CLOUD SERVICES, NETWORKING, AND MANAGEMENT

IEEE Press
445 Hoes Lane
Piscataway, NJ 08854

CLOUD SERVICES, NETWORKING, AND MANAGEMENT

Edited by

Nelson L. S. da Fonseca

Raouf Boutaba

Thomas Plevyak and
Veli Sahin, *Series Editors*

Library of Congress Cataloging-in-Publication Data

Fonseca, Nelson L. S. da.
Cloud services, networking, and management / Nelson L. S. da Fonseca, Raouf Boutaba.
pages cm
ISBN 978-1-118-84594-3 (cloth)
1. Cloud computing. I. Boutaba, Raouf. II. Title.
QA76.585.F66 2015
004.67′82–dc23

2014037179

Printed in the United States of America

10 9 8 7 6 5 4 3 2 1

For our families

CONTENTS

Preface xiii
Contributors xvii

PART I BASIC CONCEPTS AND ENABLING TECHNOLOGIES 1

1 CLOUD ARCHITECTURES, NETWORKS, SERVICES, AND MANAGEMENT 3
1.1 Introduction 3
1.2 Part I: Introduction to Cloud Computing 4
1.3 Part II: Research Challenges—The Chapters in This Book 14
1.4 Conclusion 21
References 21

2 VIRTUALIZATION IN THE CLOUD 23
2.1 The Need for Virtualization Management in the Cloud 23
2.2 Basic Concepts 25
2.3 Virtualized Elements 26
2.4 Virtualization Operations 29
2.5 Interfaces for Virtualization Management 30
2.6 Tools and Systems 34
2.7 Challenges 40
References 44

3 VIRTUAL MACHINE MIGRATION 49
3.1 Introduction 49
3.2 VM Migration 51
3.3 Virtual Network Migration without Packet Loss 59
3.4 Security of Virtual Environments 61
3.5 Future Directions 66
3.6 Conclusion 68
References 68

PART II CLOUD NETWORKING AND COMMUNICATIONS 73

4 DATACENTER NETWORKS AND RELEVANT STANDARDS 75
4.1 Overview 75
4.2 Topologies 76
4.3 Network Expansion 82
4.4 Traffic 85
4.5 Routing 89
4.6 Addressing 93
4.7 Research Challenges 96
4.8 Summary 98
References 99

5 INTER-DATA-CENTER NETWORKS WITH MINIMUM OPERATIONAL COSTS 105
5.1 Introduction 105
5.2 Inter-Data-Center Network Virtualization 108
5.3 IDC Network Design with Minimum Electric Bills 115
5.4 Inter-Data-Center Network Design with Minimum Downtime Penalties 120
5.5 Overcoming Energy versus Resilience Trade-Off 123
5.6 Summary and Discussions 124
References 126

6 OPENFLOW AND SDN FOR CLOUDS 129
6.1 Introduction 129
6.2 SDN, Cloud Computing, and Virtualization Challenges 130
6.3 Software-Defined Networking 132
6.4 Overview of Cloud Computing and OpenStack 138
6.5 SDN for Cloud Computing 142
6.6 Combining OpenFlow and OpenStack with OpenDaylight 145
6.7 Software-Defined Infrastructures 149
6.8 Research Trends and Challenges 150
6.9 Concluding Remarks 151
References 151

7 MOBILE CLOUD COMPUTING 153
7.1 Introduction 153
7.2 Mobile Cloud Computing 155
7.3 Risks in MCC 163

7.4 Risk Management for MCC 177
7.5 Conclusions 184
References 186

PART III CLOUD MANAGEMENT 191

8 ENERGY CONSUMPTION OPTIMIZATION IN CLOUD DATA CENTERS 193
8.1 Introduction 193
8.2 Energy Consumption in Data Centers: Components and Models 195
8.3 Energy Efficient System-Level Optimization of Data Centers 198
8.4 Conclusions and Open Challenges 210
References 211

9 PERFORMANCE MANAGEMENT AND MONITORING 217
9.1 Introduction 217
9.2 Background Concepts 219
9.3 Related Work 221
9.4 X-Cloud Application Management Platform 222
9.5 Implementation 229
9.6 Experiments and a Case Study 232
9.7 Challenges in Management on Heterogeneous Clouds 238
9.8 Conclusion 239
References 240

10 RESOURCE MANAGEMENT AND SCHEDULING 243
10.1 Introduction 243
10.2 Basic Concepts 244
10.3 Applications 248
10.4 Problem Definition 249
10.5 Resource Management and Scheduling in Clouds 254
10.6 Challenges and Perspectives 262
10.7 Conclusion 264
References 264

11 CLOUD SECURITY 269
11.1 Introduction 270
11.2 Technical Background 273

11.3 Existing Solutions 274
11.4 Transforming to the New IDPS Cloud Security Solutions 278
11.5 FlowIPS: Design and Implementation 279
11.6 FlowIPS vs Snort/Iptables IPS 282
11.7 Network Reconfiguration 284
11.8 Performance Comparison 288
11.9 Open Issues and Future Work 290
11.10 Conclusion 291
References 291

12 SURVIVABILITY AND FAULT TOLERANCE IN THE CLOUD 295
12.1 Introduction 295
12.2 Background 296
12.3 Failure Characterization in Cloud Environments 298
12.4 Availability-Aware Resource Allocation Schemes 299
12.5 Conclusion 307
References 307

PART IV CLOUD APPLICATIONS AND SERVICES 309

13 SCIENTIFIC APPLICATIONS ON CLOUDS 311
13.1 Introduction 311
13.2 Background Information 313
13.3 Related Work 313
13.4 IWIR Workflow Model 314
13.5 Amazon SWF Background 315
13.6 RainCloud Workflow 317
13.7 IWIR-to-SWF Conversion 319
13.8 Experiments 324
13.9 Open Challenges 328
13.10 Conclusion 329
References 330

14 INTERACTIVE MULTIMEDIA APPLICATIONS ON CLOUDS 333
14.1 Introduction 333
14.2 Delivery Models for Interactive Multimedia Services 335
14.3 Cloud Gaming 339
14.4 UGC Live Streaming 345
14.5 Time-Shifting Video Streaming 351

14.6 Open Challenges 353
14.7 Conclusion 354
References 355

15 BIG DATA ON CLOUDS (BDOC) 361

15.1 Introduction 361
15.2 Historical Perspective and State of the Art 362
15.3 Clouds—Supply and Demand of Big Data 364
15.4 Emerging Business Applications 365
15.5 Cloud and Service Availability 368
15.6 BDOC Security Issues 372
15.7 BDOC Legal Issues 379
15.8 Enabling Future Success—Stem Cultivation and Outreach 384
15.9 Open Challenges and Future Directions 385
15.10 Conclusions 388
References 388

Index 393

PREFACE

With the wide availability of high-bandwidth, low-latency network connectivity the Internet has enabled the delivery of rich services such as social networking, content delivery, and e-commerce at unprecedented scales. This technological trend has led to the development of cloud computing, a paradigm that harnesses the massive capacities of data centers to support the delivery of online services in a cost-effective manner. The National Institute of Standards and Technology (NIST) provided a relatively complete and widely accepted definition of cloud computing as follows: "cloud computing is a model for enabling ubiquitous, convenient, on-demand network access to a shared pool of configurable computing resources (e.g., networks, servers, storage, applications, and services) that can be rapidly provisioned and released with minimal management effort or service provider interaction." NIST further defined five essential characteristics as follows: (1) on-demand self-service, which states that a consumer can acquire resources based on service demand; (2) broad network access, which states that cloud services can be accessed remotely from heterogeneous client platforms (e.g., mobile phones); (3) resource pooling, where resources are pooled and shared by consumers in a multitenant fashion; (4) rapid elasticity, which states that cloud resources can be rapidly provisioned and released with minimal human involvement; (5) measured service, which states that resources are controlled (and possibly priced) by leveraging a metering capability (e.g., pay per use) that is appropriate to the type of the service.

These characteristics provide a relatively accurate picture of how cloud computing systems should look like. Furthermore, in a cloud computing environment, the traditional role of service providers is divided into two: *cloud providers* who own the physical data centers and lease resources (e.g., virtual machines) to service providers; and *service providers* who use resources leased from cloud providers to execute applications. By leveraging the economies of scale of data centers, cloud computing can provide significant reduction in operational expenditure. At the same time, it also supports new applications such as big data analytics (e.g., MapReduce) that process massive volumes of data in a scalable and efficient fashion. The rise of cloud computing has made a profound impact on the development of the IT industry in recent years. While large companies like Google, Amazon, Facebook, and Microsoft have developed their own cloud platforms and technologies, many small companies are also embracing cloud computing by leveraging open-source software and deploying services in public clouds.

This wide adoption of cloud computing is largely driven by successful deployment of a number of enabling technologies currently subject to extensive research including data center virtualization, cloud networking, data storage and management, MapReduce programming model, resource management, energy management, security, and privacy.

Data Center Virtualization—One of the main characteristics of cloud computing is that the infrastructure (e.g., data centers) is often shared by multiple tenants (e.g., service providers) running applications with different resource requirements and performance objectives. Hence, there is an emerging trend toward virtualizing physical infrastructures, that is virtualizing not only servers but also data center networks. Similar to server virtualization, network virtualization aims at creating multiple virtual networks on top of a shared physical network, allowing each tenant to implement and manage his virtual network independently from the others. This raises the question regarding how virtualized data center resources should be allocated and managed by each tenant.

Cloud Networking—to ensure predictable performance over the cloud, it is of utmost importance to design efficient networks that are able to provide guaranteed performance and to scale with the ever-growing traffic volumes in the cloud. Therefore, extensive research work is needed on designing new data center network architectures that enhance performance, fault tolerance, and scalability. Furthermore, the advent of software-defined networking (SDN) technology brings new opportunities to redesign cloud networks. Thanks to the programmability offered by this technology it is now possible to dynamically adapt the configuration of the network based on the workload in order to achieve potential cloud providers' objectives in terms of performance, utilization, survivability, and energy efficiency

Data Storage and Management—As mentioned previously one of the key driving forces for cloud computing is the need to process large volumes of data in a scalable and efficient manner. As cloud data centers typically consist of commodity servers with limited storage and processing capacities, it is necessary to develop distributed storage systems that support efficient retrieval of desired data. At the same time, as failures are common in commodity machine-based data centers, the distributed storage system must also be resilient to failures. This usually implies each file block must be replicated on multiple machines. This raises challenges regarding how the distributed storage system should be designed to achieve availability and high performance, while ensuring file replicas remain consistent over time.

MapReduce Programming Model—Cloud computing has become the most cost-effective technology for hosting Internet-scale applications. Companies like Google and Facebook generate enormous volumes of data on a daily basis that need to be processed in a timely manner. To meet this requirement, cloud providers use computational models such as MapReduce. However, despite its success, the adoption of MapReduce has implications on the management of cloud workload and cluster resources, which is still largely unstudied. In particular, many challenges pertaining to MapReduce job scheduling, task and data placement, resource allocation, and sharing require further exploration.

Resource Management—Resource management has always been a central theme of cloud computing. Given the large variety of applications running in the cloud, it is a challenging problem to determine how each application should be scheduled and managed in a scalable and dynamic manner. The scheduling of individual application component can be formulated as a variant of the multidimensional vector bin-packing problem, which is NP-hard in the general case. Furthermore, different applications may have different scheduling needs. Therefore, finding a scheduling scheme that satisfy diverse application scheduling requirement is a challenging problem.

Energy Management—Data centers consume tremendous amount of energy not only for powering up the servers and network devices but also for cooling down these components to prevent overheating conditions. It has been reported that energy cost accounts for 15% of the average data center operation expenditure. At the same time, such large energy consumption also raises environmental concerns regarding the carbon emissions for energy generation. As a result, improving data center energy efficiency has become a primary challenge for today's data center operators.

Security and Privacy—Security is another major concern of cloud computing. While security is not a critical concern in many private clouds, it is often a key barrier to the adoption of cloud computing in public clouds. Specifically, since service providers typically do not have access to the physical security system of data centers, they must rely on cloud providers to achieve full data security. The cloud provider, in this context, must provide solutions to achieve the following objectives: (1) confidentiality for secure data access and transfer and (2) auditability for attesting whether security setting of applications has been tampered or not.

Despite the wide adoption of cloud computing in the industry the current cloud technologies are still far from unleashing their full potential. In fact, cloud computing was known as a buzzword for several years and many IT companies were uncertain about how to make successful investment in cloud computing. With the recent adoption in industry and academia, cloud computing is evolving rapidly with advancements in almost all aspects, ranging from data center architectural design, scheduling and resource management, server and network virtualization, data storage, programming frameworks, energy management, pricing, and service connectivity to security and privacy

The goal of this book is to provide a general introduction to cloud services, networking, and management. We first provide an overview of cloud computing, describing its key driving forces, characteristics, and enabling technologies. Then we focus on the different characteristics of cloud computing systems and key research challenges that are covered in the subsequent fourteen chapters of this book. Specifically, the chapters delve into several topics related to cloud services, networking, and management including virtualization and SDN technologies, intra- and interdata center network architectures, resource, performance and energy management in the cloud, survivability, fault tolerance and security mobile cloud computing, and cloud applications notably big data, scientific, and multimedia applications. We hope that the readers find this journey through *Cloud Services, Networking, and Management* inspirational and informative.

Nelson L. S. da Fonseca
Raouf Boutaba

CONTRIBUTORS

Hadi Bannazadeh, Department of Electrical and Computer Engineering, University of Toronto, Toronto, Ontario, Canada

Marinho P. Barcellos, Institute of Informatics, Federal University of Rio Grande do Sul, Porto Alegre, Brazil

Joseph Betser, The Aerospace Corporation, El Segundo, CA, USA

Luiz F. Bittencourt, Institute of Computing, State University of Campinas, Campinas, São Paulo, Brazil

Raouf Boutaba, D.R. Cheriton School of Computer Science, University of Waterloo, Waterloo, Ontario, Canada

Pascal Bouvry, Faculty of Science, Technology and Communications, University of Luxembourg, Luxembourg City, Luxembourg

Otto Carlos M. B. Duarte, Grupo de Teleinformática e Automação (GTA/UFRJ), PEE/COPPE - DEL/Poli, Universidade Federal do Rio de Janeiro, Rio de Janeiro, Brazil

Rafael Pereira Esteves, Institute of Informatics, Federal University of Rio Grande do Sul, Porto Alegre, Brazil

Thomas Fahringer, Institute for Computer Science, University of Innsbruck, Innsbruck, Austria

Lyno Henrique G. Ferraz, Grupo de Teleinformática e Automação (GTA/UFRJ), PEE/COPPE - DEL/Poli, Universidade Federal do Rio de Janeiro, Rio de Janeiro, Brazil

Nelson L. S. da Fonseca, Institute of Computing, State University of Campinas, Campinas, São Paulo, Brazil

Luciano P. Gaspary, Institute of Informatics, Federal University of Rio Grande do Sul, Porto Alegre, Brazil

Fabrizio Granelli, Department of Information Engineering and Computer Science, University of Trento, Trento, Trentino, Italy

Lisandro Zambenedetti Granville, Institute of Informatics, Federal University of Rio Grande do Sul, Porto Alegre, Brazil

Simon Gwendal, Telecom Bretagne, Institut Mines-Telecom, Paris, France

Myron Hecht, The Aerospace Corporation, El Segundo, CA, USA

Dijiang Huang, School of Information Technology and Engineering, Arizona State University, Tempe, AZ, USA

Matthias Janetschek, Institute for Computer Science, University of Innsbruck, Innsbruck, Austria

B. Kantarci, Department of Electrical and Computer Engineering, Clarkson University, Potsdam, New York, USA

Dzmitry Kliazovich, Interdisciplinary Centre for Security, Reliability and Trust, University of Luxembourg, Luxembourg City, Luxembourg

Alberto Leon-Garcia, Department of Electrical and Computer Engineering, University of Toronto, Toronto, Ontario, Canada

Marin Litoiu, School of Information Technology, York University, Toronto, Ontario, Canada

Seng W. Loke, Department of Computer Science and Computer Engineering, Latrobe University, Melbourne, Australia

Hongbin Lu, Department of Computer Science and Engineering, York University, Toronto, Ontario, Canada

Edmundo R. M. Madeira, Institute of Computing, State University of Campinas, Campinas, São Paulo, Brazil

Daniel S. Marcon, Institute of Informatics, Federal University of Rio Grande do Sul, Porto Alegre, Brazil

Diogo M. F. Mattos, Grupo de Teleinformática e Automação (GTA/UFRJ), PEE/COPPE - DEL/Poli, Universidade Federal do Rio de Janeiro, Rio de Janeiro, Brazil

Deep Medhi, Computer Science and Electrical Engineering Department, University of Missouri-Kansas City, Kansas City, MO, USA

H. T. Mouftah, School of Information Technology and Engineering, University of Ottawa, Ottawa, Ontario, Canada

Rodrigo R. Oliveira, Institute of Informatics, Federal University of Rio Grande do Sul, Porto Alegre, Brazil

Simon Ostermann, Institute for Computer Science, University of Innsbruck, Innsbruck, Austria

Karine Pires, Telecom Bretagne, Institut Mines-Telecom, Paris, France

Radu Prodan, Institute for Computer Science, University of Innsbruck, Innsbruck, Austria

Haiyang Qian, China Mobile Technology, Milpitas, CA, USA

Karl Reed, Department of Computer Science and Computer Engineering, Latrobe University, Melbourne, Australia

Javeria Samad, Department of Computer Science and Computer Engineering, Latrobe University, Melbourne, Australia

Mark Shtern, Department of Computer Science and Engineering, York University, Toronto, Ontario, Canada

Bradley Simmons, School of Information Technology, York University, Toronto, Ontario, Canada

Michael Smit, School of Information Management, Dalhousie University, Halifax, Nova Scotia, Canada

Juliano Araujo Wickboldt, Institute of Informatics, Federal University of Rio Grande do Sul, Porto Alegre, Brazil

Tianyi Xing, School of Computing, Informatics, and Decision systems Engineering, Arizona State University, Tempe, AZ, USA

Zhengyang Xiong, School of Computing, Informatics, and Decision systems Engineering, Arizona State University, Tempe, AZ, USA

Qi Zhang, Department of Electrical and Computer Engineering, University of Toronto, Toronto, Ontario, Canada

Mohamed Faten Zhani, Department of Software and IT Engineering, École de technologie supérieure, University of Quebec Montreal, Canada

PART I

BASIC CONCEPTS AND ENABLING TECHNOLOGIES

1

CLOUD ARCHITECTURES, NETWORKS, SERVICES, AND MANAGEMENT

Raouf Boutaba[1] and Nelson L. S. da Fonseca[2]

[1]*D.R. Cheriton School of Computer Science, University of Waterloo, Waterloo, Ontario, Canada*

[2]*Institute of Computing, State University of Campinas, Campinas, São Paulo, Brazil*

1.1 INTRODUCTION

With the wide availability of high-bandwidth, low-latency network connectivity, the Internet has enabled the delivery of rich services such as social networking, content delivery, and e-commerce at unprecedented scales. This technological trend has led to the development of cloud computing, a paradigm that harnesses the massive capacities of data centers to support the delivery of online services in a cost-effective manner. In a cloud computing environment, the traditional role of service providers is divided into two: *cloud providers* who own the physical data center and lease resources (e.g., virtual machines or VMs) to service providers; and *service providers* who use resources leased by cloud providers to execute applications. By leveraging the economies-of-scale of data centers, cloud computing can provide significant reduction in operational expenditure. At the same time, it also supports new applications such as big-data analytics (e.g., MapReduce [1]) that process massive volumes of data in a scalable and efficient fashion. The rise of cloud computing has made a profound impact on the development of the IT industry in recent years. While large companies like Google, Amazon, Facebook,

Cloud Services, Networking, and Management, First Edition.
Edited by Nelson L. S. da Fonseca and Raouf Boutaba.

and Microsoft have developed their own cloud platforms and technologies, many small companies are also embracing cloud computing by leveraging open-source software and deploying services in public clouds.

However, despite the wide adoption of cloud computing in the industry, the current cloud technologies are still far from unleashing their full potential. In fact, cloud computing was known as a buzzword for several years, and many IT companies were uncertain about how to make successful investment in cloud computing. Fortunately, with the significant attraction from both industry and academia, cloud computing is evolving rapidly, with advancements in almost all aspects, ranging from data center architectural design, scheduling and resource management, server and network virtualization, data storage, programming frameworks, energy management, pricing, service connectivity to security, and privacy.

The goal of this chapter is to provide a general introduction to cloud networking, services, and management. We first provide an overview of cloud computing, describing its key driving forces, characteristics and enabling technologies. Then, we focus on the different characteristics of cloud computing systems and key research challenges that are covered in the subsequent 14 chapters of this book. Specifically, the chapters delve into several topics related to cloud services, networking and management including virtualization and software-defined network technologies, intra- and inter- data center network architectures, resource, performance and energy management in the cloud, survivability, fault tolerance and security, mobile cloud computing, and cloud applications notably big data, scientific, and multimedia applications.

1.2 PART I: INTRODUCTION TO CLOUD COMPUTING

1.2.1 What Is Cloud Computing?

Despite being widely used in different contexts, a precise definition of cloud computing is rather elusive. In the past, there were dozens of attempts trying to provide an accurate yet concise definition of cloud computing [2]. However, most of the proposed definitions only focus on particular aspects of cloud computing, such as the business model and technology (e.g., virtualization) used in cloud environments. Due to lack of consensus on how to define cloud computing, for years cloud computing was considered a buzz word or a marketing hype in order to get businesses to invest more in their IT infrastructures. The National Institute of Standards and Technology (NIST) provided a relatively standard and widely accepted definition of cloud computing as follows: "cloud computing is a model for enabling ubiquitous, convenient, on-demand network access to a shared pool of configurable computing resources (e.g., networks, servers, storage, applications, and services) that can be rapidly provisioned and released with minimal management effort or service provider interaction." [3]

NIST further defined five essential characteristics, three service models, and four deployment models, for cloud computing. The five essential characteristics include the following:

1. On-demand self-service, which states that a consumer (e.g., a service provider) can acquire resources based on service demand;

2. Broad network access, which states that cloud services can be accessed remotely from heterogeneous client platforms (e.g., mobile phones);
3. Resource pooling, where resources are pooled and shared by consumers in a multi-tenant fashion;
4. Rapid elasticity, which states that cloud resources can be rapidly provisioned and released with minimal human involvement;
5. Measured service, which states that resources are controlled (and possibly priced) by leveraging a metering capability (e.g., pay-per-use) that is appropriate to the type of the service.

These characteristics provide a relatively accurate picture of what cloud computing systems should look like. It should be mentioned that not every cloud computing system exhibits all five characteristics listed earlier. For example, in a private cloud, where the service provider owns the physical data center, the metering capability may not be necessary because there is no need to limit resource usage of the service unless it is reaching data center capacity limits. However, despite the definition and aforementioned characteristics, cloud computing can still be realized in a large number of ways, and hence one may argue the definition is still not precise enough. Today, cloud computing commonly refers to a computing model where services are hosted using resources in data centers and delivered to end users over the Internet. In our opinion, since cloud computing technologies are still evolving, finding the precise definition of cloud computing at the current moment may not be the right approach. Perhaps once the technologies have reached maturity, the true definition will naturally emerge.

1.2.2 Why Cloud Computing?

In this section, we present the motivation behind the development of cloud computing. We will also compare cloud computing with other parallel and distributed computing models and highlight their differences.

1.2.2.1 Key Driving Forces. There are several driving forces behind the success of cloud computing. The increasing demand for large-scale computation and big data analytics and economics are the most important ones. But other factors such as easy access to computation and storage, flexibility in resource allocations, and scalability play important roles.

Large-scale computation and big data: Recent years have witnessed the rise of Internet-scale applications. These applications range from social networks (e.g., facebook, twitter), video applications (e.g., Netflix, youtube), enterprise applications (e.g., SalesForce, Microsoft CRM) to personal applications (e.g., iCloud, Dropbox). These applications are commonly accessed by large numbers of users over the Internet. They are extremely large scale and resource intensive. Furthermore, they often have high performance requirements such as response time. Supporting these applications requires extremely large-scale infrastructures. For instance, Google has hundreds of compute clusters deployed worldwide with hundreds of thousands of servers. Another salient characteristic is that these applications also require access to huge volumes of data. For instance, Facebook stores tens of petabytes of data and processes over a hundred terabytes per day. Scientific applications (e.g., brain image processing, astrophysics, ocean monitoring, and DNA analysis) are more and more deployed in the cloud. Cloud computing emerged in this context as a computing model designed for running large applications in a scalable and cost-efficient manner by harnessing massive resource capacities in data centers and by sharing the data center resources among applications in an on-demand fashion.

Economics: To support large-scale computation, cloud providers rely on inexpensive commodity hardware offering better scalability and performance/price ratio than supercomputers. By deploying a very large number of commodity machines, they leverage economies of scale bringing per unit cost down and allowing for incremental growth. On the other hand, cloud customers such as small and medium enterprises, which outsource their IT infrastructure to the cloud, avoid upfront infrastructure investment cost and instead benefit from a pay-as-you-go pricing and billing model. They can deploy their services in the cloud and make them quickly available to their own customers resulting in short time to market. They can start small and scale up and down their infrastructure based on their customers demand and pay based on usage.

Scalability: By harnessing huge computing and storage capabilities, cloud computing gives customers the illusion of infinite resources on demand. Customers can start small and scale up and down resources as needed.

Flexibility: Cloud computing is highly flexible. It allows customers to specify their resource requirements in terms of CPU cores, memory, storage, and networking capabilities. Customers are also offered the flexibility to customize the resources in terms of operating systems and possibly network stacks.

Easy access: Cloud resources are accessible from any device connected to the Internet. These devices can be traditional workstations and servers or less traditional devices such as smart phones, sensors, and appliances. Applications running in the cloud can be deployed or accessed from anywhere at anytime.

1.2.2.2 Relationship with Other Computing Models. Cloud computing is not a completely new concept and has many similarities with existing distributed and parallel computing models such as Grid computing and Cluster computing. But cloud computing also has some distinguishing properties that explain why existing models are not used and justify the need for a new one. These can be explained according to two dimensions: scale and service-orientation. Both parallel computing and cloud, computing are used to solve large-scale problems often by subdividing these problems into smaller parts and carrying out the calculations concurrently on different processors. In the cloud, this is achieved using computational models such as MapReduce. However, while parallel computing relies on expensive supercomputers and massively parallel multi-processor machines, cloud computing uses cheap, easily replaceable commodity hardware. Grid computing uses supercomputers but can also use commodity hardware, all accessible through open, general-purpose protocols and interfaces, and distributed management and job scheduling middleware. Cloud computing differs from Grid computing in that it provides high bandwidth between machines, that is more suitable for I/O-intensive applications such as log analysis, Web crawling, and big-data analytics. Cloud computing also differs from Grid computing in that resource management and job scheduling is centralized under a single administrative authority (cloud provider) and, unless this evolves differently in the future, provides no standard application programming interfaces (APIs). But perhaps the most distinguishing feature of cloud computing compared to previous computing models is its extensive reliance on virtualization technologies to allow for efficient sharing of resources while guaranteeing isolation between multiple cloud tenants. Regarding the second dimension, unlike other computing models designed for supporting applications and are mainly application-oriented, cloud computing extensively leverages service orientation providing everything (infrastructure, development platforms, software, and applications) as a service.

1.2.3 Architecture

Generally speaking, the architecture of a cloud computing environment can be divided into four layers: the hardware/datacenter layer, the infrastructure layer, the platform layer, and the application layer, as shown in Figure 1.1. We describe each of them in detail in the text that follows:

The hardware layer: This layer is responsible for managing the physical resources of the cloud, including physical servers, routers, and switches, and power, and cooling systems. In practice, the hardware layer is typically implemented in data centers. A data center usually contains thousands of servers that are organized in racks and interconnected through switches, routers, or other fabrics. Typical issues at hardware layer include hardware configuration, fault-tolerance, traffic management, and power and cooling resource management.

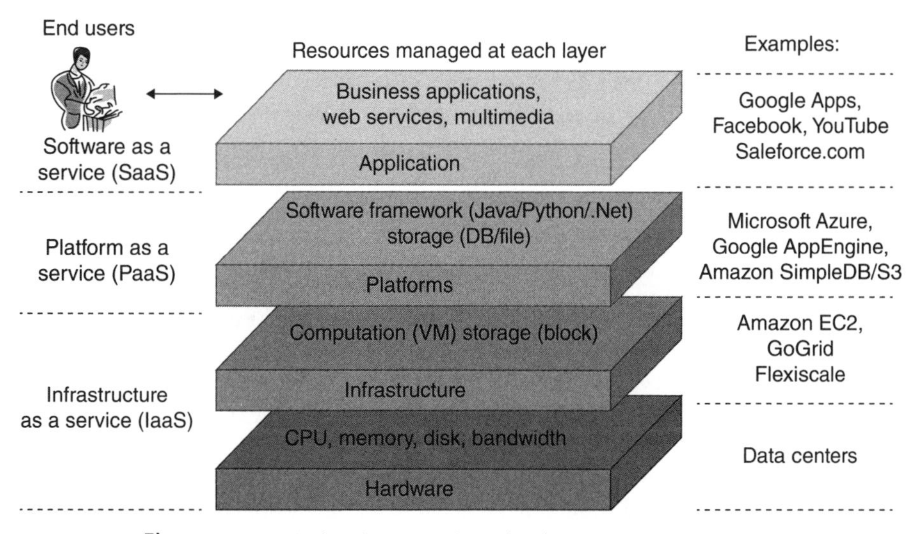

Figure 1.1. Typical architecture in a cloud computing environment.

The infrastructure layer: Also known as the virtualization layer, the infrastructure layer creates a pool of storage and computing resources by partitioning the physical resources using virtualization technologies such as Xen [4], KVM [5], and VMware [6]. The infrastructure layer is an essential component of cloud computing, since many key features, such as dynamic resource assignment, are only made available through virtualization technologies.

The platform layer: Built on top of the infrastructure layer, the platform layer consists of operating systems and application frameworks. The purpose of the platform layer is to minimize the burden of deploying applications directly into VM containers. For example, Google App Engine operates at the platform layer to provide API support for implementing storage, database, and business logic of typical Web applications.

The application layer: At the highest level of the hierarchy, the application layer consists of the actual cloud applications. Different from traditional applications, cloud applications can leverage the automatic-scaling feature to achieve better performance, availability, and lower operating cost. Compared to traditional service hosting environments such as dedicated server farms, the architecture of cloud computing is more modular. Each layer is loosely coupled with the layers above and below, allowing each layer to evolve separately. This is similar to the design of the protocol stack model for network protocols. The architectural modularity allows cloud computing to support a wide range of application requirements while reducing management and maintenance overhead.

1.2.4 Cloud Services

Cloud computing employs a service-driven business model. In other words, hardware and platform-level resources are provided as services on an on-demand basis. Conceptually, every layer of the architecture described in the previous section can be implemented as a service to the layer above. Conversely, every layer can be perceived as a customer of the layer below. However, in practice, clouds offer services that can be grouped into three categories: software as a service (SaaS), platform as a service (PaaS), and infrastructure as a service (IaaS).

1. *Infrastructure as a service*: IaaS refers to on-demand provisioning of infrastructural resources, usually in terms of VMs. The cloud owner who offers IaaS is called an IaaS provider.
2. *Platform as a service*: PaaS refers to providing platform layer resources, including operating system support and software development frameworks.
3. *Software as a service*: SaaS refers to providing on-demand applications over the Internet.

The business model of cloud computing is depicted in Figure 1.2. According to the layered architecture of cloud computing, it is entirely possible that a PaaS provider runs its cloud on top of an IaaS providers cloud. However, in the current practice, IaaS and

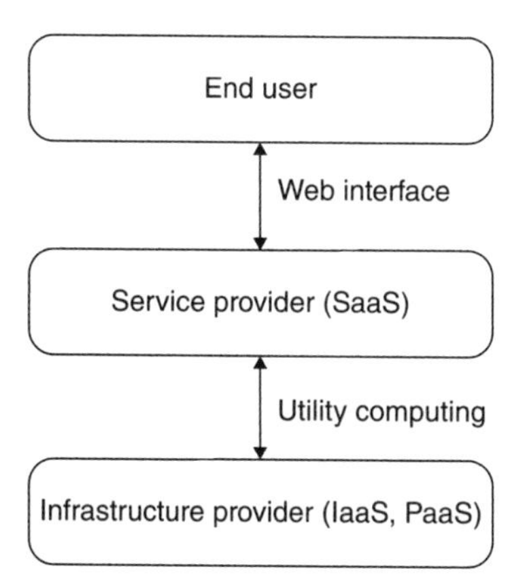

Figure 1.2. Cloud computing business model.

PaaS providers are often parts of the same organization (e.g., Google). This is why PaaS and IaaS providers are often called cloud providers [7].

1.2.4.1 *Type of Clouds.* There are many issues to consider when moving an enterprise application to the cloud environment. For example, some enterprises are mostly interested in lowering operation cost, while others may prefer high reliability and security. Accordingly, there are different types of clouds, each with its own benefits and drawbacks:

- *Public clouds*: A cloud in which cloud providers offer their resources as services to the general public. Public clouds offer several key benefits to service providers, including no initial capital investment on infrastructure and shifting of risks to cloud providers. However, current public cloud services still lack fine-grained control over data, network and security settings, which hampers their effectiveness in many business scenarios.
- *Private clouds*: Also known as internal clouds, private clouds are designed for exclusive use by a single organization. A private cloud may be built and managed by the organization or by external providers. A private cloud offers the highest degree of control over performance, reliability, and security. However, they are often criticized for being similar to traditional proprietary server farms and do not provide benefits such as no up-front capital costs.
- *Hybrid clouds*: A hybrid cloud is a combination of public and private cloud models that tries to address the limitations of each approach. In a hybrid cloud, part of the service infrastructure runs in private clouds while the remaining part runs in public clouds. Hybrid clouds offer more flexibility than both public and private clouds. Specifically, they provide tighter control and security over application data compared to public clouds, while still facilitating on-demand service expansion

and contraction. On the down side, designing a hybrid cloud requires carefully determining the best split between public and private cloud components.

- *Community clouds*: A community cloud refers to a cloud infrastructure that is shared between multiple organizations that have common interests or concerns. Community clouds are a specific type of cloud that relies on the common interest and limited participants to achieve efficient, reliable, and secure design of the cloud infrastructure.

Private cloud has always been the most popular type of cloud. Indeed, the development of cloud computing was largely due to the need of building data centers for hosting large-scale online services owned by large private companies, such as Amazon and Google. Subsequently, realizing the cloud infrastructure can be leased to other companies for profits, these companies have developed public cloud services. This development has also led to the creation of hybrid clouds and Community clouds, which represent different alternatives to share cloud resources among service providers. In the future, it is believed that private cloud will remain to be the dominant cloud computing model. This is because as online services continue to grow in scale and complexity, it becomes increasingly beneficial to build private cloud infrastructure to host these services. In this case, private clouds not only provide better performance and manageability than public clouds but also reduced operation cost. As the initial capital investment on a private cloud can be amortized across large number of machines over many years, in the long-term private cloud typically has lower operational cost compared to public clouds.

1.2.4.2 SME's Survey on Cloud Computing. The European Network and Information Security Agency (ENISA) has conducted a survey on the adaption of the cloud computing model by small to medium enterprises (SMEs). The survey provides an excellent overview of the benefits and limitations of today's cloud technologies. In particular, the survey has found that the main reason for adopting cloud computing is to reduce total capital expenditure on software and hardware resources. Furthermore, most of the enterprises prefer a mixture of cloud computing models (public cloud, private cloud), which comes with no surprise as each type of cloud has own benefits and limitations. Regarding the type of cloud services, it seems that IaaS, PaaS, and SaaS all received similar scores, even though SaaS is slightly in favor compared to the other two. Last, it seems that data availability, privacy, and confidentiality are the main concerns of all the surveyed enterprises. As a result, it is not surprising to see that most of the enterprises prefer to have a disaster recovery plan when considering migration to the cloud. Based on these observations, cloud providers should focus more on improving the security and reliability aspect of cloud infrastructures, as they represent the main obstacles for adopting the cloud computing model by today's enterprises.

1.2.5 Enabling Technologies

The success of cloud computing is largely driven by successful deployment of its enabling technologies. In this section, we provide an overview of cloud enabling technologies and describe how they contribute to the development of cloud computing.

1.2.5.1 Data Center Virtualization. One of the main characteristics of cloud computing is that the infrastructure (e.g., data centers) is often shared by multiple tenants (e.g., service providers) running applications with different resource requirements and performance objectives. This raises the question regarding how data center resources should be allocated and managed by each service provider. A naive solution that has been implemented in the early days is to allocate dedicated servers for each application. While this "bare-metal" strategy certainly worked in many scenarios, it also introduced many inefficiencies. In particular, if the server resource is not fully utilized by the application running on the server, the resource is wasted as no other application has the right to acquire the resource for its own execution. Motivated by this observation, the industry has adopted virtualization in today's cloud data centers. Generally speaking, virtualization aims at partitioning physical resources into virtual resources that can be allocated to applications in a flexible manner. For instance, server virtualization is a technology that partitions the physical machine into multiple VMs, each capable of running applications just like a physical machine. By separating logical resources from the underlying physical resources, server virtualization enables flexible assignment of workloads to physical machines. This not only allows workload running on multiple VMs to be consolidated on a single physical machine, but also enables a technique called VM migration, which is the process of dynamically moving a VM from one physical machine to another. Today, virtualization technologies have been widely used by cloud providers such as Amazon EC2, Rackspace, and GoGrid. By consolidating workload using fewer machines, server virtualization can deliver higher resource utilization and lower energy consumption compared to allocating dedicated servers for each application.

Another type of data center virtualization that has been largely overlooked in the past is network virtualization. Cloud applications today are becoming increasingly data-intensive. As a result, there is a pressing need to determine how data center networks should be shared by multiple tenants with diverse performance, security and manageability requirements. Motivated by these limitations, there is an emerging trend towards virtualizing data center networks in addition to server virtualization. Similar to server virtualization, network virtualization aims at creating multiple VNs on top of a shared physical network substrate allowing each VN to be implemented and managed independently. By separating logical networks from the underlying physical network, it is possible to implement network resource guarantee and introduce customized network protocols, security, and management policies. Combining with server virtualization, a fully virtualized data centers support the allocation in the form of virtual infrastructures or VIs (also known as virtual data centers (VDC)), which consist of VMs inter-connected by virtual networks. The scheduling and management of VIs have been studied extensively in recent years. Commercial cloud providers are also pushing towards this direction. For example, the Amazon Virtual Private Cloud (VPC) already provides limited features to support network virtualization in addition to server virtualization.

1.2.5.2 Cloud Networking. To ensure predictable performance over the cloud, it is of utmost importance to design efficient networks that are able to provide guaranteed performance and to scale with the ever-growing traffic volumes in the cloud. Traditional data center network architectures suffer from many limitations that may hinder the performance of large-scale cloud services. For instance, the widely-used tree-like topology does not provide multiple paths between the nodes, and hence limits the scalability of the network and the ability to mitigate node and link congestion and failures. Moreover, current technologies like Ethernet and VLANs are not well suited to support cloud computing requirements like multi-tenancy or performance isolation between different tenants/applications. In recent years, several research works have focused on designing new data center network architectures to overcome these limitations and enhance performance, fault tolerance and scalability (e.g., VL2 [38], Portland [9], NetLord [10]). Furthermore, the advent of software-defined networking (SDN) technology brings new opportunities to redesign cloud networks [11]. Thanks to the programmability offered by this technology, it is now possible to dynamically adapt the configuration of the network based on the workload. It also makes it easy to implement policy-based network management schemes in order to achieve potential cloud providers' objectives in terms of performance, utilization, survivability, and energy efficiency.

1.2.5.3 Data Storage and Management. As mentioned previously, one of the key driving forces for cloud computing is the need to process large volumes of data in a scalable and efficient manner. As cloud data centers typically consist of commodity servers with limited storage and processing capacities, it is necessary to develop distributed storage systems that support efficient retrieval of desired data. At the same time, as failures are common in commodity machine-based data centers, the distributed storage system must also be resilient to failures. This usually implies each file block must be replicated on multiple machines. This raises challenges regarding how the distributed storage system should be designed to achieve availability and high performance, while ensuring file replicas remain consistent over time. Unfortunately, the famous CAP theorem [12] states that simultaneously achieving all three objectives (consistency, availability, and robustness to network failures) is not a viable task. As result, recently many file systems such Google File System [13], Amazon Dynamo [14], Cassandra [15] are trying to explore various trade-offs among the three objectives based on applications' needs. For example, Amazon Dynamo adopts an eventual consistency model that allow replicas to be temporary out-of-sync. By sacrificing consistency, Dynamo is able to achieve significant improvement in server response time. It is evident that these storage systems provide the foundations for building large-scale data-intensive applications that are commonly found in today's cloud data centers.

1.2.5.4 MapReduce Programming Model. Cloud computing has become the most cost-effective technology for hosting Internet-scale applications. Companies like Google and Facebook generate enormous volumes of data on a daily basis that need to be processed in a timely manner. To meet this requirement, cloud providers use computational models such as MapReduce [1] and Dryad [16]. In these models, a job spawns many small tasks that can be executed concurrently on multiple machines, resulting in significant reduction in job completion time. Furthermore, to cope with software and hardware exceptions frequent in large-scale clusters, these models provide built-in fault tolerance features that automatically restart failed tasks when exceptions occur. As a result, these computational models are very attractive not only for running data-intensive jobs but also for computation-intensive applications. The MapReduce model, in particular, is largely used nowadays in cloud infrastructures for supporting a wide range of applications and has been adapted to several computing and cluster environments. Despite this success, the adoption of MapReduce has implications on the management of cloud workload and cluster resources, which is still largely unstudied. In particular, many challenges pertaining to MapReduce job scheduling, task and data placement, resource allocation, and sharing are yet to be addressed.

1.2.5.5 Resource Management. Resource management has always been a central theme of cloud computing. Given the large variety of applications running in the cloud, it is a challenging problem to determine how each application should be scheduled and managed in a scalable and dynamic manner. The scheduling of individual application component can be formulated as a variant of the multi-dimensional vector bin-packing problem, which is already NP-hard in the general case. Furthermore, different applications may have different scheduling needs. For example, individual tasks of a single MapReduce job can be scheduled independently over time, whereas the servers of a three-tier Web application must be scheduled simultaneously to ensure service availability. Therefore, finding a scheduling scheme that satisfy diverse application scheduling requirement is a challenging problem. The recent work on multi-framework scheduling (e.g., MESOS [17]) provides a platform to allow various scheduling frameworks, such as MapReduce, Spark, and MPI to coexist in a single cloud infrastructure. The work on distributed schedulers (e.g., Omega [18] and Sparrow [19]) also aim at improving the scalability of schedulers by having multiple schedulers perform scheduling in parallel. These technologies will provide the functionality to support a wide range of workload in the cloud data center environments.

1.2.5.6 Energy Management. Data centers consume tremendous amount of energy, not only for powering up the servers and network devices, but also for cooling down these components to prevent overheating conditions. It has been reported that energy cost accounts for 15% of the average data center operation expenditure. At the same time, such large energy consumption also raises environmental concerns regarding the carbon emissions for energy generation. As a result, improving data center energy efficiency has become a primary concern for today's data center operators. A widely used metric for measuring energy efficiency of data centers is power usage effectiveness (PUE), which is computed as the ratio between the computer infrastructure usage and the total data center power usage. Even though none of the existing data centers can achieve the ideal PUE value of 1.0, many cloud data centers today have become very energy efficient with PUE less than 1.1.

There are many techniques for improving data center energy efficiency. At the infrastructure level, many cloud providers leverage nearby renewable energy source (i.e., solar and wind) to reduce energy cost and carbon footprint. At the same time, it is also possible to leverage environmental conditions (e.g., low temperature conditions) to reduce cooling cost. For example, Facebook recently announced the construction of a cloud data center in Sweden, right on the edge of the arctic circle, mainly due to the low air temperature that can reduce cooling cost. The Net-Zero Energy Data Center developed by HP labs leverages locally generated renewable energy and workload demand management techniques to significantly reduce the energy required to operate data centers. We believe the rapid development of cloud energy management techniques will continue to push the data center energy efficiency towards the ideal PUE value of 1.0.

1.2.5.7 Security and Privacy. Security is another major concern of cloud computing. While security is not a critical concern in many private clouds, it is often a key barrier to the adoption of cloud computing in public clouds. Specifically, since service providers typically do not have access to the physical security system of data centers, they must rely on cloud providers to achieve full data security. The cloud provider, in this context, must achieve the following objectives: (1) confidentiality, for secure data access and transfer, and (2) auditability, for attesting whether security setting of applications has been tampered or not. Confidentiality is usually achieved using cryptographic protocols, whereas auditability can be achieved using remote attestation techniques. Remote attestation typically requires a trusted platform module (TPM) to generate nonforgeable system summary (i.e., system state encrypted using TPM private key) as the proof of system security. However, in a virtualized environment like the clouds, VMs can dynamically migrate from one location to another, hence directly using remote attestation is not sufficient. In this case, it is critical to build trust mechanisms at every architectural layer of the cloud. First, the hardware layer must be trusted using hardware TPM. Second, the virtualization platform must be trusted using secure VM monitors. VM migration should only be allowed if both source and destination servers are trusted. Recent work has been devoted to designing efficient protocols for trust establishment and management.

1.3 PART II: RESEARCH CHALLENGES—THE CHAPTERS IN THIS BOOK

This book covers the fundamentals of cloud services, networking and management and focuses on most prominent research challenges that have drawn the attention of the IT community in the past few years. Each of the 14 chapters of this book provides an overview of some of the key architectures, features, and technologies of cloud services, networking and management systems and highlights state-of-the-art solutions and possible research gaps. The chapters of the book are written by knowledgeable authors that were carefully selected based on their expertise in the field. Each chapter went through a rigorous review process, including external reviewers, the book editors Raouf Boutaba and Nelson Fonseca, and the series editors Tom Plevyak and Veli Sahin. In the following, we briefly describe the topics covered by the different chapters of this book.

1.3.1 Virtualization in the Cloud

Virtualization is one of the key enabling technologies that made cloud computing model a reality. Initially, virtualization technologies have allowed to partition a physical server into multiple isolated environments called VMs that may eventually host different operating systems and be used by different users or applications. As cloud computing evolved, virtualization technologies have matured and have been extended to consider not only the partitioning of servers but also the partitioning of the networking resources (e.g., links, switches and routers). Hence, it is now possible to provide each cloud user with a VI encompassing VMs, virtual links, and virtual routers and switches. In this context, several challenges arise especially regarding the management of the resulting virtualized environment where different types of resources are shared among multiple users.

In this chapter, the authors outline the main characteristics of these virtualized infrastructures and shed light on the different management operations that need to be implemented in such environments. They then summarize the ongoing efforts towards defining open standard interfaces to support virtualization and interoperability in the cloud. Finally, the chapter provides a brief overview of the main open-source cloud management platforms that have recently emerged.

1.3.2 VM Migration

One of the powerful features brought by virtualization is the ability to easily migrate VMs within the same data center or even between geographically distributed data centers. This feature provides an unprecedented flexibility to network and data center operators allowing them to perform several management tasks like dynamically optimizing resource allocations, improving fault tolerance, consolidating workloads, avoiding server overload, and scheduling maintenance activities. Despite all these benefits, VM migration induces several costs, including higher utilization of computing and networking resources, inevitable service downtime, security risks, and more complex management challenges. As a result, a large number of migration techniques have been recently proposed in the literature in order to minimize these costs and make VM migration a more effective and secure tool in the hand of cloud providers.

This chapter starts by providing an overview of VM migration techniques. It then presents, XenFlow, a tool based on Xen and OpenFlow, and allowing to deploy, isolate and migrate VIs. Finally, the authors discuss potential security threats that can arise when using VM migration.

1.3.3 Data Center Networks and Relevant Standards

Today's cloud data centers are housing hundreds of thousands of machines that continuously need to exchange tremendous amounts of data with stringent performance requirements in terms of bandwidth, delay, jitter, and loss rate. In this context, the data center network plays a central role to ensure a reliable and efficient communication between machines, and thereby guarantee continuous operation of the data center and effective delivery of the cloud services. A data center network architecture is typically defined by the network topology (i.e., the way equipment are inter-connected) as well as the adopted switching, routing, and addressing schemes and protocols (e.g., Ethernet and IP).

Traditional data center network architectures suffer from several limitations and are not able to satisfy new application requirements spawned by cloud computing model in terms of scalability, multitenancy and performance isolation. For instance, the widely used tree-like topology does not provide multiple paths between the nodes, and hence limits the ability to survive node and link failures. Also, current switches have limited forwarding table sizes, making it difficult for traditional data center networks to handle the large number of VMs that may exist in virtualized cloud environments. Another issue is with the performance isolation between tenants as there is no bandwidth allocation mechanism in place to ensure predictable network performance for each of them.

In order to cope with these limitations, a lot of attention has been devoted in the past few years to study the performance of existing architectures and to design better solutions. This chapter dwells on these solutions covering data center network architectures, topologies, routing protocols and addressing schemes that have been recently proposed in the literature.

1.3.4 Interdata Center Networks

In recent years, cloud providers have largely relied on large-scale cloud infrastructures to support Internet-scale applications efficiently. Typically, these infrastructures are composed of several geographically distributed data centers connected through a backbone network (i.e., an inter-data center network). In this context, a key challenge facing cloud providers is to build cost-effective backbone networks while taking into account several considerations and requirements including scalability, energy efficiency, resilience, and reliability. To address this challenge, many factors should be considered. The scalability requirement is due to the fact that the volume of data exchanged between data centers is growing exponentially with the ever-increasing demand in cloud environments. The energy efficiency requirement concerns how to minimize the energy consumption of the infrastructure. Such a requirement is not only crucial to make the infrastructure more green and environmental-friendly but also essential to cut down operational expenses. Finally, the resilience of the interdata center network requirement is fundamental to maintain a continuous and reliable cloud services.

This chapter investigates the different possible alternatives to design and manage cost-efficient cloud backbones. It then presents mathematical formulations and heuristic solutions that could be adopted to achieve desired objectives in terms of energy efficiency, resilience and reliability. Finally, the authors discuss open issues and key research directions related to this topic.

1.3.5 OpenFlow and SDN for Clouds

The past few years have witnessed the rise of SDN, a technology that makes it possible to dynamically configure and program networking elements. Combined with cloud computing technologies, SDN enables the design of highly dynamic, efficient, and cost-effective shared application platforms that can support the rapid deployment of Internet applications and services.

This chapter discusses the challenges faced to integrate SDN technology in cloud application platforms. It first provides a brief overview of the fundamental concepts of SDN including OpenFlow technology and tools like Open vSwitch. It also introduces the cloud platform OpenStack with a focus on its Networking Service (i.e., Neutron project), and shows how cloud computing environments can benefit from SDN technology to provide guaranteed networking resources within a data center and to interconnect data centers. The authors also review major open source efforts that attempt to integrate SDN technology in cloud management platforms (e.g., OpenDaylight open source project) and discuss the notion of software-defined infrastructure (SDI).

1.3.6 Mobile Cloud Computing

Mobile cloud computing has recently emerged as a new paradigm that combines cloud computing with mobile network technology with the goal of putting the scalability and limitless resources of the cloud into the hands of mobile service and application providers. However, despite of its potential benefits, the growth of mobile cloud computing in recent years was hampered by several technical challenges and risks. These challenges and risks are mainly due to the inherent limitations of mobile devices such as the scarcity of resources, the limited energy supply, the intermittent connectivity in wireless networks, security risks, and legal/environmental risks.

This chapter starts by providing an overview of mobile cloud computing application models and frameworks. It also defines risk management and identifies and analyzes prevalent risk factors found in mobile cloud computing environments. The authors also present an analysis of mobile cloud frameworks from a risk management perspective and discusses the effectiveness of traditional risk approaches to address mobile cloud computing risks.

1.3.7 Resource Management and Scheduling

Resource allocation and scheduling are two crucial functions in cloud computing environments. Generally speaking, cloud providers are responsible for allocating resources (e.g., VMs) with the goal of satisfying the promised service-level agreement (SLA) while increasing their profit. This can be achieved by reducing operational costs (e.g., energy costs) and sharing resources among the different users. At the opposite side, cloud users are responsible for application scheduling that aims at mapping tasks from applications submitted by users to computational resources in the system. The goals of scheduling include maximizing the usage of the leased resources, and minimizing costs by dynamically adjusting the leased resources to the demand while maintaining the required quality of service.

Resource allocation and scheduling are both vital to cloud users and providers, but they both have their own specifics, challenges and potentially conflicting objectives. This chapter starts by a review of the different cloud types and service models and then discusses the typical objectives of cloud providers and their clients. The chapter provides also mathematical formulations to the problems, VM allocation, and application scheduling. It surveys some of the existing solutions and discusses their strengths and weaknesses. Finally, it points out the key research directions pertaining to resource management in cloud environments.

1.3.8 Autonomic Performance Management for Multi-Clouds

The growing popularity of the cloud computing model have led to the emergence of multiclouds or clouds of clouds where multiple cloud systems are federated together to further improve and enhance cloud services. Multiclouds have several benefits that range from improving availability, to reducing lock-in, and optimizing costs beyond what can be achieved within a single cloud. At the same time, multi-clouds bring new challenges in terms of the design, development, deployment, monitoring, and management of multi-tier applications able to capitalize on the advantages of such distributed infrastructures. As a matter of fact, the responsibility for addressing these challenges is shared among cloud providers and cloud users depending on the type of service (i.e., IaaS, PaaS, and SaaS) and SLAs. For instance, from an IaaS cloud provider's perspective, management focuses mainly on maintaining the infrastructure, allocating resources requested by clients and ensuring their high availability. By contrast, cloud users are responsible for implementing, deploying and monitoring applications running on top of resources that are eventually leased from several providers. In this context, a compelling challenge that is currently attracting a lot of attention is how to develop sophisticated tools that simplify the process of deploying, managing, monitoring, and maintaining large-scale applications over multi-clouds.

This chapter focuses on this particular challenge and provides a detailed overview of the design and implementation of XCAMP, the X-Cloud Application Management Platform that allows to automate application deployment and management in multitier clouds. It also highlights key research challenges that require further investigation in the context of performance management and monitoring in distributed cloud environments.

1.3.9 Energy Management

Cloud computing environments mainly consist of data centers where thousands of servers and other systems (e.g, power distribution and cooling equipment) are consuming tremendous amounts of energy. Recent reports have revealed that energy costs represent more than 12% of the total data center operational expenditures, which translates into millions of dollars. More importantly, high energy consumption is usually synonymous of high carbon footprint, raising serious environmental concerns and pushing governments to put in place more stringent regulations to protect the environment. Consequently, reducing energy consumption has become one of the key challenges facing today's data center managers. Recently, a large body of work has been dedicated to investigate possible techniques to achieve more energy-efficient and environment-friendly infrastructures. Many solutions have been proposed including dynamic capacity provisioning and optimal usage of renewable sources of energy (e.g., wind power and solar).

This chapter further details the trends in energy management solutions in cloud data centers. It first surveys energy-aware resource scheduling and allocation schemes aiming at improving energy efficiency, and then provides a detailed description of GreenCloud, an energy-aware cloud data center simulator.

1.3.10 Survivability and Fault Tolerance in the Cloud

Despite the success of cloud computing, its widespread and full-scale adoption have been hampered by the lack of strict guarantees on the reliability and availability of the offered resources and services. Indeed, outages, failures and service disruption can be fatal for many businesses. Not only they incur significant revenue loss—as much as hundreds of thousands of dollars per minute for some services—but they may also hurt the business reputation in the long term and impact on customers' loyalty and satisfaction. Unfortunately, major cloud providers like Amazon EC2, Google, and Rackspace are not yet able to satisfy the high availability and reliability levels required for such critical services.

Consequently, a growing body of work has attempted to address this problem and to propose solutions to improve the reliability of cloud services and eventually provide more stringent guarantees to cloud users. This chapter provides a comprehensive literature survey on this particular topic. It first lays out cloud computing and survivability-related concepts, and then covers recent studies that analyzed and characterized the types of failures found in cloud environments. Subsequently, the authors survey and compare the solutions put forward to enhance cloud services' fault-tolerance and to guarantee high availability of cloud resources.

1.3.11 Cloud Security

Security has always been a key issue for cloud-based services and several solutions have been proposed to protect the cloud from malicious attacks. In particular, intrusion detection systems (IDS) and intrusion prevention systems (IPS) have been widely deployed to improve cloud security and have been recently empowered with new technologies like SDN to further enhance their effectiveness. For instance, the SDN technology has been leveraged to dynamically reconfigure the cloud network and services and better protect them from malicious traffic. In this context, this chapter introduces FlowIPS, an OpenFlow-based IPS solution for intrusion prevention in cloud environments. FlowIPS implements SDN-based control functions based on Open vSwitch (OVS) and provides novel Network Reconfiguration (NR) features by programming POX controllers. Finally, the chapter presents the performance evaluation of FlowIPS that demonstrates its efficiency compared to traditional IPS solutions.

1.3.12 Big Data on Clouds

Big data has emerged as a new term that describes all challenges related to the manipulation of large amounts of data including data collection, storage, processing, analysis, and visualization.

This chapter articulates some of the success enablers for deploying Big Data on Clouds (BDOC). It starts by providing some historical perspectives and by describing emerging Internet services and applications. It then describes some legal issues related to big data on clouds. In particular, it highlights emerging hybrid big data management roles, the development and operations (DevOps), and Site Reliability Engineering (SRE). Finally, the chapter discusses science, technology, engineering, and mathematics (STEM) talent cultivation and engagement, as an enabler to technical succession and future success for global enterprises of big data on clouds.

1.3.13 Scientific Applications on Clouds

In order to cope with the requirements of scientific applications, cloud providers have recently proposed new coordination and management tools and services (e.g., Amazon Simple WorkFlow or SWF) in order to automate and streamline task processes executed by the cloud applications. Such services allow to specify the dependencies between the tasks, their order of execution and make it possible to track their progress and the current state of each of them. In this context, a compelling challenge is to ensure the compatibility between existing workflow systems and to provide the possibility to reuse scientific legacy code.

This chapter presents a software engineering solution that allows the scientific workflow community to use Amazon cloud via a single front-end converter. In particular, it describes a wrapper service for executing legacy code using Amazon SWF. The chapter also describes the experimental results demonstrating that the automatically SWF application generated by the wrapper provides a performance comparable to the native manually optimized workflow.

1.3.14 Interactive Multimedia Applications on Clouds

The booming popularity of cloud computing has led to the emergence of a large array of new applications such as social networking, gaming, live streaming, TV broadcasting, and content delivery. For instance, cloud gaming allows direct on-demand access to games whose content is stored in the cloud and streamed directly to end users through thin clients. As a result, less powerful game consoles or computers are needed as most of the processing is carried out in the hosting cloud, leveraging its seemingly unlimited resources. Another prominent cloud application is the Massive user-generated content (UGC) live streaming that allows each simple Internet user to become a TV or content provider. A similar application that has become extremely popular is time-shifting on-demand TV as many services like catch-up TV (i.e., the content of a TV channel is recorded for many days and can be requested on demand) and TV surfing (i.e., the possibility of pausing, forwarding or rewinding of a video stream) have recently became widely demanded. Naturally, the cloud is the ideal platform to host such services as it provides the processing and storage capacity required to ensure a high quality of service. However, several challenges are not addressed yet especially because of the stringent performance requirements (e.g., delay) of such multimedia applications and the increasing amounts of traffic they generate.

This chapter discusses the deployment of these applications over the cloud. It starts by laying out content delivery models in general, and then provides a detailed study of the performance of three prominent multimedia cloud applications, namely cloud gaming, massive user-generated content live streaming and time-shifting on-Demand TV.

1.4 CONCLUSION

Editing and preparing a book on such an important topic is a challenging task requiring a lot of effort and time. As the editors of this book, we are grateful to many individuals who contributed to its successful completion. We would like to thank the chapters' authors for their high-quality contributions, the reviewers for their insightful comments and feedback, and the book series editors for their support and guidance. Finally, we hope that the reader finds the topics and the discussions presented in this book informative, interesting, and inspiring and pave the way for designing new cloud platforms able to meet the requirements of future Internet applications and services.

REFERENCES

1. J. Dean and S. Ghemawat, "MapReduce: Simplified data processing on large clusters," *Communications of the ACM*, vol. 51, no. 1, pp. 107–113, 2008.
2. L. M. Vaquero, L. Rodero-Merino, J. Caceres, and M. Lindner, "A break in the clouds: Towards a cloud definition," *ACM SIGCOMM Computer Communication Review*, vol. 39, no. 1, pp. 50–55, 2008.
3. P. Mell and T. Grance, "The NIST definition of cloud computing (draft)," *NIST Special Publication*, vol. 800, no. 145, p. 7, 2011.
4. P. Barham, B. Dragovic, K. Fraser, S. Hand, T. Harris, A. Ho, R. Neugebauer, I. Pratt, and A. Warfield, "Xen and the art of virtualization," *ACM SIGOPS Operating Systems Review*, vol. 37, no. 5, pp. 164–177, 2003.
5. A. Kivity, Y. Kamay, D. Laor, U. Lublin, and A. Liguori, "KVM: The linux virtual machine monitor," in *Proceedings of the Linux Symposium*, vol. 1, Dttawa, Dntorio, Canada, 2007, pp. 225–230.
6. F. Guthrie, S. Lowe, and K. Coleman, *VMware vSphere Design*. John Wiley & Sons, Indianapolis, IN, 2013.
7. A. Fox, R. Griffith, A. Joseph, R. Katz, A. Konwinski, G. Lee, D. Patterson, A. Rabkin, and I. Stoica, "Above the clouds: A berkeley view of cloud computing," *Department of Electrical Engineering and Computer Sciences, University of California, Berkeley, CA, Rep. UCB/EECS*, vol. 28, p. 13, 2009.
8. A. Greenberg, J. Hamilton, N. Jain, S. Kandula, C. Kim, P. Lahiri, D. Maltz, P. Patel, and S. Sengupta, "VL2: A scalable and flexible data center network," in *Proceedings ACM SIGCOMM*, Barcelona, Spain, August 2009.
9. R. Mysore, A. Pamboris, N. Farrington, N. Huang, P. Miri, S. Radhakrishnan, V. Subramanya, and A. Vahdat, "PortLand: A scalable fault-tolerant layer 2 data center network fabric," in *Proceedings ACM SIGCOMM*, Barcelona, Spain, August 2009.

10. J. Mudigonda, P. Yalagandula, B. Stiekes, and Y. Pouffary, "NetLord: A scalable multi-tenant network architecture for virtualized datacenters," in *Proceedings ACM SIGCOMM*, Toronto, Dntorio, Canada, August 2011.
11. N. McKeown, T. Anderson, H. Balakrishnan, G. Parulkar, L. Peterson, J. Rexford, S. Shenker, and J. Turner, "Openflow: Enabling innovation in campus networks," *SIGCOMM Computer Communnication Review*, vol. 38, no. 2, pp. 69–74, March 2008.
12. S. Gilbert and N. Lynch, "Brewer's conjecture and the feasibility of consistent, available, partition-tolerant web services," *ACM SIGACT News*, vol. 33, no. 2, pp. 51–59, 2002.
13. S. Ghemawat, H. Gobioff, and S.-T. Leung, "The Google file system," in *ACM SIGOPS Operating Systems Review*, vol. 37, no. 5, pp. 29–43, 2003.
14. G. DeCandia, D. Hastorun, M. Jampani, G. Kakulapati, A. Lakshman, A. Pilchin, S. Sivasubramanian, P. Vosshall, and W. Vogels, "Dynamo: Amazon's highly available key-value store," in *ACM SIGOPS Operating Systems Review*, vol. 41, no. 6, pp. 205–220, 2007.
15. A. Lakshman and P. Malik, "Cassandra: A decentralized structured storage system," *ACM SIGOPS Operating Systems Review*, vol. 44, no. 2, pp. 35–40, 2010.
16. M. Isard, M. Budiu, Y. Yu, A. Birrell, and D. Fetterly, "Dryad: Distributed data-parallel programs from sequential building blocks," *ACM SIGOPS Operating Systems Review*, vol. 41, no. 3, pp. 59–72, 2007.
17. B. Hindman, A. Konwinski, M. Zaharia, A. Ghodsi, A. D. Joseph, R. Katz, S. Shenker, and I. Stoica, "MESOS: A platform for fine-grained resource sharing in the data center," in *Proceedings of the 8th USENIX Conference on Networked Systems Design and Implementation*, Boston, MA, 2011, pp. 22–22.
18. M. Schwarzkopf, A. Konwinski, M. Abd-El-Malek, and J. Wilkes, "Omega: Flexible, scalable schedulers for large compute clusters," in *Proceedings of the 8th ACM European Conference on Computer Systems*, Prague, Czech Republic. ACM, New York, 2013, pp. 351–364.
19. K. Ousterhout, P. Wendell, M. Zaharia, and I. Stoica, "Sparrow: Distributed, low latency scheduling," in *Proceedings of the Twenty-Fourth ACM Symposium on Operating Systems Principles*, Farmington, PA. ACM, New York, 2013, pp. 69–84.

2

VIRTUALIZATION IN THE CLOUD

Lisandro Zambenedetti Granville, Rafael Pereira Esteves, and
Juliano Araujo Wickboldt

*Institute of Informatics, Federal University of Rio Grande do Sul,
Porto Alegre, Brazil*

2.1 THE NEED FOR VIRTUALIZATION MANAGEMENT IN THE CLOUD

Cloud infrastructures are aggregates of computing, storage, and networking resources deployed along centralized or distributed data centers devoted to support companies' applications or, in the case of services being offered through the Internet, to support cloud customers' applications. Companies such as Google, Amazon, Facebook, and Microsoft rely on cloud infrastructures to support various services such as Web search, e-mail, social networking, and e-commerce. By leasing physical infrastructure to external customers, cloud providers encourage the development of novel services and, at the same time, generate revenue to cover deployment and operation costs of clouds. Cloud resource sharing is then critical for the cloud computing model.

To allow multiple customers, cloud providers rely on virtualization technologies to build *virtual infrastructures* (VIs) comprising logical instances of physical resources (e.g., servers, network, and storage). The provisioning of VIs must consider requirements of both cloud providers *and* customers. While the main objective of cloud providers is

Cloud Services, Networking, and Management, First Edition.
Edited by Nelson L. S. da Fonseca and Raouf Boutaba.

to generate revenue by accommodating a large number of VIs, customers, in their turn, have specific needs, such as storage capacity, high availability, processing power (usually represented by the number of leased virtual machines or VMs), guaranteed bandwidth among VMs, and load balancing. Inefficiencies in the provisioning process can lead to negative consequences for cloud providers, including customer defection, financial penalties when service-level agreements (SLAs) are not satisfied, and low utilization of the physical infrastructure. In summary, management of physical and VIs is vital to enabling proper cloud resource sharing.

Current cloud provisioning systems allow customers to select among different resource configurations (e.g., CPU, memory, and disk) to build a VI. Customers are the main responsible for choosing the resources that will better fit their application's needs. The cloud provider, in turn, either (a) allocates resources for the VI on physical data centers, or (b) rejects the allocation if there are not enough resources to satisfy the customers' requirements. Cloud providers run allocation algorithms to find the best way to map VIs onto the physical substrate according to well-defined objectives, such as minimizing the allocation cost, reducing energy consumption, or maximizing residual capacity of the infrastructure. Mapping virtual to physical resources is commonly referred to as *embedding* and has been extensively studied in the context of network virtualization [1–3].

Embedding is an example of a network virtualization aspect that needs to be properly managed. Choosing the appropriate embedding algorithm, and deciding when it should be triggered (e.g., when new VI requests arrive or when on-the-fly optimizations of the physical substrate are needed) is a management activity that needs to be consciously performed by the cloud management operator or team. Other virtualization management aspects encompass operations such as: monitoring, to detect abusing applications/customers; configuration, to tune VI and physical substrate; and discovery, to identify collaborating VIs that would be better placed closer to one another. In addition to management operations, virtualization management requires understanding of the diversity of target elements because both VI and physical substrate are quite heterogeneous in regard to the resources they use/offer. That impacts the management operations themselves, since, for example, monitoring and configuring a physical server can be quite different than monitoring and configuring network devices and traffic. Operations and target elements are thus two important dimensions of cloud virtualization management.

In this chapter, we cover the management of virtualization in the cloud. Our observations primarily take the perspective of cloud providers who need to manage their substrate and hosted VIs to guarantee that the services offered to customers are operating properly. Virtualization management is a quite new discipline because virtualization itself, at least as it has been employed these days, is also quite recent. Management is achieved by borrowing techniques from other areas, such as network management and parallel processing. We concentrate our discussion on the two management dimensions mentioned before, i.e., management operations and target elements. Although other dimensions do exist, we will focus on the operations and elements because they are the essential dimensions a cloud manager needs to take into account in the first place.

The remaining of this chapter is organized as follows. In Section 2.2, we review some basic concepts of virtualization in cloud computing environments. In Section 2.3, we describe the main elements of a virtualized cloud environment. In Section 2.4, we list the main virtualization-related operations that need to be supported by a cloud platform. In Section 2.5, we review some of the most important efforts towards the definition of open standard interfaces to support virtualization and interoperability in the cloud. In Section 2.6, we list some of the most important efforts currently targeted to build tools and systems for virtualization and cloud resource management. Finally, in Section 2.7, we list key challenges that can guide future developments in the virtualization management and also mention some ongoing research in the field.

2.2 BASIC CONCEPTS

Clouds can be public, private, or hybrid. Public clouds offer resources to any interested tenant (e.g., Amazon EC2 and Windows Azure). Private clouds usually belong to a single organization and only the members of that organization have access to the resources. Hybrid clouds are combinations of different types of cloud. For example, a private cloud that needs to temporarily extend its capacity can borrow resources from a public cloud, thus forming a hybrid cloud.

Cloud services are organized according to three basic business models. In infrastructure as a service (IaaS), cloud providers offer logical instances of physical resources, such as VMs, virtual storage, and virtual links to interested tenants. In platform as a service (PaaS), tenants can request a computing platform including an operating system and a development environment. The software as a service (SaaS) model offers end-applications (e.g., Google Drive and Dropbox) to customers. Other models, such as network as a service (NaaS) are also possible in the cloud, but are not found so frequently in the literature [4].

Virtualization is the key technology to enable cloud computing. Virtualization abstracts the internal details of physical resources and enables resource sharing. Using virtualization, a physical resource (e.g., server, router, link) can be shared among different users or applications. The core of cloud computing environments is based on virtualized data centers, which are facilities consisting of computing servers, storage, network devices, cooling, and power systems.

Virtualization can be accomplished by different technologies according to the target element. Server virtualization, for example, relies on a layer of software called hypervisor, also known as VM monitor. The hypervisor is the component responsible for actually creating, removing, copying, migrating, and running VMs. Virtual links, in turn, can be created by configuring Ethernet VLANs between the physical nodes hosting the virtual ones. Multiprotocol label switching (MPLS) label switched paths (LSPs) and generic routing encapsulation (GRE) tunnels are other candidates to establish virtual links.

Participants in the cloud comprise two main roles: the cloud provider, also known as infrastructure provider, owns the physical resources that can be leased to one or more tenants, also known as service providers, who build VIs composed of virtual instances of computing, storage, and networking resources. A VI can be also referred to as a cloud slice. After the instantiation of a VI, tenants can deploy a variety of applications that will rely on these virtual resources.

A cloud platform is a software that allows tenants to request and instantiate VIs. Tenants can specify the amount of resources to build their VIs and the specific characteristics of each resource, such as CPU and memory for computing, disk size for storage, and bandwidth capacity for links. The cloud platform then interacts with the underlying virtualization software (hypervisor) to create and configure the VI. In order to facilitate resource management and allow interoperability, cloud platforms offer specific interfaces for applications running in VIs. Such interfaces define operations that can be executed in the cloud platform.

2.3 VIRTUALIZED ELEMENTS

As stated in previous sections, virtualization plays a key role in modern cloud computing environments by improving resource utilization and reducing costs. Typical elements that can be virtualized in a cloud computing environment include computing and storage. Recently, virtualization has been extended also to the networking domain and can overcome limitations of current cloud environments such as poor isolation and increased security risks [5]. In this section, we describe the main elements that can be virtualized in a cloud computing environment.

2.3.1 Computing

The virtualization of computing resources (e.g., CPU and memory) is achieved by server virtualization technologies (e.g., VMWare, Xen, and QEMU) that allow multiple virtual machines (VMs) to be consolidated in a single physical one. The benefits of server virtualization for cloud computing include performance isolation, improved application performance, and enhanced security.

Cloud computing providers deploy their infrastructure in data centers comprising several virtualized servers interconnected by a network. In the IaaS model, VMs are instantiated and allocated to customers (i.e., tenants) on-demand. Server virtualization adds flexibility to the cloud because VMs can be dynamically created, terminated, copied, and migrated to different locations without affecting existing tenants. In addition, the capacity of a VM (i.e., CPU, memory, and disk) can be adjusted to reflect changes in tenants' requirements without hardware changes.

Cloud operators have flexibility to decide where to allocate VMs in the physical servers considering diverse criteria such as cost, energy consumption, and performance. In this regard, several VM allocation schemes have been proposed in the literature that leverage VM flexibility to optimize resource utilization [6–8, 10–12, 20].

2.3.2 Storage

Storage virtualization consists of grouping multiple (possibly heterogeneous) storage devices that are seen as a single virtual storage space. There are two main abstractions to represent storage virtualization in clouds: virtual volumes and virtual data objects. The virtualization of storage devices as virtual volumes is important in this context because it simplifies the task of assigning disks to VMs. Furthermore, many implementations also include the notion of virtual volume pools, which represent different sources of available virtualizable storage spaces to allocate virtual volumes from (e.g., separate local physical volumes or a remote Network File System or NFS). On the other hand, cloud storage of virtual data objects enables scalable and redundant creation and retrieval of data objects directly into/from the cloud. This abstraction is also often accompanied by the concept of containers, which in general serve to create a hierarchical structure of data objects similar to files and folders on any operating system.

Storage virtualization for both volumes and data objects is of utmost importance to enable the elasticity property of cloud computing. For example, VMs can have their disk space adjusted dynamically to support changes in cloud application requirements. Such adjustment is too complex and dynamic to be performed manually and, by virtualizing storage, cloud providers offer a uniform view to their users and reduce the need for manual provisioning. Also, with storage virtualization, cloud users do not need to know exactly where their data are stored. The details of which disks and partitions contain which objects or volumes are transparent to users, which also facilitates storage management for cloud providers.

2.3.3 Networking

Cloud infrastructures rely on local and wide area networks to connect the physical resources (i.e., servers, switches, and routers) of their data centers. Such networks are still based on the current IP architecture that has a number of problems. These problems are mainly related to the lack of isolation, which can allow that one VI or application interferes with another, resulting in poor performance or, even worse, in security problems. Another issue is the limited support for innovation, which hinders the development of new architectures that could suit better cloud applications.

To overcome the limitations of current network architectures, virtualization can also be extend to the cloud networks. ISP network virtualization has been a hot topic of investigation in recent years [13, 14] and is now being considered in other contexts, such as cloud networking. Similar to virtualized ISP networks, in virtualized cloud networks, multiple virtual networks (VNs) share a physical network and run isolated protocol stacks. A VN is part of a VI that comprises VN nodes (i.e., switches and routers) and virtual links.

The advantages of virtualization of cloud networks include network performance isolation, improved security, and the possibility to introduce new protocols and addressing schemes without disrupting production services. Figure 2.1 shows how virtualization can be tackled in cloud network infrastructures. In the substrate layer, physical nodes and links from different network administrative domains serve as a substrate for the deployment of VNs. Physical nodes, at the core of the physical networks, represent network devices (e.g., switches and routers) that internally run virtual (or logical) routers instantiated to serve VNs' routing necessities.

In the virtualization layer, virtual nodes and links are created on the top of the substrate and combined to build VNs. A VN can use resources from different sources,

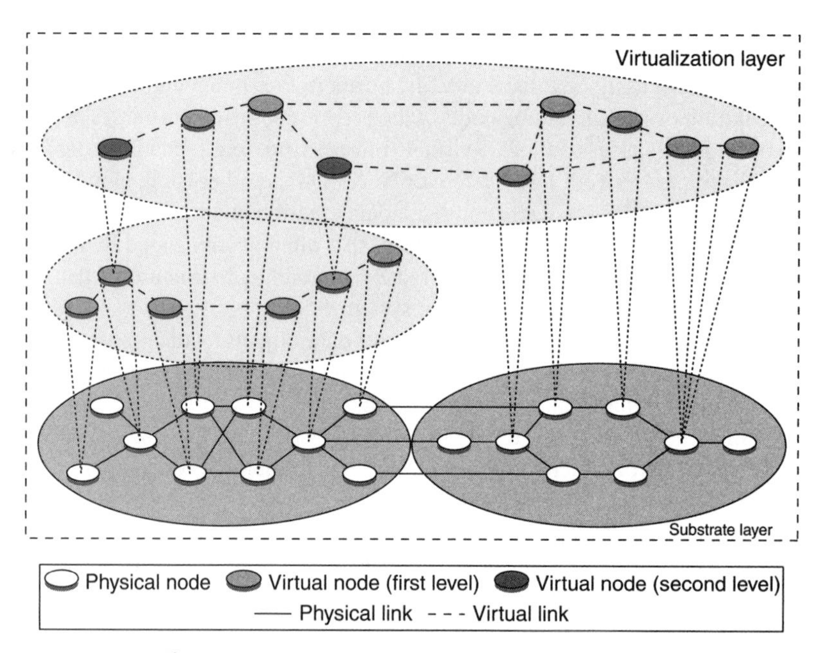

Figure 2.1. Virtualized cloud network infrastructure.

including resources from other VNs, which in this case results in a hierarchy of VNs. VNs can also be entirely placed into a single physical node (e.g., physical end-host). In this case, since virtual links are not running on top of any physical counterpart, isolation, and performance guarantees should be offered, for example, through memory isolation and scheduling mechanisms. In another setup, VNs can spread across different adjacent physical infrastructures (i.e., different administrative domains). In this case, network operators, at the substrate layer, must cooperate to provide a consistent view of the underlying infrastructure used by networks from the virtualization layer.

2.3.4 Management

The management of cloud infrastructures plays a key role to allow cloud providers to efficiently use the resources and increase revenue. At the virtual level, each VI can operate its own management protocols, resource allocation schemes, and monitoring tools. For example, one tenant can use Simple Network Management Protocol (SNMP) to manage his/her VIs, while other can use NETCONF or Web services.

Different resource allocation schemes tailored for specific cloud applications define how virtual resources are mapped in the data center. Adaptive, application-driven resource provisioning allows multiple tenants and a large diversity of applications to efficiently share a cloud infrastructure.

Monitoring is another management aspect that can be virtualized. Once a new VI is created, a set of monitoring tools need to be configured [15] in order to start monitoring the computing resources that form the VI. The set of monitoring tool configurations and the corresponding monitored metrics is referred to as a monitoring slice. Every VI is coupled with a monitoring slice [16]. To monitor the computing resources that form VIs, cloud operators generally use in their monitoring slices tools with native support to cloud platforms.

2.4 VIRTUALIZATION OPERATIONS

Virtualization operations are structured according to the components described in the previous section: *computing*, *storage*, *networking*, and *management*. A non-exhaustive list of the main virtualization operations derived from existing cloud platforms [9, 17–19, 21] is described next.

- Computing (Virtual Machines)
  - *Create/Remove*: defines/undefines the internal representation of a VM with its specified characteristics (*e.g.*, CPU, memory, and guest image).
  - *Deploy/Undeploy*: defines/undefines a VM within the hypervisor of a node of the cloud infrastructure, including the transfer/removal of the image file.
  - *Start/Stop/Suspend/Resume*: basic operations to handle the state of the guest operating system.
  - *Migrate*: undefines a VM in one node and defines it on another. The destination node needs to be specified.
  - *Modify*: modifies the attributes of a VM.
  - *Snapshot*: creates a snapshot of a VM.
  - *Restore*: restores a VM from a snapshot.
  - *List*: lists currently deployed VMs.
- Computing (Images)
  - *Create/Remove*: defines/undefines a guest operating system image at the main repository of the platform, including the transfer/removal of the file.
- Storage (Virtual Volumes)
  - *Create/Remove*: allocates/deletes chunks of storage on nodes.
  - *Attach Volume to VM*: attaches a volume file to a given VM.
- Storage (Virtual Volume Pools)
  - *Create/Remove*: defines/undefines a pool for storing virtual volumes (typically a local or remote/NFS directory).
  - *Add Volume*: adds a virtual volume to a volume pool.
- Storage (Virtual Data Objects)
  - *Create/Remove*: allocates/deletes storage for data objects on nodes.
  - *Upload/Download*: transfers the actual data in and out of the cloud environment.
  - *Stream*: sends the content out to the general public, sometimes even adopting massive scale distribution employing concepts of Content Delivery Networks (CDNs).

- Storage (Virtual Containers)
  - *Create/Remove*: defines/undefines a container for storing data objects.
  - *Add Data Object*: adds a virtual data object to a container.
- Networking (Virtual Links)
  - *Create/Remove*: defines/undefines the internal representation of a virtual link that connects point-to-point virtual interfaces of two virtual devices (i.e., VMs or virtual routers).
  - *Establish/Disable*: establishes/disables the virtual link within the network, enabling/disabling traffic to flow between the connected devices.
  - *Configure*: configures additional parameters of a virtual link (e.g., bandwidth).
- Networking (Virtual Routers)
  - *Create/Remove*: defines/undefines the internal representation of a virtual router that has multiple virtual ports to interconnect multiple virtual interfaces of virtual devices.
  - *Deploy/Undeploy*: deploys/undeploys the virtual router into a node of the infrastructure.
  - *Add/Edit/Remove Routes*: defines/modifies/undefines routes for a virtual router.
- Management (Virtual Devices)
  - *Monitor/Unmonitor*: deploys/undeploys the monitoring infrastructure required to monitor a given virtual device.
  - *Get Monitoring Information*: fetches monitoring information within the monitoring system for a given virtual device.
- Management (Events)
  - *Create/Remove*: defines/undefines the internal representation of an event that belongs to a specific slice or operates in global scope.
  - *Deploy/Undeploy*: deploys/undeploys the event on the monitoring infrastructure to be triggered on demand.
- Management (Physical)
  - *Discover Resources*: this is actually a collection of operations to discover nodes and network topology available on the infrastructure. This collection also retrieves information about resource allocation on these physical elements.
  - *Get Monitoring Information*: fetches monitoring information within the monitoring system for a given physical device (e.g., node or switch).

2.5 INTERFACES FOR VIRTUALIZATION MANAGEMENT

Today, there are many heterogeneous cloud platforms that support the provisioning of virtualized infrastructures under a plethora of different specifications and technologies. Each cloud provider chooses the platform that suits it better or designs its own platforms to provide differentiated services to its tenants. The problem with this heterogeneity is that it hinders interoperability and causes vendor lock-in for tenants. In order to allow the remote management of virtual elements, many platforms already offer specific interfaces (e.g., Amazon EC2/S3, Elastic Hosts, Flexiscale, Rackspace Cloud Servers, and VMware vSphere) to communicate with external applications.

To cope with this variety of technologies and support the development of platform-agnostic cloud applications, some proposals use basically two different approaches: (1) employing proxy-style APIs in order to communicate with multiple providers using a set of technology-specific adapters and (2) creating standardized generic interfaces to be implemented by cloud platforms. The first approach has a drawback of introducing an additional layer of software in cloud systems, which results in overhead and increased latency. Nevertheless, there are libraries and tools that are widely employed, such as Apache Deltacloud and Libcloud, which are further discussed in the next section. The second approach, on the other hand, represents a more elegant solution to the problem by proposing some sort of *lingua franca* to communicate among cloud systems. The problem with standardization is to make participants to agree onto the same standard [22]. Ideally, a standardized interface should be open and extensible to allow widespread adoption by cloud management platforms and application developers. In this section, we review some of the most important efforts towards the definition of open standard interfaces to support virtualization and interoperability in the cloud.

2.5.1 Open Cloud Computing Interface

The Open Cloud Computing Interface (OCCI) [23] introduces a set of open, community-driven specifications to deal with cloud service resource management [24]. OCCI is supported by the Open Grid Forum and was originally conceived to create a remote management API for IaaS platforms allowing interoperability for common tasks, such as deployment, scaling, and monitoring virtual resources. Besides the definition of an open application programming interface (API), this specification also introduces a RESTful Protocol for exchanging management information and actions. The current release of OCCI is not anymore focused only in IaaS and includes other cloud business models, such as PaaS and SaaS.

The current version of the specification[1] is designed to be modular and extensible, thus it is split in three complementary documents. The OCCI Core document (GFD.183) describes the formal definition of the OCCI Core Model. This document also describes how the core model can be interacted with renderings (including associated behaviors) and expanded through extensions. The second document is OCCI Infrastructure (GFD.184), which contains the definition of the OCCI infrastructure extension for the IaaS domain. This document also defines additional resource types, their attributes, and actions that each resource type can perform. The third document, OCCI HTTP Rendering (GFD.185), defines means of interacting with the OCCI Core Model through the RESTful OCCI API. Moreover, this document defines how the OCCI Core Model can be communicated and serialized over HTTP.

[1] As of the ending of 2013, the current version of OCCI is v1.1 (release date April 7, 2011)

The OCCI Infrastructure document describes the modeling of virtual resources in IaaS as three basic element types: (1) compute that are information processing resources, (2) storage that are intended to handle information recording, and (3) network representing L2 networking elements (e.g., virtual switches). Also, there is an abstraction for creation of links between resources. Links can be of two types: i.e., Network Interface or Storage Link, depending on the type of resource they connect. It is also possible to use this specification to define Infrastructure Templates, which are predefined virtual resource specifications (e.g., small, medium, and large VM configurations). Moreover, the OCCI HTTP Rendering document complements these definitions by specifying management operations, such as creating, retrieving, updating, and deleting virtual resources. The document also details general requirements for the transmission of information over HTTP, such as security and authentication.

OCCI is currently implemented in many popular cloud management platforms, such as OpenStack, OpenNebula, and Eucalyptus. There are also base implementations in programming languages, such as rOCCI in Ruby and jclouds in Java, and automated compliance tests with doyouspeakOCCI. One particular effort aims to improve the intercloud networking standardization by proposing an extension to OCCI, called Open Cloud Networking Interface (OCNI) [25]. There is also a reference implementation of OCNI called pyOCNI, written as a Python framework including JSON serialization for resource representation.

2.5.2 Open Virtualization Format

The Open Virtualization Format (OVF) [26], currently in version 2.0.1, was introduced late in 2008 within the Virtualization Management initiative of the Distributed Management Task Force (DMTF), aiming to provide an open and extensible standard for packaging and distribution of software to be run in VMs. Its main virtue is to allow portability of virtual appliances onto multiple platforms through so-called OVF Packages, which may contain one or more virtual systems. The OVF standard is not tied to any particular hypervisor or processor architecture. Nevertheless, it is easily extensible through the specification of vendor-specific metadata included in OVF Packages.

OVF Packages are a core concept of the OVF specification, which consist of several files placed into one directory describing the structure of the packed virtual systems. An OVF Package includes one OVF Descriptor, which is an XML document containing metadata about the package contents, such as product details, virtual hardware requirements, and licensing. The OVF Package may also include certificates, disk image files, or ISO images to be attached to virtual systems.

Within an OVF Package, an Envelope Element describes all metadata for the VMs included in the package. Among this metadata, a detailed Virtual Hardware Description (based on CIM classes) can specify all types of virtual hardware resources required by a virtual system. This specification can be abstract or incomplete, allowing the virtualization platform to decide how to better satisfy the resource requirements, as long as the required virtual devices are deployed. Moreover, OVF Environment information can be added to define how the guest software and the deployment platform interact.

This environment allows the guest software to access information about the deployment platform, such as the values specified for the properties defined in the OVF Descriptor.

This standard is present in many hypervisor implementations and has shown to be very useful for migrating virtual systems information among many hypervisors or platforms, since it allows precise description of VMs and virtual hardware requirements. However, it is not within the objectives of OVF to provide detailed specification for complete VIs (i.e., detailing interconnections, communication requirements, and network elements).

2.5.3 Cloud Infrastructure Management Interface

The Cloud Infrastructure Management Interface (CIMI) [27] standard is another DMTF proposal within the context of the Cloud Management initiative. This standard defines a model and a protocol for managing interactions between cloud IaaS providers and tenants. CIMI's main objective is to provide tenants with access to basic management operations on IaaS resources (VMs, storage, and networking), facilitating portability between different cloud implementations that support this standard. CIMI also specifies a RESTful protocol over HTTP using both JSON or XML formats to represent information and transmit management operations.

The model defined in CIMI includes basic types of virtualized resources, where Machine Resources are used to represent VMs, Volume Resources for storage, and Network Resources for VN devices and ports. Besides, CIMI also defines a Cloud Entry Point type of resource, which represents a catalog of virtual resources that can be queried by a tenant. A System Resource in this standard gathers one or more Network, Volume, or Machine Resources, and can be operated as a single resource. Finally, a Monitoring Resource is also defined to track progress of operations, metering, and monitoring of other virtual resources.

The protocol relies on basic HTTP operations (i.e., PUT, GET, DELETE, HEAD, and POST) and uses either JSON or XML to transmit the message body. To manipulate virtual resources, there are four basic create, read, update, and delete (CRUD) operations. It is also possible to extend the protocol by creating or customizing operations to manipulate the state of each particular resource. Moreover, the CIMI specification can also be integrated with OVF, in which case VMs represented as OVF Packages can be used to create Machine Resources or System Resources.

Today, implementations of the CIMI standard are not so commonly found as OCCI or OVF are. One specific implementation that is worth noting is found within the Apache Deltacloud[2] project, which exposes a CIMI REST API to communicate with external applications supporting manipulation of Machine and Volume Resources abstractions.

2.5.4 Cloud Data Management Interface

The Cloud Data Management Interface (CDMI) [28] is a standard specifically targeted to define an interface to access cloud storage and to manage data objects. CDMI is

[2] http://deltacloud.apache.org/cimi-rest.html

comparable to Amazon's S3 [29], with the fundamental difference that it is conceived by the Storage Networking Industry Association (SNIA) to be an open standard targeted for future ANSI and ISO certification. This standard also includes a RESTful API running over HTTP to allow accessing capabilities of cloud storage systems, allocating and managing storage containers and data objects, handling users and group access rights, among other operations.

The CDMI standard defines a JSON serializable interface to manage data stored in clouds based on several abstractions. Data objects are fundamental storage components analogous to files within a file system, which include metadata and value (contents). Container objects are intended to represent grouping of data, analogous to directories in regular file systems; this abstraction links together zero or more Data objects. Domain objects represent the concept of administrative ownership of data stored within cloud systems. This abstraction is very useful to facilitate billing, to restrict management operations to groups of objects, and to represent hierarchies of ownership. Queue objects provide first-in, first-out access to store or retrieve data from the cloud system. Queuing provides a simple mechanism for controlling concurrency when reading and writing Data objects in a reliable way. To facilitate interoperability, this standard also includes mechanisms for exporting data to other network storage platforms, such as iSCSI, NFS, and WebDAV.

Regarding implementations, CDMI is also not so commonly deployed in most popular cloud management platforms. SNIA's Cloud Storage Technical Working Group (TWG) provides a Reference Implementation for the standard, which is currently a working draft and provides support only for version 1.0 of the specification. Some independent projects, such as CDMI add-on for OpenStack Swift and the CDMI-Serve in Python, have implemented basic support for the CDMI standard but do not present much recent activity.

Besides all the aforementioned efforts to create new standardized interfaces for virtual resource management in cloud environments, other approaches, protocols, and methods have been studied and may be of interest in particular situations [30]. Moreover, many organizations, such as OASIS, ETSI, ITU, NIST, and ISO, are currently engaged with their cloud and virtualization related working groups on developing standards and recommendations. We recommend the interested reader to look at DMTF's maintained wiki page Cloud-Standards.org[3] to keep track of future standardization initiatives.

2.6 TOOLS AND SYSTEMS

In this section, we list some of the most important efforts currently targeted to build tools and systems for virtualization and cloud resource management. Initially, we describe open source cloud management platforms, which are in fact complete solutions to deploy and operate private, public, or hybrid clouds. Afterwards, we discuss some tools and

[3] http://cloud-standards.org/

libraries to perform specific operations for virtual resource management and cloud integration.

2.6.1 Open Source Cloud Management Platforms

2.6.1.1 Eucalyptus. Eucalyptus started as a research project in the Computer Science Department at the University of California, Santa Barbara, in 2007, within a project called Virtual Grid Application Development Software Project (VGrADS) funded by the National Science Foundation. This is one of the first open source initiatives to build cloud management platforms that allow users to deploy their own private clouds [18]. Currently, Eucalyptus is in version 3.4 and comprises full integration with Amazon Web Services (AWS)—including EC2, S3, Elastic Block Store (EBS), Identity and Access Management (IAM), Auto Scaling, Elastic Load Balancing (ELB), and CloudWatch—enabling both private and hybrid cloud deployments.

Eucalyptus architecture is based on four high-level components: (1) Node Controller executes at hosts and is responsible for controlling the execution of VM instances; (2) Cluster Controller works as a front-end at the cluster-level (i.e., Availability Zone) managing VM execution and scheduling on Node Controllers, controls cluster-level SLAs, and also manages VNs; (3) Storage Controller exists both at cluster-level and at cloud-level (Walrus) and implements a put/get SaaS solution based on Amazon's S3 interface, providing a mechanism for storing and accessing VM images and user data; and (4) Cloud Controller is the entry-point into the cloud for users and administrators, it implements an EC2-compatible interface and coordinates other components to perform high-level tasks, such as authentication, accounting, reporting, and quota management.

For networking, Eucalyptus offers four operating modes: (1) Managed, in which the platform manages layers 2 and 3 VM isolation, employing a built-in DHCP service. This mode requires a switch to forward a configurable range of VLAN-tagged packets; (2) Managed (no VLAN), in which only layer 3 VM isolation is possible; (3) Static, where there is no VM isolation, employs a built-in DHCP service for static IP assignment; and (4) System, where there is also no VM isolation and, in this case, no automatic address handling since Eucalyptus will rely on an existing external DHCP service. In version 4.0, released in April 2014, Eucalyptus has introduced new functionality for networking support through a new Edge Networking Mode.

The main technical characteristics of the Eucalyptus platform as the following:

- *Programming Language*: Written mostly in C and Java
- *Compatibility/Interoperability*: Fully integrated with AWS
- *Supported Hypervisors*: vSphere, ESXi, KVM, any AWS-compatible clouds
- *Identity Management*: Role-Based Access Control mechanisms with Microsoft Active Directory or LDAP systems
- *Resource Usage Control*: resource quotas for users and groups
- *Networking*: Basic support with four operating modes
- *Monitoring*: CloudWatch

- *Version/Release*: 3.4.1 (Released on December 16, 2013)
- *License*: GPL v3.0

2.6.1.2 OpenNebula. In its early days OpenNebula was a research project at the Universidad Complutense de Madrid. The first version of the platform was released under an open source license in 2008 within the European Union's Seventh Framework Programme (FP7) project called RESERVOIR—Resources and Services Virtualization without Barriers (2008–2011). Nowadays, OpenNebula (version 4.4 Retina released December 3, 2013) is a feature-rich platform used mostly for the deployment of private clouds, but is also capable of interfacing with other systems to work as hybrid or public cloud environment.

OpenNebula is conceptually organized in a three-layered architecture [31]. At the top, the Tools layer comprises higher level functions, such as cloud-level VM scheduling, providing CLI and GUI access for both users and administrators, managing and supporting multi-tier services, elasticity and admission control, and exposing interfaces to external clouds through AWS and OCCI. At the Core layer, vital functions are performed, such as accounting, authorization, and authentication, as well as resource management for computing, storage, networking, and VM images. Also at this layer, the platform implements resource monitoring by retrieving information available from hypervisors to gather updated status of VMs and manages federations, enabling access to remote cloud infrastructures, which can be either partner infrastructures governed by a similar platform or public cloud providers. At the bottom, the Drivers layer implements infrastructure and cloud drivers to provide an abstraction to communicate with the underlying devices or to enable access to remote cloud providers.

OpenNebula allows administrators to set up multiple zones and create federated VIs considering different federation paradigms (e.g., cloud aggregation, bursting, or brokering), in which case each zone operates their network configurations independently. From the user's viewpoint, setting up a network in the OpenNebula platform is restricted to the creation of a DHCP IP range that will be automatically configured in each VM. The administrator can change the way VMs connect to the physical ports of the host machine using one of many options, that is, VLAN 802.1Q to allow isolation, EBtables and Open vSwitch to permit implementation of traffic filtering, and VMware VLANs, which isolate VMs running over VMware hypervisor. It is also possible to deploy Virtual Routers from OpenNebula's Marketplace to work as an actual router, DHCP, or DNS server.

The main technical characteristics of the OpenNebula platform as the following:

- *Programming Language*: C++ (Integration APIs in Ruby, JAVA, and Python)
- *Compatibility/Interoperability*: AWS, OCCI, and XML-RPC API
- *Supported Hypervisors*: KVM, Xen, and VMWare
- *Identity Management*: Sunstone, EC2, OCCI, SSH, x509 certificates, and LDAP
- *Resource Usage Control*: resource quotas for users and groups
- *Networking*: IP/DHPC ranges customizable by users, many options for administrator require manual configuration

- *Monitoring*: Internal, gathers information from hypervisors
- *Version/Release*: 3.4.1 (Released on December 3, 2013)
- *License*: Apache v2.0

2.6.1.3 OpenStack. OpenStack started as a joint project between Rackspace Hosting and NASA around mid 2010, aiming to provide a cloud-software solution to run over commodity hardware [19]. Right after the first official release (beginning of 2011), OpenStack was quickly adopted and packed within many Linux distributions, such as Ubuntu, Debian, and Red Hat. Today, it is the cloud management platform with the most active community counting on more than 13,000 registered people from over 130 countries. OpenStack is currently developed in nine parallel core projects (plus four incubated) all coordinated by the OpenStack Foundation, which is embodied by 9,500 individuals and 850 different organizations.

The OpenStack architecture consists of a myriad of interconnected components, each one developed under a separate project, to deliver a complete cloud infrastructure management solution. Initially, only two components were present, Compute (Nova) and Object Storage (Swift), which respectively provide functionality for handling VMs and a scalable redundant object storage system. Adopting an incremental approach, incubated/community projects were gradually included in the core architecture, such as Dashboard (Horizon) to provide administration GUI access, Identity Service (Keystone) to support a central directory of users mapped to services, and Image Service (Glance) to allow discovery, registration, and delivery of disk and server images. The current release of OpenStack (Havana) includes advanced network configuration with Neutron, persistent block-level storage with Cinder, a single point of contact for billing systems through Ceilometer, and a service to orchestrate multiple composite cloud applications via Heat.

As for networking, a community project called Quantum started in April 2011 and was targeted to further develop the networking support of OpenStack by employing VN overlays in a Connectivity as a Service perspective. From release Folsom on, Quantum was added as a core project and renamed Neutron. Currently, this component lets administrators to employ from basic networking configuration of IP addresses, allowing both dedicated static address assignment and DHCP, to complex configuration with software-defined networking (SDN) technology like OpenFlow. Moreover, Neutron allows the addition of plug-ins to introduce more complex functionality to the platform, such as quality of service, intrusion detection systems, load balancing, firewalls, and virtual private networks.

- *Programming Language*: Python
- *Compatibility/Interoperability*: Nova and Swift are feature-wise compatible to EC2 and S3 (applications need to be adapted though), OCCI support (under development)
- *Supported Hypervisors*: QEMU/KVM over libvirt (fully supported), VMware and XenAPI (partially supported), many others at nonstable development stages
- *Identity Management*: Local database, EC2/S3, RBAC, token-based, SSL, x509 or PKI certificates, and LDAP

- *Resource Usage Control*: configurable quotas per user (tenant) defined by each project
- *Networking*: several options via Neutron component, extensible with plug-ins
- *Monitoring*: simple customizable dashboard relies on information provided by other components
- *Version/Release*: Havana (Released on October 17, 2013)
- *License*: Apache v2.0

2.6.1.4 CloudStack. CloudStack started as a project from a startup company called VMOps in 2008, later renamed Cloud.com, and was first released as open source in mid 2010. After Cloud.com was acquired by Citrix, CloudStack was relicensed to Apache 2.0 and incubated by the Apache Software Foundation in April 2012. Ever since, the project has developed a powerful cloud platform to orchestrate resources in highly distributed environments for both private and public cloud deployments [21].

CloudStack deployments are organized into two basic building blocks, a Management Server and a Cloud Infrastructure. The Management Server is a central point of configuration for the cloud (these servers might be clustered for reliability reasons). It provides a Web user interface and API access, manages the assignment of guest VMs to hosts, allocates public and private IP addresses to particular accounts, manages images, among other tasks. A Cloud infrastructure comprises distributed Zones (typically, data centers) hierarchically organized into Pods, Clusters, Hosts, Primary and Secondary Storage. A CloudStack Cloud Infrastructure may also optionally include Regions (perhaps geographically distributed), to aggregate multiple Zones, and each Region is controlled by a different set of Management Servers, turning the platform into a highly distributed and reliable system. Moreover, a separate Python tool called CloudMonkey is available to provide CLI and shell environments for interacting with CloudStack-based clouds.

CloudStack offers two types of networking configurations: (1) Basic, which is an AWS-style networking providing a single network where guest isolation can be achieved through layer 3 means, such as security groups and (2) Advanced, where more sophisticated network topologies can be created. CloudStack also offers a variety of NaaS features, such as creation of VPNs, firewalls, and load balancers. Moreover, this tool provides the ability to create a Virtual Private Cloud, which is a private, isolated part of CloudStack that can have its own VN topology. VMs in this VN can have any private addresses since they are completely isolated from others.

- *Programming Language*: Mostly Java
- *Compatibility/Interoperability*: CloudStack REST API (XML or JSON)
- *Supported Hypervisors*: XenServer/XCP, KVM, and/or VMware ESXi with vSphere
- *Identity Management*: Internal or LDAP
- *Resource Usage Control*: Usage server separately installed provides records for billing, resource limits per project
- *Networking*: two operating modes, several networking as a service options in advanced configurations

- *Monitoring*: some performance indicators available through the API are displayed to users and administrators
- *Version/Release*: 4.2.0 (Released on October 1, 2013)
- *License*: Apache v2.0

2.6.2 Specific Tools and Libraries

The following describes some tools and libraries mainly designed to deal with the diversity of technologies involved in cloud virtualization. Unlike cloud platforms, these tools do not intend to offer a complete solution for cloud providers. Nevertheless, they play a key role in integration and allow applications to be written in a more generic manner in terms of virtual resource management.

2.6.2.1 Libcloud. Libcloud is a client Python library for interacting with the most popular cloud management platforms [32]. This library originally started being developed within Cloudkick (extinct cloud monitoring software project, now part of Rackspace) and today is an independent free software project licensed under the Apache License 2.0. The main idea behind Libcloud is to create a programming environment to facilitate developers on the task of building products that can be ported across a wide range of cloud environments. Therefore, much of the library is about providing a long list of drivers to communicate with different cloud platforms. Currently, Libcloud supports more than 26 different providers, including Amazon's AWS, OpenStack, OpenNebula, and Eucalyptus, just to mention a few.

Moreover, this library also provides a unified Python API, offering a set of common operations to be mapped to the appropriate calls to the remote cloud system. These operations are divided into four abstractions: (1) Compute, which enables operations for handling VMs (e.g., list/create/reboot/destroy VMs) and its extension Block Storage to manage volumes attached to VMs (e.g., create/destroy volumes, attach volume to VM); (2) Load Balancer, which includes operations for the management of load balancers as a service (e.g., create/list members, attach/detach member or compute node) and is available in some providers; (3) Object Storage, which offers operations for creating an environment for handling data objects in a cloud (list/create/delete containers or objects, upload/download/stream object) and its extension for CDNs to assist providers that support these operations (e.g., enable CDN container or object, get CDN container or object URL); and (4) Domain Name System (DNS), which allows management operations for DNS as a service (e.g., list zones or records, create/update zone or record) in providers that support it, such as Rackspace Cloud DNS.

2.6.2.2 Deltacloud. Deltacloud follows a very similar philosophy as compared to Libcloud. It is also an Apache Software Foundation project—left incubation in October 2011 and is now a top-level project—and is similarly targeted to provide an intermediary layer to let applications communicate with several different cloud management platforms. Nevertheless, instead of providing a programming environment through a specific programming language, Deltacloud enables management of resources in different clouds by the use of one of three supported RESTful APIs [33]: (1) Deltacloud classic, (2) DMTF CIMI, (3) Amazon's EC2.

Deltacloud implements drivers for more than 20 different providers and offers several operations divided into two main abstractions: (1) Compute Driver, which includes operations for managing VMs, such as create/start/stop/reboot/destroy VM instances, list all/get details about hardware profiles, realms, images, and VM instances; and (2) Storage Driver, providing operations similar to Amazon S3 to manage data objects stored in clouds, such as create/update/delete buckets (analogous to folders), create/update/delete blobs (analogous to data files), and read/write blobs data and attributes.

2.6.2.3 Libvirt. Libvirt is a toolkit for interacting with multiple virtualization providers/hypervisors to manage virtual compute, storage, and networking resources. It is a free collection of software available under GNU LGPL and is not particularly targeted to cloud systems. Nevertheless, Libvirt has shown to be very useful to handle low level virtualization operations and is actually used under the hood by cloud platforms like OpenStack to interface with some hypervisors. Libvirt supports several hypervisors (e.g., KVM/QEMU, Xen, VirtualBox, and VMware), creation of VNs (e.g., bridging or NAT), and storage on IDE, SCSI, and USB disks and LVM, iSCSI, and NFS file systems. It also provides remote management using TLS encryption, x509 certificates, and authentication through Kerberos or SASL.

Libvirt provides a C/C++ API with bindings to several other languages, such as Python, Java, PHP, Ruby, and C#. This API includes operations for managing virtual resources as well as retrieving information and capabilities from physical hosts and hypervisors. Virtual resource management operations are divided into three abstractions: (1) Domains, which are common VM-related operations, such as create, start, stop, and migrate; (2) Storage, for managing block storage volumes or pools; and (3) Network, which includes operations such as, creating bridges, connecting VMs to these bridges, enabling NAT and DHCP. Note that network operations are all performed within the scope of a single physical host, that is, it is not possible to connect two VMs in separate hosts to the same bridged network, for example.

2.7 CHALLENGES

Because virtualization management in the cloud is still in its infant days, important challenges are in place. In this section, we list key challenges that can guide future developments in the virtualization management area. We also mention some ongoing research in the field.

2.7.1 Scalability

Although the benefits of virtualization enables the cloud model, from the management perspective, virtualization impacts the scalability of management solutions. The transition from the traditional management of physical infrastructures to virtual one is not smooth in terms of scale because few physical devices can host a much larger number of virtual device, each one requiring management actions. The number of management elements immediately explodes because such number not only duplicates but is proportional to the number of virtual devices each physical one supports. Traditional management applications have not been conceived to support a so drastic increase in the number of elements, and as a consequence, such solutions do not scale.

Novel management approaches need to be considered, or traditional approaches need to be adapted (if possible) to the cloud context and scale. The problem is also exacerbated because the managed environments (i.e., clouds) are much more dynamic, having new elements created very quickly, while older elements can be destroyed frequently too. Virtual servers can go up and down (even forever) quite fast, which is unusual for traditional management solutions. Adaptation is then required not only because of the new scales of cloud environments but also because they are much more dynamic than traditional IT infrastructures. Very distributed solutions have to be investigated, like the usage of peer-to-peer for management [34]. Autonomic management also becomes an alternative, in order to reduce human intervention as much as possible [35].

2.7.2 Monitoring

Monitoring is a permanent challenging task in the cloud because of the large number of resources in production cloud data centers. Centralized monitoring approaches suffer from low scalability and resilience. Cooperative monitoring [36] and gossiping [37] aim to overcome these limitations by enabling distributed and robust monitoring solutions for large scale environments. The goal is to minimize the negative impact of management traffic on the performance of the cloud. At the same time, finding a scalable solution for aggregating relevant monitoring information without hurting accuracy is a challenge that needs to be tackled by monitoring tools designed specifically for cloud data centers.

Usually, monitoring generates more management data than other activities. With virtualization in the cloud, and the aforementioned scalability issues, the overwhelming amount of monitoring data can hinder proper observation of the cloud environment. As such, monitoring considering big data techniques may be a possible path to follow. Compressing of data structures [38], for example, can be convenient to find a reasonable balance between amount of data and analysis precision.

2.7.3 Management Views

Since cloud computing creates an environment with different actors with particular management roles, such different actors need different management views. Operators of a cloud infrastructure need to have a broader, possibly complete view of the physical infrastructure, but should be prevented of accessing management information that are solely related to a tenant application, because of privacy issues. Cloud tenants also need to have access to management information related to their rented VI, but must also be isolated from accessing management information of both other tenants and physical infrastructure operator.

Different management views are already supported in traditional solutions, but in the case of cloud environments, trust relationships between cloud provider and tenants become more apparent. Since the management software runs in the cloud itself, tenants accessing their management view need to trust the cloud provider assuming that sensitive information is not available to the cloud operator. Tenants can also employ their own management system operating at his/her local IT infrastructure. In this case, management interfaces and protocols that connect the tenant management solution and remote managed virtual elements need to be present.

2.7.4 Energy Efficiency

Efficient energy management aims to reduce the operational cost of cloud infrastructures. A challenge in optimal energy consumption is to design energy-proportional data center architectures, where energy consumption is determined by server and network utilization [39, 40]. ElasticTree [39], for example, attempts to achieve energy proportionality by dynamically powering off switches and links. In this respect, cloud network virtualization can further contribute to reduce power consumption through network consolidation (e.g., through VN migration [41]).

Minimizing energy consumption, however, usually comes with the price of performance degradation. Energy efficiency and performance is often conflicting, representing a tradeoff. Thus, designing energy-proportional data center architectures factoring in cloud virtualization, and finding good balance between energy consumption and performance are interesting research questions.

2.7.5 Fault Management

Detection and handling of failures are requirements of any cloud, especially because in cloud environments failures of a single physical resource can potentially affect multiple customers' virtual resources. Because failures also tend to propagate, the damage caused by a faulty cloud physical device impacts much more severely the cloud business. In additional, the lack of faults in physical devices is not always a synonym that there is not faulty virtual devices. As such, the traditional fault management needs to be expanded to consider faulty virtual devices too.

Most existing architectures rely on reactive failure handing approaches. One drawback is the potentially long response time, which can negatively impact application performance. Ideally, fault management should be implemented in a proactive manner, where the management system predicts the occurrence of failures and acts before they occur. In practice, proactive fault management is often ensured by means of redundancy, for example, provisioning backup paths. As such, offering high reliability without incurring excessive costs or energy consumption is a problem requiring further exploration.

2.7.6 Security

Security issues are challenging in the context of cloud virtualization because of the complex interactions between tenants and cloud providers. Although the virtualization of both servers and networks can improve security (e.g., limiting information leakage, avoiding the existence of side channels, and minimizing performance interference attacks), today's virtualization technologies are still in their infancy in terms of security. In particular, various vulnerabilities in server virtualization technologies, such as VMWare [42], Xen [43], and Microsoft Virtual PC and Virtual Server [44] have been revealed in the literature. Similar vulnerabilities are likely to occur in programmable network components too. Thus, not only network virtualization techniques give no guaranteed protection from existing attacks and threats to physical and VNs, but also lead to new security vulnerabilities. For example, an attack against a VM may lead to an attack against a hypervisor of a physical server hosting the VM, subsequent attacks against other VMs hosted on that server, and eventually, all VNs sharing that server [45]. This raises the issue of designing secure virtualization architectures immune to these security vulnerabilities.

In addition to mitigating security vulnerabilities related to virtualization technologies, there is a need to provide monitoring and auditing infrastructures, in order to detect malicious activities from both tenants and cloud providers. It is known that data center network traffic exhibits different characteristics than the traffic of traditional data networks [46]. Thus, appropriate mechanisms may be required to detect network anomalies. On the other hand, auditability in cloud virtualization should be mutual between tenants and cloud providers to prevent malicious behaviors from either party. However, there is often an overhead associated with such infrastructures, especially in large-scale clouds. In Ref. [47], the authors showed that it is a challenge to audit Web services in cloud environments without deteriorating application performance. Much work remains to be done on designing scalable and efficient mechanisms for monitoring and auditing cloud virtualization.

2.7.7 Cloud Federations

The federation of virtualized infrastructures from multiple cloud providers enables access to larger scale infrastructures. This is already happening with virtualized network testbeds, allowing researchers to conduct realistic network experiments at large scale, which would not have been possible otherwise. ProtoGENI [48] is an example of federation that allows cooperation among multiple organizations. However, guaranteeing predictable performance for participating entities through SLA enforcement has not been properly addressed by current solutions and remains an open issue.

Cloud federations cannot be considered a wide reality in the cloud marketplace. Competition possibly prevents cloud providers to cooperate among one another in federated environments, but the lack of proper technologies devoted to materialize federations of clouds certainly does not improve the current situation either. As in other areas, solutions to federate different resources already exist, but an integrated, global solution to support federating heterogeneous resources between heterogeneous cloud providers also needs further investigation.

2.7.8 Standard Management Protocols and Information Models

The VR-MIB module [49] described a set of SNMP management variables for the management of physical routers with virtualization support. However, it did not progress in the IETF standardization track. More recently, the VMM-MIB module [50] is progressing, but it is limited to manage virtualization of servers (network devices are not explicitly considered); VMM-MIB is also devoted mainly for monitoring, and configuration is weakly supported. In general, the situation of SNMP-based management solutions for cloud environments are still weak.

Other existing management protocols are considered. NETCONF [51], for example, would be more appropriate for configuration aspects, while NetFlow/IPFIX [52] could be expanded for virtual router monitoring. The WS-Management [53] suite, in turn, is more appropriate for server management. A myriad of proprietary solutions is also present in the market, but the large diversity of management interfaces and protocols forces cloud operators to deal with too many different technologies. Although a protocol that fits every need is unlikely to exist or be largely accepted/adopted, there is a clear lack in this area today, which represents an interesting opportunity for research and standardization.

REFERENCES

1. M. Chowdhury, M.R. Rahman, and R. Boutaba. ViNEYard: Virtual Network Embedding Algorithms with Coordinated Node and Link Mapping. *IEEE/ACM Transactions on Networking*, 20(1):206–219, February. 2012.
2. M. Yu, Y. Yi, J. Rexford, and M. Chiang. Rethinking Virtual Network Embedding: Substrate Support for Path Splitting and Migration. *ACM Computer Communication Review*, 38(2):17–29, April 2008.
3. X. Cheng, S. Su, Z. Zhang, K. Shuang, F. Yang, Y. Luo, and J. Wang. Virtual Network Embedding Through Topology Awareness and Optimization. *Computer Networks*, 56(6):1797–1813, 2012.
4. A. Lenk, M. Klems, J. Nimis, S. Tai, and T. Sandholm. What's Inside the Cloud? An Architectural Map of the Cloud Landscape. In *Proceedings of the 2009 ICSE Workshop on Software Engineering Challenges of Cloud Computing (CLOUD '09)*, pages 23–31, Washington, DC, 2009. IEEE Computer Society.
5. M. F. Bari, R. Boutaba, R. Esteves, L. Granville, M. Podlesny, M. Rabbani, Q. Zhang, and F. Zhani. Data Center Network Virtualization: A Survey. *IEEE Communications Surveys and Tutorials*, 15(2):909–928, 2012.
6. Q. Zhu and G. Agrawal. Resource Provisioning with Budget Constraints for Adaptive Applications in Cloud Environments. In *Proceedings HPDC 2010*, Chicago, IL, 2010.
7. M. E. Frincu and C. Craciun. Multi-objective Meta-heuristics for Scheduling Applications with High Availability Requirements and Cost Constraints in Multi-Cloud Environments. In *Proceedings of the Fourth IEEE International Conference on Utility and Cloud Computing (UCC)*, pages 267–274, Victoria, NSW, December 2011.

8. J. Rao, X. Bu, K. Wang, and C.-Z. Xu. Self-adaptive Provisioning of Virtualized Resources in Cloud Computing. In *Proceedings SIGMETRICS 2011*, 2011.
9. OpenNebula. The Open Source Solution for Data Center Virtualization, 2008. http://www.opennebula.org. Accessed on November 2013.
10. J. Rao, Y. Wei, J. Gong, and C.-Z. Xu. DynaQoS: Model-free Self-Tuning Fuzzy Control of Virtualized Resources for QoS Provisioning. In *19th IEEE International Workshop on Quality or Service (IWQoS)*, pages 1–9, San Jose, CA, June 2011.
11. J. Rao, X. Bu, C.-Z. Xu, and K. Wang. A Distributed Self-learning Approach for Elastic Provisioning of Virtualized Cloud Resources. In *Proceedings IEEE MASCOTS 2011*, pages 45–54, Singapore, July 2011.
12. J. Z. W. Li, M. Woodside, J. Chinneck, and M. Litoiu. CloudOpt: Multi-goal Optimization of Application Deployments across a Cloud. In *Proceedings CNSM 2011*, pages 1–9, October 2011.
13. A. Khan, A. Zugenmaier, D. Jurca, and W. Kellerer. Network Virtualization: A Hypervisor for the Internet? *IEEE Communications Magazine*, 50(1):136–143, January 2012.
14. N. Chowdhury and R. Boutaba. Network Virtualization: State of the Art and Research Challenges. *IEEE Communications Magazine*, 47(7):20–26, July 2009.
15. J. Montes, A. Sánchez, B. Memishi, M. S. Pérez, and G. Antoniu. GMonE: A Complete Approach to Cloud Monitoring. *Future Generation Computer Systems*, 29(8):2026–2040, 2013.
16. M. Carvalho, R. Esteves, G. Rodrigues, L. Z. Granville, and L. M. R. Tarouco. A Cloud Monitoring Framework for Self-Configured Monitoring Slices Based on Multiple Tools. In *9th International Conference on Network and Service Management 2013 (CNSM 2013)*, pages 180–184, Zürich, Switzerland, October 2013.
17. Amazon. Amazon elastic compute cloud (Amazon EC2), 2013. http://aws.amazon.com/ec2/. Accessed on May. 2013.
18. Eucalyptus. The Open Source Cloud Platform, 2009. http://open.eucalyptus.com. Accessed on November 2013.
19. Rackspace Cloud Computing. OpenStack Cloud Software, 2010. http://openstack.org. Accessed on December 2013.
20. S. Islam, J. Keung, K. Lee, and A. Liu. Empirical Prediction Models for Adaptive Resource Provisioning in the Cloud. *Future Generation Comp. Syst.*, 28(1):155–162, January 2012.
21. Apache Software Foundation. Apache CloudStack: Open Source Cloud Computing, 2012. http://cloudstack.apache.org. Accessed on December 2013.
22. S. Ortiz. The Problem with Cloud-Computing Standardization. *IEEE Computer*, 44(7):13–16, 2011.
23. Open Grid Forum. Open Cloud Computing Interface, 2012. http://occi-wg.org/. Accessed on September 2012.
24. A. Edmonds, T. Metsch, A. Papaspyrou, and A. Richardson. Toward an Open Cloud Standard. *Internet Computing, IEEE*, 16(4):15–25, 2012.
25. H. Medhioub, B. Msekni, and D. Zeghlache. OCNI—Open Cloud Networking Interface. In *22nd International Conference on Computer Communications and Networks (ICCCN)*, pages 1–8, Nassau, Bahamas, 2013.
26. Distributed Management Task Force (DMTF). Open Virtualization Format (OVF) Specification—Version 2.0.1, Ago 2013. http://dmtf.org//standards/cloud. Accessed on December 2013.

27. Distributed Management Task Force (DMTF). Cloud Infrastructure Management Interface (CIMI)—Version 1.0.0, 2013. http://dmff.org/standards/cloud. Accessed on May 2013.
28. Storage Networking Industry Association (SNIA). Cloud Data Management Interface (CDMI)—version 1.0.2, June 2012. http://www.snia.org/cdmi. Accessed on December 2013.
29. Storage Networking Industry Association (SNIA). S3 and CDMI: A CDMI Guide for S3 Programmers—version 1.0, May 2013. http://www.snia.org/cdmi. Accessed on December 2013.
30. R. P. Esteves, L. Z. Granville, and R. Boutaba. On the Management of Virtual Networks. *IEEE Communications Magazine*, 51(7):80–88, 2013.
31. R. Moreno-Vozmediano, R. S. Montero, and I. M. Llorente. IaaS Cloud Architecture: From Virtualized Datacenters to Federated cloud infrastructures. *IEEE Computer*, 45(12):65–72, 2012.
32. Apache Software Foundation. Apache Libcloud a Unified Interface to the Cloud, 2012. http://libcloud.apache.org/. Accessed on December 2013.
33. Apache Software Foundation. Apache DeltaCloud an API that Abstracts the Differences between Clouds, 2011. http://deltacloud.apache.org/. Accessed on December 2013.
34. L. Z. Granville, D. M. da Rosa, A. Panisson, C. Melchiors, M. J. B. Almeida, and L. M. Rockenbach Tarouco. Managing Computer Networks Using Peer-to-Peer Technologies. *Communications Magazine, IEEE*, 43(10):62–68, 2005.
35. C. C. Marquezan and L. Z. Granville. On the Investigation of the Joint Use of Self-* Properties and Peer-to-Peer for Network Management. In *2011 IFIP/IEEE International Symposium on Integrated Network Management (IM)*, pages 976–981, 2011.
36. K. Xu and F. Wang. Cooperative Monitoring for Internet Data Centers. In *IEEE International Performence, Computing and Communications Conference (IPCC)*, pages 111–118, Austin, TX, December 2008.
37. F. Wuhib, M. Dam, R. Stadler, and A. Clemm. Robust Monitoring of Network-Wide Aggregates through Gossiping. *IEEE Transactions on Network and Service Management*, 6(2): 95–109, 2009.
38. L. Quan, J. Heidemann, and Y. Pradkin. Trinocular: Understanding Internet Reliability Through Adaptive Probing. In *Proceedings of the ACM SIGCOMM 2013 Conference on SIGCOMM (SIGCOMM '13)*, pages 255–266, Hong Kong, China, 2013. ACM, New York.
39. B. Heller, S. Seetharaman, P. Mahadevan, Y. Yiakoumis, P. Sharma, S. Banerjee, and N. McKeown. ElasticTree: Saving Energy in Data Center Networks. In *Proceedings USENIX NSDI*, April 2010.
40. H. Yuan, C. C. J. Kuo, and I. Ahmad. Energy Efficiency in Data Centers and Cloud-based Multimedia Services: An Overview and Future Directions. In *Proceedings IGCC*, Chicago, IL, August 2010.
41. Y. Wang, E. Keller, B. Biskeborn, J. van der Merwe, and J. Rexford. Virtual Routers on the Move: Live Router Migration as a Network-Management Primitive. *ACM Computer Communication Review*, 38:231–242, August 2008.
42. VMware. VMware Shared Folder Bug Lets Local Users on the Guest OS Gain Elevated Privileges on the Host OS. VMWare vulnerability. http://securitytracker.com/alerts/2008/Feb/1019493.html, 2008.
43. Xen. Xen Multiple Vulnerabilities. Xen vulnerability. http://secunia.com/advisories/26986, 2007.

44. Microsoft. Vulnerability in Virtual PC and Virtual Server Could Allow Elevation of Privilege. Virtual PC Vulnerability. http://technet.microsoft.com/en-us/security/bulletin/MS07-049, 2007.
45. J. Szefer, E. Keller, R. B. Lee, and J. Rexford. Eliminating the Hypervisor Attack Surface for a More Secure Cloud. In *Proceedings of the 18th ACM conference on Computer and Communications Security (CSS)*, pages 401–412, Chicago, IL, October 2011.
46. T. Benson, A. Anand, A. Akella, and M. Zhang. Understanding Data Center Traffic Characteristics. *ACM SIGCOMM Computer Communication Review*, 40(1):92–99, 2010.
47. A. Chukavkin and G. Peterson. Logging in the Age of Web Services. *IEEE Security and Privacy*, 7(3):82–85, June 2009.
48. ProtoGENI. ProtoGENI, Dec 2013. Available at: http://www.protogeni.net/trac/protogeni (Dec. 2013).
49. E. Stelzer, S. Hancock, B. Schliesser, and J. Laria. Virtual Router Management Information Base Using SMIv2. *Internet-Draft draft-ietf-ppvpn-vr-mib-05 (obsolete)*, June, 2013. http://tools.ieff.org//html/draft-ietf-ppvpn-vr.mib-05. Accessed on December 2014.
50. H. Asai, M. MacFaden, J. Schoenwaelder, Y. Sekiya, K. Shima, T. Tsou, C. Zhou, and H. Esaki. Management Information Base for Virtual Machines Controlled by a Hypervisor. *Internet-Draft draft-asai-vmm-mib-05 (work in progress)*, October 13, 2013.
51. R. Enns. RFC 4741: NETCONF Configuration Protocol, December 2006. http://tools.ietf.org//html/rfc4741. Accessed on December 10, 2014.
52. B. Claise, B. Trammell, and P. Aitken. RFC 7011: Specification of the IPFIX Protocol for the Exchange of Flow Information, September 2013. http://tools.ieff.org//html/rfctoll. Accessed on December 2014.
53. Distributed Management Task Force (DMTF). Web Services for Management (WS-Management) Specification. DMTF, Ago 2012.

3

VIRTUAL MACHINE MIGRATION

Diogo M. F. Mattos, Lyno Henrique G. Ferraz, and
Otto Carlos M. B. Duarte

Grupo de Teleinformática e Automação (GTA/UFRJ), PEE/COPPE - DEL/Poli, Universidade Federal do Rio de Janeiro, Rio de Janeiro, Brazil

3.1 INTRODUCTION

Cloud computing is experiencing an extraordinary growth [1–5]. Besides, virtualization technologies are widely adopted by companies to manage flexible computing environments and to run isolated virtual environments for each customer [6]. Virtualization also provides the means to accomplish efficient allocation of resources and to improve management, reducing operating costs, improving application performance and increasing reliability. Virtualization logically slices physical resources into virtual environments, which have the illusion of accessing the entire available physical resource. Hence, the physical machine resources are shared between multiple VMs, which run their own isolated environment with an operating system and applications. By decoupling VMs from their underlying physical realization, virtualization allows flexible allocation of VMs over physical resources. To this end, virtualization introduces a new management primitive: VM migration [7]. VM migration is the relocation of virtual machines over the underlying physical machines, even if the VM is still running.

Cloud Services, Networking, and Management, First Edition.
Edited by Nelson L. S. da Fonseca and Raouf Boutaba.

The VM migration primitive enhances user mobility, load balancing, fault management, and system management [8]. The migration that occurs without the interruption of services running is called live migration.

Virtual machine migration is similar to the process migration, but it migrates a complete operating system and its applications. Process migration moves a running process from one machine to another. Process migration is very difficult, or even impossible to accomplish, because processes are strongly bound to operating systems, by means of open sockets, pointers, file descriptors and other resources [8]. Unlike process migration, VM migration moves the entire operating system along with all the running processes. Migrating an entire operating system with its applications is a more manageable procedure, and is facilitated by the hypervisor, which exposes an interface between physical machine and the VM operating system. The details of what is happening inside the VM can be ignored during migration. The VM migration also has challenges inherent security to transfer the state of a VM across physical machines and to establish a trustworthy computing environment on the destination physical machine.

In the context of virtualization, it is necessary to ensure that virtual environments are secure and trustworthy. Thus, the hypervisor, which is a software layer responsible for creating the hardware abstraction to the virtual environment, must implement a trusted computing base (TCB) [9, 10]. Indeed, the TCB is divided into two parts: the hypervisor and an administrative domain, see (Figure 3.1). The hypervisor controls the hardware directly and executes at the highest privilege level of the processor. The administrative domain is a privileged VM that controls and monitors other VMs. The administrative domain have privileges to start and stop VMs, to run guest VM configuration, to use and to monitor physical resources, and to run I/O operation directly on

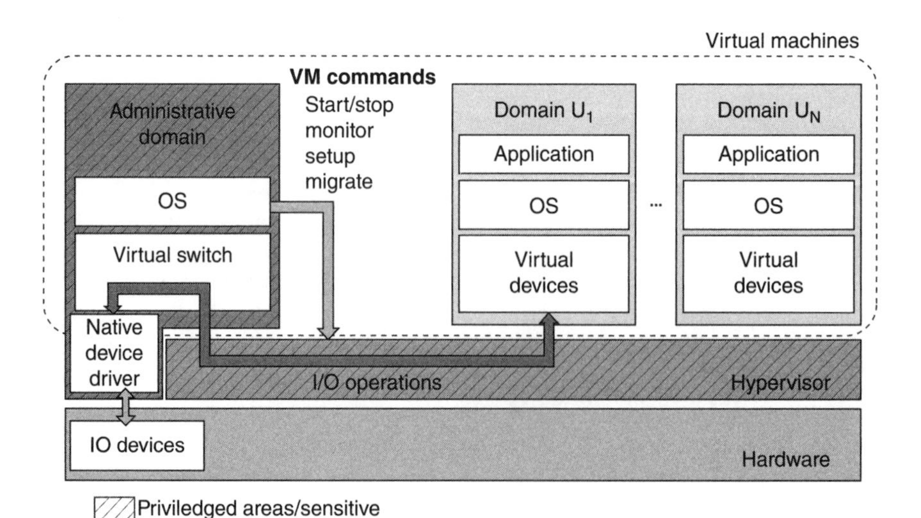

Figure 3.1. General Xen-based virtualization architecture. The hachured areas, administrative domain and hypervisor, indicate the most sensitive software modules because they run on highest privilege level.

the physical devices for the virtualized domains. This common architecture for virtualized systems creates, however, security challenges, such as lack of privacy of guest VMs. Administrative domain runs in a privileged level to inspect the state of guest VMs, such as the contents of its registers into memory and vCPUs. This privilege can be usurped by attacks on the software stack in the administrative domain and by malicious system administrators [11]. Therefore, it is necessary to establish a trusted computing base (TCB) on the hypervisor and on the administrative domain to ensure the security of virtualized environments.

Specific relevance is given to a hybrid virtualization system based on Xen and OpenFlow platforms, called XenFlow [12], which focuses on router virtualization, especially on the virtual router migration without packet losses. XenFlow provides migration of virtual topologies over the physical realization, performing both migration of virtual routers to another physical host and remapping virtual links on one or more physical links. This feature allows to extent virtual-router migration when compared to the other proposals in literature [13–15], because routes are remapped to any destination physical node by means of OpenFlow network.

This chapter presents the major VM migration techniques. This work highlights the benefits, costs and challenges for the realization of the live migration of VMs. We highlight I/O virtualization techniques and discuss how to migrate VM even if they directly access I/O devices or use I/O virtualization techniques. The chapter sets out the main security requirements to be ensured during the migration of virtual environments. Then, we examine various schemes of VM migration and discuss research directions in virtualization security. The ultimate goal is to provide a deep understanding of the developments and the future directions regarding virtualized environments migration primitive.

The rest of this chapter is organized as follows. Section 3.2 sets a background for understanding VM migration and its challenges. Virtual network migration is explained on Section 3.3, in which we also present a proposal for migrating virtual routers without packet losses. The main security requirements and proposals for virtualized environments are identified on Section 3.4. Future research directions and open challenges are discussed in Section 3.5. Section 3.6 concludes this chapter.

3.2 VM MIGRATION

The procedure of migrating the operating system and applications from a physical machine to another physical machine is an important feature in a virtualized environment. VM migrations encompass four main resource transferring: processor, memory, network, and storage [15]. During the migration process, the VM is paused on source host and is resumed on the destination host only when all resources have already been migrated and configured into the new host. The VM stays offline during a period of time, called downtime, which corresponds to time when the VM is paused until its resumption at the destination. The downtime period varies according to the resources available to the VM, to the workload submitted to VM, and to the migration technique: offline or live migration.

3.2.1 Offline and Live Migration

Offline migration transfers the VM to destination physical host while the VM is off. The offline migration introduces a great delay in services of VM, but it is the easiest to accomplish because it does not require the VM state preservation. As the VM is off, there is no network connections to preserve, and it is neither necessary to transfer the processor state nor the RAM content. The offline migration procedure just comprises shutting down and restarting the VM into another location.

The storage migration, or disk migration, is accomplished by standard data transfer tools and is the only network traffic generated. It takes a long time and a lot of network bandwidth to transfer a whole disk. As a matter of fact, VM migration is usually accomplished within a LAN with a network-attached storage (NAS) device that allows a VM to access its disk from anywhere in the network, which makes unnecessary to migrate the disk.

Live migration transfers the VM while it still runs. The live migration should not cause a perceptible downtime to the VM user. Assuming the source and destination physical machines in the same LAN with a NAS, live migration only should transfer the state of the processor, the state of the memory and network connections.

The processor live migration consists of creating a virtual CPU (vCPU) for the VM at the destination and copying the vCPU state from source to destination physical machines. Nevertheless, this task becomes complex when the source and the destination host processors are different. Migrations between different processors of the same manufacturer require the same instruction sets to work properly. In these cases, as a consequence, it is necessary to limit the instruction set of the virtual CPU to a common instructions set of both processors. This operation is called CPU mask.

The network live migration procedure should maintain the Internet Protocol (IP) address of the source VM to preserve all open transmission control protocol (TCP) connections. To keep the same IP address at the destination, it is very simple when the source and destination physical machines are in the same local area network (LAN). In this case, the destination physical machine generates gratuitous Address Resolution Protocol (ARP) replies to advertise the new physical location of the VM, that is, only the advertisement of the medium access control (MAC) address of the migrated VM is required. Otherwise, when the source and destination machines are not in the same LAN, network redirection mechanisms would be required due to the localization semantic of the IP address.

Live memory migration is the transfer of memory contents from the source to the destination host taking into account memory changes during the migration procedure, called retransmission of dirty pages. Venkat [16] divided the memory migration into three phases:

1. Push phase: While the source VM is running, its memory pages are transferred to the destination VM. If a page is modified after being transferred, it is necessary to resend this page to avoid failures.
2. Stop-and-copy phase: As the name suggests, the source VM is stopped, then the memory pages are transferred.

3. Pull phase: The destination VM is started and generates a page fault when it tries to access a page that was not copied yet. This fault requests the page to be transferred from the source to the destination.

Two live migration strategies [16]: pre-copy and post-copy, only use a combination of two of the above-mentioned phases.

The pre-copy live migration strategy applies the phases: push and stop-and-copy. First, an empty VM is created at the destination physical host and the migrating VM memory pages are copied to the VM at destination physical machine, while the VM still runs on the source host. During this process, the running VM rewrite the memory pages which are resent to destination host. This push phase ends when one of the two conditions are reached: (1) The number of dirty pages per iteration are small enough to cause a short downtime period and (2) The push phase reaches a maximum number of iterations. After the push phase, it comes the stop-and-copy phase, in which the VM is suspended at the source host, the remaining dirty pages are transferred to the destination host, and the VM is resumed on the destination host. The downtime varies according the workload from tens or hundreds of milliseconds to a few seconds [15]. It is important to notice that determining when to stop the push phase and start the stop-and-copy phase is not trivial. Stopping the push phase too soon can result in longer downtime, as more data will be transferred after suspending the VM. On the other hand, stopping too late results in longer total migration time and network bandwidth occupation, as more time will be spent re-sending dirty pages. Therefore, there is a trade-off between total migration time and downtime. The pre-copy procedure requires the verification of memory pages to send them to the destination through the network. These CPU and bandwidth consumption should be monitored to minimize service degradation. Xen uses pre-copy as its live migration strategy [14].

The post-copy live migration strategy use the phase stop-and-copy first and then the pull phase. First, the VM is suspended at the source and few VM execution states are transferred to the destination host, namely CPU registers and non-paged memory. The VM is resumed at the destination despite the absence of many memory pages, which still are at the source host. The source host begins to send the remaining memory pages. The destination host generates faulty memory accesses when the VM tries to access memory pages that were not transferred yet. These faulty memory accesses are sent back to the source host, which prioritizes the requested memory pages to send. This process can degrade memory intensive application performance, but cause minimal downtime. There are some ways of handling page fetching in order to increase performance, such as the following:

- *Active pushing*: the pages are proactively pushed from the source to the destination. Page faults are handled with priority over noncritical pages.
- *Pre-paging*: an estimation of memory access pattern is generated to allow the active pushing of the pages that are most likely to generate faults.

Table 3.1 compares offline and live migrations.

TABLE 3.1. Comparison of offline and live migration techniques

Characteristic / Technique	Storage migration	Memory migration	Network migration	Downtime	Total migration time
Offline migration. Shutdown VM and restart at destination host.	Standard copying tools, if migrated.	Not transfered. Loss of volatile data.	Reconfiguration at destination host. Network connections not migrated.	Long period of time. VM and services restart (if storage is not migrated).	Equal to downtime (if storage is not migrated).
Live migration. Transfer running VM to destination host.	Not migrated. Storage is accessable through the network.	Transfered. Pre- and post-copy and retransmission of updates.	Reconfiguration at destination host. Transfer of network state. Network connections preserved.	Short period of time. VM pause/resume.	Equal to downtime plus memory transfer time. Vary according to the workload that requires retransmission of dirty pages.

3.2.2 I/O Virtualization and Migration of Pass-Through Devices

Input/Output (I/O) virtualization of network devices is challenging because current network interface controllers are unable to distinguish which specific VM is writing to or reading from the shared memory space. Therefore, a controller or a hypervisor must redirect (multiplexing or demultiplexing) data to/from specific memory area in an administrative domain from/to different VM shared memory areas. This procedure negatively impacts the performance, since it introduces extra memory copies, it centralizes the interruption handling at administrative domain processing time slice, and it demands execution of software instructions for multiplexing data in administrative domain, such as virtual bridges, as shown in Figure 3.2a. Thus, a technique to improve I/O device performance is the use of pass-through technologies to avoid the centralization and memory copies by providing direct I/O procedures to/from the virtual domain from/to the physical device. Although the pass-through technology improves I/O virtualization performance, the pass-through device belongs to a single VM and cannot be shared by other VMs, as shown in Figure 3.2b.

The main technique to provide direct I/O virtualization is single root I/O virtualization (SR-IOV) for Peripheral Component Interconnect Express (PCIe) [17]. The specification SR-IOV stands for how PCIe devices can share a single root I/O device with multiple VMs. Indeed, a SR-IOV enabled hardware provides several PCIe virtual functions to the hypervisor, which can be assigned directly to VMs as pass-through devices, as shown in Figure 3.3a. Besides SR-IOV, Intel also proposes VM Device Queues (VMDq) [4] for network I/O virtualization. VMDq technology enabled network device has separated queues for VMs. The network interface classifies received packets to the queue of a VM and fairly sends packets of all queues in round robin manner. As VMDq applies a paravirtualized device driver, it uses shared pages to avoid packet copying between the virtual network interface in the VM and the physical network queue. The

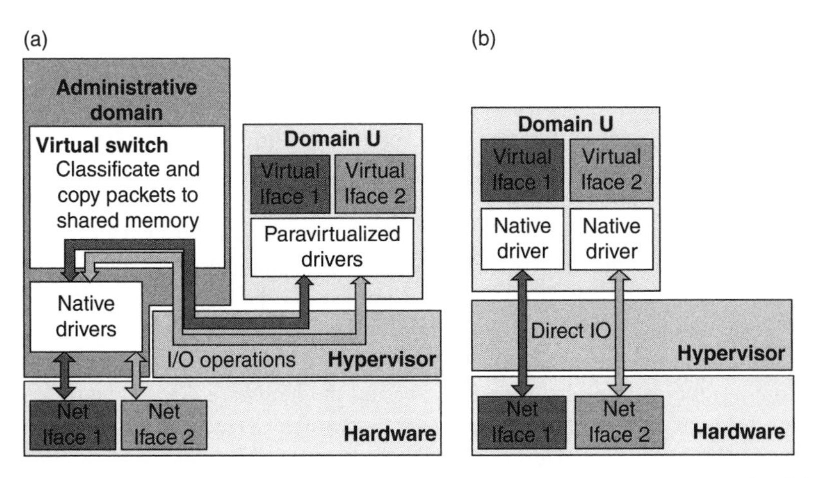

Figure 3.2. I/O virtualization modes. (a) Network I/O virtualization with paravirtualized drivers. Administrative domain centralizes all I/O operations. (b) Direct I/O network virtualization. A network interface card is directly connected to virtual machine.

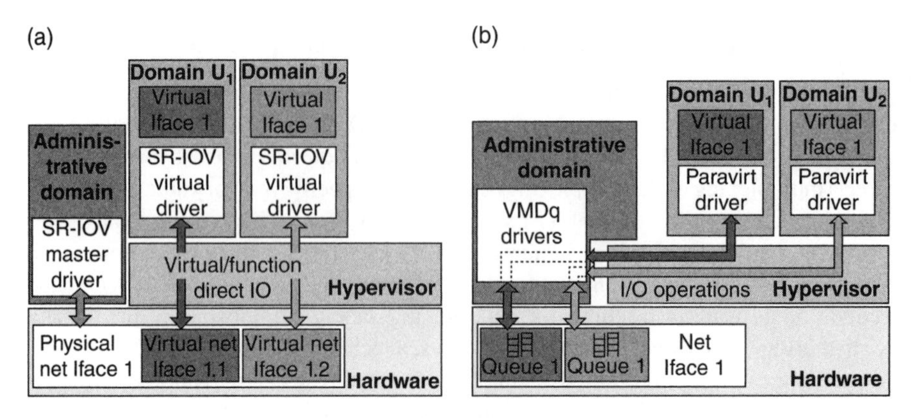

Figure 3.3. Hardware-assisted network I/O virtualization modes. (a) Network I/O virtualization with SR-IOV. Virtual machines directly access NIC virtual functions. (b) Network I/O virtualization with VMDq. Virtual machines access device queues through a paravirtualized driver.

VM benefits from faster classification and a paravirtualized device driver, while SR-IOV technology exposes a unique device interface to the VM. The implementation of VMDq paravirtualized driver assures better performance than paravirtualized network drivers. Besides, VMDq paravirtualized driver support live migration in a similar way than when using common paravirtualized drivers [4], illustrated in Figure 3.3b.

VMWare and Intel propose Network Plug-In Architecture (NPA/NPIA) [4] to live migrate pass-through devices. The proposal creates a new driver for VM, which allows the online switching between SR-IOV and paravirtualized devices. This technology designs two new software modules: a kernel shell and a plug-in for the VM. Kernel shell acts as an intermediate layer to manage pass-through devices, which implements a device driver for the SR-IOV device. Plug-in, in its turns, implements virtual functions of the device, as a device driver, but interfaces with kernel shell instead of directly controlling the device, exposing a virtualized network interface card to the virtual domain. The kernel shell provides a hardware abstraction layer and the plug-in implements hardware communication through the kernel shell. Plug-in may be plugged or unplugged on the fly. To reduce migration downtime, while performing plugging/unplugging actions, the hypervisor employs an emulated network interface. This technology trivially supports live migration because a virtual network interface can be unplugged while running the VM. On the other hand, a drawback of the this approach is the need for rewriting all the network device drivers, which may limit its adoption [4].

Pass-through I/O virtualization technology improves virtualized device performance by making a tight coupling between the VM and the hardware device. Thus, VM live migration becomes more difficult because pass-through devices are totally controlled by VM and the hypervisor does not access the internal states of the device. Indeed, in pass-through I/O virtualization the hypervisor does not interfere into the communication between the physical device and the VM. Therefore, the internal states of the physical device must be migrated with VM, in order to accomplish a successful live VM migration [4].

A way to migrate VM with pass-through devices is to let user stop everything using a pass-through device, and then migrate and restore the VM into the destination physical host. Although this method works, it is not generic enough to fit all operating systems, it involves a greater downtime, it needs to be inside the VM, and it needs a lot of intervention of the user [18]. A generic solution to suspend the VM before migrating is Advanced Configuration and Power Interface (ACPI)[1] S3 [18]. Sleep state S3 stands for the sleep or suspend state of a machine, in which the operating system freezes all process, suspends all I/O devices, and then goes to the sleep state, but the RAM remains powered. It is worth noting that in sleep state all context is lost, except for the system volatile memory. The major drawback of this approach is that whole system is affected, inducing a long service downtime besides disabling the target device.

Migration of a pass-through I/O device may also be accomplished by the PCI hotplug mechanism [18]. Migrating a VM using PCI hotplug work as follows. Before live migrating, in the source host, the entity responsible for the migration triggers an event of hot unplugging the virtual PCI pass-through device against the guest VM. Then, the migrating VM responds to the hot unplugging event, and stops using the device after unloading its driver. Without running any pass-through device, the VM can be safely live migrated to the destination host. After the live migration, in the destination host, it triggers an event of hot plugging a virtual PCI pass-through device against the VM. Eventually, the guest VM loads the appropriate driver and starts using the new pass-through device. As the guest reinitializes a new device, that has nothing to do with the old one, it should reconfigure it as the previous one.

CompSC proposes a live migration mechanism for VM using pass-through I/O virtualization [4]. The key idea of CompSC is to change as less as possible the code of drivers and prevent the hypervisor to have any specific knowledge about the migratig device. The hypervisor examines the list of registers of the network device and saves them into the sharedpt memory area. The hypervisor does not know the list of registers *a priori*. For this reason, the hypervisor gets this list of registers also from the shared memory area, where the device driver places it during the boot process. The device driver completes the state transferring between hosts. Every time before the driver releases a read lock, it stores enough information about the latest operations or set of operations to achieve a successful resume. In the resume procedure, the device triggers the target hardware using the same saved state information. The proposal also provides a layer of self-emulation, which can be placed in the hypervisor or in the device driver. Placing the self-emulation layer in hypervisor, the hypervisor intercepts all accesses to emulated registers and returns the correct value. A layer of self-emulation in the driver processes the fetched value and corrects it after the access. A layer of self-emulation in hypervisor requires only the list of emulated registers and requires few code changes to the driver, but the performance degrades due to interception of I/O operations. A layer of self-emulation in device driver requires less overhead, but produces more code changes [4]. Table 3.2 summarizes the migration proposals of main pass-through I/O virtualization techniques.

[1] Advanced Configuration and Power Interface (ACPI) specification is an open standard for device configuration and power management by the operating system. This standard replaces some other standards bringing power management under the control of the operating system instead of BIOS control as stated by the replaced standards.

TABLE 3.2. Comparison of migrating I/O virtualization techniques

Technique \ Characteristic	Pros	Cons	Summary
SR-IOV	Good performance. VM direct access to device.	Hard to migrate. Hypervisor cannot access device state.	Hardware provides multiple virtual functions. Hypervisor sets virtual functions to VMs. VM interacts directly with hardware devices.
VMDq	Good performance and easy to migrate. Packet classification by hardware and conventional paravirtualized driver.	Slightly performance degradation. Minor driver domain participation in I/O.	VMDq driver writes and reads packets directly on shared pages in driver domain which avoid packet classification and extra packet copies.
NPA-NPIA	Good performance and easy to migrate. Hotplug of SR-IOV virtual functions and paravirtualized drivers.	Hard to deploy. New virtual network device drivers in VMs.	It creates a pair "Kernel Shell and Plug-in," which allows Plug-in to be migrated carrying all device states, while Kernel Shell implements virtual function into the driver.
Pause/resume	Easy to deploy. It uses current technologies.	Hard to migrate and loss of volatile data. It depends on users' interaction. VM suspension.	VM is suspent on source host and, after, it is resumed on destination host.
PCI hotplug	Good performance and easy to deploy. It uses current technologies.	Loss of volatile data. Pass-through devices hotplugging.	Source host unplugs the virtual PCI pass-through device of VM. After migration, a new pass-through device is loaded and reconfigured on the migrated VM.
CompSC	Good performance and easy to migrate. VM uses current technologies. Easy to deploy. Hypervisor uses new live migration software.	Slightly performance degradation during migration. It uses emulated virtual network device driver.	Hypervisor saves pass-through device states before migration, and restores the device state after migration.

3.3 VIRTUAL NETWORK MIGRATION WITHOUT PACKET LOSS

Network virtualization is the technique that decouples network functions from their physical substrate, enabling virtual networks to run logically separated and over the a physical network topology [19]. The logical separation enables virtual network migration, which allows online physical topology changes avoiding reconfiguration, traffic disruption and long convergence delays [13]. The virtual network migration consists of migrating the virtual network element, also called virtual router, to another physical location, without packet losses or losing connectivity. The key idea to avoid packet losses is the separation of control and data planes, the former responsible for performing control operations, such as running routing protocols and defining QoS parameters, and the latter responsible for the packet forwarding [13, 14]. As the virtual router should always forward the traffic, the data plane is copied to the physical host while the virtual router migrates. After the migration, the data plane in source host is deactivated, so the virtual router completely runs in the new location.

Both Wang *et al.* and Pisa *et al.* use plane separation paradigm to migrate virtual routers without packet losses [13, 14]. They assume an external mechanism for link migrations to preserve neighborhood after migration, such as maintaining the same set of neighbors or tunneling. Pisa *et al.* assume all physical routers connect to the same local area network (LAN) to facilitate link migration [14]. On the other hand, flow migration on the OpenFlow platform is easy. Pisa *et al.* present an algorithm that is based on the redefinition of a flow path in the OpenFlow network [14]. This proposal has zero packet losses and low overhead of network control messages. Although, this migration proposal is limited to OpenFlow switched networks, and it is not applicable to router virtualization systems.

Mattos and Duarte present XenFlow [12], a hybrid network virtualization system based on plane separation paradigm with Xen and OpenFlow platforms [20, 21] to both migrate virtual routers and virtual links. VM act as the routers control plane running routing protocols, and data planes of all virtual routers run centrally in the Xen administrative domain Domain 0. Physical machines have an OpenFlow switch to connect Xen VMs to the physical network, and each Xen VM acts as generator of rules to these switches. The remapping of the virtual topologies is orchestrated by a network controller capable of acting on the OpenFlow switches and of triggering the migration of VMs on any network node. Figure 3.4 presents this architecture. The architecture allows to migrate virtual routers beyond a local area network, because routes are remapped to any destination physical node by means of OpenFlow network. However, the architecture forces all virtual networks to share the same data plane, violating the requirement of isolation between virtual environments. Thus, XenFlow isolates virtual networks by two mechanisms: Address space isolation among virtual networks, which ensures VMs only access VMs that belong to the same virtual network; and virtual network resources sharing isolation, which prevents virtual networks against using resources of other virtual networks [22]. The system also offers quality of service through mapping parameters of service-level agreements, defined as control plane directives, to parameters of the data plane. It controls the basic resources of virtual networks: processing, memory, and bandwidth, as those are the resources that can be locally controlled [23].

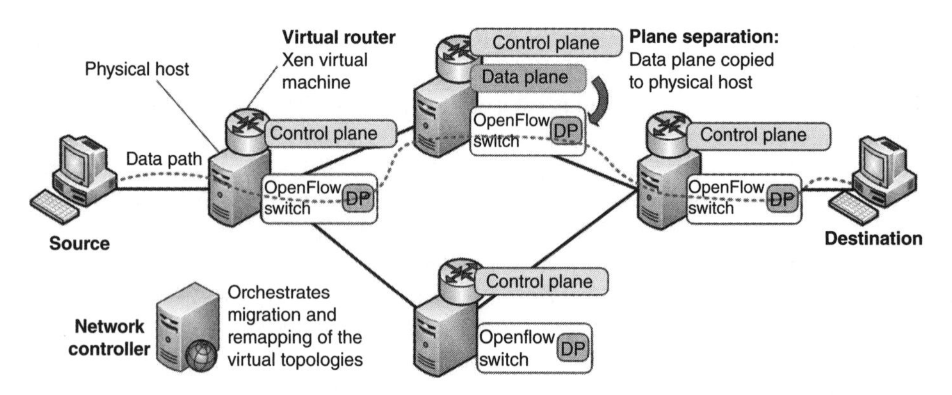

Figure 3.4. XenFlow architecture overview. Xen virtual router data plane is copied to physical host OpenFlow switch. Network controller orchestrates virtual router and link migration.

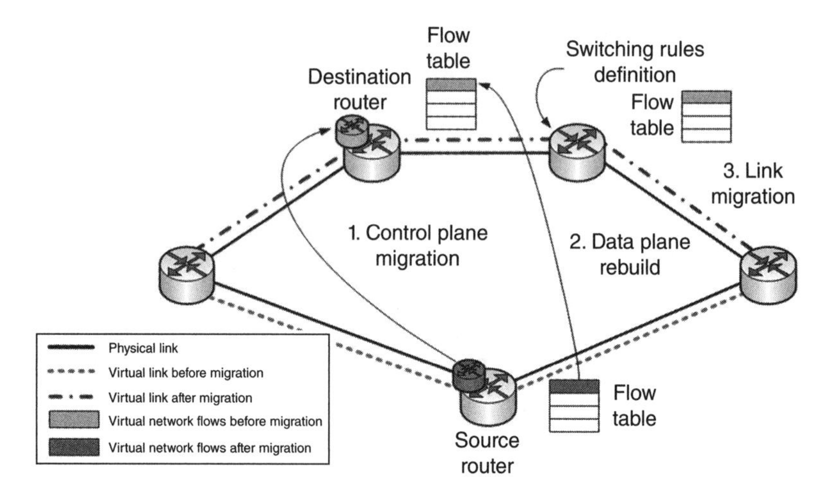

Figure 3.5. XenFlow virtual topology migration. (1) Virtual machine and all running routing protocol migration. (2) Data plane reconstruction based on control plane information. (3) Link migration by sending a predefined `ARP Reply` message.

The XenFlow routing function is performed by a flow table dynamically controlled by POX, an OpenFlow network controller [24]. Migration of virtual routers, shown in Figure 3.5, consists of three steps: migration of control plane, reconstruction of data plane, and migration of virtual links. The control plane is migrated between two physical network nodes through the live-migration mechanism of conventional Xen VMs [15]. Then, the reconstruction of data plane is performed as follows. The virtual router sends all routes to the Domain 0. When the virtual router detects a connection disruption caused by the migration, it reconnects to the Domain 0 in new physical host and sends all information about the routing and ARP tables. Upon receiving such information, Domain 0 reconfigures the data plane according to the control plane of the migrated virtual router. After migration of the control plane and reconstruction of the data plane, links are migrated. The links migration occurs in the OpenFlow switches instantiated in Domain 0 and other OpenFlow hardware switches. Link migration creates a switched path between the neighbors of the migrated virtual router to the physical host of virtual router after migration. The migrated virtual router sends an ARP reply packet with a predefined destination MAC address (AA:AA:AA:AA:AA:AA), which the network controller captures and reconfigures the paths. This procedure updates the location of a virtual router after the migration procedure, hence, the source physical host forward packets until the migration is complete, which results in a migration primitive of virtual routers without packet loss or interruption of packet-forwarding services.

XenFlow ensures the virtual router migration without packet loss, but the new path in the underlying substrate may introduce a greater or a smaller delay when compared to the original path. XenFlow does not control delay in forwarding nodes and also the new path may comprise non-XenFlow nodes. Therefore, during virtual network migration, packets may be out of order or may be received after a bigger delay of the new path. We assume that this is not a constrain because transport protocols are resilient to delay variation, as currently occurs due to changes in routing path or network congestion.

3.4 SECURITY OF VIRTUAL ENVIRONMENTS

There are several vulnerabilities that are disclosed in the current implementation of live migration of well know hypervisors, such as Xen and VMWare [25]. The biggest issue is that transferred data is not encrypted during migration procedure. Kernel memory, application state, sensitive data such as passwords and keys, and other migration data are transferred clearly, resulting in no confidentiality. Other vulnerabilities are: no guarantees that the VM is migrating to a trusted destination platform, no authentication and no authorization of operations, no integrity guarantees of VM data, and bugs in the hypervisor and migration module code that introduce security vulnerabilities. In this section, we argue about the main security issues of machine virtualization, and we expose the main security requirements for a secure virtualization platform. We focus on securing VM migration, but we also highlight security issues that affect cloud computing environment based on machine virtualization.

3.4.1 Requirements for a Secure Virtual Environment

A secure virtualization environment must ensure that processor, RAM, storage, and network, the main resources of a VM, are invulnerable against other VMs or against infrastructure attacks. Therefore, we establish six security requirements that summarize the needs of a secure virtualization environment. We also highlight that a secure live migration should provide confidentiality, to guarantee that any VM data are not accessed by others while they are transferred from one host to another, and auditability, to secure that sensitive data have not been exposed or damaged [26]. The six secure virtualization requirements are following: availability and isolation; integrity; confidentiality; access control, authentication, and authorization; nonrepudiation; and replay resistance.

Availability and isolation stands for the fact that any VM should be neither capable to access nor interfere other VMs. Even though several VMs share the same infrastructure, one VM is not able to access other VMs data or change computing results [1]. Thus, a secure hypervisor ensures strong isolation between running VMs, running each VM into a protected domain [27]. It is worth noting that isolation is achieved with confidentiality, integrity, and protection against denial of service.

Integrity aims that a virtual environment must provide the means to verify and prove the integrity and, therefore, it must be possible to identify if its processing, memory, and storage were modified. Attacks against integrity intend to modify information from virtual environments or to modify running programs in a virtual environment. The migration process should also be protected against integrity violation, because it clearly exposes the VM memory through the network to attacks, such as man in the middle attack [28]. In addition, a hardware module can run cryptographic functions to perform integrity verification and attestation. Attestation cryptographically ensures that a computing environment is trustworthy and the running application are not compromised [27]. Attestation may also assure that a remote environment is trustworthy because it has the same cryptographic signature of an integer environment. Attestation is also important to assure that after a VM migration, the destination machine is trustworthy and the migrated VM keeps its integrity as its cryptographic signature remains the same of the one before migration.

VM atomicity ensures that only one instance of the VM runs at a time [10]. Therefore, VM migration should neither add new VMs nor eliminate anyone. Thus, after successful migration, the system removes the VM instance in source host, and in case of migration failure, the system removes the VM instance in target host. The atomicity is crucial to ensure the integrity of the infrastructure for disaster recovery and to avoid generating duplicated copy of the same VM.

Confidentiality ensures that an attacker should not be able to intercept, to access or to modify the content of data transfer during the migration of a virtual machine. Therefore, system may use secure communication channel to transfer data between peer hosts. Moreover, the peers of the secure communication channel should be able to negotiate unique cryptographic keys and ensure that they are known only by the peers [10].

Access control, authentication, and authorization define that the system must ensure that a VM migration is performed between two secure authenticated platforms, which both are authorized to perform the migration, and there is no one else between them (man in the middle). Authentication ensures the true identity of an entity, hence, other security requirements depend on successful authentication. Authentication is a key feature because other security requirements depend on the authentication such as authorization, to distinguish legitimate and authorized from illegitimate participants based on authentication. Authorization ensures that only authorized entities perform operations such as VM migration. Besides, the VM should be neither migrated to unauthorized host nor from one [29].

Non-repudiation stands for the peers involved in migration cannot deny the migration participation [10]. The system must guarantee the provision of conclusive evidences of the migration event and peers participation, even when peers do not cooperate.

Replay resistance aims that an attacker cannot reproduce the migration procedure without being detected. Hence, all migration packets are unique and lose validity after migration.

3.4.2 Vulnerabilities

In a virtual environment, multiple VMs running on top of the same physical machine increase the efficiency of the system, but it also introduces software on sensitive areas of the system, which increases vulnerabilities. These vulnerabilities can be exploited by malicious users to obtain sensitive information, such as passwords and encryption keys, or perform other types of attacks such as denial of service.

In internal attacks, the system administrator performs attacks on the VMs. In this case, the system is completely vulnerable, because the administrator is authenticated and authorized to perform actions, neither cryptographic nor integrity techniques prevent the attacks. A malicious user who gains super-user privileges, via flaws in the authentication and authorization modules, performs an internal attack.

In other attacks, the attacker exploits the flaws of the virtualization system source code to inject malicious code and modify the system modules. This attack is possible due to the complexity of virtualization systems that end up having security flaws [2].

The attack can also be originated from an infected VM (or a legitimate machine with a malicious user) targeting other VMs sharing the same system. This type of attack requires that the attacker and the target VM are in the same physical machine. Due to the sharing of resources (e.g., CPU data cache), the attacker can steal cryptographic keys using techniques such as covert channel. This attack is facilitated when the network infrastructure indirectly allows the user to map the virtual networks and verify co-residence with the target VM [1]. These procedures are facilitated when static IPs are used for virtual networks, associating them with the physical IPs, but it can also be checked with IP common tools, such as traceroute.

The side channel attack is any attack that information, obtained to break the system, relies on information leaked by the hardware that are obtained by physical measurements as a "side" or an alternative channel [30]. The attack only concerns the implementation of a cryptosystem, rather than cryptanalysis of the math of the algorithm or brute force. Examples of physical measurements used to build a side channel can be: time took for performing different computations [31], varying power consumption [32], or leaked electromagnetic radiation provided by the hardware during computations, and even sound produced by the hardware. Therefore, assuming side attacks, the weakness of the security system is not the algorithm but its implementation. Brumley and Boneh [33] have shown that they succeeded to extract private keys from an OpenSSL-based Web server running on a machine in the local network. They run a timing attack in which an attacker machine measures the decryption queries response time of an OpenSSL server, in order to extract the private key stored on the server. They successfully performed the attack between two VMs, then, their results invalidate the announced isolation provided by the hypervisor. As mentioned before, side channel attacks only concern the crypto algorithm implementation and, thus, a virtualized system does not interfere on the weakness or strengthen of an implementation. Otherwise, virtualization is a shared operating hardware environment and actions of one VM may cause effects in another VM. Therefore, a virtualized system should not facilitate the access to physical measurements and should fully isolate one virtual environment from another virtual environment to prevent side channels attacks.

Covert channel is a type of security attack that creates and conveys information through a hidden communication channel, which is able to transfer information between processes that violate the security policy. A covert channel is not a legitimate channel and, therefore, it depends upon an ingenious mechanism, which is a program scheme to hide the way used to transfer the information from the source to the destination and requires access to the file system. Hence, different from side channel attack, covert channel are illegitimate communication channel built on already compromised systems. Covert channel requires viral infection of the system or a programming effort accomplished by the administrator or other authorized user of the system. Covert channels are usually difficult to detect and low detectability, the capacity to stay hidden, is often the assumed measurement of effectiveness of a covert channel attack. The usual hardware-based security mechanisms that underlie ultra-high-assurance secure operating systems cannot detect or control covert channels because they do not employ legitimate data transfer mechanisms of the computer system such as read and write. Thus, the covert channel must not interfere into legitimate operations to not be detected by security systems.

Intruders have limited options to get the data out of secured systems with Intrusion Detection Systems, Packet Anomaly Detection systems, and firewalls [34]. In this scenario, the intruder creates a covert channel. The communication media often used are ordinary actions unnoticed by administrator and legitimate users such as use of header- or payload-embedded information, altering a store location, performing operations that modify the real response time, using of packet inter-arrival times, and so on. Adding data to the payload section of Ping packets or encoding data in the unused fields of packet headers. A covert channel attack, which is the most difficult to detect, is to use inter-packet delay times to encode data. This means that the intruder does not necessarily have to create new traffic because he encodes the data by modulating the time between packets of regular legitimate communication. Data exfiltration can be an indication that a computer has been compromised even when other intrusion detection schemes have failed to detect a successful attack.

During the process of live migration, vulnerabilities may be exploited by attackers. Such vulnerabilities include authorization, integrity, and isolation failures.

Inappropriate access control policy: If access control policies are not defined properly or the module responsible for regulating them does not act effectively, an attacker can acquire undue control of systems to perform internal attacks. When the attacker controls the migration operation, the attacker can cause a denial of service by migrating multiple VMs to one physical machine to overload the communication link and the physical machine itself. The attacker may also migrate a malicious VM to a target physical machine, or migrating a target VM to a malicious physical machine. In both cases, after migration, the attacker gains full control of the target machine (physical or virtual).

Unprotected channel transmission: If the migration channel does not guarantee the confidentiality of the data, an attacker can steal or modify sensitive information, such as passwords and encryption keys. Attacks can be done passively (sniffing) or actively (man-in-the-middle) using techniques such as ARP spoofing, DNS poisoning and route hijacking. Active attacks are usually more problematic since they violate integrity, and may include modifications in the authentication services of the VM (sshd/login) and manipulation of kernel memory.

Loopholes in the migration module: The contemporary virtualization software such as Xen, VMware and KVM, have an extensive and complex code base, which tend to have bugs. Perez-Botero *et al.* identified 59 vulnerabilities in Xen and 38 in KVM until July 15, 2012, according to reports of CVE security vulnerability database [35]. These results confirm the existence of vulnerabilities, which an attacker can exploit to obstruct or access VMs.

3.4.3 Isolation, Access Control, and Availability

Several proposals aim to improve virtualization isolation, QoS provisioning, and virtual topologies migration. Besides, some proposals use software-defined networking (SDN) to manage network migrations. There are proposals for developing security applications on OpenFlow network infrastructures, as there are others that seek to ensure the security of the infrastructure itself [36].

NetLord [37] introduces a software agent on each physical server, which encapsulates packets of VMs with a new IP header. The new IP header whose semantics of addresses of layers 2 and 3 are overloaded to indicate to which virtual network the frames belong to. Similarly, VL2 [38] encapsulates IP packets of a virtual network with another IP header. In this case, the semantics of the IP addresses indicate both the virtual network and the localization of the physical host.

Distributed Overlay Virtual Ethernet (DOVE) [39] is a proposal of network virtualization that provides address space isolation by using a network identifier field of the envelop DOVE header, creating an overlay network. Address space isolation is also achieved using Virtnal extended Local Area Network (VXLAN), encapsulation [73]. VXLAN also adds to each Ethernet frame an outer Ethernet header, followed by an external IP, UDP, and VXLAN headers. *Network Virtualization Generic Routing Encapsulation* (NVGRE) [41] also encapsulates to allow multi-tenancy in public or private clouds. Both VXLAN and NVGRE use 24 bits to identify the virtual network that a frame belongs to. Nevertheless, these proposals create an overlay network that interconnects the nodes of the virtual network.

Houidi *et al.* propose an adaptive system that provides resources on demand for virtual networks [42]. It provides more resources for virtual networks as soon it detects service degradation or after a resource failure. The system uses a distributed multi-agent mechanism in physical infrastructure to negotiate requests, to fit the resources to the network needs, and to synchronize supplier nodes and virtual networks. Another proposal, OpenFlow Management Infrastructure (OMNI) [43] provides QoS to OpenFlow networks [21]. OMNI manages all flows of the network and define QoS parameters to each one. Besides, OMNI migrates flows to different paths without any packet losses. Kim *et al.* map QoS parameters of the virtual networks with different workloads on resources available on OpenFlow switches, such as queues and rate limiters [44]. The proposal's main goal is to provide QoS to scenarios in which the physical infrastructure belongs to a cloud multi-tenant provider. Nevertheless, the control of QoS parameters and QoS mapping are centralized on the OpenFlow controller node. McIlroy and Sventek provide QoS to virtual networks with a new router architecture [45]. The router comprises multiple VMs, called *Routelets*. Each *Routelet* is isolated from others and their resources are limited and guaranteed. *Routelets* that route QoS sensitive flows have access priority to substrate resources. Nevertheless, packet forwarding is performed by VMs, which limits the forwarding performance of *Routelets*.

Wang *et al.* propose a load balancer based on programming low cost OpenFlow switches to multiplex requests among different server replicas [46]. The proposed solution weightily fragments the IP address space of clients between server replicas. Thus, according to the client IP, it identifies the replica that serves a client. The proposal, however, does not guarantee the reservation of resources, nor QoS of flows. Hao *et al.* present the infrastructure *Virtually Clustered Open Router* (VICTOR) which is based on creating a cluster of datacenters via a virtualized network infrastructure [47]. The central idea of this approach is to use the OpenFlow as the basic network infrastructure of datacenters to allow moving a virtual machine from one physical location to another, as it is possible to reconfigure network paths. This proposal optimizes the datacenter network usage performing server migrations, but it does not guarantee QoS of each flow, and also does not isolate the use of resources from different virtualized servers.

3.5 FUTURE DIRECTIONS

The most important performance goal in VM live migration is a short VM downtime. Current migration approaches apply a combination of push and stop-and-copy strategies for VM live migration. The combined push and stop-and-copy strategy reduces the VM downtime at the cost of increasing the total migration time and network traffic due to migration. When transferring the VM storage during migration, total migration time is also affected. Therefore, a main research topic is to decrease the total downtime, keeping memory and storage consistence and reducing network bandwidth. Downtime directly impacts on the virtualization performance and compromises the deployment of VM migration on different scenarios.

Virtual Network Migration is another research topic. When moving a VM, its network connections should follow accordingly. VM migration between different Local Area Networks demands mechanisms for IP address migration or for networks traffic redirection. Migration within the same datacenter can also present performance problems when datacenters are globally distributed in a wide geographical area. Current research efforts focus on tunneling network traffic between source and destination host [22]. In this direction, there are proposal, such as NVGRE [41], VXLAN [73], and DOVE (Distributed Overlay Virtual Ethernet) [39], that creates tunnels to maintain virtual network connectivity even in scenarios that sites are separated by a WAN. Moreover, NetLord [37] and VL2 [38] change IP semantics for isolating and creating virtual networks within a datacenter. Other proposals for handling VM mobility across the Internet is to use Locator/Identifier Separation Protocol (LISP) [48, 49]. LISP uses two IP headers, one for the locator and other for identifier of the host. LISP maintains a globally reachable service that maps locator into identifier, and vice versa, in order to ensure the correct location of

VM no matter where it is hosted. After the VM migration, only the locator is changed, and all services remain online and reachable. Future trends also point to OpenFlow [21] as a possible approach for managing virtual network. Nevertheless, all aforementioned approaches require adaptations or more sophisticated deployments to be fully functional. To achieve a seamless network migration, we believe that new standards should take place to define a common way to migrate virtual network.

Storage is an important resource to virtualized servers, because it must be always available and present high performance. When migrating a VM, its storage should be also available at migration destination. Therefore, both source and destination sites share the storage service, or all VM storage must be sent over the network to destination host. EMC², one of the world's leader storage provider enterprise, provides a storage facility focused on a distributed federation of data, which allows data to be accessed among locations over synchronous distances. The EMC² distributed storage service is called VPLEX.[2] Moreover, Ceph is an open source project that aims to provide a distributed and redundant file system [50]. We agree that there are several initiatives for providing distributed storage service that are a step ahead for an available file system for VM migration. Nevertheless, these initiatives are new and immature. The proprietary ones have a higher maturity grade, but still are expensive and demand large infrastructure. Providing a distributed and available storage service, that requires low investment into infrastructure and is backward compatible, is a key research area.

Automated migration is also a key research topic, because the VMs allocation into physical servers is a np-hard problem. This scenario is aggravated considering big datacenters and multiple datacenters in a cloud provider's environment, due to the size and unmanageability of the scenario. There are proposed heuristics [51]-based optimization and others based on system modeling [52, 53], aiming to better use physical resources. An important factor to be considered in the use of optimization algorithms is the convergence time of the algorithm, which will directly interfere into the dynamics of the system. Proposals for optimization of use of physical resources are complementary to automatic migration systems and can be used to manage migrations. Trends show that a key research theme is matching the tradeoff between optimizing physical resource usage and limiting the number of migrations into the network.

A major research topic that arises is securing VM migration. Our studies show that there is no proposal that achieves a complete secure live migration primitive. Security must be deployed all long the development of a virtualization system. It must be present since the hypervisor, which should be reliable, trustworthy, and should provide secure virtualized environment, till the migration procedure, which should authenticate peers, check the trust of the foreign peer, and ensure a confidential channel between peers for transferring the VM. Security must also be ensured for all resources used by a VM. Isolation is a key challenge for network virtualization, as availability is another key challenge for storage virtualization. Confidentiality is an open topic while virtualizing memory. Trust warranting is a trend of research, in which we identified some works proposing protocols and new approaches [54]. We believe that providing security for

[2]http://www.emc.com/campaign/global/vplex/index.htm.

virtualized environments is a hot research topic, in which the proposals still are initial and immature. Therefore, trends show that new security mechanisms should be proposed for guaranteeing a securer virtualizing system.

3.6 CONCLUSION

VM migration is one of the most useful primitive introduced by virtualization technique. VM migration stands for the relocation of virtual computing environments over the physical infrastructure. The main idea of the migration primitive is to remap virtual resources into physical resources without disrupting the function of the virtual resources. We consider Virtual Machine Migration of particular interest for cloud computing environments and for network virtualization approaches. We claim that migration is a powerful tool for fitting computer capacity to dynamic workloads, facilitating user mobility, improving energy savings, and managing failures. In a network virtualization scenario, VM migration plays the whole of flexibly changing network topologies without constraining the physical realization of the virtual topology. Nevertheless, VM migration is both challenging in its realization and in its security guarantees.

In this chapter, we explained that live migration is the key migration mechanism of most of current hypervisor. We identified that the key resource to migrate is the VM memory, as it is constantly updated during the migration process. We also discussed how to migrate storage service of VMs through WANs. Moreover, we present a network virtualization approach, called XenFlow, which focuses on migrating virtual networks, without losing packets or disrupting network services. Besides the technical difficulties of migrating a VM, while it is running, we also highlighted how to assure that a VM migration occurs in a secure environment.

REFERENCES

1. T. Ristenpart, E. Tromer, H. Shacham, and S. Savage, "Hey, you, get off of my cloud: Exploring information leakage in third-party compute clouds," in *Proceedings of the 16th ACM Conference on Computer and Communications Security*, ser. CCS '09, 2009, pp. 199–212. [Online]. Available: http://doi.acm.org/10.1145/1653662.1653687. Accessed November 20, 2014.
2. Z. Wang and X. Jiang, "Hypersafe: A lightweight approach to provide lifetime hypervisor control-flow integrity," in *2010 IEEE Symposium on Security and Privacy (SP)*, May 2010, pp. 380–395.
3. M. Pearce, S. Zeadally, and R. Hunt, "Virtualization: Issues, security threats, and solutions," *ACM Computing Survey*, vol. 45, no. 2, pp. 17:1–17:39, 2013. [Online]. Available: http://doi.acm.org/10.1145/2431211.2431216. Accessed November 20, 2014.
4. Z. Pan, Y. Dong, Y. Chen, L. Zhang, and Z. Zhang, "Compsc: Live migration with pass-through devices," *SIGPLAN Notices*, vol. 47, no. 7, pp. 109–120, 2012. [Online]. Available: http://doi.acm.org/10.1145/2365864.2151040. Accessed November 20, 2014.

5. L. H. G. Ferraz, D. M. F. Mattos, and O. C. M. B. Duarte, "A two-phase multipathing scheme with genetic algorithm for data center network," *IEEE Global Communications Conference - GLOBECOM*, Austin, TX, December 2014.
6. O. C. M. B. Duarte and G. Pujolle, *Virtual Networks: Pluralistic Approach for the Next Generation of Internet*. Hoboken, NJ: John Wiley & Sons, Inc. 2013.
7. I. M. Moraes, D. M. Mattos, L. H. G. Ferraz, M. E. M. Campista, M. G. Rubinstein, L. H. M. Costa, M. D. de Amorim, P. B. Velloso, O. C. M. Duarte, and G. Pujolle, "FITS: A flexible virtual network testbed architecture," *Computer Networks*, Vol. 63, pp. 221–237, 2014.
8. H. T. Mouftah, H. T. Mouftah, and B. Kantarci, *Communication Infrastructures for Cloud Computing*, 1st ed. Hershey, PA: IGI Global, 2013.
9. S. Berger, R. Cáceres, K. A. Goldman, R. Perez, R. Sailer, and L. van Doorn, "vtpm: Virtualizing the trusted platform module," in *Proceedings of the 15th Conference on USENIX Security Symposium - Volume 15*, ser. USENIX-SS'06, 2006.
10. X. Wan, X. Zhang, L. Chen, and J. Zhu, "An improved vtpm migration protocol based trusted channel," in *2012 International Conference on Systems and Informatics (ICSAI)*, May 2012, pp. 870–875.
11. M. Aslam, C. Gehrmann, and M. Bjorkman, "Security and trust preserving VM migrations in public clouds," in *2012 IEEE 11th International Conference on Trust, Security and Privacy in Computing and Communications (TrustCom)*, June 2012, pp. 869–876.
12. D. M. F. Mattos and O. C. M. B. Duarte, "XenFlow: Seamless migration primitive and quality of service for virtual networks," *IEEE Global Communications Conference - GLOBECOM*, Austin, TX, December 2014.
13. Y. Wang, E. Keller, B. Biskeborn, J. van der Merwe, and J. Rexford, "Virtual routers on the move: Live router migration as a network-management primitive," in *Proceedings of the ACM SIGCOMM 2008 Conference on Data Communication*, ser. SIGCOMM '08, 2008, pp. 231–242. [Online]. Available: http://doi.acm.org/10.1145/1402958.1402985. Accessed November 20, 2014.
14. P. Pisa, N. Fernandes, H. Carvalho, M. Moreira, M. Campista, L. Costa, and O. Duarte, "Openflow and xen-based virtual network migration," in *Communications: Wireless in Developing Countries and Networks of the Future*, ser. IFIP Advances in Information and Communication Technology, A. Pont, G. Pujolle, and S. Raghavan, Eds. Boston, MA: Springer, 2010, vol. 327, pp. 170–181.
15. C. Clark, K. Fraser, S. Hand, J. Hansen, E. Jul, C. Limpach, I. Pratt, and A. Warfield, "Live migration of virtual machines," in *Proceedings of the 2nd conference on Symposium on Networked Systems Design & Implementation-Volume 2*, 2005, pp. 273–286.
16. S. Venkatesha, S. Sadhu, and S. Kintali, "Survey of virtual machine migration techniques," Department of Computer Science - University of California, Santa Barbara, CA, Technical Report, March 2009.
17. J. Suzuki, Y. Hidaka, J. Higuchi, T. Baba, N. Kami, and T. Yoshikawa, "Multi-root share of single-root i/o virtualization (sr-iov) compliant pci express device," in *2010 IEEE 18th Annual Symposium on High Performance Interconnects (HOTI)*, August 2010, pp. 25–31.
18. E. Zhai, G. D. Cummings, and Y. Dong, "Live migration with pass-through device for linux vm," in *OLS'08: The 2008 Ottawa Linux Symposium*, 2008, pp. 261–268.
19. N. Fernandes, M. Moreira, I. Moraes, L. Ferraz, R. Couto, H. Carvalho, M. Campista, L. Costa, and O. Duarte, "Virtual networks: Isolation, performance, and trends," *Annals of Telecommunications*, vol. 66, pp. 1–17, 2010.

20. N. Egi, A. Greenhalgh, M. Handley, M. Hoerdt, F. Huici, and L. Mathy, "Towards high performance virtual routers on commodity hardware," in *Proceedings of the 2008 ACM CoNEXT Conference*, 2008, pp. 1–12.
21. N. McKeown, T. Anderson, H. Balakrishnan, G. Parulkar, L. Peterson, J. Rexford, S. Shenker, and J. Turner, "OpenFlow: Enabling innovation in campus networks," *SIGCOMM Computer Communication Review*, vol. 38, pp. 69–74, 2008.
22. M. Bari, R. Boutaba, R. Esteves, L. Granville, M. Podlesny, M. Rabbani, Q. Zhang, and M. Zhani, "Data center network virtualization: A survey," *Communications Surveys Tutorials, IEEE*, vol. 15, no. 2, pp. 909–928, 2013.
23. R. Sherwood, G. Gibb, K. Yap, G. Appenzeller, M. Casado, N. McKeown, and G. Parulkar, "Flowvisor: A network virtualization layer," Technical Report OPENFLOW-TR-2009-01, OpenFlow Consortium, 2009.
24. M. Casado, T. Koponen, R. Ramanathan, and S. Shenker, "Virtualizing the network forwarding plane," in *Proceedings of the Workshop on Programmable Routers for Extensible Services of Tomorrow*, 2010, p. 8.
25. V. Melvin, "Dynamic load balancing based on live migration of virtual machines: Security threats and effects," Master's thesis, B. Thomas Golisano College of Computing and Information Sciences (GCCIS) - Rochester Institute of Technology, Rochester, NY, 2011.
26. Q. Zhang, L. Cheng, and R. Boutaba, "Cloud computing: State-of-the-art and research challenges," *Journal of Internet Services and Applications*, vol. 1, no. 1, pp. 7–18, 2010. [Online]. Available: http://dx.doi.org/10.1007/s13174-010-0007-6. Accessed November 20, 2014.
27. T. Garfinkel, B. Pfaff, J. Chow, M. Rosenblum, and D. Boneh, "Terra: A virtual machine-based platform for trusted computing," in *Proceedings of the Nineteenth ACM Symposium on Operating Systems Principles*, ser. SOSP '03, 2003, pp. 193–206. [Online]. Available: http://doi.acm.org/10.1145/945445.945464. Accessed November 20, 2014.
28. J. Oberheide, E. Cooke, and F. Jahanian, "Empirical exploitation of live virtual machine migration," in *Proceedings of BlackHat DC convention*, 2008.
29. B. Danev, R. J. Masti, G. O. Karame, and S. Capkun, "Enabling secure VM-vTPM migration in private clouds," in *Proceedings of the 27th Annual Computer Security Applications Conference*, ser. ACSAC '11, 2011, pp. 187–196. [Online]. Available: http://doi.acm.org/10.1145/2076732.2076759.Accessed November 20, 2014.
30. D. Agrawal, B. Archambeault, J. R. Rao, and P. Rohatgi, "The EM side—channel(s)," in *Cryptographic Hardware and Embedded Systems - CHES 2002*, ser. Lecture Notes in Computer Science, B. S. Kaliski, c. K. Koç, and C. Paar, Eds. Springer Berlin Heidelberg, 2003, vol. 2523, pp. 29–45. [Online]. Available: http://dx.doi.org/10.1007/3-540-36400-5_4. Accessed November 20, 2014.
31. P. C. Kocher, "Timing attacks on implementations of diffie-hellman, rsa, dss, and other systems," in *Advances in Cryptology—CRYPTO'96*, ser. Lecture Notes in Computer Science, N. Koblitz, Ed. Berlin: Springer, 1996, vol. 1109, pp. 104–113. [Online]. Available: http://dx.doi.org/10.1007/3-540-68697-5_9. Accessed November 20, 2014.
32. P. Kocher, J. Jaffe, and B. Jun, "Differential power analysis," in *Advances in Cryptology—CRYPTO'99*, ser. Lecture Notes in Computer Science, M. Wiener, Ed. Berlin: Springer, 1999, vol. 1666, pp. 388–397. [Online]. Available: http://dx.doi.org/10.1007/3-540-48405-1_25. Accessed November 20, 2014.
33. D. Brumley and D. Boneh, "Remote timing attacks are practical," *Computer Networks*, vol. 48, no. 5, pp. 701–716, 2005. [Online]. Available: http://www.sciencedirect.com/science/article/pii/S1389128605000125. Accessed November 20, 2014.

34. C. Fung, D. Lam, and R. Boutaba, "RevMatch: An efficient and robust decision model for collaborative malware detection," in *IEEE/IFIP Network Operation and Management Symposium (NOMS14)*, 2014.
35. D. Perez-Botero, J. Szefer, and R. B. Lee, "Characterizing hypervisor vulnerabilities in cloud computing servers," in *Proceedings of the 2013 International Workshop on Security in Cloud Computing*, ser. Cloud Computing'13, 2013, pp. 3–10. [Online]. Available: http://doi.acm.org/10.1145/2484402.2484406. Accessed November 20, 2014.
36. D. Kreutz, F. M. Ramos, and P. Verissimo, "Towards secure and dependable software-defined networks," in *Proceedings of the Second ACM SIGCOMM Workshop on Hot Topics in Software Defined Networking*, ser. HotSDN '13, 2013, pp. 55–60.
37. J. Mudigonda, P. Yalagandula, J. Mogul, B. Stiekes, and Y. Pouffary, "Netlord: A scalable multi-tenant network architecture for virtualized datacenters," in *Proceedings of the ACM SIGCOMM 2011*, ser. SIGCOMM '11, 2011, pp. 62–73.
38. A. Greenberg, J. R. Hamilton, N. Jain, S. Kandula, C. Kim, P. Lahiri, D. A. Maltz, P. Patel, and S. Sengupta, "Vl2: A scalable and flexible data center network," in *Proceedings of the ACM SIGCOMM 2009*, ser. SIGCOMM'09, 2009, pp. 51–62.
39. K. Barabash, R. Cohen, D. Hadas, V. Jain, R. Recio, and B. Rochwerger, "A case for overlays in DCN virtualization," in *Proceedings of the 3rd Workshop on Data Center—Converged and Virtual Ethernet Switching*, ser. DC-CaVES '11, 2011, pp. 30–37.
40. Y. Nakagawa, K. Hyoudou, and T. Shimizu, "A management method of IP multicast in overlay networks using openflow," in *Proceedings of the First Workshop on Hot Topics in Software Defined Networks*, ser. HotSDN '12, 2012, pp. 91–96. [Online]. Available: http://doi.acm.org/10.1145/2342441.2342460. Accessed November 20, 2014.
41. M. Sridharan, K. Duda, I. Ganga, A. Greenberg, G. Lin, M. Pearson, and P. Thaler, "NVGRE: Network Virtualization using Generic Routing Encapsulation," NVGRE, Internet Engineering Task Force, February 2013. [Online]. Available: http://tools.ietf.org/html/draft-sridharan-virtualization-nvgre-02. Accessed November 20, 2014.
42. I. Houidi, W. Louati, D. Zeghlache, P. Papadimitriou, and L. Mathy, "Adaptive virtual network provisioning," in *Proceedings of the Second ACM SIGCOMM Workshop on Virtualized Infrastructure Systems and Architectures*, 2010, pp. 41–48.
43. D. M. F. Mattos, N. C. Fernandes, V. T. da Costa, L. P. Cardoso, M. E. M. Campista, L. H. M. K. Costa, and O. C. M. B. Duarte, "OMNI: Openflow management infrastructure," in *2011 International Conference on the Network of the Future (NOF)*, 2011, pp. 52–56.
44. W. Kim, P. Sharma, J. Lee, S. Banerjee, J. Tourrilhes, S. Lee, and P. Yalagandula, "Automated and scalable QoS control for network convergence," *Proceedings of the INM/WREN*, April 2010.
45. R. McIlory and J. Sventek, "Resource virtualisation of network routers," in *2006 Workshop on High Performance Switching and Routing*, 2006, 6 pp.
46. R. Wang, D. Butnariu, and J. Rexford, "Openflow-based server load balancing gone wild," in *Proceedings of the 11th USENIX Conference on Hot Topics in Management of Internet, Cloud, and Enterprise Networks and Services*, 2011, pp. 12.
47. F. Hao, T. Lakshman, S. Mukherjee, and H. Song, "Enhancing dynamic cloud-based services using network virtualization," in *Proceedings of the 1st ACM Workshop on Virtualized Infrastructure Systems and Architectures*, 2009, pp. 37–44.
48. D. Phung, S. Secci, D. Saucez, and L. Iannone, "The openlisp control plane architecture," *IEEE Network*, vol. 28, no. 2, pp. 34–40, 2014.

49. P. Raad, S. Secci, D. C. Phung, A. Cianfrani, P. Gallard, and G. Pujolle, "Achieving subsecond downtimes in large-scale virtual machine migrations with lisp," *IEEE Transactions on Network and Service Management,* vol. 11, no. 2, pp. 133–143, 2014.
50. S. A. Weil, S. A. Brandt, E. L. Miller, D. D. E. Long, and C. Maltzahn, "Ceph: A scalable, high-performance distributed file system," in *Proceedings of the 7th Symposium on Operating Systems Design and Implementation*, ser. OSDI '06., 2006, pp. 307–320. [Online]. Available: http://dl.acm.org/citation.cfm?id=1298455.1298485. Accessed November 20, 2014.
51. I. Fajjari, N. Aitsaadi, G. Pujolle, and H. Zimmermann, "Vne-ac: Virtual network embedding algorithm based on ant colony metaheuristic," in *2011 IEEE International Conference on Communications (ICC)*, 2011, pp. 1–6.
52. E. Rodriguez, G. Alkmim, D. Batista, and N. da Fonseca, "Live migration in green virtualized networks," in *2013 IEEE International Conference on Communications (ICC)*, 2013, pp. 2262–2266.
53. G. P. Alkmim, D. M. Batista, and N. L. S. da Fonseca, "Mapping virtual networks onto substrate networks," *Journal of Internet Services and Applications*, vol. 4, no. 1, 2013. [Online]. Available: http://dx.doi.org/10.1186/1869-0238-4-3. Accessed November 20, 2014.
54. L. H. G. Ferraz, P. B. Velloso, and O. C. M. Duarte, "An accurate and precise malicious node exclusion mechanism for ad hoc networks," *Ad Hoc Networks*, vol. 19, pp. 142–155, 2014. [Online]. Available: http://www.sciencedirect.com/science/article/pii/S1570870514000468. Accessed November 20, 2014.

PART II

CLOUD NETWORKING AND COMMUNICATIONS

4

DATACENTER NETWORKS AND RELEVANT STANDARDS

Daniel S. Marcon, Rodrigo R. Oliveira, Luciano P. Gaspary, and Marinho P. Barcellos

Institute of Informatics, Federal University of Rio Grande do Sul, Porto Alegre, Brazil

4.1 OVERVIEW

Datacenters are the core of cloud computing, and their network is an essential component to allow distributed applications to run efficiently and predictably [1]. However, not all datacenters provide cloud computing. In fact, there are two main types of datacenters: production and cloud. Production datacenters are often shared by one tenant or among multiple (possibly competing) groups, services, and applications, but with low rate of arrival and departure. They run data analytics jobs with relatively little variation in demands, and their size varies from hundreds of servers to tens of thousands of servers. Cloud datacenters, in contrast, have high rate of tenant arrival and departure (churn) [2], run both user-facing applications and inward computation, require elasticity (since application demands are highly variable), and consist of tens to hundreds of thousands of physical servers [3]. Moreover, clouds can comprise several datacenters spread around the world. As an example, Google, Microsoft, and Amazon (three of the biggest players in the market) have datacenters in four continents; and each company has over 900,000 servers.

Cloud Services, Networking, and Management, First Edition.
Edited by Nelson L. S. da Fonseca and Raouf Boutaba.

This chapter presents an in-depth study of datacenter networks (DCNs), relevant standards, and operation. Our goal here is three-fold: (i) provide a detailed view of the networking infrastructure connecting the set of servers of the datacenter via high-speed links and commodity off-the-shelf (COTS) switches [4]; (ii) discuss the addressing and routing mechanisms employed in this kind of network; and (iii) show how the nature of traffic may impact DCNs and affect design decisions.

Providers typically have three main goals when designing a DCN [5]: scalability, fault tolerance, and agility. First, the infrastructure must scale to a large number of servers (and preferably allow incremental expansion with commodity equipment and little effort). Second, a DCN should be fault tolerant against failures of both computing and network resources. Third, a DCN ideally needs to be agile enough to assign any virtual machine or, in short, VM (which is part of a service or application) to any server [6]. As a matter of fact, DCNs should ensure that computations are not bottlenecked on communication [7].

Currently, providers attempt to meet these goals by implementing the network as a multi-rooted tree [1], using LAN technology for VM addressing and two main strategies for routing: equal-cost multipath (ECMP) and valiant load balancing (VLB). The shared nature of DCNs among a myriad of applications and tenants and high scalability requirements, however, introduce several challenges for architecture design, protocols and strategies employed inside the network. Furthermore, the type of traffic in DCNs is significantly different from traditional networks [8]. Therefore, we also survey recent proposals in the literature to address the limitations of technologies used in today's DCNs.

We structure this chapter as follows. First, we begin by examining the typical multi-rooted tree topology used in current datacenters and discuss its benefits and drawbacks. Then, we take a look at novel topologies proposed in the literature, and how network expansion can be performed in a cost-efficient way for providers. After addressing the structure of the network, we look into the traffic characteristics of these high-performance, dynamic networks and discuss proposals for traffic management on top of existing topologies. Based on the aspects discussed so far, we present layer-2 and layer-3 routing, its requirements and strategies typically employed to perform such task. We also examine existing mechanisms used for VM addressing in the cloud platform and novel proposals to increase flexibility and isolation for tenants. Finally, we discuss the most relevant open research challenges and close this chapter with a brief summary of DCNs.

4.2 TOPOLOGIES

In this section, we present an overview of datacenter topologies. The topology describes how devices (routers, switches and servers) are interconnected. More formally, this is represented as a graph, in which switches, routers and servers are the nodes, and links are the edges.

4.2.1 Typical Topology

Figure 4.1 shows a canonical three-tiered multi-rooted tree-like physical topology, which is implemented in current datacenters [1, 9]. The three tiers are: (1) the access (edge) layer, comprising the top-of-rack (ToR) switches that connect servers mounted on every rack; (2) the aggregation (distribution) layer, consisting of devices that interconnect ToR switches in the access layer; and (3) the core layer, formed by routers that interconnect switches in the aggregation layer. Furthermore, every ToR switch may be connected to multiple aggregation switches for redundancy (usually 1+1 redundancy) and every aggregation switch is connected to multiple core switches. Typically, a three-tiered network is implemented in datacenters with more than 8000 servers [4]. In smaller datacenters, the core and aggregation layers are collapsed into one tier, resulting in a two-tiered datacenter topology (flat layer-2 topology) [9].

This multitiered topology has a significant amount of oversubscription, where servers attached to ToR switches have significantly more (possibly an order of magnitude) provisioned bandwidth between one another than they do with hosts in other racks [3]. Providers employ this technique in order to reduce costs and improve resource utilization, which are key properties to help them achieve economies of scale.

This topology, however, presents some drawbacks. First, the limited bisection bandwidth[1] constrains server-to-server capacity, and resources eventually get fragmented (limiting agility) [11, 12]. Second, multiple paths are poorly exploited (e.g., only a single path is used within a layer-2 domain by spanning tree protocol), which may potentially cause congestion on some links even though other paths exist in the network and have available capacity. Third, the rigid structure hinders incremental expansion [13].

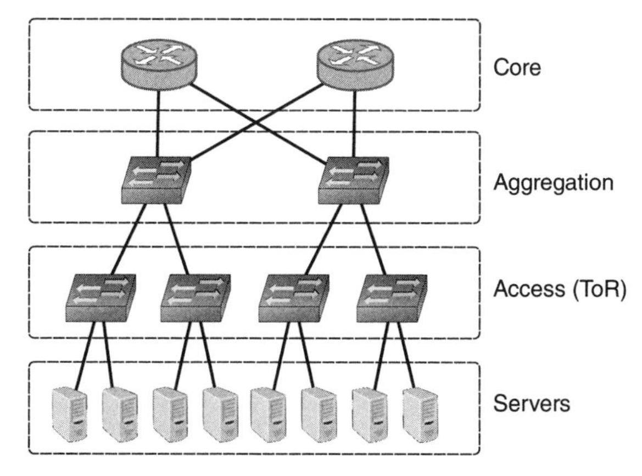

Figure 4.1. A canonical three-tiered tree-like datacenter network topology.

[1]The bisection bandwidth of the network is the worst-case segmentation (i.e., with minimum bandwidth) of the network in two equally-sized partitions [10].

Fourth, the topology is inherently failure-prone due to the use of many links, switches and servers [14]. To address these limitations, novel network architectures have been recently proposed; they can be organized in three classes [15]: switch-oriented, hybrid switch/server and server-only topologies.

4.2.2 Switch-Oriented Topologies

These proposals use commodity switches to perform routing functions, and follow a clos-based design or leverage runtime reconfigurable optical devices. A clos network [16] consists of multiple layers of switches; each switch in a layer is connected to all switches in the previous and next layers, which provides path diversity and graceful bandwidth degradation in case of failures. Two proposals follow the Clos design: VL2 [6] and Fat-Tree [4]. VL2, shown in Figure 4.2a, is an architecture for large-scale datacenters and provides multiple uniform paths between servers and full bisection bandwidth (i.e., it is non-oversubscribed). Fat-Tree, in turn, is a folded Clos topology. The topology, shown in Figure 4.2b, is organized in a non-oversubscribed k-ary tree-like structure, consisting of k-port switches. There are k two-layer pods with $k/2$ switches. Each $k/2$ switch in the lower layer is connected to $k/2$ servers, and the remaining ports are connected to $k/2$ aggregation switches. Each of the $(k/2)^2 k$-port core switches has one port connected to each of k pods. In general, a fat-tree built with k-port switches supports $k^3/4$ hosts. Despite the high capacity offered (agility is guaranteed), these architectures increase wiring costs (because of the number of links).

Optical switching architecture (OSA) [17], in turn, uses runtime reconfigurable optical devices to dynamically change physical topology and one-hop link capacities (within 10 milliseconds). It employs hop-by-hop stitching of multiple optical links to provide all-to-all connectivity for the highly dynamic and variable network demands of cloud applications. This method is shown in the example of Figure 4.3. Suppose that demands change from the left table to the right table in the figure (with a new highlighted entry). The topology must be adapted to the new traffic pattern, otherwise there will be at least one congested link. One possible approach is to increase capacity of link F–G (by reducing capacity of links F–D and G–C), so congestion can be avoided. Despite the flexibility achieved, OSA suffers from scalability issues, since it is designed to connect only a few

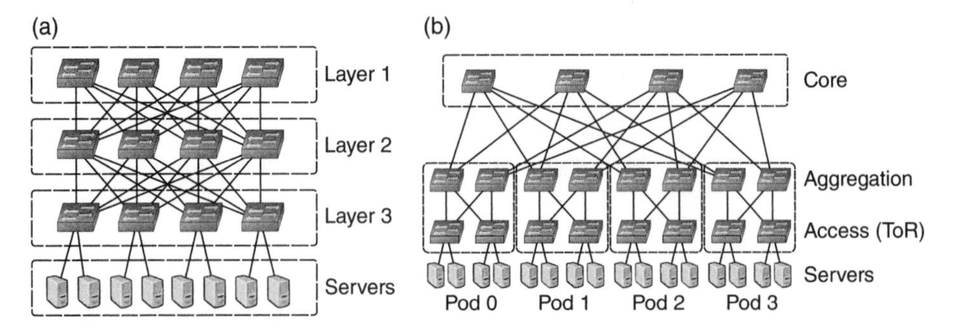

Figure 4.2. Clos-based topologies. (a) VL2 and (b) Fat-tree.

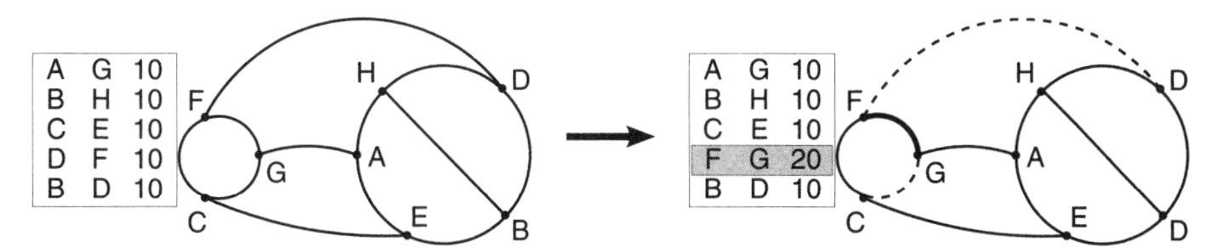

Figure 4.3. OSA adapts according to demands (adapted from Ref. [17]).

thousands of servers in a container, and latency-sensitive flows may be affected by link reconfiguration delays.

4.2.3 Hybrid Switch/Server Topologies

These architectures shift complexity from network devices to servers, i.e., servers perform routing, while low-end mini-switches interconnect a fixed number of hosts. They can also provide higher fault-tolerance, richer connectivity and improve innovation, because hosts are easier to customize than commodity switches. Two example topologies are DCell [5] and BCube [18], which can arguably scale up to millions of servers.

DCell [5] is a recursively built structure that forms a fully connected graph using only commodity switches (as opposed to high-end switches of traditional DCNs). DCell aims to scale out to millions of servers with few recursion levels (it can hold 3.2 million servers with only four levels and six hosts per cell). A DCell network is built as follows. A level-0 DCell ($DCell_0$) comprises servers connected to a n-port commodity switch. $DCell_1$ is formed with n + 1 $DCell_0$; each $DCell_0$ is connected to all other $DCell_0$ with one bidirectional link. In general, a level-k DCell is constructed with n + 1 $DCell_{k-1}$ in the same manner as $DCell_1$. Figure 4.4a shows an example of a two-level DCell topology. In this example, a commodity switch is connected with four servers ($n = 4$) and, therefore, a $DCell_1$ is constructed with 5 $DCell_0$. The set of $DCell_0$ is interconnected in the following way: each server is represented by the tuple (a_1, a_0), where a_1 and a_0 are level 1 and 0 identifiers, respectively; and a link is created between servers identified by the tuples $(i, j-1)$ and (j, i), for every i and every $j > i$.

Similarly to DCell, BCube [18] is a recursively built structure that is easy to design and upgrade. Additionally, BCube provides low latency and graceful degradation of bandwidth upon link and switch failure. In this structure, clusters (a set of servers interconnected by a switch) are interconnected by commodity switches in a hypercube-based topology. More specifically, BCube is constructed as follows: $BCube_0$ (level-0 BCube) consists of n servers connected by a n-port switch; $BCube_1$ is constructed from n $BCube_0$ and n n-port switches; and $BCube_k$ is constructed from n $BCube_{k-1}$ and n^k n-port switches. Each server is represented by the tuple (x_1, x_2), where x_1 is the cluster number and x_2 is the server number inside the cluster. Each switch, in turn, is represented by a tuple (y_1, y_2), where y_1 is the level number and y_2 is the switch number inside the level. Links are created by connecting the level-k port of the i-th server in the j-th $BCube_{k-1}$ to the j-th port of the i-th level-k switch. An example of two-level BCube with $n = 4$ (4-port switches) is shown in Figure 4.4b.

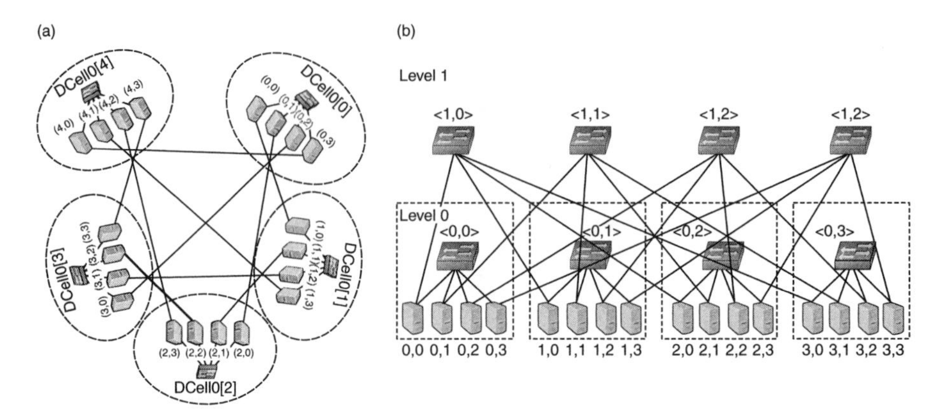

Figure 4.4. Hybrid switch/server topologies. (a) Two-level DCell and (b) two-level BCube.

Despite the benefits, DCell and BCube require a high number of NIC ports at end-hosts — causing some overhead at servers — and increase wiring costs. In particular, DCell results in non-uniform multiple paths between hosts, and level-0 links are typically more utilized than other links (creating bottlenecks). BCube, in turn, provides uniform multiple paths, but uses more switches and links than DCell [18].

4.2.4 Server-Only Topology

In this kind of topology, the network comprises only servers that perform all network functions. An example of architecture is CamCube [19], which is inspired in Content Addressable Network (CAN) [20] overlays and uses a 3D torus (k-ary 3-cube) topology with k servers along each axis. Each server is connected directly to 6 other servers, and the edge servers are wrapped. Figure 4.5 shows a 3-ary CamCube topology, resulting in 27 servers. The three most positive aspects of CamCube are (1) providing robust fault-tolerance guarantees (unlikely to partition even with 50% of server or link failures); (2) improving innovation with key-based server-to-server routing (content is hashed to a location in space defined by a server); and (3) allowing each application to define specific routing techniques. However, it does not hide topology from applications, has higher network diameter $O(\sqrt[3]{N})$ (increasing latency and traffic in the network) and hinders network expansion.

4.2.5 Summary of Topologies

Table 4.1 summarizes the benefits and limitations of these topologies by taking four properties into account: scalability, resiliency, agility and cost. The typical DCN topology has limited scalability (even though it can support hundreds of thousands of servers), as COTS switches have restricted memory size and need to maintain an entry in their Forwarding Information Base (FIB) for each VM. Furthermore, it presents low resiliency, since it provides only 1+1 redundancy, and its oversubscribed nature hinders agility.

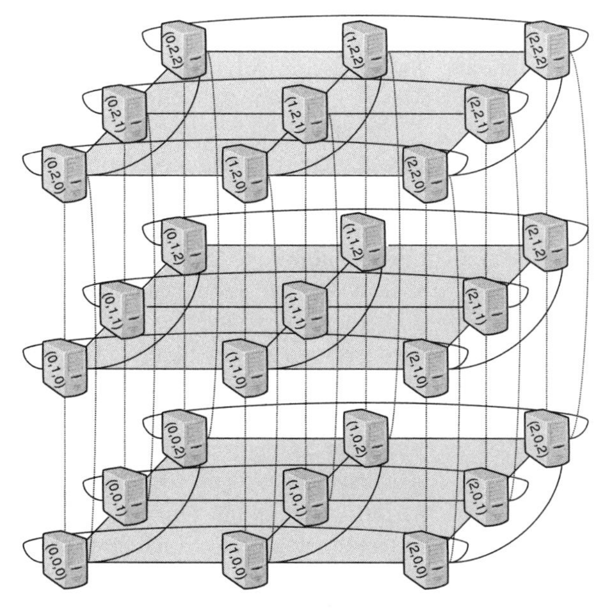

Figure 4.5. Example of 3-ary CamCube topology (adapted from Ref. [21]).

TABLE 4.1. Comparison among datacenter network topologies

Proposal	Properties			
	Scalability	Resiliency	Agility	Cost
Typical DCN	Low	Low	No	Low
Fat-Tree	High	Average	Yes	Average
VL2	High	High	Yes	High
OSA	Low	High	No	High
DCell	Huge	High	No	High
BCube	Huge	High	Yes	High
CamCube	Low	High	No	Average

Despite the drawbacks, it can be implemented with only commodity switches, resulting in lower costs.

Fat-Tree and VL2 are both instances of a Clos topology with high scalability and full bisection bandwidth (guaranteed agility). Fat-Tree achieves average resiliency, as ToR switches are connected only to a subset of aggregation devices, and has average overall costs (mostly because of increased wiring). VL2 scales through packet encapsulation, maintaining forwarding state only for switches in the network, achieves high resiliency by providing multiple shortest paths and by relying on a distributed lookup entity for handling address queries. As a downside, its deployment has increased costs (due to wiring, significant amount of exclusive resources for running the lookup system and the need of switch support for IP-in-IP encapsulation).

OSA was designed taking flexibility into account in order to improve resiliency (i.e., by using runtime reconfigurable optical devices to dynamically change physical topology and one-hop link capacities). However, it has low scalability (up to a few thousands of servers), no agility (as dynamically changing link capacities may result in congested links) and higher costs (devices should support optical reconfiguration).

DCell and BCube aim at scaling to millions of servers while ensuring high resiliency (rich connectivities between end-hosts). In contrast to BCube, DCell does not provide agility, as the set set of non-uniform multiple paths may be bottlenecked by links at level-0. Finally, their deployment costs may be significant, since they require a lot of wiring and more powerful servers in order to efficiently perform routing.

CamCube, in turn, is unlikely to partition even with 50% of server or link failures, thus achieving high resiliency. Its drawback, however, is related to scalability and agility; both properties can be hindered because of high network diameter, which indicates that, on average, more resources are needed for communication between VMs hosted by different servers. CamCube also has average deployment costs, mainly due to wiring and the need of powerful servers (to perform network functions).

As we can see, there is no perfect topology, since each proposal focus on specific aspects. Ultimately, providers are cost-driven: they choose the topology with the lowest costs, even if it cannot achieve all properties desired for a datacenter network running heterogenous applications from many tenants.

4.3 NETWORK EXPANSION

A key challenge concerning datacenter networks is dealing with the harmful effects that their ever-growing demand causes on scalability and performance. Because current DCN topologies are restricted to 1+1 redundancy and suffer from oversubscription, they can become underprovisioned quite fast. The lack of available bandwidth, in turn, may cause resource fragmentation (since it limits VM placement) [11] and reduce server utilization (as computations often depend on the data received from the network) [2]. In consequence, the DCN can loose its ability to accommodate more tenants (or offer elasticity to the current ones); even worse, applications using the network may start performing poorly, as they often rely on strict network guarantees[2].

These fundamental shortcomings have stimulated the development of novel DCN architectures (seen in Section 4.2) that provide large amounts of (or full) bisection bandwidth for up to millions of servers. Despite achieving high bisection bandwidth, their deployment is hindered by the assumption of homogeneous sets of switches (with the same number of ports). For example, consider a Fat-Tree topology, where the entire structure is defined by the number of ports in switches. These homogeneous switches limit the structure in two ways: full bisection bandwidth can only be achieved with

[2]For example, user-facing applications, such as Web services, require low-latency for communication with users, while inward computation (e.g., Map-Reduce) requires reliability and bisection bandwidth in the intra-cloud network.

specific numbers of servers (e.g., 8,192 and 27,648) and incremental upgrade may require replacing every switch in the network [13].

In fact, most physical datacenter designs are unique; hence, expansions and upgrades must be custom-designed and network performance (including bisection bandwidth, end-to-end latency and reliability) must be maximized while minimizing provider costs [11, 12]. Furthermore, organizations need to be able to incrementally expand their networks to meet the growing demands of tenants [13]. These facts have motivated recent studies [7, 11–13] to develop techniques to expand current DCNs to boost bisection bandwidth and reliability with heterogeneous sets of devices (i.e., without replacing every router and switch in the network). They are discussed next.

4.3.1 Legup

Focused on tree-like networks, Legup [12] is a system that aims at maximizing network performance at the design of network upgrades and expansions. It utilizes a linear model that combines three metrics (agility, reliability and flexibility), while being subject to the cloud provider's budget and physical constraints. In an attempt to reduce costs, the authors of Legup develop the *Theory of Heterogeneous Clos Networks* to allow modern and legacy equipment to coexist in the network. Figure 4.6 depicts an overview of the system. Legup assumes an existing set of racks and, therefore, only needs to determine aggregation and core levels of the network (more precisely, the set of devices, how they interconnect, and how they connect to ToR switches). It employs a branch and bound optimization algorithm to explore the solution space only for aggregation switches, as core switches in a heterogeneous Clos network are restricted by aggregation ones. Given a set of aggregation switches in each step of the algorithm, Legup performs three actions. First, it computes the minimum cost for mapping aggregation switches to racks. Second, it finds the minimum cost distribution of core switches to connect to the set of aggregation switches. Third, the candidate solution is bounded to check its optimality and feasibility (by verifying if any constraint is violated, including provider's budget and physical restrictions).

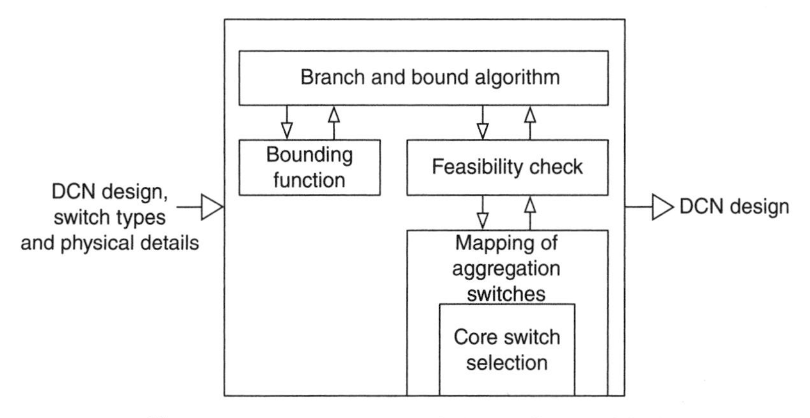

Figure 4.6. Legup's overview (adapted from Ref. [12]).

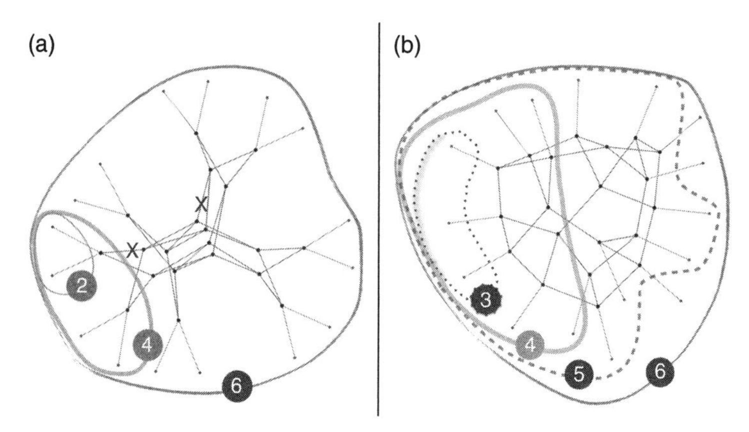

Figure 4.7. Comparison between (a) Fat-Tree and (b) Jellyfish with identical equipment (adapted from Ref. [13]).

4.3.2 Rewire

Recent advancements in routing protocols may allow DCNs to shift from a rigid tree to a generic structure [11, 22–25]. Based on this observation, Rewire [11] is a framework that performs DCN expansion on arbitrary topologies. It has the goal of maximizing network performance (i.e., finding maximum bisection bandwidth and minimum end-to-end latency), while minimizing costs and satisfying user-defined constraints. In particular, Rewire adopts a different definition of latency: while other studies model it by the worst-case hop-count in the network, Rewire also considers the speed of links and the processing time at switches (because unoptimized switches can add an order of magnitude more processing delay). Rewire uses simulated annealing (SA) [26] to search through candidate solutions and implements an approximation algorithm to efficiently compute their bisection bandwidth. The simulated annealing, however, does not take the addition of switches into account; it only optimizes the network for a given set of switches. Moreover, the process assumes uniform queuing delays for all switch ports, which is necessary because Rewire does not possess knowledge of network load.

4.3.3 Jellyfish

End-to-end throughput of a network is quantitatively proved to depend on two factors: (1) the capacity of the network and (2) the average path length (i.e., throughput is inversely proportional to the capacity consumed to deliver each byte) [13]. Furthermore, as noted earlier, rigid DCN structures hinder incremental expansion. Consequently, a degree-bounded[3] random graph topology among ToR switches, called Jellyfish [13], is introduced, with the goal of providing high bandwidth and flexibility. It supports device heterogeneity, different degrees of oversubscription and easy incremental expansion (by naturally allowing the addition of heterogeneous devices). Figure 4.7 shows a comparison

[3]Degree-bounded, in this context, means that the number of connections per node is limited by the number of ports in switches.

of Fat-Tree and Jellyfish with identical equipment and same diameter (i.e., 6). Each ring in the figure contains servers reachable within the number of hops in the labels. We see that Jellyfish can reach more servers in fewer hops, because some links are not useful from a path-length perspective in a Fat-Tree (e.g., links marked with "x"). Despite its benefits, Jellyfish's random design brings up some challenges, such as routing and the physical layout. Routing, in particular, is a critical feature needed, because it allows the use of the topology's high capacity. However, results show that the commonly used ECMP does not utilize the entire capacity of Jellyfish, and the authors propose the use of k-shortest paths and MultiPath TCP [25] to improve throughput and fairness.

4.3.4 Random Graph-Based Topologies

Singla et al. [7] analyze the throughput achieved by random graphs for topologies with both homogeneous and heterogeneous switches, while taking optimization into account. They obtain the following results: random graphs achieve throughput close to the optimal upper-bound under uniform traffic patterns for homogeneous switches, and heterogeneous networks with distinct connectivity arrangements can provide nearly identical high throughput. Then, the acquired knowledge is used as a building block for designing large-scale random networks with heterogeneous switches. In particular, they utilize the VL2 deployed in Microsoft's datacenters as a case study, showing that its throughput can be significantly improved (up to 43%) by only rewiring the same devices.

4.4 TRAFFIC

Proposals of topologies for datacenter networks presented in Sections 4.2 and 4.3 share a common goal: provide high bisection bandwidth for tenants and their applications. It is intuitive that a higher bisection bandwidth will benefit tenants, since the communication between VMs will be less prone to interference. Nonetheless, it is unclear how strong is the impact of the bisection bandwidth. This section addresses this question by surveying several recent measurement studies of DCNs. Then, it reviews proposals for dealing with related limitations. More specifically, it discusses traffic patterns—highlighting their properties and implications for both providers and tenants—and shows how literature is using such information to help designing and managing DCNs.

Traffic can be divided in two broad categories: north/south and east/west communication. North/south traffic (also known as extra-cloud) corresponds to the communication between a source and a destination host where one of the ends is located outside the cloud platform. By contrast, east/west traffic (also known as intra-cloud) is the communication in which both ends are located inside the cloud. These types of traffic usually depend on the kind and mix of applications: user-facing applications (e.g., web services) typically exchange data with users and, thus, generate north/south communication, while inward computation (i.e., MapReduce) requires coordination among its VMs, generating east/west communication. Studies [27] indicate that north/south and east-west traffic correspond to around 25% and 75% of traffic volume, respectively. They also point that both are increasing in absolute terms, but east/west is growing on a larger scale [27]. Towards understanding traffic characteristics and how it influences the proposal of novel mechanisms, we first discuss traffic properties defined by measurement studies in the literature [9, 28–30] and, then, examine traffic management and its most relevant proposals for large-scale cloud datacenters.

4.4.1 Properties

Traffic in the cloud network is characterized by flows; each flow is identified by sequences of packets from a source to a destination node (i.e., a flow is defined by a set packet header fields, such as source and destination addresses and ports and transport protocol). Typically, a bimodal flow classification scheme is employed, using elephant and mice classes. Elephant flows comprise a large number of packets injected in the network over a short amount of time, are usually long-lived and exhibit bursty behavior. In comparison, mice flows have a small number of packets and are short-lived [3]. Several measurement studies [9, 28–31] were conducted to characterize network traffic and its flows. We summarize their findings as follows:

- *Traffic asymmetry*. Requests from users to cloud services are abundant, but small in most occasions. Cloud services, however, process these requests and typically send responses that are comparatively larger.
- *Nature of traffic*. Network traffic is highly volatile and bursty, with links running close to their capacity at several times during a day. Traffic demands change quickly, with some transient spikes and other longer ones (possibly requiring more than half the full-duplex bisection bandwidth) [32]. Moreover, traffic is unpredictable at long time scales (e.g., 100 seconds or more). However, it can be predictable on shorter timescales (at 1 or 2 seconds). Despite the predictability over small timescales, it is difficult for traditional schemes, such as statistical multiplexing, to make a reliable estimate of bandwidth demands for VMs [33].
- *General traffic location and exchange*. Most traffic generated by servers (on average 80%) stays within racks. Server pairs from the same rack and from different racks exchange data with a probability of only 11% and 0.5%, respectively. Probabilities for intra- and extra-rack communication are as follows: servers either talk with fewer than 25% or to almost all servers of the same rack; and servers communicate with less than 10% or do not communicate with servers located outside its rack.
- *Intra- and inter-application communication*. Most volume of traffic (55%) represents data exchange between different applications. However, the communication matrix between them is sparse; only 2% of application pairs exchange data, with the top 5% of pairs accounting for 99% of inter-application traffic volume. Consequently, communicating applications form several highly connected components, with few applications connected to hundreds of other applications in star-like topologies. In comparison, intra-application communication represents 45% of the total traffic, with 18% of applications generating 99% of this traffic volume.

- *Flow size, duration, and number.* Mice flows represent around 99% of the total number of flows in the network. They usually have less than 10 kilobytes and last only a few hundreds of milliseconds. Elephant flows, in turn, represent only 1% of the number of flows, but account for more than half of the total traffic volume. They may have tens of megabytes and last for several seconds. With respect to flow duration, flows of up to 10 seconds represent 80% of flows, while flows of 200 seconds are less than 0.1% (and contribute to less than 20% of the total traffic volume). Further, flows of 25 seconds or less account for more than 50% of bytes. Finally, it has been estimated that a typical rack has around 10,000 active flows per second, which means that a network comprising 100,000 servers can have over 25,000,000 active flows.
- *Flow arrival patterns.* Arrival patterns can be characterized by heavy-tailed distributions with a positive skew. They best fit a log-normal curve having ON and OFF periods (at both 15 and 100 milliseconds granularities). In particular, inter arrival times at both servers and ToR switches have periodic modes spaced apart by approximately 15 milliseconds, and the tail of these distributions is long (servers may experience flows spaced apart by 10 seconds).
- *Link utilization.* Utilization is, on average, low in all layers but the core; in fact, in the core, a subset of links (up to 25% of all core links) often experience high utilization. In general, link utilization varies according to temporal patterns (time of day, day of week and month of year), but variations can be an order of magnitude higher at core links than at aggregation and access links. Due to these variations and the bursty nature of traffic, highly utilized links can happen quite often; 86% and 15% of links may experience congestion lasting at least 10 and 100 seconds, respectively, while longer periods of congestion tend to be localized to a small set of links.
- *Hot spots.* They are usually located at core links and can appear quite frequently, but the number of hot spots never exceeds 25% of core links.
- *Packet losses.* Losses occur frequently even at underutilized links. Given the bursty nature of traffic, an underutilized network (e.g., with mean load of 10%) can experience lots of packet drops. Measurement studies found that packet losses occur usually at links with low average utilization (but with traffic bursts that go beyond 100% of link capacity); more specifically, such behavior happens at links of the aggregation layer and not at links of the access and core layers. Ideally, topologies with full bisection bandwidth (i.e., a Fat-Tree) should experience no loss, but the employed routing mechanisms cannot utilize the full capacity provided by the set of multiple paths and, consequently, there is some packet loss in such networks as well [28].

4.4.2 Traffic Management

Other set of papers [34–37] demonstrate that available bandwidth for VMs inside the datacenter can vary by a factor of five or more in the worst-case scenario. Such variability results in poor and unpredictable network performance and reduced overall application performance [1, 38, 39], since VMs usually depend on the data received from the network to execute the subsequent computation.

The lack of bandwidth guarantees is related to two main factors. First, the canonical cloud topology is typically oversubscribed, with more bandwidth available in leaf nodes than in the core. When periods of traffic bursts happen, the lack of bandwidth up the tree (i.e., at aggregation and core layers) results in contention and, therefore, packet discards at congested links (leading to subsequent retransmissions). Since the duration of the timeout period is typically one or two orders of magnitude more than the round-trip time, latency is increased, becoming a significant source of performance variability [3]. Second, TCP congestion control does not provide robust isolation among flows. Consequently, elephant flows can cause contention in congested links shared with mice flows, leading to discarded packets from the smaller flows [2].

Recent proposals address this issue either by employing proportional sharing or by providing bandwidth guarantees. Most of them use the hose model [40] for network virtualization and take advantage of rate-limiting at hypervisors [41], VM placement [42] or virtual network embedding [43] in order to increase their robustness.

Proportional sharing. Seawall [2] and NetShare [44] allocate bandwidth at flow-level based on weights assigned to entities (i.e., VMs or services running inside these VMs) that generate traffic in the network. While both assign weights based on administrator specified policies, NetShare also supports automatic weight assignment. Both schemes are work-conserving (i.e., available bandwidth can be used by any flow that needs more bandwidth), provide max–min fair sharing and achieve high utilization through statistical multiplexing. However, as bandwidth allocation is performed per flow, such methods may introduce substantial management overhead in large datacenter networks (with over 10,000 flows per rack per second [9]). FairCloud [45] takes a different approach and proposes three allocation policies to explore the trade-off among network proportionality, minimum guarantees and high utilization. Unlike Seawall and NetShare, FairCloud does not allocate bandwidth along congested links at flow-level, but in proportion to the number of VMs of each tenant. Despite the benefits, FairCloud requires customized hardware in switches and is designed specifically for tree-like topologies.

Bandwidth guarantees. SecondNet [46], Gatekeeper [47], Oktopus [1], Proteus [48], and Hadrian [49] provide minimum bandwidth guarantees by isolating applications in virtual networks. In particular, SecondNet is a virtualization architecture that distributes the virtual-to-physical mapping, routing and bandwidth reservation state in server hypervisors. Gatekeeper configures each VM virtual NIC with both minimum and maximum bandwidth rates, which allows the network to be shared in a work-conserving manner. Oktopus maps tenants' virtual network requests (with or without oversubscription) onto the physical infrastructure and enforces these mappings in hypervisors. Proteus is built based on the observation that allocating the peak bandwidth requirements for applications leads to underutilization of resources. Hence, it quantifies the temporal bandwidth demands of applications and allocates each one of them in a different virtual network. Hadrian extends previous schemes by also taking inter-tenant communication into account and allocating applications according to a hierarchical hose model (i.e., per VM minimum bandwidth for intra-application communication and per tenant minimum guarantees for inter-tenant traffic). By contrast, a group of related proposals attempt to provide some level bandwidth sharing among applications of distinct tenants [50–52]. The approach introduced by Marcon et al. [51] groups applications in virtual networks, taking mutually trusting relationships between tenants into account when allocating each application. It provides work-conserving network sharing, but assumes that trust relationships are determined in advance. ElasticSwitch [52] assumes there exists an allocation method in the cloud platform and focuses on providing minimum bandwidth guarantees with a work-conserving sharing mechanism (when there is spare capacity in the network). Nevertheless, it requires two extra management layers for defining the amount of bandwidth for each flow, which may add overhead. Finally, EyeQ [50] leverages high bisection bandwidth of DCNs to support minimum and maximum bandwidth rates for VMs. Therefore, it provides work-conserving sharing, but depends on the core of the network to be congestion-free. None of these approaches can be readily deployed, as they demand modifications in hypervisor source code.

4.5 ROUTING

Datacenter networks often require specially tailored routing protocols, with different requirements from traditional enterprise networks. While the latter presents only a handful of paths between hosts and predictable communication patterns, DCNs require multiple paths to achieve horizontal scaling of hosts with unpredictable traffic matrices [4, 6]. In fact, datacenter topologies (i.e., the ones discussed in Section 4.2) typically present path diversity, in which multiple paths exist between servers (hosts) in the network. Furthermore, many cloud applications (ranging from Web search to MapReduce) require substantial (possibly full bisection) bandwidth [53]. Thus, routing protocols must enable the network to deliver high bandwidth by using all possible paths in the structure. We organize the discussion according to the layer involved, starting with the network layer.

4.5.1 Layer 3

To take advantage of the multiple paths available between a source and its destination, providers usually employ two techniques: ECMP [54] and VLB [6, 55, 56]. Both strategies use distinct paths for different flows. ECMP attempts to load balance traffic in the network and utilize all paths which have the same cost (calculated by the routing protocol) by uniformly spreading traffic among them using flow hashing. VLB randomly selects an intermediate router (occasionally, a L3 switch) to forward the incoming flow to its destination.

Recent studies in the literature [46, 53, 57, 58] propose other routing techniques for DCNs. As a matter of fact, the static flow-to-path mapping performed by ECMP does not take flow size and network utilization into account [59]. This may result in saturating commodity switch L3 buffers and degrading overall network performance [53]. Therefore, a system called Hedera [53] is introduced to allow dynamic flow scheduling for general multi-rooted trees with extensive path diversity. Hedera is designed to maximize

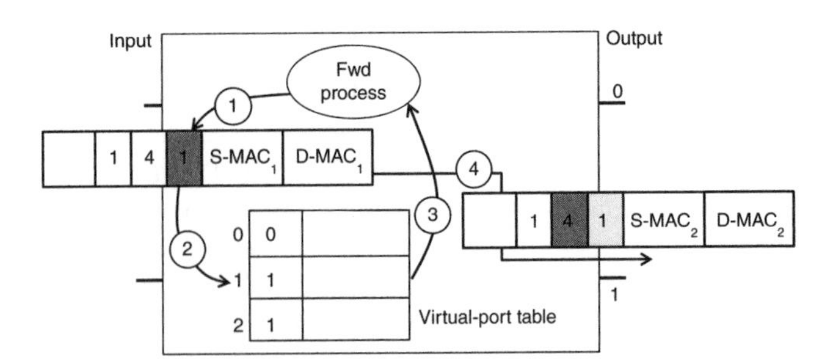

Figure 4.8. PSSR overview (adapted from Ref. [46]).

network utilization with low scheduling overhead of active flows. In general, the system performs the following steps: (1) detects large flows at ToR switches; (2) estimates network demands of these large flows (with a novel algorithm that considers bandwidth consumption according to a max–min fair resource allocation); (3) invokes a placement algorithm to compute paths for them; and (4) installs the set of new paths on switches.

Hedera uses a central OpenFlow controller[4] [60] with a global view of the network to query devices, obtain flow statistics and install new paths on devices after computing their routes. With information collected from switches, Hedera treats the flow-to-path mapping as an optimization problem and uses a simulated annealing metaheuristic to efficiently look for feasible solutions close to the optimal one in the search space. SA reduces the search space by allowing only a single core switch to be used for each destination. Overall, the system delivers close to optimal performance and up to four times more bandwidth than ECMP.

Port-switching based source routing (PSSR) [46] is proposed for the SecondNet architecture with arbitrary topologies and commodity switches. PSSR uses source routing, which requires that every node in the network knows the complete path to reach a destination. It takes advantage of the fact that a datacenter is administered by a single entity (i.e., the intra-cloud topology is known in advance) and represents a path as a sequence of output ports in switches, which is stored in the packet header. More specifically, the hypervisor of the source VM inserts the routing path in the packet header, commodity switches perform the routing process with PSSR and the destination hypervisor removes PSSR information from the packet header and delivers the packet to the destination VM. PSSR also introduces the use of virtual ports, because servers may have multiple neighbors via a single physical port (e.g., in DCell and BCube topologies). The process performed by a switch is shown in Figure 4.8. Switches read the pointer field in the packet header to get the exact next output port number (step 1), verify the next port number in the lookup virtual-port table (step 2), get the physical port number (step 3) and, in step 4, update the pointer field and forward the packet. This routing method introduces some overhead (since routing information must be included in the packet header), but, according to the authors, can be easily implemented on commodity switches using Multi-Protocol Label Switching (MPLS) [61].

[4]We will not focus our discussion in OpenFlow in this chapter. It is discussed in Chapter 6.

Bounded Congestion Multicast Scheduling (BCMS) [57], introduced to efficiently route flows in Fat-trees under the hose traffic model, aims at achieving bounded congestion and high network utilization. By using multicast, it can reduce traffic, thus minimizing performance interference and increasing application throughput [62]. BCMS is an online multicast scheduling algorithm that leverages OpenFlow to (1) collect bandwidth demands of incoming flows; (2) monitor network load; (3) compute routing paths for each flow; and (4) configure switches (i.e., installing appropriate rules to route flows). The algorithm has three main steps, as follows. First, it checks the conditions of uplinks out of source ToR switches (as flows are initially routed towards core switches). Second, it carefully selects a subset of core switches in order to avoid congestion. Third, it further improves traffic load balance by allowing ToR switches to connect to core switches with most residual bandwidth. Despite its advantages, BCMS relies on a centralized controller, which may not scale to large datacenters under highly dynamic traffic patterns such as the cloud.

Like BCMS, Code-Oriented eXplicit multicast (COXcast) [58] also focuses on routing application flows through the use of multicasting techniques (as a means of improving network resource sharing and reducing traffic). COXcast uses source routing, so all information regarding destinations are added to the packet header. More specifically, the forwarding information is encoded into an identifier in the packet header and, at each network device, is resolved into an output port bitmap by a node-specific key. COXcast can support a large number of multicast groups, but it adds some overhead to packets (since all information regarding routing must be stored in the packet).

4.5.2 Layer 2

In the Spanning Tree Protocol (STP) [63], all switches agree on a subset of links to be used among them, which forms a spanning tree and ensures a loop-free network. Despite being typically employed in Ethernet networks, it does not scale, since it cannot use the high-capacity provided by topologies with rich connectivities (i.e., Fat-Trees [24]), limiting application network performance [64]. Therefore, only a single path is used between hosts, creating bottlenecks and reducing overall network utilization.

STP's shortcomings are addressed by other protocols, including Multiple Spanning Tree Protocol (MSTP) [65], Transparent Interconnect of Lots of Links (TRILL) [22] and Link Aggregation Control Protocol (LACP) [66]. MSTP was proposed in an attempt to use the path diversity available in DCNs more efficiently. It is an extension of STP to allow switches to create various spanning trees over a single topology. Therefore, different Virtual LANs (VLANs) [67] can utilize different spanning trees, enabling the use of more links in the network than with a single spanning tree. Despite its objective, implementations only allow up to 16 different spanning trees, which may not be sufficient to fully utilize the high-capacity available in DCNs [68].

TRILL is a link-state routing protocol implemented on top of layer-2 technologies, but bellow layer-3, and is designed specifically to address limitations of STP. It discovers and calculates shortest paths between TRILL devices (called routing bridges or, in short, RBridges), which enables shortest path multihop routing in order to use all available paths in networks with rich connectivities. RBridges run Intermediate System to Intermediate System (IS-IS) routing protocol (RFC 1195) and handle frames in the following manner: the first RBridge (ingress node) encapsulates the incoming frame with a TRILL header (outer MAC header) that specifies the last TRILL node as the destination (egress node), which will decapsulate the frame.

Link Aggregation Control Protocol (LACP) is another layer-2 protocol used in DCNs. It transparently aggregates multiple physical links into one logical link known as Link Aggregation Group (LAG). LAGs only handle outgoing flows; they have no control over incoming traffic. They provide flow-level load balancing among links in the group by hashing packet header fields. LACP can dynamically add or remove links in LAGs, but requires that both ends of a link run the protocol.

There are also some recent studies that propose novel strategies for routing frames in DCNs, namely Smart Path Assignment in Networks (SPAIN) [24] and Portland [64]. SPAIN [24] focuses on providing efficient multipath forwarding using COTS switches over arbitrary topologies. It has three components: (1) path computation; (2) path setup; and (3) path selection. The first two components run on a centralized controller with global network visibility. The controller first pre-computes a set of paths to exploit the rich connectivities in the DCN topology, in order to use all available capacity of the physical infrastructure and to support fast failover. After the path computation phase, the controller combines these multiple paths into a set of trees, with each tree belonging to a distinct VLAN. Then, these VLANs are installed on switches. The third component (path selection) runs at end-hosts for each new flow; it selects paths for flows with the goals of spreading load across the pre-computed routes (by the path setup component) and minimizing network bottlenecks. With this configuration, end-hosts can select different VLANs for communication (i.e., different flows between the same source and destination can use distinct VLANs for routing). To provide these functionalities, however, SPAIN requires some modification to end-hosts, adding an algorithm to choose among pre-installed paths for each flow.

PortLand [64] is designed and built based on the observation that Ethernet/IP protocols may have some inherent limitations when designing large-scale arbitrary topologies, such as limited support for VM migration, difficult management and inflexible communication. It is a layer-2 routing and forwarding protocol with plug-and-play support for multi-rooted Fat-Tree topologies. PortLand uses a logically centralized controller (called fabric manager) with global visibility and maintains soft state about network configuration. It assigns unique hierarchical Pseudo MAC (PMAC) addresses for each VM to provide efficient, provably loop-free frame forwarding; VMs, however, do not have the knowledge of their PMAC and believe they use their Actual MAC (AMAC). The mapping between PMAC and AMAC and the subsequent frame header rewriting is performed by edge (ToR) switches. PMACs are structured as *pod.position.port.vmid*, where each field respectively corresponds to the pod number of the edge switch, its position inside the pod, the port number in which the physical server is connected to and the identifier of the VM inside the server. With PMACs, PortLand transparently provides location-independent addresses for VMs and requires no modification in commodity switches. However, it has two main shortcomings (1) it requires a Fat-Tree topology (instead of the traditional multi-rooted oversubscribed tree) and (2) at least half of the

ToR switch ports should be connected to servers (which, in fact, is a limitation of Fat-Trees) [69].

4.6 ADDRESSING

Each server (or, more specifically, each VM) must be represented by a unique canonical address that enables the routing protocol to determine paths in the network. Cloud providers typically employ LAN technologies for addressing VMs in datacenters, which means there is a single address space to be sliced among tenants and their applications. Consequently, tenants have neither flexibility in designing their application layer-2 and layer-3 addresses nor network isolation from other applications.

Some isolation is achieved by the use of VLANs, usually one VLAN per tenant. However, VLANs are ill-suited for datacenters for four main reasons [51, 70–72]: (1) they do not provide flexibility for tenants to design their layer-2 and layer-3 address spaces; (2) they use the spanning tree protocol, which cannot utilize the high-capacity available in DCN topologies (as discussed in the previous section); (3) they have poor scalability, since no more than 4094 VLANs can be created, and this is insufficient for large datacenters; and *iv*) they do not provide location-independent addresses for tenants to design their own address spaces (independently of other tenants) and for performing seamless VM migration. Therefore, providers need to use other mechanisms to allow address space flexibility, isolation and location independence for tenants while multiplexing them in the same physical infrastructure. We structure the discussion in three main topics: emerging technologies, separation of name and locator and full address space virtualization.

4.6.1 Emerging Technologies

Some technologies employed in DCNs are: Virtual eXtensible Local Area Network (VXLAN) [73], Amazon Virtual Private Cloud (VPC) [74] and Microsoft Hyper-V [75]. VXLAN [73] is an Internet draft being developed to address scalability and multipath usage in DCNs when providing logical isolation among tenants. VXLAN works by creating overlay (virtual layer-2) networks on top of the actual layer-2 or on top of UDP/IP. In fact, using MAC-in-UDP encapsulation abstracts VM location (VMs can only view the virtual layer-2) and, therefore, enables a VXLAN network to be composed of nodes within distinct domains (DCNs), increasing flexibility for tenants using multi-datacenter cloud platforms. VXLAN adds a 24-bit segment ID field in the packet header (allowing up to 16 million different logical networks), uses ECMP to distribute load along multiple paths and requires Internet Group Management Protocol (IGMP) for forwarding frames to unknown destinations, or multicast and broadcast addresses. Despite the benefits, VXLAN header adds 50 bytes to the frame size, and multicast and network hardware may limit the usable number of overlay networks in some deployments.

Amazon VPC [74] provides full IP address space virtualization, allowing tenants to design layer-3 logical isolated virtual networks. However, it does not virtualize layer-2, which does not allow tenants to send multicast and broadcast frames [71]. Microsoft Hyper-V [75] is a hypervisor-based system that provides virtual networks for tenants to design their own address spaces; Hyper-V enables IP overlapping in different virtual networks without using VLANs. Furthermore, Hyper-V switches are software-based layer-2 network switches with capabilities to connect VMs among themselves, with other virtual networks and with the physical network. Hyper-V, nonetheless, tends to consume more resources than other hypervisors with the same load [76].

4.6.2 Separation of Name and Locator

VL2 [6] and Crossroads [70] focus on providing location independence for VMs, so that providers can easily grow or shrink allocations and migrate VMs inside or across datacenters. VL2 [6] uses two types of addresses: location-specific addresses (LAs), which are the actual addresses in the network, used for routing; and application-specific addresses (AAs), permanent address assigned to VMs that remain the same even after migrations. VL2 uses a directory system to enforce isolation among applications (through access control policies) and to perform the mapping between names and locators; each server with an AA is associated with the LA from the ToR it is connected to. Figure 4.9 depicts how address translation in VL2 is performed: the source hypervisor encapsulates the AA address with the LA address of the destination ToR for each packet sent; packets are forwarded in the network through shortest paths calculated by the routing protocol, using both ECMP and VLB; when packets arrive at the destination ToR switch, LAs are removed (packets are decapsulated) and original packets are sent to the correct VMs using AAs. To provide location-independent addresses, VL2 requires that hypervisors run a shim layer (VL2 agent) and that switches support IP-over-IP.

Crossroads [70], in turn, is a network fabric developed to provide layer agnostic and seamless VM migration inside and across DCNs. It takes advantage of

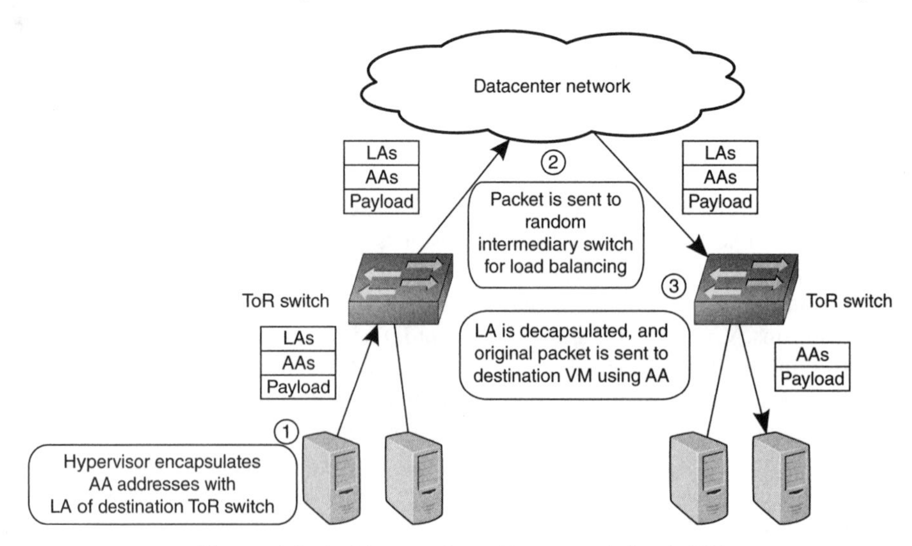

Figure 4.9. Architecture for address translation in VL2.

the Software-Defined Networking (SDN) paradigm [77] and extends an OpenFlow controller to allow VM location-independence without modifications to layer-2 and layer-3 network infrastructure. In Crossroads, each VM possess two addresses: a PMAC and a Pseudo IP (PIP), both with location and topological information embedded in them. The first one ensures that traffic originated from one datacenter and en route to a second datacenter (to which the VM was migrated) can be maintained at layer-2, while the second guarantees that all traffic destined to a migrated VM can be routed across layer-3 domains. Despite its benefits, Crossroads introduces some network overhead, as nodes must be identified by two more addresses (PMAC and PIP) in addition to the existing MAC and IP.

4.6.3 Full Address Space Virtualization

Cloud datacenters typically provide limited support for multi-tenancy, since tenants should be able to design their own address spaces (similar to a private environment) [71]. Consequently, a multi-tenant virtual datacenter architecture to enable specific-tailored layer-2 and layer-3 address spaces for tenants, called NetLord [71], is proposed. At hypervisors, NetLord runs an agent that performs Ethernet+IP (L2+L3) encapsulation over tenants' layer-2 frames and transfers them through the network using SPAIN [24] for multipathing, exploring features of both layers. More specifically, the process of encapsulating/decapsulating is shown in Figure 4.10 and occurs in three steps, as follows: (1) the agent at the source hypervisor creates L2 and L3 headers (with source IP being a tenant-assigned MAC address space identifier, illustrated as MAC_AS_ID) in order to direct frames through the L2 network to the correct edge switch; (2) the edge switch forwards the packet to the correct server based on the IP destination address in the virtualized layer-3 header; (3) the hypervisor at the destination server removes the virtual L2 and L3 headers and uses the IP destination address to deliver the original packet from the source VM to the correct VM. NetLord can be run on commodity switches and scale the network to hundreds of thousands of VMs. However, it requires an agent running on hypervisors (which may add some overhead) and support for IP forwarding on commodity edge (ToR) switches.

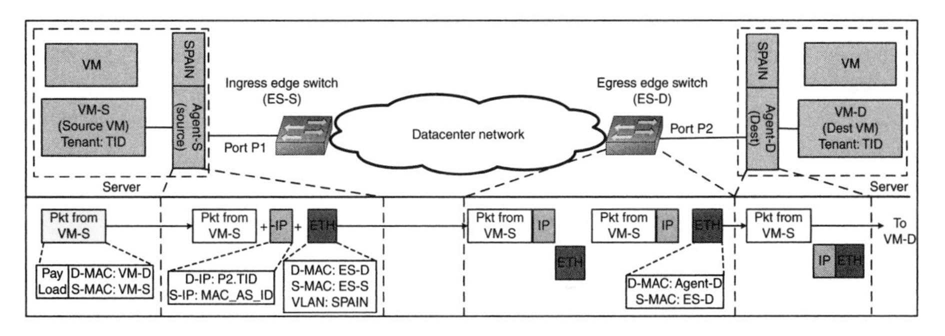

Figure 4.10. NetLord's encapsulation/decapsulation process (adapted from Ref. [71]).

4.7 RESEARCH CHALLENGES

In this section, we analyze and discuss open research challenges and future directions regarding datacenter networks. As previously mentioned, DCNs (i) present some distinct requirements from traditional networks (e.g., high scalability and resiliency); (ii) have significantly different (often more complex) traffic patterns; and (iii) may not be fully utilized, because of limitations in current deployed mechanisms and protocols (for instance, ECMP). Such aspects introduce some challenges, which are discussed next.

4.7.1 Heterogeneous and Optimal DCN Design

Presently, many Internet services and applications rely on large-scale datacenters to provide availability while scaling in and out according to incoming demands. This is essential in order to offer low response time for users, without incurring excessive costs for owners. Therefore, datacenter providers must build infrastructures to support large and dynamic numbers of applications and guarantee quality of service (QoS) for tenants. In this context, the network is an essential component of the whole infrastructure, as it represents a significant fraction of investment and contributes to future revenues by allowing efficient use of datacenter resources [15]. According to Zhang et al. [78], network requirements include (i) scalability, so that a large number of servers can be accommodated (while allowing incremental expansion); (ii) high server-to-server capacity, to enable intensive communication between any pair of servers (i.e., at full speed of their NICs); (iii) agility, so applications can use any available server when they need more resources (and not only servers located near their current VMs); (iv) uniform network utilization to avoid bottlenecks; and (v) fault tolerance to cope with server, switch and link failures. In fact, guaranteeing such requirements is a difficult challenge. Looking at these challenges from the providers viewpoint make them even more difficult to address and overcome, since reducing the cost of building and maintaining the network is seen as a key enabler for maximizing profits [15].

As discussed in Section 4.2, several topologies (e.g., Refs. [4–6, 17, 18]) have been proposed to achieve the desired requirements, with varying costs. Nonetheless, they (i) focus on homogeneous networks (all devices with the same capabilities); and (ii) do not provide theoretical foundations regarding optimality. Singla et al. [7], in turn, take an initial step towards addressing heterogeneity and optimality, as they (i) measure the upper-bound on network throughput for homogeneous topologies with uniform traffic patterns; and (ii) show an initial analysis of possible gains with heterogeneous networks. Despite this fact, a lot remains to be investigated in order to enable the development of more efficient, robust large-scale networks with heterogeneous sets of devices. In summary, very little is known about heterogeneous DCN design, even though current DCNs are typically composed of heterogenous equipment.

4.7.2 Efficient and Incremental Expansion

Providers need to be constantly expanding their datacenter infrastructures to accommodate ever-growing demands. For instance, Facebook has been expanding its datacenters for some years [79–82]. This expansion is crucial for business, as the increase of demand may negatively impact scalability and performance (e.g., by creating bottlenecks in the network). When the whole infrastructure is upgraded, the network must be expanded accordingly, with a careful design plan, in order to allow efficient utilization of resources and to avoid fragmentation. To address this challenge, some proposals in the literature [7, 11–13] have been introduced to enlarge current DCNs without replacing legacy hardware. They aim at maximizing high bisection bandwidth and reliability. However, they often make strong assumptions (e.g., Legup [12] is designed for tree-like networks, and Jellyfish [13] requires new mechanisms for routing). Given the importance of datacenters nowadays (as home of hundreds of thousands of services and applications), the need for efficient and effective expansion of large-scale networks is a key challenge for improving provider profit, QoS offered to tenant applications and quality of experience (QoE) provided for users of these applications.

4.7.3 Network Sharing and Performance Guarantees

Datacenters host applications with diverse and complex traffic patterns and different performance requirements. Such applications range from user-facing ones (i.e., Web services and online gaming) that require low latency communication to inward computation (e.g., scientific computing) that need high network throughput. To gain better understanding of the environment, studies [1, 9, 30, 49, 83] conducted measurements and concluded that available bandwidth for VMs inside the cloud platform can vary by a factor of five or more during a predefined period of time. They demonstrate that such variability ends up impacting overall application execution time (resulting in poor and unpredictable performance). Several strategies (including Refs. [2, 47, 48, 52, 84]) have been proposed to address this issue. Nonetheless, they have one or more of the following shortcomings: (i) require complex mechanisms, which, in practice, cannot be deployed; (ii) focus on network sharing among VMs (or applications) in a homogeneous infrastructure (which simplifies the problem [85]); (iii) perform static bandwidth reservations (resulting in underutilization of resources); or (iv) provide proportional sharing (no strict guarantees). In fact, there is an inherent trade-off between providing strict guarantees (desired by tenants) and enabling work-conserving sharing in the network (desired by providers to improve utilization), which may be exacerbated in a heterogenous network. We believe this challenge requires further investigation, since such high-performance networks ideally need simple and efficient mechanisms to allow fair bandwidth sharing among running applications in a heterogeneous environment.

4.7.4 Address Flexibility for Tenants

While network performance guarantees require quantitative performance isolation, address flexibility needs qualitative isolation [71]. Cloud DCNs, however, typically provide limited support for multi-tenancy, as they have a single address space divided among applications (according to their needs and number of VMs). Thereby, tenants have no flexibility in choosing layer-2 and layer-3 addresses for applications. Note that, ideally, tenants should be able to design their own address spaces (i.e., they should have similar flexibility to a private environment), since already developed applications may necessitate a specific set of addresses to correctly operate without source code modification. Some proposals in the literature [6, 70, 71] seek to address this challenge either by identifying end-hosts with two addresses or by fully virtualizing layer-2 and layer-3. Despite adding flexibility for tenants, they introduce some overhead (e.g., hypervisors need a shim layer to manage addresses, or switches must support IP-over-IP) and require resources specifically used for address translation (in the case of VL2). This is an important open challenge, as the lack of address flexibility may hinder the migration of applications to the cloud platform.

4.7.5 Mechanisms for Load Balancing Across Multiple Paths

DCNs usually present path diversity (i.e., multiple paths between servers) to achieve horizontal scaling for unpredictable traffic matrices (generated from a large number of heterogeneous applications) [6]. Their topologies can present two types of multiple paths between hosts: uniform and non-uniform ones. ECMP is the standard technique used for splitting traffic across equal-cost (uniform) paths. Nonetheless, it cannot fully utilize the available capacity in these multiple paths [59]. Non-uniform multiple paths, in turn, complicate the problem, as mechanisms must take more factors into account (i.e., path latency and current load). There are some proposals in the literature [46, 53, 57, 58] to address this issue, but they either cannot achieve the desired response times (e.g., Hedera) [86] or are developed for specific architectures (e.g., PSSR for SecondNet). Chiesa et al. [87] have taken an initial approach towards analyzing ECMP and propose algorithms for improving its performance. Nevertheless, further investigation is required for routing traffic across both uniform and non-uniform parallel paths, considering not only tree-based topologies, but also newer proposals such as random graphs [7, 13]. This investigation should lead to novel mechanisms and protocols that better utilize available capacity in DCNs (e.g., eliminating bottlenecks at level-0 links in DCell).

4.8 SUMMARY

In this chapter, we have presented basic foundations of datacenter networks and relevant standards, as well as recent proposals in the literature that address limitations of current mechanisms. We began by studying network topologies in Section 4.2. First, we examined the typical topology utilized in today's datacenters, which consists of a multi-rooted tree with path diversity. This topology is employed by providers to allow rich connectivity with reduced operational costs. One of its drawbacks, however, is the lack of full bisection bandwidth, which is the main motivation for proposing novel topologies. We used a three-class taxonomy to organize the state-of-the-art datacenter topologies: switch-oriented, hybrid switch/server and server-only topologies. The distinct characteristic is the use of switches and/or servers: switches only (Fat-Tree, VL2 and OSA), switches and servers (DCell and BCube) and only servers (CamCube) to perform packet routing and forwarding in the network.

These topologies, however, usually present rigid structures, which hinders incremental network expansion (a desirable property for the ever-growing cloud datacenters). Therefore, we took a look at network expansion strategies (Legup, Rewire and Jellyfish) in Section 4.3. All of these strategies have the goal of improving bisection bandwidth to increase agility (the ability to assign any VM of any application to any server). Furthermore, the design of novel topologies and expansion strategies must consider the nature of traffic in DCNs. In Section 4.4, we summarized recent measurement studies about traffic and discussed some proposals that deal with traffic management on top of a DCN topology.

Then, we discussed routing and addressing in Sections 4.5 and 4.6, respectively. Routing was divided in two categories: layer-3 and layer-2. While layer-3 routing typically employs ECMP and VLB to utilize the high-capacity available in DCNs through the set of multiple paths, layer-2 routing uses the spanning tree protocol. Despite the benefits, these schemes cannot efficiently take advantage of multiple paths. Consequently, we briefly examined proposals that deal with this issue (Hedera, PSSR, SPAIN and Portland). Addressing, in turn, is performed by using LAN technologies, which does not provide robust isolation and flexibility for tenants. Towards solving these issues, we examined the proposal of a new standard (VXLAN) and commercial solutions developed by Amazon (VPC) and Microsoft (Hyper-V). Furthermore, we discussed proposals in the literature that aim at separating name and locator (VL2 and Crossroads) and at allowing full address space virtualization (NetLord).

Finally, we analyzed open research challenges regarding datacenter networks: (i) the need to design more efficient DCNs with heterogeneous sets of devices, while considering optimality; (ii) strategies for incrementally expanding networks with general topologies; (iii) network schemes with strict guarantees and predictability for tenants, while allowing work-conserving sharing to increase utilization; (iv) address flexibility to make the migration of applications to the cloud easier; and (v) mechanisms for load balancing traffic across different multiple parallel paths (using all available capacity).

Having covered the operation and research challenges of intra-datacenter networks, the next three chapters inside the networking and communications part discuss the following subjects: inter-datacenter networks, an important topic related to cloud platforms composed of several datacenters (e.g., Amazon EC2); the emerging paradigm of SDN, its practical implementation (OpenFlow) and how these can be applied to intra- and inter-datacenter networks to provide fine-grained resource management; and mobile cloud computing, which seeks to enhance capabilities of resource-constrained mobile devices using cloud resources.

REFERENCES

1. Hitesh Ballani, Paolo Costa, Thomas Karagiannis, and Ant Rowstron. Towards predictable datacenter networks. In *ACM SIGCOMM*, 2011.
2. Alan Shieh, Srikanth Kandula, Albert Greenberg, Changhoon Kim, and Bikas Saha. Sharing the data center network. In *USENIX NSDI*, 2011.

3. Dennis Abts and Bob Felderman. A guided tour of data-center networking. *Communication of the ACM*, 55(6):44–51, 2012.
4. Mohammad Al-Fares, Alexander Loukissas, and Amin Vahdat. A scalable, commodity data center network architecture. In *ACM SIGCOMM*, 2008.
5. Chuanxiong Guo, Haitao Wu, Kun Tan, Lei Shi, Yongguang Zhang, and Songwu Lu. Dcell: A scalable and fault-tolerant network structure for data centers. In *ACM SIGCOMM*, 2008.
6. Albert Greenberg, James R. Hamilton, Navendu Jain, Srikanth Kandula, Changhoon Kim, Parantap Lahiri, David A. Maltz, Parveen Patel, and Sudipta Sengupta. VL2: A scalable and flexible data center network. In *ACM SIGCOMM*, 2009.
7. Ankit Singla, P. Brighten Godfrey, and Alexandra Kolla. High throughput data center topology design. In *USENIX NSDI*, 2014.
8. Jian Guo, Fangming Liu, Xiaomeng Huang, John C.S. Lui, Mi Hu, Qiao Gao, and Hai Jin. On efficient bandwidth allocation for traffic variability in datacenters. In *IEEE INFOCOM*, 2014a.
9. Theophilus Benson, Aditya Akella, and David A. Maltz. Network traffic characteristics of data centers in the wild. In *ACM IMC*, 2010.
10. Nathan Farrington, Erik Rubow, and Amin Vahdat. Data Center Switch Architecture in the Age of Merchant Silicon. In *IEEE HOTI*, 2009.
11. Andrew R. Curtis, Tommy Carpenter, Mustafa Elsheikh, Alejandro Lopez-Ortiz, and S. Keshav. Rewire: An optimization-based framework for unstructured data center network design. In *IEEE INFOCOM*, 2012.
12. Andrew R. Curtis, S. Keshav, and Alejandro Lopez-Ortiz. Legup: Using heterogeneity to reduce the cost of data center network upgrades. In *ACM Co-NEXT*, 2010.
13. Ankit Singla, Chi-Yao Hong, Lucian Popa, and P. Brighten Godfrey. Jellyfish: Networking data centers randomly. In *USENIX NSDI*, 2012.
14. Yang Liu and Jogesh Muppala. Fault-tolerance characteristics of data center network topologies using fault regions. In *IEEE/IFIP DSN*, 2013.
15. Lucian Popa, Sylvia Ratnasamy, Gianluca Iannaccone, Arvind Krishnamurthy, and Ion Stoica. A cost comparison of datacenter network architectures. In *ACM Co-NEXT*, 2010.
16. Charles Clos. A Study of non-blocking switching networks. *BellSystem Technical Journal*, 32:406–424, 1953.
17. Kai Chen, Ankit Singlay, Atul Singhz, Kishore Ramachandranz, Lei Xuz, Yueping Zhangz, Xitao Wen, and Yan Chen. Osa: An optical switching architecture for data center networks with unprecedented flexibility. In *USENIX NSDI*, 2012.
18. Chuanxiong Guo, Guohan Lu, Dan Li, Haitao Wu, Xuan Zhang, Yunfeng Shi, Chen Tian, Yongguang Zhang, and Songwu Lu. BCube: A high performance, server-centric network architecture for modular data centers. In *ACM SIGCOMM*, 2009.
19. Hussam Abu-Libdeh, Paolo Costa, Antony Rowstron, Greg O'Shea, and Austin Donnelly. Symbiotic routing in future data centers. In *ACM SIGCOMM*, 2010.
20. Sylvia Ratnasamy, Paul Francis, Mark Handley, Richard Karp, and Scott Shenker. A scalable content-addressable network. In *ACM SIGCOMM*, 2001.
21. Paolo Costa, Thomas Zahn, Ant Rowstron, Greg O'Shea, and Simon Schubert. Why should we integrate services, servers, and networking in a data center? In *ACM WREN*, 2009.
22. Transparent Interconnection of Lots of Links (TRILL): RFCs 5556 and 6325, 2013. Available at: http://tools.ietf.org/rfc/index. Accessed November 20, 2014.

23. Changhoon Kim, Matthew Caesar, and Jennifer Rexford. Floodless in seattle: A scalable ethernet architecture for large enterprises. In *ACM SIGCOMM*, 2008.
24. Jayaram Mudigonda, Praveen Yalagandula, Mohammad Al-Fares, and Jeffrey C. Mogul. SPAIN: COTS data-center Ethernet for multipathing over arbitrary topologies. In *USENIX NSDI*, 2010.
25. Damon Wischik, Costin Raiciu, Adam Greenhalgh, and Mark Handley. Design, implementation and evaluation of congestion control for multipath tcp. In *USENIX NSDI*, 2011.
26. Scott Kirkpatrick, C. Daniel Gelatt, and Mario P. Vecchi. Optimization by simulated annealing. *Science*, 220(4598):671–680, 1983.
27. Renato Recio. The coming decade of data center networking discontinuities. ICNC, August 2012. keynote speaker.
28. Theophilus Benson, Ashok Anand, Aditya Akella, and Ming Zhang. MicroTE: Fine grained traffic engineering for data centers. In *ACM CoNEXT*, 2011.
29. Peter Bodík, Ishai Menache, Mosharaf Chowdhury, Pradeepkumar Mani, David A. Maltz, and Ion Stoica. Surviving failures in bandwidth-constrained datacenters. In *ACM SIGCOMM*, 2012.
30. Srikanth Kandula, Sudipta Sengupta, Albert Greenberg, Parveen Patel, and Ronnie Chaiken. The nature of data center traffic: Measurements & analysis. In *ACM IMC*, 2009.
31. Xiaoqiao Meng, Vasileios Pappas, and Li Zhang. Improving the scalability of data center networks with traffic-aware virtual machine placement. In *IEEE INFOCOM*, 2010.
32. Chi H. Liu, Andreas Kind, and Tiancheng Liu. Summarizing data center network traffic by partitioned conservative update. *IEEE Communications Letters*, 17(11):2168–2171, 2013.
33. Meng Wang, Xiaoqiao Meng, and Li Zhang. Consolidating virtual machines with dynamic bandwidth demand in data centers. In *IEEE INFOCOM*, 2011.
34. Alexandrm-Dorin Giurgiu. Network performance in virtual infrastrucures, February 2010. Available at: http://staff.science.uva.nl/~delaat/sne-2009-2010/p29/presentation.pdf. Accessed November 20, 2014.
35. Dave Mangot. Measuring EC2 system performance, May 2009. Availabel at: http://bit.ly/48Wui. Accessed November 20, 2014.
36. Jörg Schad, Jens Dittrich, and Jorge-Arnulfo Quiané-Ruiz. Runtime measurements in the cloud: Observing, analyzing, and reducing variance. *Proceedings of the VLDB Endowment*, 3(1–2):460–471, 2010.
37. Guohui Wang and T. S. Eugene Ng. The impact of virtualization on network performance of amazon ec2 data center. In *IEEE INFOCOM*, 2010.
38. Haiying Shen and Zhuozhao Li. New bandwidth sharing and pricing policies to achieve A win-win situation for cloud provider and tenants. In *IEEE INFOCOM*. 2014.
39. Eitan Zahavi, Isaac Keslassy, and Avinoam Kolodny. Distributed adaptive routing convergence to non-blocking DCN routing assignments. *IEEE Journal on Selected Areas in Communications*, 32(1):88–101, 2014.
40. Nick G. Duffield, Pawan Goyal, Albert Greenberg, Partho Mishra, K. K. Ramakrishnan, and Jacobus E. van der Merive. A flexible model for resource management in virtual private networks. In *ACM SIGCOMM*, 1999.
41. Barath Raghavan, Kashi Vishwanath, Sriram Ramabhadran, Kenneth Yocum, and Alex C. Snoeren. Cloud control with distributed rate limiting. In *ACM SIGCOMM*, 2007.

42. Joe W. Jiang, Tian Lan, Sangtae Ha, Minghua Chen, and Mung Chiang. Joint VM placement and routing for data center traffic engineering. In *IEEE INFOCOM*, 2012.
43. Minlan Yu, Yung Yi, Jennifer Rexford, and Mung Chiang. Rethinking virtual network embedding: Substrate support for path splitting and migration. *SIGCOMM Compuer. Communication Review*, 38:17–29, 2008.
44. Vinh The Lam, Sivasankar Radhakrishnan, Rong Pan, Amin Vahdat, and George Varghese. Netshare and stochastic netshare: Predictable bandwidth allocation for data centers. *ACM SIGCOMM CCR*, 42(3), 2012.
45. Lucian Popa, Gautam Kumar, Mosharaf Chowdhury, Arvind Krishnamurthy, Sylvia Ratnasamy, and Ion Stoica. FairCloud: Sharing the network in cloud computing. In *ACM SIGCOMM*, 2012.
46. Chuanxiong Guo, Guohan Lu, Helen J. Wang, Shuang Yang, Chao Kong, Peng Sun, Wenfei Wu, and Yongguang Zhang. SecondNet: A data center network virtualization architecture with bandwidth guarantees. In *ACM CoNEXT*, 2010.
47. Henrique Rodrigues, Jose Renato Santos, Yoshio Turner, Paolo Soares, and Dorgival Guedes. Gatekeeper: Supporting bandwidth guarantees for multi-tenant datacenter networks. In *USENIX WIOV*, 2011.
48. Di Xie, Ning Ding, Y. Charlie Hu, and Ramana Kompella. The only constant is change: Incorporating time-varying network reservations in data centers. In *ACM SIGCOMM*, 2012.
49. Hitesh Ballani, Keon Jang, Thomas Karagiannis, Changhoon Kim, Dinan Gunawardena, and Greg O'Shea. Chatty tenants and the cloud network sharing problem. In *USENIX NSDI*, 2013a.
50. Vimalkumar Jeyakumar, Mohammad Alizadeh, David Mazières, Balaji Prabhakar, Changhoon Kim, and Albert Greenberg. EyeQ: Practical network performance isolation at the edge. In *USENIX NSDI*, 2013.
51. Daniel Stefani Marcon, Rodrigo Ruas Oliveira, Miguel Cardoso Neves, Luciana Salete Buriol, Luciano Paschoal Gaspary, and Marinho Pilla Barcellos. Trust-based Grouping for Cloud Datacenters: Improving security in shared infrastructures. In *IFIP Networking*, 2013.
52. Lucian Popa, Praveen Yalagandula, Sujata Banerjee, Jeffrey C. Mogul, Yoshio Turner, and Jose Renato Santos. ElasticSwitch: Practical work-conserving bandwidth guarantees for cloud computing. In *ACM SIGCOMM*, 2013.
53. Mohammad Al-Fares, Sivasankar Radhakrishnan, Barath Raghavan, Nelson Huang, and Amin Vahdat. Hedera: dynamic flow scheduling for data center networks. In *USENIX NSDI*, 2010.
54. C. Hopps. Analysis of an equal-cost multi-path algorithm, 2000. RFC 2992.
55. Albert Greenberg, Parantap Lahiri, David A. Maltz, Parveen Patel, and Sudipta Sengupta. Towards a next generation data center architecture: Scalability and commoditization. In *ACM PRESTO*, 2008.
56. Leslie G. Valiant and Gordon J. Brebner. Universal schemes for parallel communication. In *ACM STOC*, 1981.
57. Zhiyang Guo, Jun Duan, and Yuanyuan Yang. On-line multicast scheduling with bounded congestion in Fat-Tree data center networks. *IEEE Journal on Selected Areas in Communications*, 32(1):102–115, 2014b.
58. Wen-Kang Jia. A scalable multicast source routing architecture for data center networks. *IEEE Journal on Selected Areas in Communications*, 32(1):116–123, 2014.

59. Sivasankar Radhakrishnan, Malveeka Tewari, Rishi Kapoor, George Porter, and Amin Vahdat. Dahu: Commodity switches for direct connect data center networks. In *ACM/IEEE ANCS*, 2013.
60. Nick McKeown, Tom Anderson, Hari Balakrishnan, Guru Parulkar, Larry Peterson, Jennifer Rexford, Scott Shenker, and Jonathan Turner. Openflow: Enabling innovation in campus networks. *SIGCOMM Computer Communication Review*, 38:69–74, 2008.
61. E. Rosen, A. Viswanathan, and R. Callon. Multiprotocol label switching architecture, 2001. RFC 3031.
62. Dan Li, Mingwer Xu, Ying Liu, Xia Xie, Yong Cui, Jingyi Wang, and Gihai Chen. Reliable multicast in data center networks., 2014. IEEE Transactions Computers, 63: 2011-2024, 2014.
63. 802.1D - MAC Bridges, 2013. Available at: http://www.ieee802.org/1/pages/802.1D.html. Accessed November 20, 2014.
64. Radhika Niranjan Mysore, Andreas Pamboris, Nathan Farrington, Nelson Huang, Pardis Miri, Sivasankar Radhakrishnan, Vikram Subramanya, and Amin Vahdat. PortLand: A scalable fault-tolerant layer 2 data center network fabric. In *ACM SIGCOMM*, 2009.
65. 802.1s - Multiple Spanning Trees, 2013. Available at: http://www.ieee802.org/1/pages/802.1s.html. Accessed November 20, 2014.
66. 802.1ax - Link Aggregation Task Force, 2013. Available at: http://ieee802.org/3/axay/. Accessed November 20, 2014.
67. 802.1Q - Virtual LANs, 2013. Available at: http://www.ieee802.org/1/pages/802.1Q.html. Accessed November 20, 2014.
68. Understanding Multiple Spanning Tree Protocol (802.1s), 2007. Available at: http://www.cisco.com/en/US/tech/tk389/tk621/technologies_white_paper09186a0080094cfc.shtml. Accessed November 20, 2014.
69. Md. Faizal Bari, Raonf Boutaba, Rafael Esteves, Lisandro Z. Granville, Maxim Podlesny, Md. Golam Rabbani, Qi Zhang, and Mohamed F. Zhani. Data center network virtualization: A survey. *IEEE Communications Surveys Tutorials*, 15(2):909–928, 2013.
70. Vijay Mann, Ailkumar Vishnoi, Kalapriya Kannan, and Shivkumar Kalyanaraman. CrossRoads: Seamless VM mobility across data centers through software defined networking. In *IEEE/IFIP NOMS*, 2012.
71. Jayaram Mudigonda, Praveen Yalagandula, Jeff Mogul, Bryan Stiekes, and Yanick Pouffary. NetLord: A scalable multi-tenant network architecture for virtualized datacenters. In *ACM SIGCOMM*, 2011.
72. Brent Stephens, Alan Cox, Wes Felter, Colin Dixon, and John Carter. PAST: Scalable ethernet for data centers. In *ACM CoNEXT*, 2012.
73. VXLAN: A Framework for Overlaying Virtualized Layer 2 Networks over Layer 3 Networks, 2013. Available at: http://tools.ietf.org/html/draft-mahalingam-dutt-dcops-vxlan-02. Accessed November 20, 2014.
74. Amazon Virtual Private Cloud, 2013. Available at: http://aws.amazon.com/vpc/. Accessed November 20, 2014.
75. Microsoft Hyper-V Server 2012, 2013a. Available at: http://www.microsoft.com/en-us/server-cloud/hyper-v-server/. Accessed November 20, 2014.
76. Hyper-V Architecture and Feature Overview, 2013b. Available at: http://msdn.microsoft.com/en-us/library/dd722833(v=bts.10).aspx. Accessed November 20, 2014.

77. Teemu Koponen, Martin Casado, Natasha Gude, Jeremy Stribling, Leon Poutievski, Min Zhu, Rajiv Ramanathan, Yuichiro Iwata, Hiroaki Inoue, Takayuki Hama, et al. Onix: A distributed control platform for large-scale production networks. In *USENIX OSDI*, 2010.
78. Yan Zhang and Nirwan Ansari. On architecture design, congestion notification, tcp incast and power consumption in data centers. *Communications Surveys Tutorials, IEEE*, 15(1):39–64, 2013.
79. Facebook to Expand Prineville Data Center, 2010. Available at: https://www.facebook.com/notes/prineville-data-center/facebook-to-expand-prineville-data-center/411605058132. Accessed November 20, 2014.
80. Tad Andersen. Facebook's Iowa expansion plan goes before council, 2014. Available at: http://www.kcci.com/news/facebook-just-announced-new-expansion-plan-in-iowa/25694956#!0sfWy. Accessed November 20, 2014.
81. David Cohen. Facebook eyes expansion of oregon data center, 2012. Available at: http://allfacebook.com/prineville-oregon-data-center-expansion_b97206. Accessed November 20, 2014.
82. John Rath. Facebook Considering Asian Expansion With Data Center in Korea, 2013. Available at: http://www.datacenterknowledge.com/archives/2013/12/31/asian-expansion-has-facebook-looking-at-korea/. Accessed November 20, 2014.
83. Ryan Shea, Feng Wang, Haiyang Wang, and Jiangchuan Liu. A deep investigation into network performance in virtual machine based cloud environment. In *IEEE INFOCOM*. 2014.
84. Katrina LaCurts, Shuo Deng, Ameesh Goyal, and Hari Balakrishnan. Choreo: Network-aware task placement for cloud applications. In *ACM IMC*, 2013.
85. Fei Xu, Fangming Liu, Hai Jin, and A.V. Vasilakos. Managing performance overhead of virtual machines in cloud computing: A survey, state of the art, and future directions. *Proceedings of the IEEE*, 102(1):11–31, 2014.
86. Andrew R. Curtis, Jeffrey C. Mogul, Jean Tourrilhes, Praveen Yalagandula, Puneet Sharma, and Sujata Banerjee. Devoflow: Scaling flow management for high-performance networks. In *Proceedings of the ACM SIGCOMM 2011 Conference*, SIGCOMM '11, pp. 254–265, New York, 2011. ACM. Available at: http://doi.acm.org/10.1145/2018436.2018466. Accessed November 20, 2014.
87. Marco Chiesa, Guy Kindler, and Michael Schapira. Traffic engineering with equal-cost-multiPath: an algorithmic perspective. In *IEEE INFOCOM*, 2014.

5

INTER-DATA-CENTER NETWORKS WITH MINIMUM OPERATIONAL COSTS

B. Kantarci[1] and H. T. Mouftah[2]

[1]*Department of Electrical and Computer Engineering, Clarkson University, Potsdam, New York, USA*
[2]*School of Information Technology and Engineering, University of Ottawa, Ottawa, Ontario, Canada*

5.1 INTRODUCTION

Cloud computing enables users to receive infrastructure/platform/software as a service (XaaS) via a shared pool of resources based on the pay-as-you-go fashion [1]. Automated service provisioning, virtual machine migration, data security, reliability, and energy management have been pointed as the challenges faced by cloud providers [2], whereas energy management and reliability appear as two important issues that impact the operational expenditures (Opex) of the operators. As data centers are the main hosts of physical resources, they play the key role in the delivery of cloud services. Hence, interconnection of data centers over a backbone network is one of the major challenges affecting the performance of the cloud system, as well as the Opex of the service providers.

As illustrated in Figure 5.1, inter-data-center (IDC) networks are considered to be accommodated within the public telecom network that consists of heterogeneous network segments such as wireless backhaul networks, wireline local area networks (LANs), wireless sensor networks (WSNs), wireline Multiprotocol Label Switching (MPLS) networks, legacy IP networks, and so on [3]. In the Cloud era, the volume of the traffic

Cloud Services, Networking, and Management, First Edition.
Edited by Nelson L. S. da Fonseca and Raouf Boutaba.

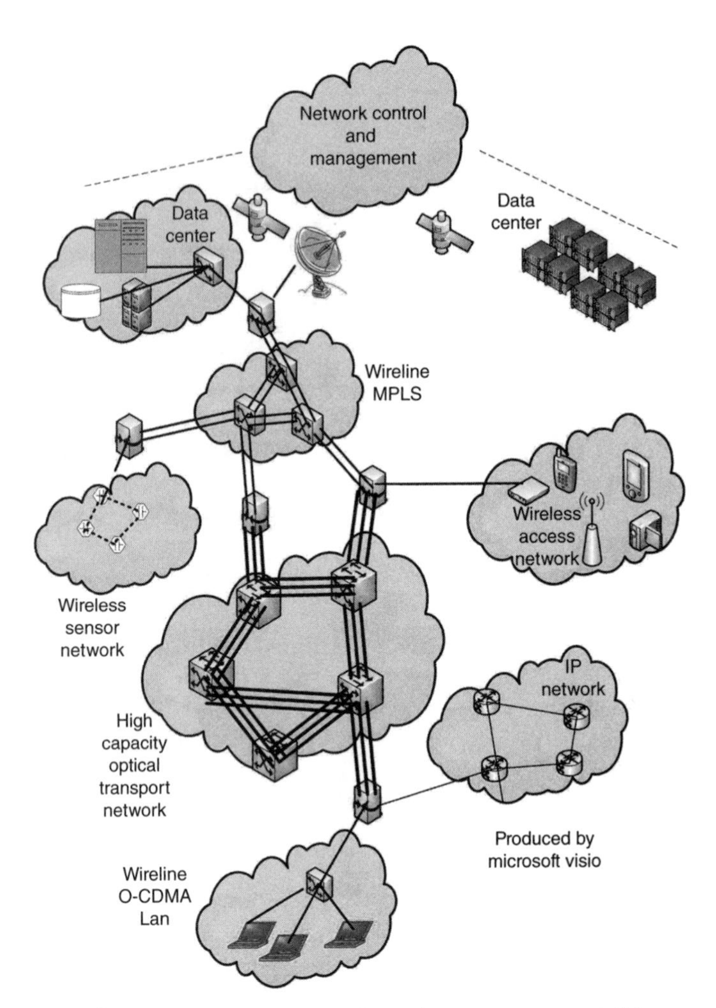

Figure 5.1. Heterogeneous inter-data-center network [3].

between data centers increases tremendously due to on-demand accessing to shared pool of resources by large number of users. This phenomenon increases the capacity demands of the IDC networks introducing challenges related to capacity scaling and operational expenses [4]. Furthermore, virtual machine migration within and between the data centers or massive arrival of new cloud resource requests can lead to frequent reconfiguration of the network between the servers in data centers, as well as the network interconnecting the data centers [5]. High bandwidth and low energy cost are reported as the two crucial requirements of IDC networks which make optical networks the leading transport technology [6, 7].

Optical IDC networks call for intelligent design schemes by considering content replicas, as well as the location and number of data centers in order to ensure survivability against failures [8, 9]. Furthermore, energy-efficient design of the IDC network is crucial to minimize the operational expenses of the network and data center providers as high energy consumption leads to increased the electric bills. Here, energy-efficiency denotes power saving design and planning of the network, as well as reducing the nonrenewable energy consumption in powering the data centers and the inter-data-center network [10].

Virtualization of the network is a key concern in network design as connectivity can be guaranteed to the cloud customers by offering network as a service (NaaS) [11]. In the same study, Baroncelli et al. define virtualization as mapping the cloud services with the corresponding end addresses where cloud requests are submitted. Besides virtualization, communication mode is another factor which affects Opex of the network and data center operators. In conventional networks, unicast and multicast communication modes are used. However, in a virtualized cloud environment, requests can be routed toward virtual resources based on anycast or manycast paradigm. In an IDC network consisting of N nodes, anycast is denoted by $<s, d \in D>$ where s and d denote source and destination addresses, respectively, whereas D is the set of candidate destination addresses. Thus, reaching at any of the candidate destination addresses is sufficient to provision the corresponding request. On the other hand, manycast is denoted by $<s, D' \subseteq D>$ where reaching at a subset of the candidate destinations is sufficient to provision a submitted request. Anycast and manycast communication modes provide the flexibility of allocating resources in different data centers; hence energy efficiency and resilience can be ensured by adopting these communication modes [12–14].

As mentioned earlier, energy efficiency impacts the electric bills of the operators; therefore design schemes considering electricity prices based on location and time are also emergent. According to recent research results, electricity price-aware design of the inter-data-center network can enable Opex savings if and only if electricity price-aware inter-data-center workload migration is enabled along with provisioning the demands in the data centers where electricity prices are low at the time of provisioning [15, 16].

An IDC network design with the objective of energy efficiency (or minimum electric bills) is different from an energy-efficient transport network design due to the difference between energy consumption levels of the network components and data centers. The most power hungry components of the transport networks are reported to be the IP router ports, while the power consumption of a cloud data center is at least ten to hundred times of that of a corporate data center. An IP router port consumes around 1 kW [17], whereas the total power consumption of a cloud data center can reach up to multi-mega-watts (MMW). Recent research reports that 61.4 MMW of total energy consumption of US data centers (according to the report in 2006 [18]) has dramatically increased by the end of 2013 [19].

Indeed, when designing a virtual IDC network, resilience is at the expenses of energy savings as reported in Ref. [20]. Therefore, in order to address this trade-off, there has been proposals such as the resilient virtual infrastructure design under 1:1 protection for lightpaths and virtual servers [20] and IDC workload migration-enabling virtual network design schemes [21].

This chapter provides a reference on the design methods for operational cost-efficient design of a cloud backbone through demand profile-based network virtualization where the data centers are located at the core nodes of a transport network. Addressing energy-efficiency in a cloud backbone helps reducing the Opex of the network and data center operators. Another factor that affects the Opex of the operators is the downtime of cloud services that can be denoted by resiliency, availability, and/or reliability. This chapter considers two major components of the operational costs of an IDC network: (1) electric bills of the operators, (2) Downtime penalties due to service unavailability, i.e., outage. Therefore, the design methods that aim at cutting the electric bills of the operators, as well as the methods that aim at reducing the outage probability, are covered in the following sections. Furthermore, the approaches that jointly consider these challenges and overcome the related challenges are studied, as well. At the end of the chapter, a brief summary of the studied schemes is complemented by a comprehensive comparison in terms of various aspects of Opex and other performance parameters affecting it.

In Section 5.2, we introduce IDC network virtualization and a generic virtualization scheme. Section 5.3 introduces virtual IDC network design with the objective of minimum electric bills by presenting mixed integer linear programming (MILP)-based optimization and heuristic solutions. Section 5.4 introduces IDC network design with the objective of minimum downtime penalties. As mentioned earlier, there exists a trade-off between these two Opex elements. Therefore, Section 5.5 presents a solution to address this trade-off. The chapter is summarized along with discussions for open issues and challenges in Section 5.6.

5.2 INTER-DATA-CENTER NETWORK VIRTUALIZATION

In the cloud dominated era, virtualization and infrastructure as a service (IaaS) enables providing several portions of the physical infrastructure as a service by different operators where infrastructure denotes computing and storage resources in data centers, as well as the communication infrastructure interconnecting the data centers [22]. Furthermore, by taking advantage of transparent optical devices, virtualization of an optical network enables bypassing IP routers that are the most power hungry components in the backbone [17].

Figure 5.2 presents a minimalist illustration of virtualization of an IDC network. Each data center is associated with a backbone node where backbone nodes are interconnected via fiber links. If a lightpath can be established between two nodes, the two backbone nodes with the allocated resources in the associated data centers are said to be virtually linked. Thus, in the virtual infrastructure, a virtual node is denoted by the virtualized resources of a data center and its associated backbone node.

In the literature, planning of the virtual infrastructure denotes mapping the virtual network onto the physical topology. The physical infrastructure is considered to be the set of data centers, optical backbone nodes, and fiber links interconnecting them, whereas the virtual infrastructure is a subset of the physical infrastructure consisting of a set of virtualized data center resources and fiber channels [23].

The objective of network virtualization can be various such as energy minimization, cost minimization, reliability maximization, and so on. In this section, we present a previously proposed energy-minimized design of an inter-data-center network [13] which adopts the multihop optical bypass-based virtualization technique in an IP over

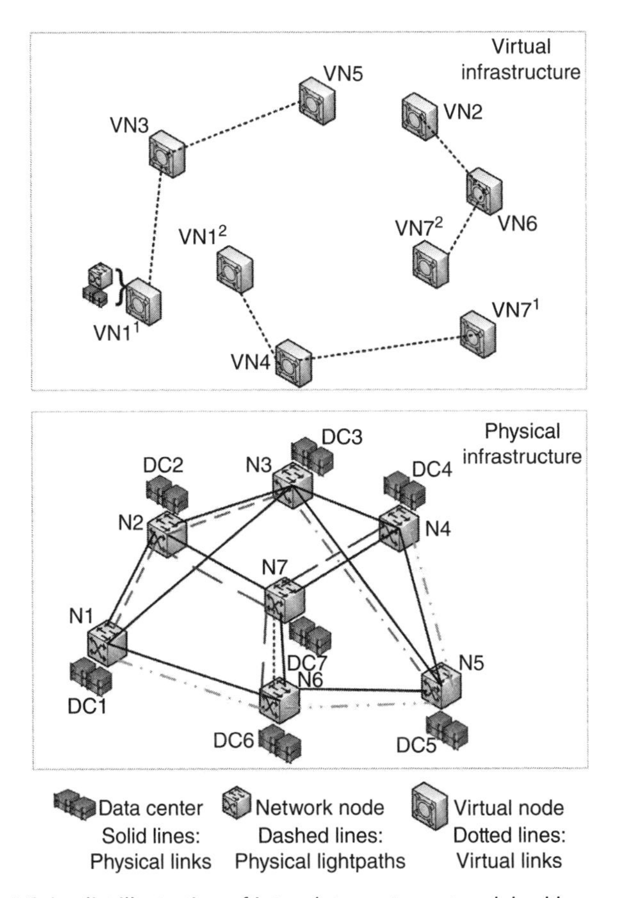

Figure 5.2. Minimalist illustration of inter-data-center network backbone virtualization.

WDM network [17]. In Ref. [13], the authors have proposed an energy efficient design of the IDC network backbone through MILP formulations, as well as heuristics. These schemes have been extended to address both inter and IDC network provisioning with the objective of energy efficiency [24]. However, since the scope of this chapter is limited to IDC network design, we refer the interested reader to the corresponding reference. In the next sections, the corresponding formulation will serve as a benchmark for the Opex-minimized design schemes.

For the sake of simplicity, let us assume that a single virtual infrastructure is mapped onto the physical infrastructure. Furthermore, the following assumptions hold in the design of the virtual infrastructure:

- Three types of demands are assumed in the network, namely downstream data center demands, upstream data center demands, and regular demands. An upstream demand is submitted from a backbone node, and it is destined to any or a number of data centers. A downstream data center demand originates from a few data centers and destined to a certain backbone node where they are aggregated and delivered to the corresponding end users. A regular demand denotes a nondata center unicast flow between two backbone nodes.

- Intensities of all types of demands in a certain time interval are forecasted in advance. Thus, virtualization of the backbone is performed in advance of the occurrence of the corresponding demand profile so that the virtualization objective can be met.
- For an incoming upstream demand of any size, the overhead of allocating resources in a given data center is known in terms of utilization, power consumption, and power usage efficiency.

5.2.1 Mathematical Formulation

In the formulation, the physical infrastructure is denoted by a directed graph G whereas the virtual infrastructure is also represented by a directed graph denoted by G'. Given an upstream data center demand originating at a source node, s, the set of data centers that are capable of provisioning the corresponding demand is denoted by D. Routing the demand towards any data center out of D is referred as anycast, whereas routing toward a subset of the eligible data centers is named as manycast. Thus, a manycast demand can be denoted by the tuple $<s, D' \subseteq D>$. Here, if $|D'| = 1$, the communication mode becomes equivalent to anycast while in case of $D' = D$, it becomes identical to multicast communication. In this design scheme, upstream data center demands are assumed to be provisioned based on the manycast communication mode.

Table 5.1 illustrates the notation used in explaining the virtualization framework in Ref. [13]. Mathematical formulation of the model is presented in Equation 5.1 and 5.2. Equation 5.1 presents the objective of the virtualization, which is minimized energy consumption throughout the IDC network. As seen in the equation, the total power consumption in the network is the sum of the power consumptions at each node location. Power consumption at each node location is a function of the power consumption of the associated data center (summation term 1), the active IP router ports (summation term-2) and transponders in the directed wavelength channels along with the erbium-doped fiber amplifiers (EDFAs) in the directed fiber links (summation term 3). The number of active IP ports is calculated by the number of outgoing virtual lightpaths at the corresponding node. Besides, in the third summation term, the number of EDFAs on a physical link, S_{ij} is set at $\lfloor Lf_{ij}/\Delta_{span} \rfloor + 1$ where Δ_{span} is the fiber span length.

$$\min \sum_{i \in N^v} \left(DC_i + \sum_{j \in N_i^v} P_r \cdot C_{ij} + \sum_{j \in N_i^p} (P_t \cdot W_{ij} + S_{ij} \cdot P_{edfa} \cdot f_{ij}) \right) \tag{5.1}$$

Before proceeding with the details, it is worthwhile to provide information on the power consumption of a data center. Based on the assumptions summarized earlier, the prospective power consumption of data center-d (DC_d) is a function of current processing and cooling power consumption in the corresponding data center and the total

TABLE 5.1. The notation used in the virtualization scheme

Notation	Explanation
P_i	Power consumption at node i
DC_i	Power consumption of the data center i
P_r	Power consumption of an IP router port
C_{ij}	Number of lightpaths in the virtual link ij
N_i^v (N_i^p)	Set of neighbors of node i in the virtual (physical) topology
P_t	Power consumption of a transponder
P_{edfa}	Power consumption of an EDFA
S_{mn}	Number of EDFAs in the physical link mn
W_{ij}	Number of wavelengths in the physical link ij
f_{ij}	Number of fibers in the physical link ij
Ω_{ds}^{DOWN}	Downstream demand from data center s to node d
Ω_s^{UP}	Upstream traffic (job submission) to data centers from to node s
$\gamma_{ij_{down}}^{ds}$	Binary variable is one if there is downstream traffic from data center s to node d traversing the virtual link-ij.
λ_{ij}^{sd}	Regular traffic demand traversing the virtual link ij and destined from node s to node d
Λ_{sd}	Regular traffic demand from node s to node d
Υ_{up}^{sd}	Possible demand from node s to data center d
γ_{ijup}^{sd}	Binary variable is one if there is traffic from node s to data center d traversing the virtual link ij
D_{max}^s	Maximum number of destinations for the upstream traffic from node s
D_{min}^s	Minimum number of destinations for the upstream traffic from node s
W_{ij}^{mn}	Number of wavelength channels on the virtual link ij traversing the physical link mn
DC_d^{cool}	Cooling power consumed at data center d
DC_d^{proc}	Processing power consumed at data center d
$\Theta_{s,d}$	Power consumption overhead introduced to data center d by the job submitted by node s
$L_{i,j}$	Shortest distance from node i to node j
$Lf_{m,n}$	Fiber length between node m to node n

additional power consumption overhead of the demands submitted from other locations and provisioned in data center-d. Equation 5.2 formulates this expression.

$$DC_d = DC_d^{cool} + DC_d^{proc} + \sum_{s \in V} \sum_{i \neq d} \Theta_{s,d} \cdot \gamma_{id_{up}}^{sd}, \forall d \in V \quad (5.2)$$

Three subsets of constraints form the constraint set of the design model. Virtualization is mainly based on a typical routing and fiber and wavelength assignment (RFWA) in an optical network. Therefore, flow conservation constraints dominate the rest of the constraint set. In the virtual topology, single-hop routing is performed in the IP layer while multihop routing in the physical topology is handled by the optical layer. Hence, flow conservation constraints for both layers have to be formulated separately.

Besides, capacity constraints and manycast constraints for upstream data center traffic are needed. It is worthwhile to note that this chapter only presents the constraints related to the upstream data center (i.e., manycast) demands. Downstream data center demands can also be considered as multiple unicast demands such as the regular demands. For detailed information on the formulation of unicast the constraints, the reader is referred to Refs. [13, 17].

Flow conservation constraints in the IP layer: An upstream data center demand requires to be provisioned in at least (D^s_{min}) and at most (D^s_{max}) data centers. A backbone node must initiate manycast traffic whose size is less than the demand size for maximum number of destination data centers and greater than the demand size for minimum number of destination data centers as shown in Equation 5.3. Manycast communication mode requires a "light-tree" in the network [25]. Thus, the total demand size on the first branches of the light-tree is the size of the manycast demand. Furthermore, an upstream data center demand has to arrive at sufficient number of destinations by modifying Equation 5.3 appropriately. Besides, flow conservation at the intermediate nodes has to be met, that is, incoming and outgoing traffic volumes for a given type of demand at an intermediate node have to be equal.

$$D^s_{min} \cdot \Omega^{UP}_s \leq \sum_{d \in V} \sum_{j \in V} \Upsilon^{sd}_{up} \cdot \gamma^{sd}_{sjup} - \sum_{d \in V} \sum_{j \in V} \Upsilon^{sd}_{up} \cdot \gamma^{sd}_{jsup} \leq D^s_{max} \cdot \Omega^{UP}_s, \quad \forall (s) \in V \qquad (5.3)$$

Flow conservation constraint in the optical layer: In the optical layer, source node of a virtual link does not have any incoming wavelength channels as formulated in Equation 5.4, while the destination node of a virtual link does not have any outgoing wavelength channels. Besides, a virtual link does not contain any loops.

$$W^{ij}_{mn} - W^{ij}_{nm} = \left\{ \begin{array}{cc} -C_{ij} & m = i \\ C_{ij} & m = j \\ 0 & else \end{array} \right\}, \quad \forall m, n, i, j \in V \qquad (5.4)$$

Capacity constraints: Total channel capacity of the fibers from *node* − *m* to *node* − *n* sets the upper bound for the number of lightpaths traversing a physical link-*mn* as shown in Equation 5.5.

$$\sum_{i \in V} \sum_{j \in V} W^{ij}_{nm} - W \cdot f_{mn} \leq 0, \quad \forall m, n \in V \qquad (5.5)$$

Furthermore, a virtual link must have sufficient capacity to accommodate the regular traffic, downstream DC traffic and the upstream DC traffic traversing it as shown in Equation 5.6, where C denotes the wavelength channel capacity.

$$\sum_{s \in V} \sum_{d \in V} \lambda^{sd}_{ij} + \Upsilon^{sd}_{up} \cdot \gamma^{sd}_{ijup} + \Upsilon^{ds}_{down} \cdot \gamma^{ds}_{ijdown} \leq C \cdot C_{ij}, \quad \forall i, j, \in V \qquad (5.6)$$

Manycast constraints: Equation 5.7 ensures that each DC upstream demand reaches to sufficient number of destinations. Furthermore, at most one virtual link can be utilized prior to reaching at a destination as shown in Equation 5.8. To be able to distribute the traffic over the branches of the light-tree, backbone nodes in the optical domain have to be multicast capable. Thus, an upstream data center demand can be accommodated by the same virtual links up to node-j where the demand is split into multiple virtual links. Equation 5.8 formulates this constraint.

$$D^{s}_{min} \leq \sum_{i}\sum_{i\neq d} \Upsilon^{sd}_{up} \cdot \gamma^{sd}_{idup} \leq D^{s}_{max} \cdot \Omega^{UP}_{s}, \quad \forall s \in V \tag{5.7}$$

$$\sum_{i\neq d} \gamma^{sd}_{idup} \leq 1, \quad \forall s, d \in V \tag{5.8}$$

$$\sum_{d\in V} \gamma^{sd}_{ijup} \leq 1, \quad \forall s, i, j \in V \tag{5.9}$$

5.2.2 Heuristic Solution for Inter-Data-Center Network Virtualization

In Ref. [13], the authors have proposed a heuristic to solve the aforementioned MILP solution. The heuristic adopts the energy-efficient virtualization of an IP over WDM network [17], and introduces data center demands, as well as data center utilization and power consumption constraints. The heuristic is named as Power Minimized Provisioning (PoMiP). Figure 5.3 illustrates a generic flowchart for virtualization steps of an IDC network. The algorithm starts with a set of demands that are sorted in decreasing order. Starting from each demand, the algorithm aims at routing the demand over virtual topology, G'. If the demand can be routed over the virtual topology, the remaining virtual link capacities are updated, and the algorithm proceeds with the next non-provisioned demand. If the demand cannot be routed over the virtual topology; a new virtual link is added between source and destination nodes, which is later routed over the physical topology. The capacity of the newly added virtual link is also updated accordingly.

Here, as seen in the figure, there are three distinguishing functions of the heuristic as follows: (1) Virtual link cost (φ^{v}_{ij}) assignment, (2) Physical link cost (φ^{phy}_{mn}) assignment, (3) Selection of destination data centers.

Equation 5.10 formulates the virtual link cost assignment. The heuristic aims at selecting the virtual links with higher remaining capacity and lower physical link costs.

$$\varphi^{v}_{ij} = \left\{ \begin{array}{cc} \dfrac{\sum\limits_{link-mn\in link-ij} \varphi^{phy}_{mn}}{C'_{ij}} & C'_{ij} > 0 \\ \infty & \text{else} \end{array} \right\} \tag{5.10}$$

Physical link cost assignment is formulated in Equation 5.11. As the power consumption of IP routers are avoided in the optical domain, power consumption of EDFAs (P_{edfa}) as a function of the distance between the two nodes forming the link, power consumption of the transponders as a function of the number of active wavelengths in the physical link are the factors that contribute the power consumption in a physical

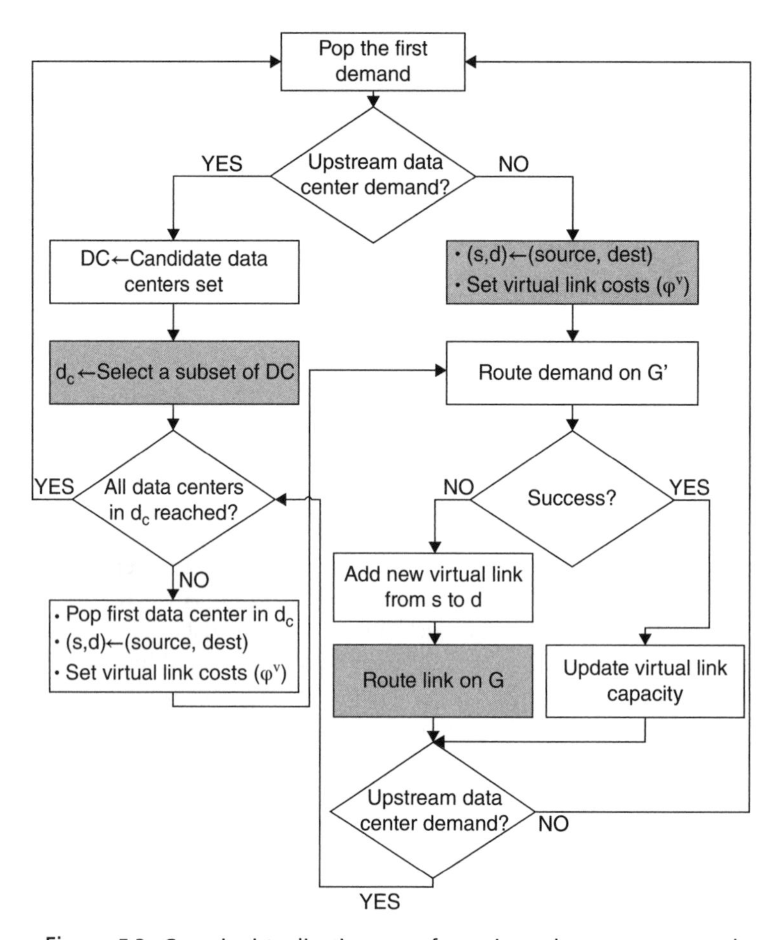

Figure 5.3. Generic virtualization steps for an inter-data-center network.

link. Thus, according to Equation 5.11, the heuristic aims at selecting the least power consuming links on a physical path.

$$\varphi_{mn}^{phy} = \left\{ \begin{array}{cc} P_{edfa} \cdot S_{mn} + P_t \cdot W_{mn} & W_{mn} > 0 \\ \infty & \text{else} \end{array} \right\} \tag{5.11}$$

Since PoMiP aims at minimum power consumption throughout the network, for an upstream data center demand initiated at the backbone node s, it ranks and sorts the data centers with respect to their prospective power consumption in increasing order, and selects the first $D_{\min}^{s}$ of them as the destinations. Thus, the heuristic maps the manycast flow provisioning problem onto multiple unicast flows provisioning problem.

Given a physical network of N backbone nodes and their associated data centers, for any demand, if a virtual path is found on the virtual topology, runtime of the algorithm is bounded above by $O(N^2)$ which is the complexity of a typical shortest path routing algorithm. If the demand cannot be routed over the virtual topology, newly added virtual link is routed over the virtual topology within $O(N^2)$. Furthermore, searching for a fiber and a lightpath throughout the physical path requires $O(F \cdot W)$ runtime on each link, where F is the number of fibers per link and W is the number of transponders (i.e., wavelengths) per node. Since fiber and wavelength search will be performed on each link with wavelength continuity constraint, fiber, and and wavelength search throughout the path can be repeated by $(N-1)^2$ times. Therefore, runtime complexity of the algorithm is $O(N^2 \cdot F \cdot W)$. Since $N \geq F$ and $N \geq W$, it can be said that the heuristic runs in $O(N^4)$ in the worst case.

5.3 IDC NETWORK DESIGN WITH MINIMUM ELECTRIC BILLS

Energy-efficient design of the IDC network reduces the Opex of the operators as they are charged due to electricity consumption. Therefore, Opex of the operators can be further reduced if energy-efficient design is consolidated with electricity price-awareness. Furthermore, taking advantage of demand response (DR) component in smart grids can help reducing Opex of the operators while keeping the power consumption fairly distributed among the network. It is worthwhile to note that DR denotes regulating power consumption through generation of varying price tariffs Three approaches, namely time-of-use (ToU) pricing, real time pricing (RTP), and critical peak Pricing (CPP) are the most popular approaches among existing time varying tariffs. Since smart grid and dynamic pricing is a new concept, customers are not willing to join RTP-based pricing tariffs as they are used to being charged by flat rates. Moreover, elasticity of RTP tariffs requires rapid adaptation of customers. Based on the analysis in [26], this chapter considers ToU pricing despite several benefits of RTP.

In Ref. [15], the authors have analyzed the impact of ToU-aware virtualization of the inter-data-center network where the network is virtualized based on the forecasted demand profile and the ToU rates in a certain timeslot with the objective of minimum electric bills for data center and network operators. In the corresponding study, the authors report that ToU-awareness enables reduction in the electric bills of the network and data center operators while introducing longer provisioning delays for the user demands that are submitted to the data centers. In Ref. [16], the authors have shown that ToU-aware IDC network virtualization is beneficial as long as IDC workload sharing is enabled during virtualization. To this end, the authors have proposed ToU-aware Provisioning (ToUP) which adopts and extends the virtualization scheme in Section 5.2.

Since data centers are the most power hungry components of an IDC network, significant reduction of the Opex by cutting the electric bills can be possible by enabling workload sharing between data centers. Therefore, ToUP re-defines the distinguishing functions of the virtualization heuristic in order to meet its objective. Furthermore, in addition to the three demand types, it also accommodates the fourth demand type, namely the IDC demands. Distinguishing functions of the heuristics are re-defined as follows:

Virtual link cost assignment: Equation 5.12 formulates the virtual link cost assignment at time, T. As seen in the equation, if there are sufficient remaining lightpaths on a virtual link, its cost is set at the total cost of physical links forming the corresponding virtual link.

$$\varphi_{ij}^{v}(T) = \left\{ \begin{array}{cc} \sum\limits_{link-mn \in link-ij} \varphi_{mn}^{phy}(T) & C'_{ij} > 0 \\ \infty & \text{else} \end{array} \right\} \tag{5.12}$$

Physical link cost assignment: Besides, cost of the physical link *mn* is set at its contribution to the electric bill per unit time. Thus, the product of the ToU price at the location of the destination end node of the physical link *mn* ($Price_n(T)$) and the total energy consumption on the corresponding link per unit time is the unit contribution to the electric bill of the network operator.

$$\varphi_{mn}^{phy}(T) = \left\{ \begin{array}{cc} Price_n(T) \cdot \left[Lf_{mn} \cdot (P_{edfa} \cdot S_{mn} + P_t \cdot W_{mn}) \right] & W_{mn} > 0 \\ \infty & \text{else} \end{array} \right\} \tag{5.13}$$

Data center subset selection for upstream data center demands: For an upstream data center demand, ToUP computes the prospective contribution of the corresponding demand to the electric bill of each data center in the network. As mentioned before, it is assumed that the network virtualization manager knows the power consumption and resource utilization overhead of an upstream data center demand on any data center. Therefore, contribution of an upstream data center demand to the electric bill of the data center operator is calculated by the product of the prospective energy consumption in the data center and the ToU rate at the time of virtualization at the location of the corresponding data center as seen in Equation 5.14.

$$R_i(T) = Price_i(T) \cdot \sum_{j \neq i} \Theta_{s,i} \cdot \gamma_{jiup}^{si} \forall i, \quad s \in V \tag{5.14}$$

IDC workload sharing: As mentioned before, ToUP enables accompanying backbone network virtualization IDC workload migration so that workloads are hosted in those data centers that experience lower ToU prices during the corresponding period. Here, a new workload-data center mapping is aimed to be obtained. To this end, in [16], the authors have proposed a simulated annealing-based procedure which is presented in Algorithm 5.1.

Algorithm 5.1

Inter-Data-Center Workload Migration Algorithm {

Begin

Sort demands in decreasing order

$Opx_{temp} \leftarrow \sum_i O_{dc}^i$ use *Map* to calculate

$Map_{ii} \leftarrow 100, TempMap_{ii} \leftarrow 100, \forall i$

$O_{current} \leftarrow Opx_{temp}$

while (*converge* = FALSE)

{

$TempMap_{ij} \leftarrow Map_{ij}, \forall i,j$

randrow ← Select a random row in *Map*

candidates ← Count ($\Upsilon_{IDC}^{randrow,d} > 0$)

```
if(Count(Map_{randrow,i} > 0)>0)
  {
  TempMap_{randrow,randrow} ← κ_{MIN}
  Remainder ← 1 − κ_{MIN}
  }
  else
    {
    randcol ← Select a random column in TempMap
    if((randcol) is on the diagonal)
      {
      TempMap_{randrow,randrow} ← κ_{MIN}
      Remainder ← 1 − κ_{MIN}
      }
      else
      {
      Remainder ← TempMap_{randrow,randcol}
      TempMap_{randrow,randcol} ← 0
      }
    }
dest1, dest2 ← Candidate destinations out of candidates
share_{dest1} ←Migration to dest1; share_{dest1} ∈ [0, Remainder]
share_{dest2} ←Migration to dest2; share_{dest2} ∈ [0, Remainder − share_{dest1}]
TempMap[randrow][dest1] increment by share_{dest1}
TempMap[randrow][dest2] increment by share_{dest2}
TempMap[randrow][randrow] increment by Remainder − (share_{dest1} + share_{dest2})
Opx_{temp} ← ∑_i O^i_{dc} use TempMap to calculate
F ← e^{(O_{current} − Opx_{temp})/(100·B·t_{cool})}
if((F ≥ 1))
  Map[i][j] ← TempMap[i][j], O_{current} ← Opx_{temp}
  else if(F < 1)
    Map[i][j] ← TempMap[i][j], O_{current} ← Opx_{temp} with prob. F
T ← T · t_{cool}
if(T ≤ T_{ground} OR change in O_{current} ≪ 1)
  converge ← true
  }
 End
}
```

Before proceeding with the details of the algorithm, it is worthwhile to see Table 5.2 for the notation, as well as the settings. The algorithm aims at obtaining a new data center-workload mapping matrix, *Map*, and in each annealing iteration it uses a temporary mapping matrix, *TempMap*. Each data center is assumed to migrate a certain portion of its workload to at most DC_{max} data centers which is set at two in the pseudocode for the sake of simplicity. By the term Opex, the algorithm denotes the total electric bills of the data center operators. Initially, *Map* is set at $100 \cdot I$ where I is the identity matrix. Thus, each data center hosts 100% of its original workload. Until the algorithm converges, the following iteration steps are repeated: *TempMap* is set equal to *Map*, and a random row of *Map* denoting the source data center is selected along with a random column which denotes a candidate destination data center. If the candidate destination data center is the data center itself, the algorithm sets the value of the corresponding cell at κ_{MIN}, otherwise it is set at zero. Then, the remainder of the workload is aimed at being distributed

TABLE 5.2. The notation used in inter-data-center workload migration algorithm

Notation	Explanation
Map:	Workload distribution matrix
TempMap:	A temporary workload distribution matrix
DC_s:	Candidate DCs set to share the workload of the data center s
Υ_{IDC}^{sd}:	Possible workload migration demand from data center s to data center d
$Opx_{temp}(O_{current})$:	Temporary (current) Opex
T, B:	Annealing temperature, Boltzman constant
t_{cool}:	Cooling rate of the system
d_c:	List of destination data centers
κ_{MIN}:	Lower bound for the workload percentage not to be migrated

among two of the rest of the data centers. The new workload-data center mapping is stored in *TempMap*, which is used to calculate the possible electric bill contribution of the new workload distribution (Opx_{temp}). The newly computed workload distribution map is accepted by running a function (i.e., F in the peseudocode) of the current actual Opex (O_{current}), newly computed Opex (Opx_{temp}), Boltzman constant and the cooling rate of the system. If F is greater than or equal to one, the workload that is stored in *TempMap* is accepted and assigned to *Map*. Otherwise, it is accepted with a probability of F.

At the end of each iteration, the system temperature is cooled by the cooling rate, t_{cool}. If the system temperature is equal to or less than the previously defined ground temperature, or if the change in current Opex is significantly low, the annealing system is said to have converged. At this point, the algorithm stops and accepts the new workload distribution among the data centers.

In Ref. [16], the authors evaluated the performance of ToUP under a medium-scale cloud system located in the 14-node NSFNET backbone where each backbone node is associated with a data center which is initially loaded between 0.1 and 0.7. Backbone network [13] is considered to be an IP over WDM network with 16 40 Gbps-wavelengths per fiber in which EDFAs are placed at every 80 km. Four time zones are assumed with the demand profile in Figure 5.4a, whereas the ToU rates have been synthetically derived for each location as shown in Figure 5.4b. It is worthwhile to note that the NSFNET topology four different time zones exist, namely the Eastern Standard Time (EST), Central Standard Time (CST), Mountain Standard Time (MST), and Pacific Standard time (PST) zones. An entire day is partitioned into eight equal timeslots. Network equipments, namely an EDFA, a transponder and an IP router port are assumed to consume 8, 73, and 1000 W, respectively [17]. Besides, workload placement in a data center utilizes minimizing heat recirculation [27], and a data center is assumed to consume 168 kW (100 kW) of idle IT (cooling) power and 319.2 kW (280 kW) of full utilization IT (cooling) power. An upstream data center demand is assumed to increase the data workload between 0.025 and 0.2. In the IDC workload distribution algorithm, the Boltzman constant is set at 0.01, whereas the cooling rate is 0.95. The ground temperature and the minimum temperature change are considered to be 0.005 and 0.001, respectively.

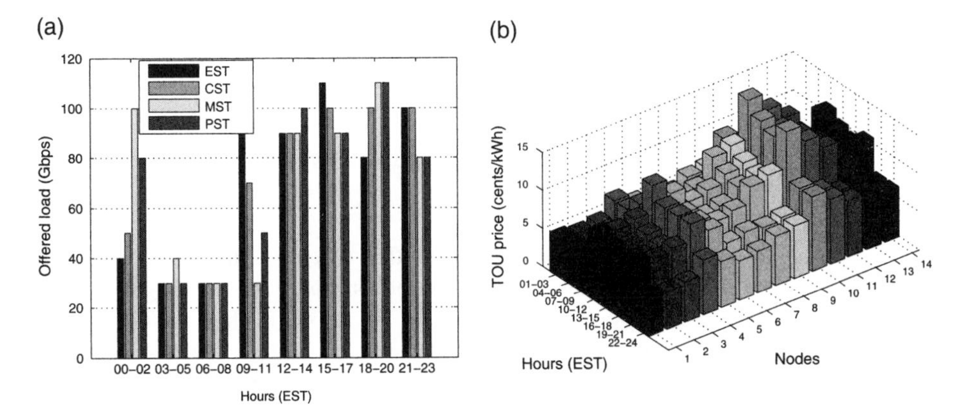

Figure 5.4. (a) Demand profile in different time zones. (b) ToU rates in different locations of the network.

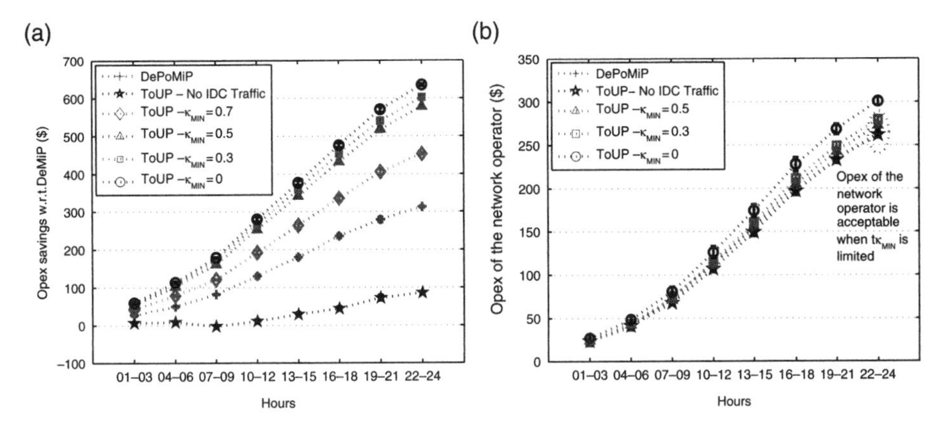

Figure 5.5. (a) Opex savings in the inter-data-center network. (b) Opex of the network equipment.

In Figure 5.5, performance evaluation of ToUP is presented in comparison to delay and power-minimized provisioning (DePoMiP) which has previously been proposed in [13]. Furthermore, ToUP is also evaluated by disabling IDC migration ($\kappa_{MIN} = 1$). Figure 5.5a illustrates overall Opex savings in the IDC network with respect to delay-minimized provisioning (DeMiP) where DeMiP aims at virtualization of the backbone network with shortest lightpaths for unicast demands and shortest light-trees for upstream data center demands. It is clearly seen that ToUP is outperformed by DePoMiP if IDC workload migration is disabled. Furthermore, enabling IDC workload migration introduces more Opex savings when compared to DePoMiP. Thus, lower κ_{MIN} values leads to higher Opex savings in the entire cloud system. However, as seen in Figure 5.5b, the smaller the κ_{MIN}, the higher the Opex of the network operator. Therefore, limiting the allowable IDC workload migration seems to be viable. As the authors report in Ref. [16], under such a scenario, enforcing around 30% of the workload to be hosted in the original data center leads to the best compromise between the Opex of the data center and the network operators.

5.4 INTER-DATA-CENTER NETWORK DESIGN WITH MINIMUM DOWNTIME PENALTIES

Besides energy efficiency and electricity bills, another type of significant operational expenses are the downtime penalties. Outage of network and/or computing resources in data centers can occur due to component failure. Therefore, resilient design of IDC network design with the objective of minimum outage probability is required to reduce Opex. In Ref. [8] the content placement along with path and content protection in an optical IDC network have been addressed via ILP formulations and heuristics. Similarly, in [9], locations of data centers are determined via ILP formulations based on anycast communication mode with the objective of maximum resilience. In Ref. [28], the authors have proposed network virtualization-aware IDC network over an elastic optical network (EON) backbone. Although the proposed architecture is transparent to the transport technology, the authors have adopted the elastic optical networking technology based on the report of the recent research. Recent research reports that energy consumption, bandwidth utilization, and deployment cost are enhanced by elastic optical networks in comparison to the conventional wavelength switched optical transport networks [29].

An outage denotes unavailability of the IDC network, and it can occur due to either network component failure or a failure in the data center. Therefore, a resilient design scheme should jointly consider the availability of network components as well as the availability of data centers. The outage probability of a virtual link ($\nu\ell^{ij}$) can be formulated by Equation 5.15 where O^i_{IP}, a_{edfa}, a_{rcvr}, and a_{rcvr} denote the outage probability of an IP router, availability of an EDFA, availability of a transceiver and the availability of a receiver, respectively. Besides, S_{mn}, represents the number of EDFAs deployed in the physical link mn ($\rho\ell^{mn}$).

$$O^{ij}_{\nu\ell} = O^i_{IP} + O^j_{IP} + \sum_{m\in G} \sum_{n\in G,\rho\ell^{mn}\in\nu\ell^{ij}} S_{mn} \cdot (1 - a_{edfa}) + (1 - a_{tran}) + (1 - a_{rcvr}) \quad (5.15)$$

Thus, a virtual link ij is said to be out of service if one of the following conditions holds:

- IP routers at the source/destination nodes of the link fails.
- An EDFA along the physical path forming the virtual link ij fails.
- Transmitter at node i fails.
- Receiver at node j fails.

Once the outage probability of the virtual link is formulated, outage probability of a virtual path (VP) can be formulated as the sum of outage probabilities of the virtual links forming the path. If a data center is located at the end of the path, its outage probability is also added to the outage probability of the virtual path.

Outage probability of an upstream data center demand submitted at node s can be formulated as shown in Equation 5.16 by simply assuming that the workload is replicated in two data centers where D^s_{list}, γ^{sd}_{ijwp}, and O^c_{dc} denote the list of selected data centers by node-s, a binary variable to denote if data center-d utilizes the virtual link ij and the outage probability of data center c, respectively. In order to ensure resilience of a given upstream data center demand; at least one lightpath toward the destination node and its corresponding data center must be available (first summation term). In the first summation term, duplicates of outage probability of the virtual links and the data centers can occur, which are eliminated by the second summation term. Since the entire summation leads to the availability of the corresponding demand, one's complement of the summation is equal to the outage probability.

$$O^s_{US} = 1 - \left[\sum_{d \in D^s_{list}} \left((1 - O^{sd}_{\nu P}) \cdot (1 - O^d_{dc}) \right) - \sum_{d,c \in DC} \sum_{i \in G'} \sum_{j \in G'} \left((1 - O^{ij}_{\nu \ell}) \cdot \gamma^{sd}_{ijwp} \cdot \gamma^{sc}_{ijwp} \cdot (1 - O^c_{dc}) \cdot (1 - O^d_{dc}) \right) \right] \quad (5.16)$$

5.4.1 Minimum Outage Probability in Cloud

In Ref. [28], the authors have proposed an IDC virtual network design scheme, namely minimum outage probability in cloud (MOPIC). MOPIC computes virtual paths to the data centers in DC_s, and DC_{min} data centers are selected where the outage probability of the data centers and that of the corresponding virtual paths lead to minimum outage probability.

While routing the virtual links over the physical topology, MOPIC assigns the outage probability of each physical link as the link cost, thus, a virtual link is aimed to be routed over the most resilient lightpath in the physical topology.

5.4.2 Resource Saving Minimum Outage Probability In Cloud

Resilience requires additional resource usage will introduce additional energy consumption. Therefore, an efficient design scheme is expected to make a compromise between resource usage and resilience. To this end, resource saving minimum outage probability in cloud (RS-MOPIC) has been proposed. Although computing resource usage cannot be reduced, some savings in network resource usage is possible. Reducing the length of the path traversed from source node to the destination data center can enable resource saving. In the virtualization algorithm in Figure 5.1, virtual link cost assignment is done as follows. Each virtual link ij is assigned the product of its outage probability and the number of hops in its physical topology mapping. The same principle holds in determining the destination data centers. Thus, virtual paths to each data center in DC_s is searched by using the virtual link cost assignment as mentioned above, and selects DC_{min} data centers leading to the DC_{min}-minimum outage probabilities for the corresponding workload placement. Similarly, while routing the virtual link ij over the physical topology, each physical link-*mn* on the physical topology is assigned the product of its outage probability and the number of nodes traversed by node *n* to node *j*.

Performance of MOPIC and RS-MOPIC has been evaluated by using a benchmark approach called minimum resource provisioning in cloud (MRPIC). MRPIC mainly aims at designing the virtual network with minimum network resource usage. To this end, it sets the virtual link cost at the number of physical links forming the corresponding link while routing a demand over the virtual topology. In order to map a virtual link on the physical topology, MRPIC sets the cost of a physical link at the number of hops to the destination node of the corresponding virtual link. Similarly, while for the upstream data center demands DC_{min} data centers out of DC_s are selected based on the locality principle.

In a medium-scale simulation scenario under the 24-node US National Backbone topology [17], the demand profile in Figure 5.4 is considered where each 3 h timeslot is denoted by D_i. Fiber links interconnecting the data centers are assumed to have 1000 GHz spectrum capacity with a data rate/bandwidth ratio of 2 bps/Hz. Besides, 1 GHz subcarriers are assumed with a guard band of 10 Ghz whereas the transponder capacity is assumed to be equal to the capacity of 50 subcarriers. DC_{min} is set at two for the sake of simplicity. It is assumed that the outage probability of a router port is 10^{-6}. Further assumptions on the outage probabilities of the optical network components such as transceivers and EDFAs are taken from Ref. [30]. Besides, four-tier data centers are considered with respect to their availability values such as Tier 1, Tier 2, Tier 3, and Tier 4 with the availability levels of 99.67%, 99.74%, 99.98%, and 99.995%, respectively [31].

In Figure 5.6a, RS-MOPIC and MOPIC are compared to the benchmark scheme, MRPIC in terms of outage probability of the upstream data center demands. Introducing outage probability awareness to RSA and destination data center selection process reduces the outage probability of upstream data center demands dramatically. The outage probability under MRPIC is always at the level of 10^{-6} whereas the outage probability of an upstream data center demand is reduced to the level of 10^{-7} under MOPIC and

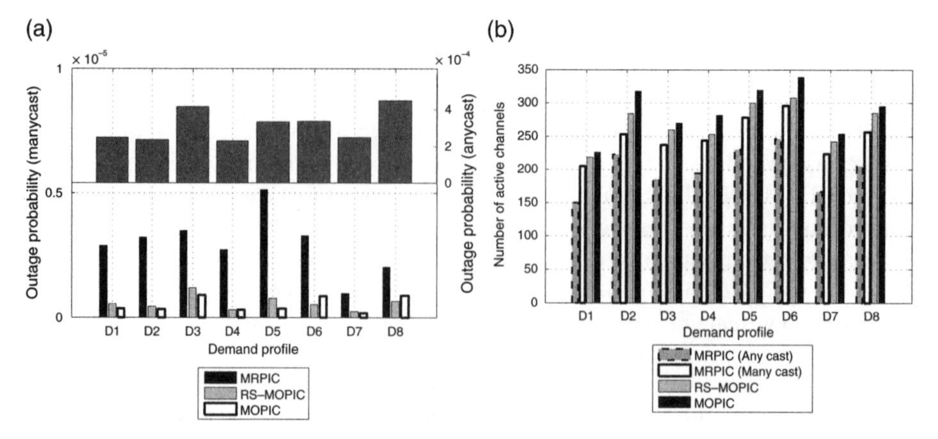

Figure 5.6. (a) Outage probability of upstream data center demands. (b) Number of active channels in the virtual inter-data-center network.

RS-MOPIC. Furthermore, joint awareness of resource consumption and outage probability does not degrade resilience of the demands as RS-MOPIC introduces similar outage probability with MOPIC. Moreover, under heavy demand profiles (e.g., D6 and D8), RS-MOPIC slightly reduces the outage probability of MOPIC by selecting shorter physical lightpaths for virtual topology mapping. The upper half of the figure shows the outage probability under anycast-based MOPIC. Instead of placing the workload on multiple data centers, provisioning on a single data center increases the outage probability up to 10^{-4}.

Besides, in Figure 5.6b, number of active channels are presented as the resource consumption of the evaluated schemes. Indeed, anycast-based implementation of MRPIC ($DC_{min} = 1$) introduces the least resource consumption due to utilizing less network resources by 13% and 35% based on the demand profile. Resource consumption awareness incorporated in outage probability-aware provisioning increases the resource consumption of MRPIC by 3.5%–12% depending on the demand profile whereas pure outage probability-aware design of the virtual IDC network increases channel utilization by 10%–25%. Thus, RS-MOPIC is more viable to be adopted in order to make a compromise between resilience and resource overhead.

5.5 OVERCOMING ENERGY VERSUS RESILIENCE TRADE-OFF

Although RS-MOPIC improves MOPIC in terms of resource consumption, it is not power-aware; hence, energy-efficient improvement over MOPIC is required in order to ensure low Opex for the operators. To this end, in [21], the authors have proposed resilient provisioning with minimum power consumption in cloud (RPMPC) which aims at making a compromise between power consumption and outage probability. RPMPC improves MOPIC in the following four ways:

(1) For upstream data center demands, RPMPC selects $\lceil DC_{min}/2 \rceil$ data centers out of D_s based on minimum power consumption, whereas the rest are selected based on those leading to minimum outage probability (see Eq. 5.16).

(2) While routing over virtual topology, RPMPC uses a two-piece function as shown in Equation 5.17. Thus, the first summation term formulates the total physical link cost forming the corresponding virtual link whereas the second term formulates the outage probability of the virtual link. In the second piece of the cost assignment function M denotes a large number to avoid the dominance of the first piece, that is, power consumption.

$$\varphi_{ij}^{v} = \left\{ \begin{array}{cc} \left(\sum_{\rho\ell^{mn} \in v\ell^{ij}} \varphi_{mn}^{phy} \right) + \left(M \cdot O_{v\ell}^{ij} \right) & A_{ij} > 0 \\ \infty & \text{else} \end{array} \right\} \tag{5.17}$$

(3) In order to route a virtual link over the physical topology, RPMPC uses power consumption and the outage probability of the corresponding physical link as formulated in Equation 5.18. It is worthwhile to note that since the backbone is

considered to be an elastic optical network, power consumption of the transponders is formulated by the term $P_t \cdot W_{mn} + \sum_{\lambda_k \in \Lambda_{mn}} P_t^c \cdot R_k^{mn}$ as P_t is the fixed power consumption and P_t^c is the bandwidth-variable power consumption of a transponder, respectively, whereas R_k^{mn} is the current bitrate on the corresponding transponder. It is worthwhile to note that selection of IP over elastic optical network as the transport medium is to enable transmission in finer granularity and flexibility in spectrum allocation [32]. However, the proposed framework is adaptable to any optical transport technology. Due to limited space, the reader is referred to Ref. [33] for the details of the transmission medium.

$$\varphi_{mn}^{phy} = \left\{ \begin{array}{ll} P_{edfa} \cdot S_{mn} + P_t \cdot W_{mn} + \sum_{\lambda_k \in \Lambda_{mn}} P_t^c \cdot R_k^{mn} + M \cdot O_{\rho\ell}^{mn} & W_{mn} > 0 \\ \infty & \text{else} \end{array} \right\} \tag{5.18}$$

(4) RPMPC enables IDC workload sharing in order to ensure energy savings and lower outage probability. To this end, it adopts the IDC workload distribution algorithm in Algorithm 5.1, and modifies it to meet both objectives. The only difference between the workload distribution algorithm of RPMPC and Algorithm 5.1 is the calculation of the temporary Opex (Opx_{temp}). The new temporary Opex calculation is performed by running Equation 5.19. Thus, the temporary Opex function consists of two pieces where the first piece denotes the power consumption overhead of the migrated workload on the destination data centers (i.e., $\hbar(Map[s][i] \cdot \Upsilon_{IDC}^{si})$ and the second piece is the outage probability of the demands destined to the selected alternate data centers. In the equation, $\hbar(\cdot)$ denotes a function which returns the additional cooling and processing power for a data center due to workload migration whereas Υ_{IDC}^{sd} is the possible workload migration demand from data center s to data center d.

$$Opx_{temp} \leftarrow \sum_i \left[\hbar(Map[s][i] \cdot \Upsilon_{IDC}^{si}) + M \cdot (O_{\nu P}^{id} + O_{dc}^{i}) \right] \tag{5.19}$$

In Ref. [21], the authors have evaluated the performance of RPMPC under the same simulation settings in Section 5.4, and compared its performance to MOPIC and PoMiP in terms of power consumption and outage probability. In Figure 5.7a, it is clearly seen that power consumption under RPMPC is similar to that under PoMiP. Furthermore, by the employment of RPMPC, up to 6.7% enhancement can be introduced to MOPIC which is purely outage probability-aware. Besides, in Figure 5.7b where MOPIC demonstrates the best performance in terms of outage probability, RPMPC improves the outage probability under PoMiP dramatically. Therefore, the trade-off between resilience and energy efficiency can be addressed by RPMPC to ensure significant Opex savings.

5.6 SUMMARY AND DISCUSSIONS

With the advent of cloud computing, users are rapidly receiving XaaS via a shared pool of resources based on the pay-as-you-go fashion. Data centers, as the hosts of physical

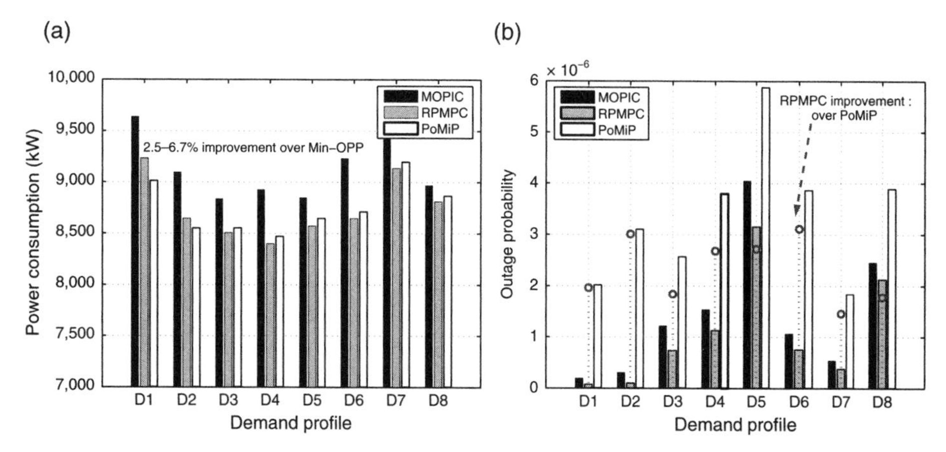

Figure 5.7. (a) Power consumption of the inter-data-center network under RPMPC, MOPIC, and POMIP, (b) Outage probability of upstream data center demands under RPMPC, MOPIC, and POMIP.

servers, play the key role in the delivery of cloud services. Therefore, interconnection of data centers over a backbone network is one of the major challenges affecting the performance of the cloud system, as well as the Opex of the service providers. This chapter has introduced recent design approaches for operational cost-efficient design of an virtual IDC network design. We have focused on energy efficiency (and electric bills) and outage probability which can be also denoted by resiliency, availability, and/or reliability to help reducing the Opex of the the network and computing services. A generic virtual IDC design framework has been introduced which forms a basis for all of the schemes studied in this chapter. Then, it has been followed by the PoMiP which aims at minimum power consumption throughout the network, ToU-aware provisioning which aims at minimum electric bills for network and data center operators, MOPIC which aims at minimum downtime for network as well as computing services, and RPMPC which aims at meeting both objectives. All schemes have been discussed with pros and cons in terms of energy consumption and resilience which has also introduced the trade-off between these two factors affecting the Opex. The chapter has been concluded by introducing the benefits of RPMPC which adopts MOPIC and PoMiP to address this trade-off. In Table 5.3, these schemes have been summarized with a comparison with respect to backbone network technology, energy efficiency, resilience, electricity price awareness, workload migration, and resource usage.

This area of research has still open issues and challenges to be addressed by the researchers working in this field. Extension of RPMPC by considering the presence of differentiated SLAs in the cloud backbone is an immediate research direction. Furthermore, the impact of the intra-data-center network on the performance of the proposed policies needs further study. Future work should also investigate the impact of using different routing and spectrum/wavelength assignment schemes on the performance of the proposed frameworks in terms of energy efficiency, outage probability, as well as resource utilization. Last but not least, communication overhead between the IDC

TABLE 5.3. Summary of the virtual inter-data-center network design schemes studied in this chapter

Scheme	Backbone network	Energy Efficiency	Resilience	Electricity Price	Workload Migration	Resource Usage
PoMiP [13]	IP/WDM	√	×	×	×	×
DePoMiP [13]	IP/WDM	√	×	×	×	√
ToUP [16]	IP/WDM	√	×	√	√	√
MOPIC [28]	EON	×	√	×	×	×
RSMOPIC [28]	EON	×	√	×	×	√
RPMPC [21]	EON	√	√	×	√	√

network and the smart grid communication network prior to virtualization needs to be studied and addressed by future research.

REFERENCES

1. Q. Zhang, L. Cheng, and R. Boutaba, "Cloud computing: State-of-the-art and research challenges," *Journal of Internet Services and Applications*, **ED-1**, 7–18 (2010).
2. R. Moreno-Vozmediano, R. S. Montero, and I. M. Llorente, "Key challenges in cloud computing to enable the future internet of services," *IEEE Internet Computing*, **ED-17/4**, 18–25 (2013).
3. S. J. B. Yoo, Y. Yin, and K. Wen, "Intra and inter datacenter networking: The role of optical packet switching and flexible bandwidth optical networking," *Proceeding of International Conference on Optical Network Design and Modeling (ONDM)*, **ED-14**, 1–6 (2012).
4. X. Zhao, V. Vusirikala, B. Koley, V. Kamalov, and T. Hofmeister, "The prospect of inter-data-center optical networks," *IEEE Communications Magazine*, **ED-51/4**, 32–38, (2013).
5. M. Gharbaoui, B. Martini, and P. Castoldi, "Anycast-based optimizations for inter-data-center interconnections," *IEEE/OSA Journal of Optical Communications and Networking*, **ED-4/11**, B168–B178 (2012).
6. Y. Li, N. Hua, H. Zhang, and X. Zheng, "Reconfigurable bandwidth service based on optical network state for inter-data center communication," *IEEE International Conference on Communications in China: Optical Networks and Systems*, **ED-1**, 282–284 (2012).
7. D. Develder, M. De Leenheer, B. Dhoedt, and M. Pickavet "Optical networks for grid and cloud computing applications," *Proceedings of the IEEE*, **ED-100/5**, 1149–1167 (2012).
8. M. F. Habib, M. Tornatore, M. De Leenheer, F. Dikbiyik, and Mukherjee, "Design of disaster-resilient optical datacenter networks," *IEEE/OSA Journal of Lightwave Technology*, **ED-30/16**, 2563–2573 (2012).
9. B. Jaumard, A. Shaikh, and C. Develder, "Selecting the best locations for data centers in resilient optical grid/cloud dimensioning," *Proceedings of the International Conference on Transparent Optical Networks (ICTON)*, **ED-14**, 1–4 (2012).
10. X. Dong, T. El-Gorashi, and J. M. H. Elmirghani, "Green IP over WDM networks with data centers," *IEEE/OSA Journal of Lightwave Technology*, **ED-29/12**, 1861–1880 (2011).

11. F. Baroncelli, B. Martini, and P. Castoldi "Network virtualization for cloud computing," *Annals of Telecommunications*, **ED-65/11-12**, 713–721 (2010).
12. J. Buysse, C. Cavdar, M. de Leenheer, B. Dhoedt, and C. Develder, "Improving energy efficiency in optical cloud networks by exploiting anycast routing," *Proceedings of the SPIE – Network Architectures, Management, and Applications*, **ED8310**, 1–6 (2011).
13. B. Kantarci and H. T. Mouftah, "Designing an energy-efficient cloud network," *IEEE/OSA Journal of Optical Communications and Networking*, **ED-4/11**, B101–B113 (2012).
14. C. Develder, M . Tornatore, M. F. Habib, and B. Jaumard, "Dimensioning resilient optical grid/cloud networks," *Communication Infrastructures for Cloud Computing*, eds. Hussein T. Mouftah and Burak Kantarci, 191 Global, Hershey, PA, 73–106 (2014).
15. B. Kantarci and H. T. Mouftah, "The impact of time of use (ToU)-Awareness in energy and Opex performance of a cloud backbone," *Proceedings of IEEE Global Communications Conference (GLOBECOM)*, pp. 3250–3255 (2012).
16. B. Kantarci and H. T. Mouftah, "Time of use (ToU)-Awareness with inter-data center workload sharing in the cloud backbone," *Proceedings of IEEE International Conference on Communications (ICC)*, pp. 4207–4211 (2013).
17. G. Shen and R. S. Tucker, "Energy-minimized design for IP over WDM networks," *IEEE/OSA Journal of Optical Communications and Networking*, **ED-1/1**, 176–186 (2009).
18. Environmental Protection Agency (EPA), "Report to Congress on server and data center energy efficiency." Environmental Protection Agency, Washington, DC [Online] http://www.energystar.gov/ia/partners/prod_development/downloads/EPA_Datacenter_Report_Congress_Final1.pdf (2007).
19. G. Sun, V. Anand, D. Liao, C. Lu, X. Zhang, and N.-H. Bao, "Power-efficient provisioning for online virtual network requests in cloud-based data centers," *IEEE Systems Journal*, accepted to appear, DOI: 10.1109/JSYST.2013.2289584.
20. A. Tzanakaki et al., "Energy efficiency considerations in integrated IT and optical network resilient infrastructures," *Proceedings of International Conference on Transparent Optical Networks (ICTON)*, **ED-13**, 1–4 (2011).
21. B. Kantarci and H. T. Mouftah, "Minimum outage probability provisioning in an energy-efficient cloud backbone," *Proceedings of IEEE Global Communications Conference (GLOBECOM)*, SAC.GDC.1–5 (2013).
22. A. Pages, J. Perello, S. Spadaro, and G. Junyent, "Strategies for virtual optical network allocation," *IEEE Communications Letters*, **ED-16/2**, 268–271 (2012).
23. K. N. Georgakilas, A. Tzanakaki, M. Anastasopoulos, and J. M. Pedersen, "Converged optical network and data center virtual infrastructure planning," *IEEE/OSA Journal of Optical Communications and Networking*, **ED-4/9**, 681–691 (2012).
24. B. Kantarci, L. Foschini, A. Corradi, and H. T. Mouftah, "Design of energy-efficient cloud systems via network and resource virtualization," *Wiley-International Journal of Network Management*, 1–16 DOI: 10.1002/nem.1838 (2013).
25. R. Lin, M. Zukerman, G. Shen, and W.-D. Zhong, "Design of light-tree based optical inter-datacenter networks," *IEEE/OSA Journal of Optical Communications and Networking*, **ED-5/12**, 1443–1455 (2013).
26. R. de Sa Ferreira, L. A. Barroso, P. R. Lino, M. M. Carvalho, and P. Valenzuela, "Time-of-use tariff design under uncertainty in price-elasticities of electricity demand: A stochastic optimization approach," *IEEE Transactions on Smart Grid*, **ED-4/4**, 2285–2295 (2013).

27. J. Moore, J. Chase, P. Ranganathan, and R. Sharma, "Making scheduling cool: Temperature-aware workload placement in data centers," *Proceedings of USENIX Annual Technical Conference (ATEC)*, 61–74, 2005.
28. B. Kantarci and H. T. Mouftah, "Resilient design of a cloud system over an optical backbone," accepted, 2014.
29. M. Klinkowski and K. Walkowiak, "On the advantages of elastic optical networks for provisioning of cloud computing traffic," *IEEE Network*, **ED-27/6**, 44–51 (2013).
30. M. Tornatore, G. Maier, and A. Pattavina, "Availability design of optical transport networks," *IEEE Journal on Selected Areas in Communications*, **ED-23/8**, 1520–1532 (2005).
31. R. Arno, A. Driedl, P. Gross, and R. J. Schuerger, "Reliability of data centers by tier classification," *IEEE Transactions on Industry Applications*, **ED-48/2**, 777–783 (2012).
32. Z. Shuqiang and B. Mukhjere, "Energy-efficient dynamic provisioning for spectrum elastic optical networks," in *IEEE International Conference on Communications (ICC)*, 1–6 (2012).
33. G. Zhang, M. De Leenheer, A. Morea, and B. Mukherjee, "A survey on OFDM-based elastic core optical networking," *IEEE Communications Surveys and Tutorials*, **ED-15/1**, 65–87 (2013).

6

OPENFLOW AND SDN FOR CLOUDS

Alberto Leon-Garcia, Hadi Bannazadeh, and Qi Zhang

Department of Electrical and Computer Engineering, University of Toronto, Toronto, Ontario, Canada

6.1 INTRODUCTION

Application platforms consist of the software and infrastructure (personal and sensor devices, wireless and wired access networks, Internet, and computing clouds) that are involved in the delivery of content and applications. Application platforms have been evolving to provide unprecedented flexibility, scalability and economies of scale. This evolution is expected to continue driven by applications that address mobility, social networking, big data, and smart infrastructures.

Service-oriented computing and virtualization are key notions in application platforms. Service-oriented computing uses services to support the rapid creation of large-scale interoperable distributed applications. Applications comprise services that can be accessed through networks. Service-oriented computing and virtualization together provide a foundation for resource management. A virtual resource reveals only the attributes that are relevant to the service or capability offered by the resource, and it hides implementation details. Virtualization therefore simplifies resource management and allows operation over infrastructures consisting of heterogeneous resources.

Cloud Services, Networking, and Management, First Edition.
Edited by Nelson L. S. da Fonseca and Raouf Boutaba.

Virtual machines (VMs) play a central role in cloud computing [1], and virtual networks (VNs) play a key role in the debate over the design of the Future Internet and in software-defined networks (SDN) [2, 3]. In combination, cloud computing and SDN can enable highly dynamic, efficient, and cost-effective shared application platforms that can support the rapid deployment of applications for a multiplicity of application providers.

In this chapter we discuss the interdependencies between cloud computing and SDN in application platforms, and we provide a sample of major open source efforts that address these interdependencies. First, we consider the basic use case of web browsing to introduce the basic issues in the interplay between cloud computing and SDN. Section 6.3 discusses the features and advantages of SDN and its most influential example, OpenFlow. Section 6.4 discusses cloud computing and introduces OpenStack focusing on the Networking Service provided by its Neutron project. Section 6.5 examines challenges and issues in combining SDN and cloud computing, and we highlight the important role of Open vSwitch in providing network connectivity to VMs. Section 6.6 introduces the OpenDaylight open-source project. Section 6.7 shows how SDN and cloud computing come together, and introduces the notion software-defined infrastructures. This chapter is focused on providing an integrated view of SDN and cloud computing. We conclude in Section 6.8 a brief discussion of research trends and challenges in SDN for cloud computing. At the end of the chapter, we provide references to various surveys and introductory articles on specifics of SDN or cloud computing.

6.2 SDN, CLOUD COMPUTING, AND VIRTUALIZATION CHALLENGES

Figure 6.1 depicts a scenario in which an end user wishes to access Web content through a handheld device. In a traditional, un-virtualized Internet service model, the request for accessing an HTML page is first received by the wireless access point. The access point forwards the request to a Web server through an access network and an Internet gateway and then a firewall across the Internet. Depending on the current load condition, the request may be forwarded by a load balancer to one of several dedicated application servers, which in turn may access a shared database (e.g., SQL) server.

While this un-virtualized model is simple and straightforward to implement, it also has several limitations. First, when the service demand is low the application servers may become underutilized, leading to resource wastage. On the other hand, when service

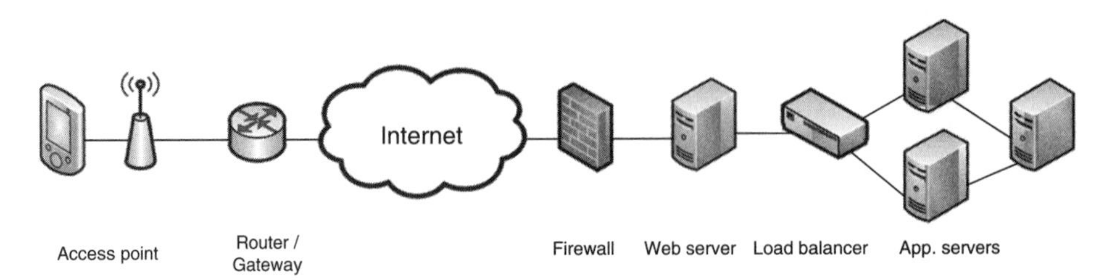

Figure 6.1. Physical infrastructure for web browsing use case.

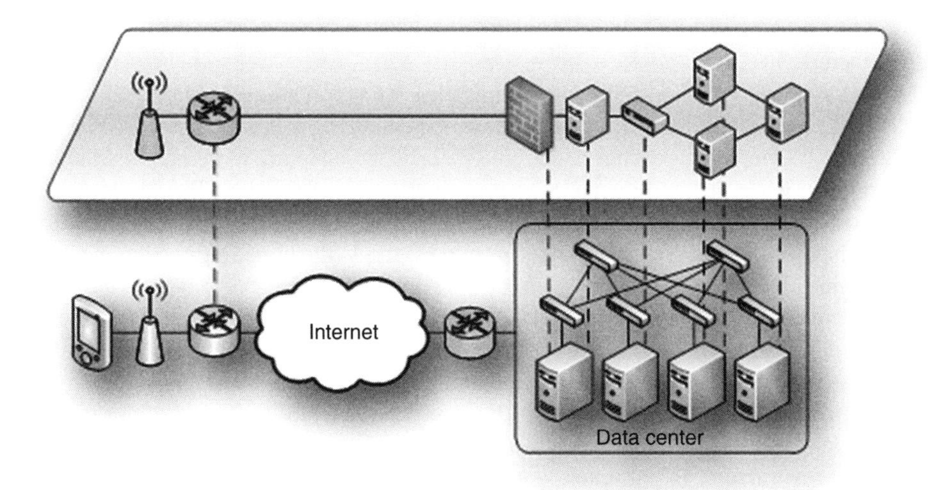

Figure 6.2. Virtual infrastructure for Web browsing use case.

demand is high, it is also difficult to scale up the service, as this requires attaching another physical server. If the demand surge is temporary, the attached server may subsequently become underutilized. Furthermore, it is difficult to provide quality-of-service (QoS) guarantees in this model, as the "best-effort" packet delivery implies customers can experience long response times and insufficient throughput when the underlying network is busy.

Motivated by these limitations, there is a trend toward building virtualized infrastructure for end-to-end service delivery. Infrastructure virtualization aims at having multiple virtual infrastructures share the same physical infrastructure. In a nutshell, a virtual infrastructure consists of VMs that are interconnected by an underlying VN, which may consist of virtual routers, virtual switches as well as virtual links that interconnect them.

For the Web browsing use case, a virtualized infrastructure is depicted in Figure 6.2. The Web content provider first specifies the topology and resource requirement of its Web content delivery service as a virtual infrastructure as shown in Figure 6.2. The firewall, load balancer, and all the servers are implemented by VMs, whereas routers and wireless access points are represented as virtual routers and virtual wireless access points. This virtual infrastructure is then embedded in the physical network infrastructure, where VMs are placed in data centers, and virtual routers and virtual switches are mapped to physical routers and switches or to software implementations of these. The notion of a *flow* is central to the virtualization of a network. The flows of packets in a VN are identified by specific values in their header fields that allow routers and switches to identify them and treat them as prescribed by their VN.

This model is beneficial for several reasons. First, by isolating VNs from each other, it is possible to achieve better QoS as network performance variability is limited. Second, by separating logical service infrastructures from the underlying physical network infrastructure, it is possible to improve resource utilization by consolidating multiple VMs on a single physical machine, or multiple virtual links on a single physical link. Virtualization also enables the scaling (up or down) and migration of virtual resources (e.g., VMs and VNs) to achieve better resource efficiency, by adapting to demand variation. For example, when demand increases, it is desirable to increase the CPU and memory allocation of a VM that hosts an application server. If the physical machine currently hosting the VM does not have sufficient resources, the VM can be migrated to another physical machine.

The creation of VNs is at the core of the above model. The benefits of the above model require an environment where multiple independent tenants are each allotted their own dedicated virtual infrastructure; but in fact, they share the same physical infrastructure, ideally completely oblivious of the presence of other tenants. From a tenant perspective, the virtual computing and networking infrastructure should behave as a physical infrastructure. In particular, a tenant requires control over connectivity, bandwidth and QoS, MAC and IP addresses, node number, and node location and mobility. Each tenant may also bring requirements for security, load balancing, caching, and application performance.

On the other hand, the infrastructure provider should be able to handle large numbers of tenants, while meeting their specific requirements for security, isolation, network control, and availability. In cases where a tenant requires presence in multiple geographic sites, a provider may be called upon to extend its VNs across multiple data centers or even to federate with other providers. In short, the provider needs controllers to manage its own network connectivity and resources while providing tenants with their own specific connectivity and networking needs.

SDN is an emerging concept for the control and management of network infrastructure that enables programmability of the network. SDN is therefore a prime candidate to address the multiplicity of challenges for creating VNs in application platforms.

6.3 SOFTWARE-DEFINED NETWORKING

6.3.1 What Is SDN?

SDN is an emerging concept that has grown from efforts to define network architectures that are flexible, evolvable and can avoid the ossification pitfalls of the current Internet [2]. The early experience with OpenFlow in particular influenced the current view on SDN [3, 4]. We refer the reader to Ref. [5] for a recent survey of SDN and programmable networks.

SDN separates the control of network functionality from the forwarding functionality in packet-forwarding devices as shown in Figure 6.3. A logically centralized network controller is responsible for the decision on how traffic from a given flow is handled. The decision of the controller is then programmed into the packet-forwarding device.

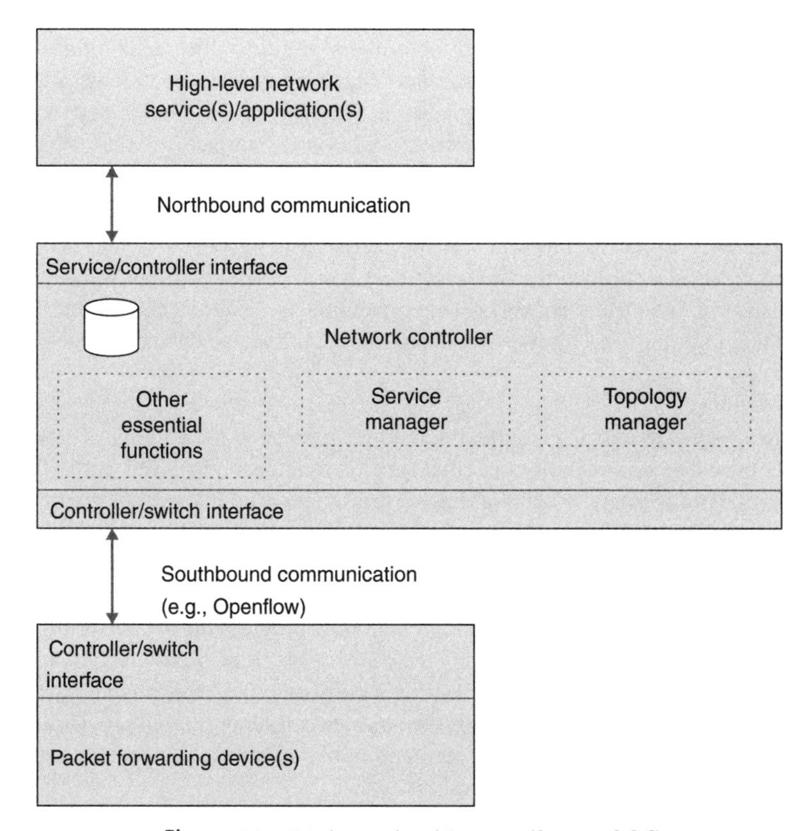

Figure 6.3. SDN layered architecture (from Ref. [5]).

Proceeding bottom up in Figure 6.3 we have the following layers:

1. Packet-forwarding devices: These devices execute the actual forwarding behavior on flows of packets. These devices can consist of actual physical switches designed to provide programmability or software-implemented switches or routers.
2. Southbound interface: This interface provides the means for network controllers to communicate and control the packet-forwarding devices. For example, OpenFlow provides a protocol for this interface.
3. Network controller: The controller provides network services to higher layers by programming the packet-forwarding devices. The network controller is positioned to make optimal decisions on resource allocations because it has a global view of the state of the overall network resources. A number of open source and proprietary network controllers are available.
4. Northbound interface: This interface provides the means for applications and high-level services to access the services provided by the network controller. In

general, the vendors of network controllers prefer to differentiate their service offerings, and so the northbound interface has not been open or standardized. However the OpenDaylight project (discussed in the following text) has been working on the development of open northbound interfaces.

5. Applications and high-level services use the services of the network controller to provide more complex functionality. For example, an application could provide the virtual machines and virtual network to support a virtual tenant. The application would invoke the services of the network controller in creating and managing its virtual networks. Other example applications could orchestrate or chain multiple services together to provide security, load balancing, caching, and so on.

SDN provides the foundation for flexible and customizable networking. The network controller provides the capability to define specific treatments for given traffic flows by installing rules in the network forwarding devices. The centralization of control in the network controller and the northbound interface allow novel applications to define the operation of the network in software. This also opens the way for faster and adaptive configuration of the network.

In the context of this chapter, the data center SDN provides the means for supporting network virtualization and automated migration of VMs. It also provides the means to achieve bandwidth optimization, as well as higher utilization of servers and higher energy efficiency. Across data centers, SDN VN capabilities can support rapid provisioning and migration of cloud services across private and public clouds in support of large-scale geographically distributed applications.

We note that the layered view in Figure 6.3 is limited in scope to the networking infrastructure, and the Application level provides an indication of the broader cloud computing context within which networking must take place. In Sections 6.4 and 6.5, we will see how SDN fits within this context in general and within OpenStack in particular.

6.3.2 OpenFlow

OpenFlow was originally presented as an approach to allow experimentation in new network protocols on campus networks [3] although its roots are in the Ethane project to enable highly flexible and secure enterprise networking [6]. OpenFlow separates the network control and packet forwarding, a concept first widely deployed in MPLS [7]. It has two basic elements as shown in Figure 6.4: (1) defining packet forwarding behavior by allowing a controller to set flow tables that associate an action with each flow in an OpenFlow switch and (2) defining an open secure protocol that enables a network controller to exchange commands and packets with an OpenFlow switch. This is essentially the SDN southbound interface in Figure 6.3. The notion of an open management interface was introduced in Ref. [9].

The Open Networking Foundation was formed to provide specifications to cover the components and basic functions of the OpenFlow switch and the OpenFlow protocol to manage the switch from a remote controller [8]. OpenFlow Switch Specification 1.0.0

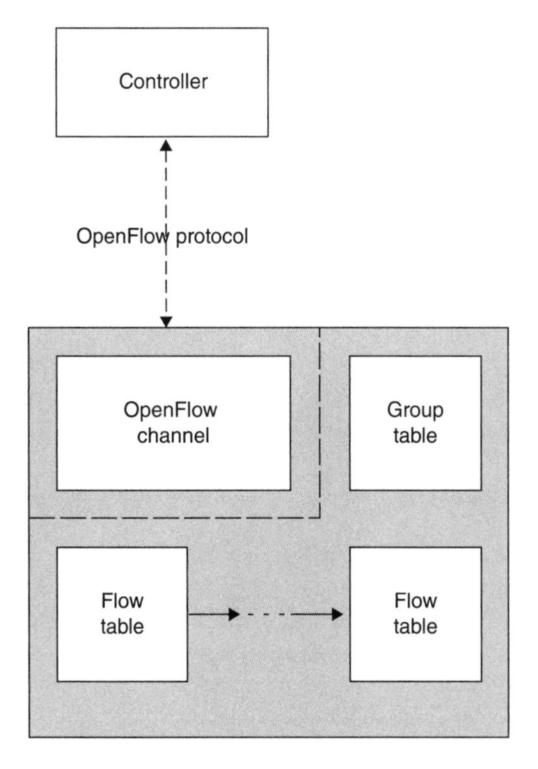

Figure 6.4. OpenFlow switch and controller (from Ref. [9]).

TABLE 6.1. Components of a flow entry in a flow table

Match fields	Priority	Counters	Instructions	Timeouts	Cookie

was released in December 2009. The latest release was 1.4.0 on October 2013. In the following, we summarize the OpenFlow specification as described in Ref. [8].

The OpenFlow switch consists of one or more flow tables and a group table that are used to perform packet lookups and forwarding. The OpenFlow protocol enables the controller to manage the OpenFlow switch. The controller can add, update, and delete flow entries in the flow tables. As shown in Table 6.1 each entry consists of match fields, counters, and instructions that are applied to matching packets. The match fields consist of ingress port and packet headers and possibly metadata specified by previous flow tables. The required match fields in OpenFlow version 1.4.0 are ingress port, Ethernet destination and source addresses with arbitrary bitmasks, Ethernet type, IPv4 and IPv6 protocol number, IPv4 source and destination addresses with subnet masks or arbitrary bitmasks, IPv6 source and destination addresses with subnet marks or arbitrary bitmasks,

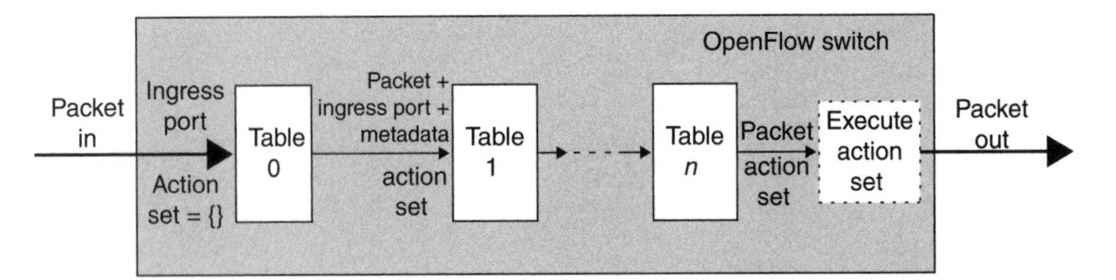

Figure 6.5. Matching of packets against tables (from Ref. [9]).

TCP and UDP source and destination port numbers. Additional optional match fields include switch physical input port, metadata between tables, VLAN ID and priority, IP DSCP and ECN, SCTP, ICMP, ARP, and MPLS and Ethernet PBB. Altogether, the match fields enable a very rich set of packet classifications spanning from physical port and through layers 2–4.

Within each table, entries are matched to packets in priority order so that the first matching entry is used. As shown in Figure 6.5, an arriving packet is first matched against the first table, and if an entry is matched, the instructions in the entry are performed. If a packet is not matched, then the table-miss flow entry is consulted to determine the appropriate action. This could include dropping the packet, forwarding it to the controller using the OpenFlow channel, or continuing to the next table.

The instructions in a flow entry can include actions such as packet forwarding, packet modification, and group table processing. The instructions may also modify the pipeline processing by directing packets, and associated metadata, to subsequent tables for additional processing. An arriving packet begins with an empty action set and this set is updated each time a match to a table entry is identified. The table pipeline processing ends when a table entry does not specify another table. At this point, the action set is executed.

In OpenFlow, a port is the network interface where packets pass between OpenFlow processing and the rest of the network. Packets that arrive on an ingress port are processed by the OpenFlow pipeline and may be forwarded to an output port. OpenFlow ports can be physical ports (e.g., Ethernet interface), logical ports that do not correspond to a hardware interface (e.g., tunnel), or reserved ports that specify generic forwarding actions (e.g., send to controller, or flooding).

The group table in Figure 6.4 contains group entries, and each entry contains a list of action buckets that are applied to packets sent to the group. The group abstraction allows flow entries to point to common output actions in a switch. For example, a group of type "all" enables the controller to implement flooding.

There are three ways to remove flow entries from tables. First, the controller can request removal of an entry. Second, the switch has a flow-expiry mechanism that removes an entry after either a hard timeout expires or after the entry has not been matched for some specified period of time. Third, a switch may evict table entries when it needs to recover resources, when eviction is enabled. If a flag is set, the switch is required to notify the controller that the entry has been removed.

TABLE 6.2. Meter entry in meter table; Meter band in meter entry

	Meter Identifier	Meter-Bands	Counters
Band Type	Rate	Counters	Type Specific Arguments

OpenFlow uses per-flow meters to implement QoS. A meter is a switch element that is used to measure and control the rate of packets. Flow entries can specify a meter in its instruction set. A meter measures and controls the aggregate rate of all flows entries to which it is attached. A meter table consists of flow entries as in Table 6.2. Each meter has a 32-bit identifier. The meter counter is updated each time a packet is processed by a meter. The meter bands specify rate bands and corresponding packet processing. The meter applies the meter band with the highest configured rate that is lower than the current measured packet rate. For example, rate limiting can be applied if the band type is "drop."

OpenFlow provides required and optional counters associated with flow tables, flow entries, ports, queues, groups, group buckets, meters, and meter bands. Counters are unsigned integers that may measure counts, such as bytes or packets, and durations such as seconds or nanoseconds. In combination, these counters can measure rates.

The OpenFlow channel in Figure 6.4 is the interface that allows the controller to configure and manage the switch, to send packets out the switch, and to receive notifications from the switch. The channel usually operates over TCP and uses transport layer security (TLS).

The OpenFlow protocol has three message types. The controller uses controller-to-switch messages to manage and monitor the switch state. For example, the Modify-State message is used to add, delete, and modify entries in the flow and group tables, and the Packet-Out message is used by the controller to send packets out of the switch. The switch uses asynchronous messages to update the controller. Thus, the Packet-In message is used to transfer control of a packet to the controller, for example, after a table-miss event. The Flow-Removed message is used to notify the controller that a flow entry has been removed. Symmetric messages are used by the switch or by the controller without solicitation, for example, to start the switch-controller connection ("Hello") or to monitor liveness in the connection ("Echo").

Altogether, the OpenFlow specifications allow the customization of the forwarding and treatment of classified traffic flows across the network. Indeed, OpenFlow can exercise tight control over which packets are admitted into the network. For example, table entries can be configured so that packets can traverse the network only after associated table entries have been established by the controller. Any flow without such entries are forwarded by the switch to the controller after a table miss, and the controller then decides whether to accept the flow.

The interest in OpenFlow has led to the availability of switches and routers that support the specification. OpenFlow is influencing the design of packet-processing chips with advanced parsing and classification capabilities. OpenFlow has also influenced the development of software-based switches. We will discuss these switches after introducing OpenStack cloud computing.

6.4 OVERVIEW OF CLOUD COMPUTING AND OPENSTACK

Cloud computing is a computational approach in which software is hosted in large data centers and where software is provided as a service [1]. The key technology in cloud computing is the VM that provides an abstraction of a physical host machine as shown in Figure 6.6. The VM is enabled by the introduction of a hypervisor that intercepts the instructions between the OS and hardware and manages the sharing of the hardware among multiple VMs. Cloud computing provides a computing utility which provides the illusion of infinite resources through the on-demand sharing of computing resources. A major advantage of cloud computing is its flexible billing model that provides access to computing without upfront cost. Cloud computing has revolutionized the delivery of applications and its tremendous potential impact has stimulated the development of an open source platform.

OpenStack is a project developing an open source cloud computing platform to provide infrastructure as a service (IaaS). OpenStack offers a set of interrelated services, each through an application programming interface (API) [10]:

- *Dashboard (Horizon project)*: A Web-based portal to interact with OpenStack services, such as launching an instance, assigning IP addresses and configuring access control.
- *Compute (Nova project)*: Manages lifecycle of compute instances: spawning, scheduling, and decommissioning of VMs on demand.
- *Networking (Neutron project)*: Enables network connectivity for other OpenStack services, such as OpenStack Compute. Provides API for users to define networks and attachments. Supports plug-ins for networking vendors and technologies.
- *Object Storage (Swift project)*: Stores and retrieves arbitrary unstructured data objects via a *RESTful*, HTTP-based API.

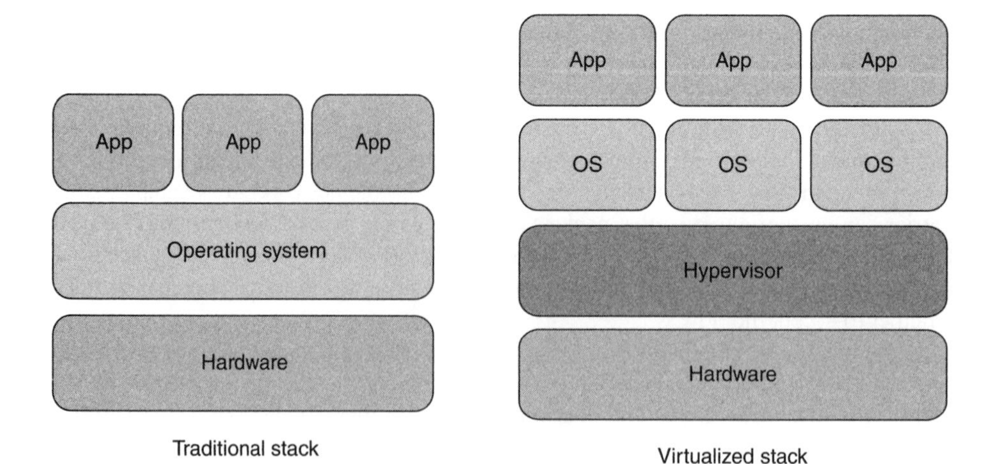

Figure 6.6. Virtual machines.

- *Block Storage (Cinder project)*: Provides persistent block storage to running instances.
- *Identity service (Keystone project)*: Provides an authentication and authorization service for other OpenStack services.
- *Image Service (Glance project)*: Stores and retrieves VM disk images, and it is used by OpenStack Compute during instance provisioning.
- *Telemetry (Ceilometer project)*: Monitors and meters the OpenStack cloud for billing, benchmarking, scalability, and statistical purposes.
- *Orchestration (Heat project)*: Orchestrates multiple composite cloud applications.
- *Database Service (Trove project)*: Provides scalable and reliable Cloud database-as-a-service functionality for both relational and nonrelational database engines.

Figure 6.7 shows the conceptual architecture of the OpenStack projects. End users can interact with OpenStack through the dashboard, command-line interfaces and APIs.

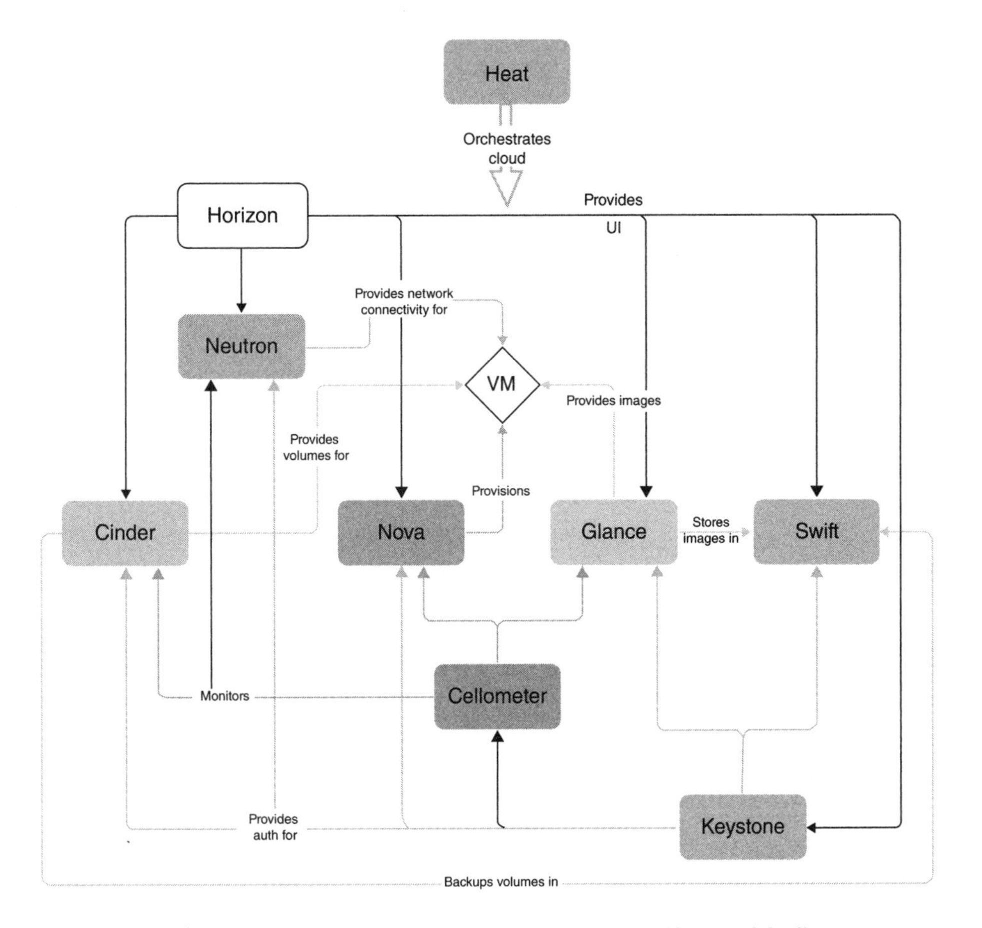

Figure 6.7. Conceptual architecture of OpenStack (from Ref. [10]).

The Identity service (Keystone) is used to authenticate services. The individual services interact through public APIs.

We focus on the Nova compute service and the Neutron networking service. Nova and neutron use a message queue as a central hub for passing messages. A message-oriented middleware, such as RabbitMQ, is used. Asynchronous calls are used for request response and a call-back is triggered once a response is received. The nova-api provides an interface for interaction with the cloud infrastructure. It supports the OpenStack Compute API as well as the Amazon EC2 API. The API server communicates with the relevant components through the Message Queue. The nova-scheduler process takes a VM instance request and decides the compute server it should run on. The nova-compute process (Compute worker) deals with instance management life cycle; it creates and terminates VM instances through hypervisor APIs. The nova-compute process requests networking tasks from the Neutron networking service.

Nova compute is designed for use by multiple tenants in a shared system. Each tenant has an individual VN, and volumes, instances, images, keys, and users. A user can specify the tenant by a tenant ID. Tenant resource limits are set by quotas on: Number of volumes that may be launched; number of processor cores and RAM that can be allocated; floating IP addresses assigned to any instance; and fixed IP addresses assigned to the same instance when it launches.

The Neutron networking service provides a VN service with connectivity between interface devices managed by OpenStack services, typically compute. Just as the Nova Compute API provides a virtual server abstraction, the Neutron API provides a VN abstraction that allows a user to create and attach interfaces to networks. The Neutron server accepts API requests and directs these to the appropriate OpenStack plug-in and agents that plug and unplug ports, create networks and subnets, and provide IP addressing. OpenStack networking has a plug-in architecture that allows it to support a variety of vendor and networking technologies.

In Neutron networking, three types of network resources are identified:

- *Network*: An isolated layer 2 segment, analogous to a VLAN in a physical network.
- *Subnet*: A block of IPv4 or IPv6 addresses and associated configuration state.
- *Port*: A connection point for attaching a single device, for example, a NIC for a virtual server, to a VN. Includes associated network configuration, for example, associated MAC and IP addressing.

Users access Neutron networking to configure network topologies and to then instruct the other OpenStack services to attach virtual devices to these networks. Tenants can create their own private networks with their own IP addressing schemes.

The typical data center deployment, shown in Figure 6.8, includes a cloud controller host, a network gateway host and a number of hypervisors for hosting VMs. The deployment includes the following physical networks:

- *Management Network*: Internal communication between OpenStack components.
- *Data Network*: Inter-VM communications; IP addressing as per plug-in used.

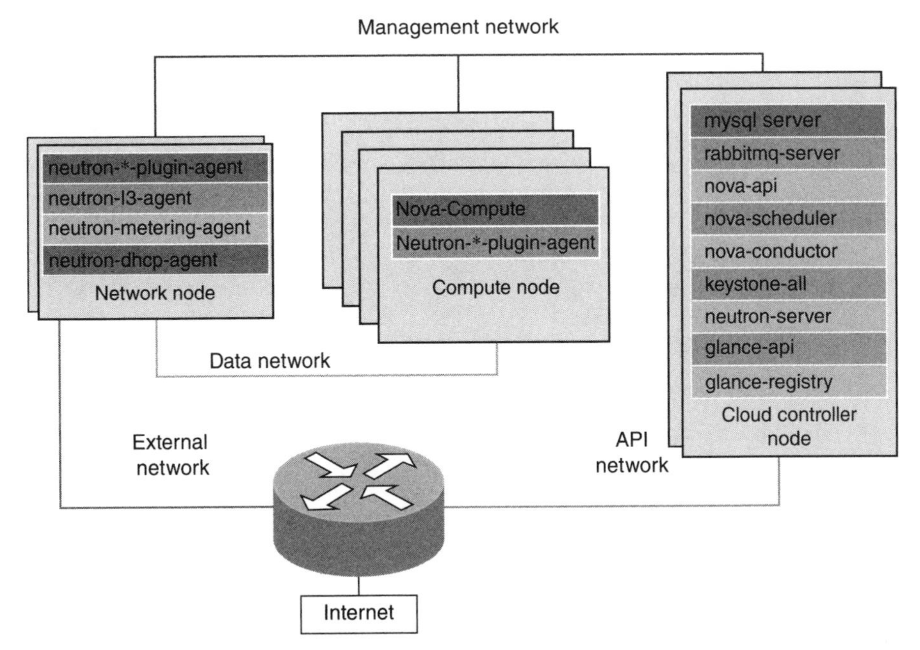

Figure 6.8. Typical physical data center networks in OpenStack (from Ref. [10]).

- *External Network*: VM access to Internet; user on Internet can reach IP addresses in external network.
- *API Network*: Exposes OpenStack APIs to tenants; IP addresses reachable from Internet.

Figure 6.9 shows the possible tenant and provider networks. Tenant networks provide connectivity and need to be isolated from other tenants. Neutron networking supports several tenant network types:

- *Flat*: All instances are on the same network, which can also be shared with the hosts. There is no VLAN tagging or other network segregation.
- *Local*: All instances reside on the local compute host and are isolated from external networks.
- *VLAN*: Users create multiple provider or tenant networks using VLAN IDs (802.1Q tagged) that correspond to VLANs present in the physical network. Instances can communicate with each other, as well as with dedicated servers, firewalls, load balancers, and other networking infrastructure on the same layer 2 VLAN.
- *VXLAN (Virtual Extensible LAN) and GRE (Generic Routing Encapsulation)*: These network overlays support private communication between instances.

Provider networks use an existing physical network in the data center. These networks may be shared among tenants. By allowing tenant networks to select their IP addresses,

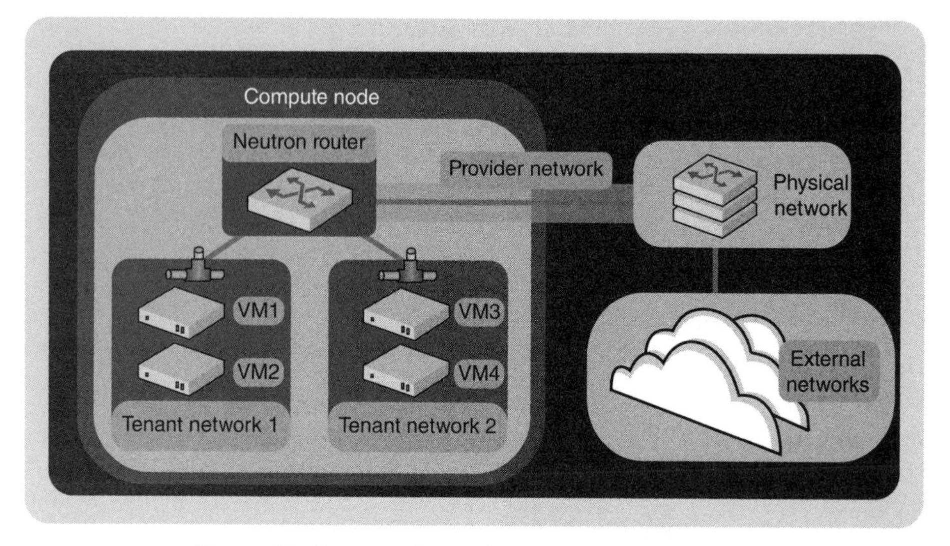

Figure 6.9. Tenant and provider networks (from Ref. [10]).

it becomes possible to migrate applications between user data centers and public data centers as required by demand or fault conditions.

6.5 SDN FOR CLOUD COMPUTING

Figure 6.9 shows that VMs require a new layer of access network *inside* the compute node. As shown in Figure 6.10 each VM now has one of more virtual interfaces (VIFs) that connect it to virtual switches that are in turn connected to physical interfaces (PIFs). Network connectivity is now required to interconnect VMs in the same host as well as in other hosts. Linux bridging is available to provide this connectivity, but it does not adequately meet new requirements that arise with VMs [11]. The migration of applications imposes new network mobility requirements. The deployment of tens of VMs in a host and hundreds of thousands of VMs in a data center poses new scalability challenges. The sharing of computing resources among multiple tenants introduces new security risks and heightened requirements for isolation. The combination of these requirements can be met by extending SDN into the networks that connect VMs, as done for example, by Open vSwitch.

6.5.1 Open vSwitch

The Open vSwitch (OVS) is a software-based virtual switch to provide intra- and inter-VM connectivity while also providing an external interface for the control of configuration state and forwarding behavior [11]. This allows OpenFlow capabilities for fine-grained control of flows to be leveraged and integrated across a multilayer network for connecting VMs. Support can then be provided for QoS, tunneling, and filtering that in turn can be used to provide isolation, security, and network mobility.

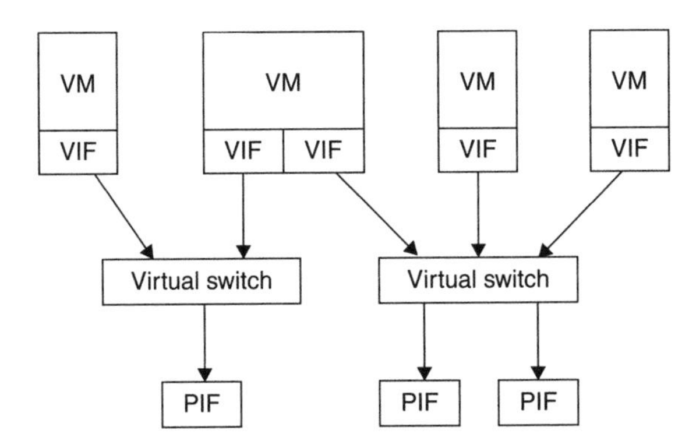

Figure 6.10. Virtual interfaces connect VMs to virtual switches and to physical interfaces (from Ref. [11]).

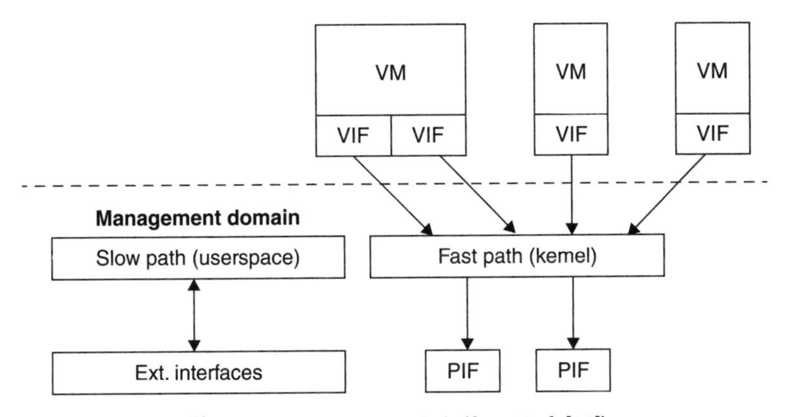

Figure 6.11. Open VSwitch (from Ref. [11]).

The virtualization environment that supports VMs can provide the new virtual switches with useful information, for example, the MAC addresses for the virtual interfaces, the movement of VMs, and the joining to multicast groups. Better coordination between the virtual computing and networking is possible because the same information that is used by the hypervisor and management layer to power VMs on/off, to migrate hosts, and to control access can also be used to manage the VN.

Open vSwitch uses Ethernet switching for VM networking with VLAN, RSPAN (Remote Switched Port Analyzer) to analyze traffic, and Access Control Lists. It also provides port bonding, GRE and IPsec tunneling, and per-VM policing.

Figure 6.11 shows the architecture of the Open vSwitch. The Open vSwitch software resides within the hypervisor. The switch has a "fast path" module in the kernel that implements the speed-critical forwarding engine and the counters for table entries. The switch has a "slow path" in user space that implements forwarding logic, and remote visibility and configuration interfaces, including OpenFlow. Thus Open vSwitch allows the forwarding path to be manipulated by writing to the forwarding table and specifying how packets are handled (forwarded, dropped, encapsulated) based on header fields.

The Open vSwitch local management interface allows the virtualization layer to manage the creation of switches and the connectivity of virtual and physical interfaces. The rule-based forwarding provided by the flow tables allows network configuration state and forwarding to be associated with specific flows, for example, a VM or group of VMs. This enables a global management process not only to have visibility of local state in the virtual switch but also to migrate the associated network configuration state corresponding to a group of VMs that is moved between servers.

In multitenant settings, it is desirable for VMs from different tenants to share the same physical server while providing strong isolation. On the other hand, it is also necessary to provide connectivity between VMs that belong to the same tenant but that reside in different hosts. Open vSwitch provides the capability to create virtual private networks to connect VMs from the same tenant while providing isolation from other tenants. Open vSwitch allows a tenant to be assigned a VLAN ID in small-scale deployments.

Connectivity in larger scale deployments is handled by Open vSwitch through the use of GRE tunnels. In GRE, an Ethernet frame is encapsulated inside an IP datagram, which is routed from the originating subnet to the destination subnet [12]. A GRE tunnel is established between any two servers that have a VM belonging to the same tenant. The MAC-to-IP mapping required for the tunnel is downloaded into the Open vSwitch table entries using OpenFlow. This approach has the advantage that no state concerning the tunnels needs to be maintained in the physical network. OpenFlow does not have a tunnel-provisioning message, so the Open vSwitch Database Management Protocol (OVSDB) was developed to construct the mesh of GRE tunnels between servers that have VMs from the same tenant.

VXLAN is an alternative tunneling method to GRE. VXLAN creates tunnels by encapsulating Ethernet frames on top of UDP and IP. The approach provides 24-bit tags to overcome the VLAN scale limitations. The reader should refer to Ref. [12] for more discussion on SDN networking issues for cloud computing.

6.5.2 Meeting Networking Requirements

We have seen that SDN provides many powerful techniques for realizing network virtualization: (i) resource allocation and bandwidth provisioning, (ii) resource isolation and addressing, and (iii) support for tenant-specific communication and routing protocols. We have seen that SDN-enabled components such as Open vSwitch (OVS) support the creation of virtual switches and interfaces within VM hypervisors, or in OVS-enabled switches. Meanwhile, SDN frameworks such as Openflow provide simple and efficient means to provision virtual links.

As for resource isolation, Openflow includes the capability to limit bandwidth usage and there are numerous proposals [13] on achieving rate limiting at different levels, including flow, ingress, and slice limiting. Thus, we anticipate that future Openflow-enabled switches will have the capability to provide guaranteed bandwidth for individual VNs. Even though address isolation is often implemented using tunneling, recent proposals use address translation supported by OpenFlow to achieve address isolation. Finally, supporting tenant-specific routing protocols can be achieved using a variety of software components, for example, using FlowVisor.

6.5.3 Inter-Data Center Networking

New challenges arise when a tenant network is deployed or migrated across different data centers [14, 15]. The VN needs to handle the addressing schemes and forwarding fabric of the data centers. The connectivity between the data centers may be shared with the public Internet and some means of allocating resources for the tenant networks is required. In addition, the live migration of VMs can impose special performance requirements on the inter-data center network.

An SDN approach to providing inter-data center connectivity is attractive for several reasons. First, it allows the extension of the network abstraction that is already in use in the individual data center. Second, the isolation techniques from intra-data center SDN can be extended. Third, the management approach of OpenFlow can be applied. An SDN approach is proposed in Ref. [15] for providing VNs on demand on loosely coupled data centers. The approach involves dynamically organizing Virtual Private LAN Service (VPLS) paths to extend VNs across data centers. In Ref. [16], an SDN abstraction and API is used to extend OpenStack VN into the WAN. This approach improves over IPSec and SSL VPNs by building on WAN services that can support QoS. The application of SDN in optical transport networks has begun to receive attention. For example, Ref. [17] presents an Open Transport Switch for bursting data between data centers using optical transport networks.

6.6 COMBINING OPENFLOW AND OPENSTACK WITH OPENDAYLIGHT

Given the large number of technologies that can implement network virtualization at various levels, it becomes increasingly important to design frameworks in the management plane to ensure these technologies can work seamlessly to achieve management including consistency, efficiency, performance, reliability, and security. These frameworks need to provide various functionalities including monitoring, scheduling, resource allocation, dynamic adaptation, and policy enforcement. For example, OpenDaylight [18] is a framework that provides functionalities for managing VNs in the context of SDN.

6.6.1 OpenDaylight Overview

OpenDaylight is an open-source project that is developing a modular, pluggable, and flexible controller platform. The controller exposes open northbound APIs to applications. These applications can then use the controller to gather network intelligence, perform analytics, and then orchestrate new rules using the controller.

As shown in Figure 6.12, the controller platform consists of dynamically pluggable modules that perform required network tasks. Base network services address basic management functions of network devices. The topology manager builds the network topology, and the Stats manager collects statistics. The Switch manager handles southbound device information, and the Forwarding Rules manager (FRM) installs flows on southbound devices. The host tracker tracks connected hosts and the ARP handler handles ARP messages. Other network services can be added to the controller platform.

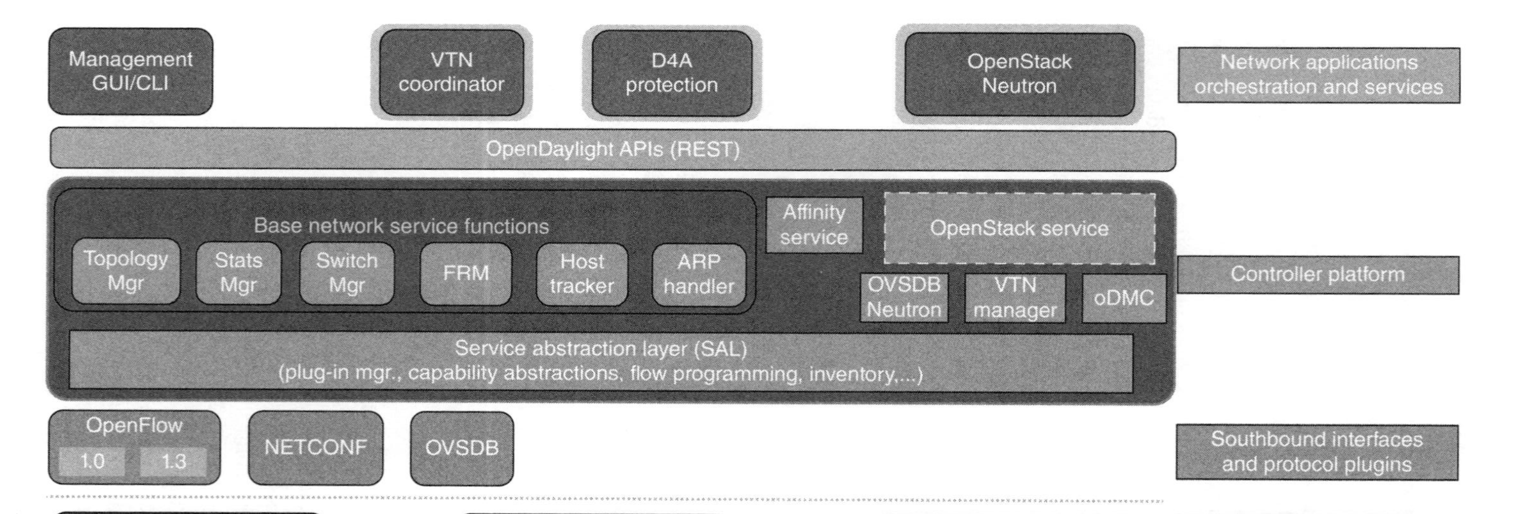

Figure 6.12. OpenDaylight architecture (from Ref. [18]). VTN, virtual tenant network; oDMC, open dove management console; D4A, defense4A# production; LISP, locator/identifier separation protocol; OCSDB, Open vSwitch data base protocol; BGP, border gateway protocol; PCEP, path computation element communication protocol; SNMP, simple network management protocol.

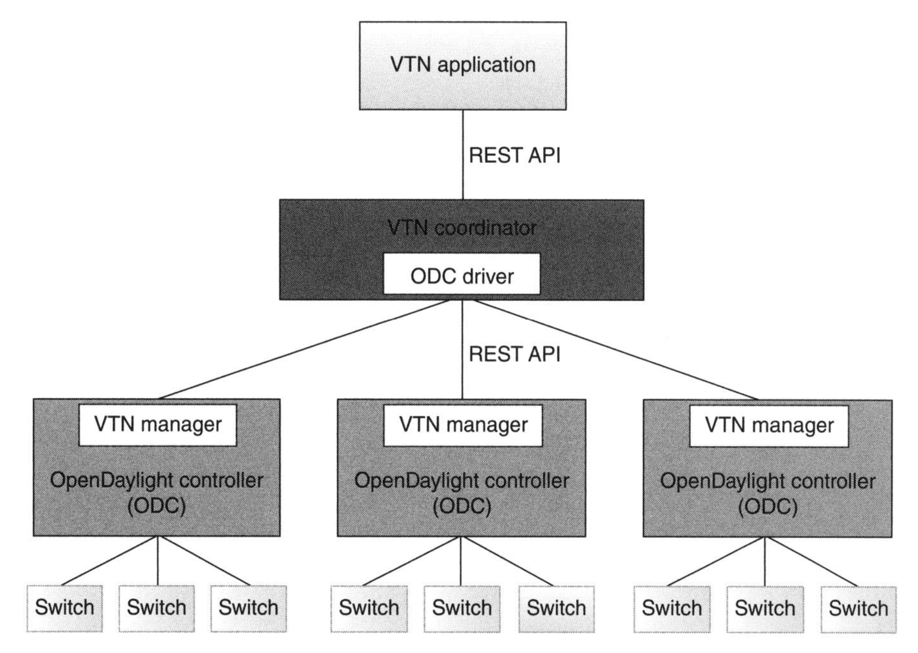

Figure 6.13. Virtual tenant network architecture (from Ref. [18]).

Figure 6.12 shows the Virtualization edition of OpenDaylight, which targets data centers. This edition includes the OVSDB protocol southbound to configure Open vSwitches in VNs. In particular, the Neutron bundle of the Virtualization edition supports VXLAN and GRE tunnels for OpenStack and CloudStack deployments.

The Virtualization edition also supports Virtual Tenant Network (VTN) service. VTN provides multitenant VN on an SDN controller. VTN allows users to design and deploy a network without requiring knowledge of the physical network. VTN maps the desired network to the underlying physical network. Figure 6.13 shows the architecture of the VTN application. The VTN coordinator is an application that allows a user to use the VTN Virtualization. The coordinator interacts with one or more VTN Managers to implement the user configuration.

Figures 6.7 and 6.12 in combination show how cloud computing and SDN interact in the deployment of virtual computing and networking resources. In Figure 6.7, the user may initiate the deployment of an application that requires support from a set of VMs with connectivity requirements. Figure 6.12 shows how the Neutron networking service in OpenStack can invoke the services of OpenFlow to provide the desired network connectivity.

In a typical scenario, a service provider (i.e., a tenant) submits a virtual infrastructure request. The request describes the topology of the virtual infrastructure, and provides resource requirement of each virtual node (i.e., VMs, virtual switches, routers, and firewalls) as well as bandwidth requirement for each virtual link. This request is sent to a scheduler that makes decisions regarding how each virtual node and link is mapped to physical resources.

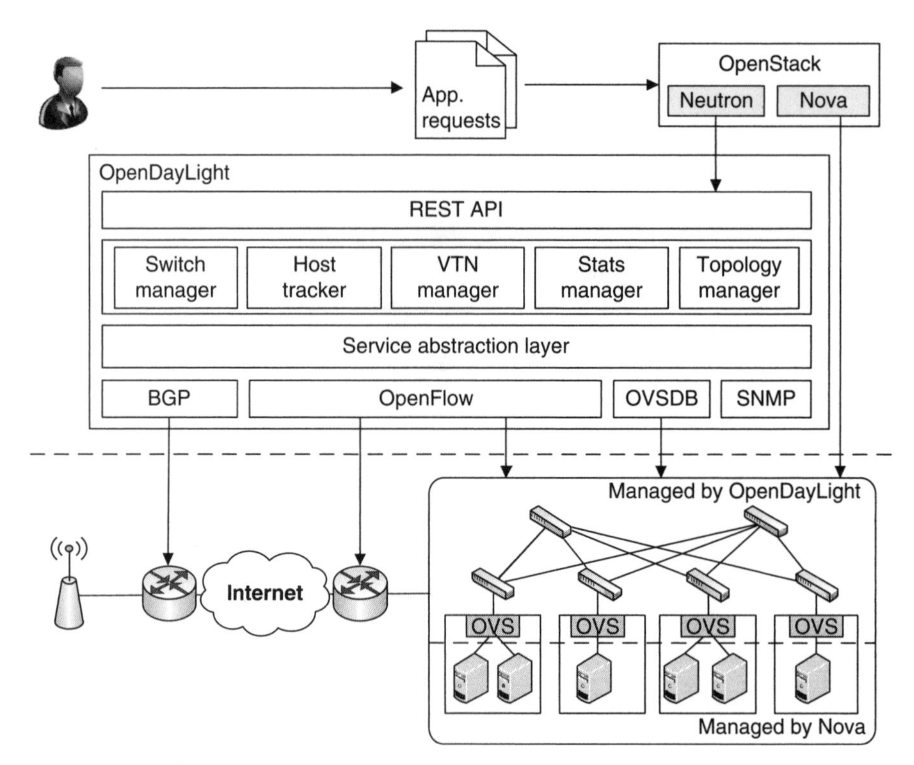

Figure 6.14. Architecture for managing virtual infrastructures.

The scheduling of virtual infrastructure is also known as the VN embedding problem [19], whose goal is to improve the acceptance rate of virtual infrastructure requests while minimizing operational costs such as energy. Once the scheduling decision is made, the scheduler allocates appropriate physical resources according to the scheduling decision. This is done by scheduling VMs and other virtual resources, creating appropriate virtual switch and routers and installing forwarding policies in Openflow-enabled switches.

In Figure 6.14, we use OpenDaylight as an example to illustrate this process. A tenant submits its virtual infrastructure request to OpenStack, which in turn, uses Nova to schedule corresponding VMs in the data center. It also delegates OpenDaylight to schedule the VN. To do so, the scheduling request is first sent to OpenDaylight manager through its REST API. OpenDaylight solves the VN embedding problem and contacts the underlying components such as OpenFlow controllers, OpenvSwitch database (OVSDB) to create VN components and install forwarding rules in Openflow switches. Once the VN is created, the VN topology information is then stored in the Topology Manager, and information of the VN components (e.g., virtual switches) is stored in the switch manager. Once the virtual infrastructure is scheduled, a stats manager will continuously monitor the status of the virtual infrastructure through the service abstraction layer. Based on the operating conditions, the allocation of each virtual infrastructure may need to be changed over time. For example, the tenant may want to scale up or down the virtual infrastructure at run-time to cope with demand fluctuation.

6.7 SOFTWARE-DEFINED INFRASTRUCTURES

Throughout this chapter, we have assumed that the resources consist of computing and networking resources. However, in slightly different contexts, for example, in virtualizing the wireless and wireless access networks in Figure 6.14, additional resources could include programmable hardware and other sharable high-performance resources. In the SAVI project [20, 21], we address the more general setting where the Software-Defined Infrastructure (SDI) includes heterogeneous virtualized resources that are managed in integrated fashion along with computing and networking. In addition to cloud and network controllers, these SDIs may require controllers for these additional resources, for example, programmable hardware resources [22].

Figure 6.15 show the architecture for the SAVI SDI resource management system (RMS). The SDI manager has overall control of resources of different types, for example, A, B, and C. The external entities request virtual resources from the SDI resource management system through open interfaces. The SDI RMS executes coordinated and integrated resource management for the heterogeneous resources through an SDI manager and a topology manager. The SDI manager performs its management functions based on the resource information provided by the topology manager. Resource-specific controllers (e.g., OpenStack or OpenFlow controllers) are responsible for managing resources of a given type. Each resource controller accepts the high-level user descriptions and manages the resources of a given type. The topology manager maintains a global view of the resources, their relationships, as well as monitoring and measurement data. It enables the SDI manager to perform state-aware resource management.

SAVI is exploring the deployment of applications in a multitier cloud that includes massive core data centers, smart edge nodes, and access networks. SAVI has designed a node cluster that provides virtualized and physical computing and networking resources,

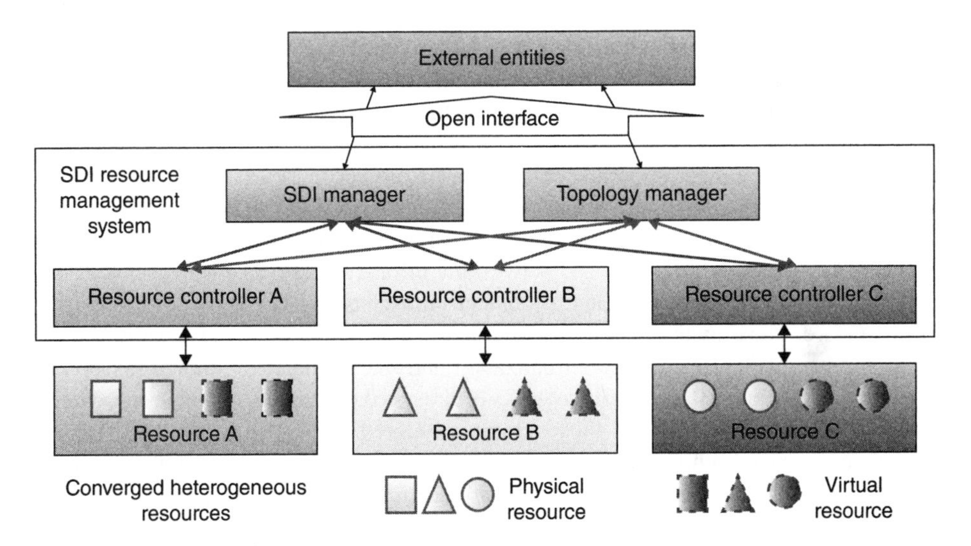

Figure 6.15. SAVI SDI resource management system.

including heterogeneous resources such as Intel Xeon servers, storage, OpenFlow switches, GPUs, NetFPGAs, Altera DE5-Net FPGAs, and ATOM servers. SAVI has implemented the Janus SDI Resource Management System to manage the heterogeneous resources provided by a SAVI node. Janus builds on top OpenStack and OpenFlow. A Canadian test bed has been deployed with nodes in the following universities: Victoria (British Columbia); Calgary and Alberta; Carleton, Toronto, York, Waterloo in Ontario; and McGill in Quebec. The SAVI test bed is supporting research on large-scale applications, multitier cloud computing, architecture of the smart edge, virtualized wireless access, and management of SDI.

6.8 RESEARCH TRENDS AND CHALLENGES

We began this chapter with a discussion of application platforms to provide a holistic view of the broad range of requirements that must be met by future SDN and computing clouds. While great progress has been made in advancing SDN and clouds, here we reiterate several major challenges that remain to be addressed: Orchestration, adaptive resource management, content distribution, and scalability.

Methods for the orchestration of the resources to support distributed applications are in a relatively early stage of development. Methods are needed for the automated determination and allocation of the computing and networking resources for applications. The Heat project in OpenStack is striving to meet this need by developing an orchestration engine for launching cloud applications [23]. A Heat template is used to describe the infrastructure resources required by an application. The Network Functions Virtualization (NFV) concept is being developed to virtualize network node functions that can serve as building blocks to create communication services [24]. Clearly, orchestration is a key element in NFV.

The automated scaling of resources allocated to support an application is essential to achieving the economies of scale that derive from cloud computing and virtualization. Methods are required for the measurement and monitoring of demand and available resources and for the autoscaling of resources. To be implemented, the rich literature in adaptive resource management requires a platform for measurement and monitoring and automated resource management. The Ceilometer project in OpenStack is developing an infrastructure to collect measurements within OpenStack to support monitoring and metering [25]. The SAVI project discussed above is exploring the use of Ceilometer in converged virtualized computing and networking infrastructures.

The collection and distribution of content represents a major driver of current IT infrastructure. The growth in video services and the emergence of Big Data applications necessitate an exploration of the virtualization and management of storage resources. The churn in demand for specific content requires striking a balance between content that is stored remotely in a few sites and content stored broadly in local sites. Various information-centric architectures need to be explored in the context of the multitier cloud infrastructures that are emerging to support application platforms. The huge volumes of content that need to be transferred also motivate the investigation of the optical transport technologies in these new architectures.

The scalability of management systems will be challenged by the continuous growth in application platforms and associated resources. The volume of messaging consumed by management and data collection must be kept to reasonable levels while providing the responsiveness, effectiveness, and reliability required of the management system. This requires further research in management system architectures.

6.9 CONCLUDING REMARKS

The potential benefits of service-oriented computing and the virtualization of resources have spurred intense activity in the advancement of cloud computing and SDN. In this chapter, we have provided an integrated view of how cloud computing and SDN, and specifically OpenFlow, OpenStack, Open vSwitch, and OpenDaylight come together. We have also introduced the SAVI project which explores the notion of SDI that encompasses both cloud computing and SDN to support large-scale applications.

REFERENCES

1. Fox, Armando, et al. "Above the clouds: A Berkeley view of cloud computing." Department of Electrical Engineering and Computer Science, University of California, Berkeley, Report UCB/EECS 28 (2009).
2. Anderson, Thomas, et al. "Overcoming the Internet impasse through virtualization." *Computer* 38.4 (2005): 34–41.
3. McKeown, Nick, et al. "OpenFlow: Enabling innovation in campus networks." *ACM SIGCOMM Computer Communication Review* 38.2 (2008): 69–74.
4. Shenker, Scott, et al. "The future of networking, and the past of protocols." *Open Networking Summit* (2011).
5. Nunes, Bruno Astuto A., et al. "A survey of software-defined networking: Past, present, and future of programmable networks." *IEEE Communications Surveys and Tutorials* 16.3 (2013): 1617–1634.
6. Casado, Martin, et al. "Ethane: Taking control of the enterprise." *ACM SIGCOMM Computer Communication Review* 37.4 (2007): 1–12.
7. Davie, Bruce, and Yakov Rekhter. *MPLS: Technology and Applications*. San Francisco, CA: Morgan Kaufmann Publishers Inc., 2000.
8. Open Networking Foundation, Open Flow Switch Specification version 1.4.0, October 14, 2013.
9. Campbell, Andrew T., et al. "Open signaling for ATM, internet and mobile networks (OPENSIG'98)." *ACM SIGCOMM Computer Communication Review* 29.1 (1999): 97–108.
10. OpenStack Foundation, OpenStack Administrator Guide, Havana, April 6, 2014.
11. Pfaff, Ben, et al. "Extending networking into the virtualization layer." Hotnets. New York, October 2009.
12. Azodolmolky, Siamak, Philipp Wieder, and Ramin Yahyapour. "SDN-based cloud computing networking." 2013 15th International Conference on Transparent Optical Networks (ICTON), IEEE, June 23–27, Cartagena, Spain, 2013.

13. http://archive.openflow.org/wk/index.php/Rate_Limiter_Proposal. Accessed November 21, 2014.
14. Wood, Timothy, et al. "The case for enterprise-ready virtual private clouds." Usenix Hot-Cloud Proceedings of the 2009 Conference on Hot Topics in Cloud Computing, June 14–19, San Diego, CA, 2009.
15. Luo, Mon-Yen, and Jun-Yi Chen. "Software defined networking across distributed datacenters over cloud." 2013 IEEE 5th International Conference on Cloud Computing Technology and Science (CloudCom), vol. 1. IEEE, December 2–5, Bristol, 2013.
16. Baucke, Stephan, et al. "Cloud Atlas: A software-defined networking abstraction for cloud to WAN virtual networking." 2013 IEEE 6th International Conference on Cloud Computing, June 28–July 3, Santa Clara, CA, pp. 895–902, 2013.
17. Sadasivarao, Abhinava, et al. "Bursting data between data centers: Case for transport SDN." 2013 IEEE 21st Annual Symposium on High Performance Interconnects, August 21–23, San Jose, CA, pp. 87–90, 2013.
18. Linux Foundation Collaborative Projects, Open Daylight Technical Overview, www.opendaylight.org/project/technical-overview. Accessed November 21, 2014.
19. Zhani, Mohamed Faten, et al. "VDC planner: Dynamic migration-aware virtual data center embedding for clouds." 2013 IFIP/IEEE International Symposium on Integrated Network Management (IM 2013), IEEE, May 27–31, Ghent, Belgium, 2013.
20. Kang, Joon-Myung, Hadi Bannazadeh and Alberto Leon-Garcia. "Software-defined infrastructure and the future CO." ICC Communications Workshops, Budapest, Hungary, June 2013.
21. Kang, Joon-Myung, Hadi Bannazadeh and Alberto Leon-Garcia. "Software-defined infrastructure and the SAVI testbed." TridentCom 2014, Guangzhou, May 2014.
22. Byma, Stuart, Hadi Bannazadeh, Alberto Leon-Garcia, J. Gregory Steffan, Paul Chow. "Virtualized reconfigurable hardware resources in the SAVI Testbed." Tridentcom 2014, Guangzhou, May 2014.
23. Heat: Openstack Orchestration, https://wiki.openstack.org/wiki/Heat. Accessed July 30, 2014.
24. Ersue, Mehmet. ETSI NFV management and orchestration, https://www.google.ca/webhp?sourceid=chrome-instant&rlz=1C5MACD_enCA568CA577&ion=1&espv=2&es_th=1&ie=UTF-8#q=nfv%20orchestratino. Accessed July 30, 2014.
25. Ceilometer: OpenStack Telemetry, https://wiki.openstack.org/wiki/Ceilometer, Accessed July 30, 2014.

7

MOBILE CLOUD COMPUTING

Javeria Samad, Seng W. Loke, and Karl Reed

Department of Computer Science and Computer Engineering, Latrobe University, Melbourne, Australia

7.1 INTRODUCTION

Cloud computing opened the doors for a paradigm shift for the ways in which systems are deployed and used. It has made possible utility computing with infinite scalability and universal availability of systems [1]. Mobile cloud computing (MCC) has taken this to a step further by enabling the users to carry on their tasks irrespective of their movement and location [2, 3]. Despite the increasing popularity and usage of MCC, there are certain issues inherent with it that still haunt the mobile cloud community, making it difficult to utilize the full potential of the clouds. These issues or "risks" span the whole structure and life cycle of mobile clouds and could be as varied as security, operations, performance, and end users.

This chapter aims at exploring MCC further and to highlight the risks related to mobile clouds, in addition to the risks normally associated with system development. While we briefly present practical solutions for most of these issues via some standard method or approaches, suitable for the respective issues, the aim is to point out the need for systematic risk analysis and management frameworks for such applications.

Cloud Services, Networking, and Management, First Edition.
Edited by Nelson L. S. da Fonseca and Raouf Boutaba.

7.1.1 Significance/Motivation

Cloud computing mainly focuses on how to best manage the computing, storage, and communication resources shared by multiple users virtually; whereas MCC works by applying cloud computing solutions using resources available in mobile environment. It allows the execution of mobile applications, data storage, and processing on external/remote resources rather than on the mobile device itself, while allowing free movement of the user/mobile device. MCC requires functional collaboration between different mobile devices. It requires the mobile devices to be aware of "presence," "status," and the "context" of other portable devices within their network, so as to provide the best possible ad hoc communication environment [2].

The complexity and dynamism of a mobile cloud system poses many risks. At system level, these include the risks of connectivity, limited resources, security, and limited power supply. As the system complexity increases, both the technical and nontechnical risks increase, and so is the need to manage these risks. The ad hoc nature and mobility [2] in MCC environments means that the development of these system need more rigorous and specialized risk management to deal with all the risks. This can further burden the developers of MCC frameworks and applications. In addition to the complexity of mobile cloud infrastructure, they also have to deal with the risks at framework/application level including but not limited to efficient job distribution, virtualization and scalability, and so on.

In the current scenario, from our review so far, we conclude there is no available formal risk management process in place to deal with the risks of MCC. As with any development and deployment activity, an effective risk management is integral to the success of any MCC system; it's a critical element while designing MCC systems. However, the literature review shows that the current work on mobile cloud systems focuses more on cost and resource savings, and there has been little progress toward the development of mobile cloud "aware" risk management methodologies. There is a need to make the mobile cloud developers and users realize the importance of an effective and efficient risk management in place. Risk management not only protect the organizations from various risks but also plays a critical role in enabling mobile cloud providers to achieve their goals by improved decision making through up-to-date risk reporting and also to help meet end users' quality-of-service requirements. An efficient risk management process can also protect the providers from risks of cost overruns during the whole system life cycle, and can also improve customer satisfaction/confidence in a delivered system.

The organization of this paper is as follows: Section 7.2 provides an overview of the MCC domain and provides a discussion on different selected mobile cloud frameworks and their categorization. Section 7.3 defines risk management and presents an analysis of risk factors currently prevalent in the MCC domain. An illustration of how these risks can affect an application is also presented in this section. Section 7.4 presents an analysis of mobile cloud frameworks (surveyed in Section 7.2) from a risk management perspective and also discusses the effectiveness of traditional risk approaches in dealing with MCC risks. Section 7.5 summarizes the review and concludes.

7.2 MOBILE CLOUD COMPUTING

Cloud computing refers to the provisioning of computing capabilities as a "service" (instead of product) to users via Internet and Web technologies. Cloud computing has been defined by different authors in various words. Some see it only as an enhancement to multiple existing technologies, whereas some others are very enthusiastic about the potential of cloud computing. A simple definition of cloud is presented in François Ragnet and Conlee [4] as "a model for enabling convenient, on-demand network access to a shared pool of configurable computing resources (e.g., networks, servers, storage, applications, and services) that can be rapidly provisioned and released with minimal management effort or service provider interaction." Vaquero et al. [5] has provided a comprehensive comparison of different cloud definitions prevailing in the literature to clarify what cloud computing really is. Based on their explanation, the cloud computing concept can be summarized as a large pool of easily usable, accessible, and dynamically scalable virtualized resources, offered as pay-per-use model, allowing for optimum resource utilization.

Basic cloud computing is an amalgam of various technologies all put together to change the ways in which IT infrastructures are built. The cloud is different from other older technologies (i.e., Internet, grid/distributed computing) in a sense that with clouds the users can use service when they need it and for as long as they need it. Cloud computing works on a mechanism of "utility-based services" where you pay only for duration and amount/type of services used. Also, unlike these technologies, cloud computing provides architectural, domain and platform independence.

The key characteristics of any cloud infrastructure are "abstraction" and "virtualization." Cloud computing must be able to allow the users to use computing services on shared resources virtually, in a dynamically scalable way, without having knowledge about location of, or the hardware and software resources involved, database design, and storage infrastructure. With cloud computing, the users can enjoy much needed elasticity (scalability), resource sharing/pooling, on-demand service access on utility basis, and broader network access and availability. Abstraction and virtualization are provided by individual cloud vendors at different levels, with some allowing total flexibility and some others offering somewhat restricted control to the users.

Cloud computing can be seen from two perspectives: (1) the way the clouds are deployed and (2) the services that are delivered by the cloud platform [6, 7]. Cloud can be deployed as either: *public cloud*—where the cloud infrastructure is available to general public and the cloud provider and consumer usually belong to different organisations; *private cloud*—where cloud infrastructure is limited to a private group (or a single organization); or *hybrid cloud*—which combines services of public and private clouds.

The services offered by cloud can be categorized in three different ways that is (1) platform as a service (PaaS), (2) software as service (SaaS), or (3) infrastructure as service (IaaS), which are self-explanatory for the type of services they offer [5–7]. In Huang et al. [8], the authors have discussed yet another approach for mobile cloud service models that classifies mobile cloud in three models, that is, mobile as service consumer (MaaSC), mobile as service provider (MaasP), and mobile as service broker

(MAASB). The authors advocate using a more user-centric approach for ensuring mobile cloud design principles.

Mobile cloud computing is an enhancement of cloud computing in which the capabilities of cloud computing are realized using mobile communication infrastructure [9]. Basic MCC is based on same techniques as that of mobile networks. A mobile cloud differs from simple cloud computing in the same manner as mobile networks differ from wireless networks.

The popularity of mobile applications has increased dramatically in past decade, allowing users to use applications plus mobility. Mobile applications provide much needed freedom to the users as it enables them to use these applications whenever they need and wherever they need. Such applications span all walks of life including but not limited to entertainment, gaming, learning, healthcare, and commerce. Despite the ease of use and popularity, mobile users still suffer from the issues such as limited power supplies, limited storage space, and limited computing resources on their mobile device [10, 11]. Answer to this problem would be to export all the complex processing and storage to some external server or "cloud" instead of mobile device itself. Cloud computing provides one such solution.

MCC is formally defined as a "model for transparent elastic augmentation of mobile device capabilities via ubiquitous wireless access to cloud storage and computing resources, with context-aware dynamic adjusting of offloading in respect to change in operating conditions, while preserving available sensing and interactivity capabilities of mobile devices" [12].

In more general terms, MCC refers to the usage of cloud computing on mobile devices, independent of the movement of user. All the storage and complicated processing is done external to the mobile device, thereby saving tremendously on computing resources and power supply for mobile device. The added benefit a mobile cloud presents is mobility; however, a mobile cloud cannot be fully advantageous, if it doesn't cater for the other functionality aspects associated with conventional clouds, that is, adaptability, scalability, and availability. The primary aim of MCC is to merge the advanced computing and communications technologies, to provide users with a seamless computing environment.

MCC provide users with a number of benefits including sharing the resources and applications without investing huge amounts of money on specialized hardware and software. Also, as most of the complex processing is done externally, the users can enjoy cost reductions for computing power as well [13]. Instead of using a remote cloud for processing, an ideal mobile cloud scenario could make use of a "local" cloud made up of surrounding mobile devices. This would also eliminate the dependence of device on remote servers, and possibly reduce data transfer latency.

As with any other wireless mobile networks, MCC faces challenges as well. A primary challenge for any mobile cloud environment is providing constant network availability, irrespective of user movement, which can be difficult or impossible at times. However, the emerging technologies are addressing this problem intensively with some systems providing "caching" facilities for mobile applications so that users can continue work seamlessly, even if connection is disrupted momentarily.

At present, most mobile cloud applications depend on remote servers for processing and storage, exposing the users still to the risk of loosing precious resources over connectivity with remote servers. This remote connectivity also presents the issues of the bandwidth, time delays, costly data services, and context ignorance. Mobile cloud users often need information services relevant to their recent/current contexts (like location and time). An ideal mobile cloud scenario should be capable of utilizing the resources from a more local cloud and accessible mobile devices, if this is better, instead of relying on remote servers [13, 14]. A temporary local mobile cloud made up of eligible mobile devices present in the surroundings at the same time can solve the problems of bandwidth, costs, and time delays, while ensuring more context aware solutions for users [12, 15]. Technology currently in use for mobile and sensor-based networks can be utilized for development and deployment of this local mobile cloud setting.

7.2.1 Types of Mobile Clouds

The mobile cloud setup can be seen in three different ways. More or less similar categorization is also proposed in Fernando et al. [16]:

1. Client server: In this approach, a mobile device works as a thin client for a remote server, where the processing for mobile applications is done on remote servers. All the public clouds like Amazon's EC2, Microsoft Azure Platform, or Google AppEngine can be examples of such client-server cloud models [6].
2. Peer to peer: In this approach, all the eligible mobile devices can act as resource servers for other eligible mobile devices within the surroundings. All these mobile devices make up a local cloud, thus eliminating the need to connect to remote servers. This is the ideal scenario, ensuring highest level of mobility [12, 14, 17].
3. Hybrid approach: As with other wired and wireless networks, a hybrid approach comprises the features of both client server and peer-to-peer approaches. This approach works by enabling a mobile device to act as a client for a local cloud, which in turn connects to a remote server. However, as a mobile device might not connect to a remote server directly, it can bypass the remote connectivity issues. Kovachev's mobile community cloud [15] can be an example of such a hybrid approach.

7.2.2 Mobile Cloud Application Models and Frameworks: A Brief Overview

There are various existing MCC frameworks and application models, each trying to provide solutions for or improve some of prevalent MCC concerns [16, 18, 19]. Here, we present a brief overview of the current approaches in MCC and provide an analysis for a subset of these methodologies, from the risk management perspective. We discuss three aspects of MCC: mobile cloud architectures, communication mechanisms or connection protocols, and inherent risk management strategies within MCC frameworks. These aspects are useful and significant for analysing the current approaches from the risk perspective, as the most common risk contributors can be the underlying structure and technology itself.

7.2.2.1 Mobile Cloud Architectures. An MCC architecture refers to the different approaches used by multiple frameworks for their respective job distribution and processing. Various approaches have been used in literature to distribute the jobs effectively between mobile device and clouds dynamically; however, we will restrict ourselves to the approaches adopted by the frameworks surveyed in this review.

The MCC frameworks and concepts surveyed in this chapter can be categorized into two groups:

1. that use application partitioning/client server such as Hyrax, Spectra, Chroma, Alfredo, CMCVR, Cuckoo, and MWSMF;
2. that use VM technology such as Cloudlets, CloneClouds, MAUI, and MobiCloud.

The first category is directly related to previously discussed client-server and peer-to-peer mobile cloud types. An interesting observation, however, is that, in some situations the *virtualization* can be mapped on to the "Hybrid" mobile cloud type as it sometimes incorporates the characteristics of client-server or peer-to-peer technologies. Moreover, some frameworks such as MAUI incorporate the characteristics of both categories collectively.

The MCC frameworks and concepts belonging to each category are discussed in the following sections.

APPLICATION PARTITIONING/CLIENT SERVER This category comprises the frameworks that use client-server approach for task offloading. Frameworks belonging to this category work by a mechanism of application partitioning in which the task is divided for processing between mobile device and remote server. In most cases, the criterion for this division is embedded in the code; however, in a few other cases, it can be decided at runtime.

Apache Hadoop is one such software framework that supports distributed applications and data-intensive processing across large sets of independent computers, and is capable of dynamic scalability for up to thousands of machines in minimal time [20]. It's claimed to be capable of automatic failure detection and handling through Hadoop's NameNode and Hadoop Distributed File System (HDFS); the task failures are handled through node-replication mechanisms. Hadoop is a free implementation of MapReduce [21]. Its significance is eminent from the fact that many of the current MCC frameworks and models are based on Hadoop and MapReduce.

One such platform based on Hadoop is Hyrax [22], which supports cloud computing on Android smartphones. Hyrax works by utilizing a resource pool of multiple mobile devices present in the surroundings. Hyrax designers have discussed how such a mobile cloud could be formed, enabling applications to utilize computational resources of all the mobile devices (making up cloud) collectively. The key processes in any Hadoop implementation are NameNode, JobTracker, DataNode, and TaskTracker. A Hadoop clusters works via master node and slave or worker node (that works as DataNode and

TaskTracker). Distributed processing is supported through Hadoop's MapReduce implementation where a job is divided into independent activities and processed separately. The Reduce functions processes the outputs from each of these tasks and produces the results collectively that are then stored in HDFS. The fault tolerance is provided in Hyrax via the fault tolerance mechanisms of Hadoop.

Another framework that uses application partitioning via a client-server technique is proposed in Jan et al. [23]. It discusses the Alfredo framework for distributed processing of applications between mobile devices and remote servers. In Alfredo, the application presentation layer or UI remains at the mobile device, but the data processing is done on servers. Alfredo framework is based on R-OSGi which is a middleware platform allowing applications to be distributed in multiple modules. R-OSGi is itself an extension of the OSGi model, allowing apps to run on multiple virtual machines instead of on one machine. When a device requests some application, the application's details and relevant services information is sent to client's (mobile device) "renderer" which in turn generates the UI accordingly. The services are run in one of two ways: either on client side, or in server in which case an ad hoc proxy is created for the client to access these services on the server. They have not discussed any risk management specifically.

The Spectra framework presented by Flinn et al. [24] is also a client-server architecture where the mobile device offloads its processing to a server via communication protocols. One major drawback of this approach is that the services have to be preinstalled over servers. Spectra is not suitable for very fast response applications, rather it targets apps that can afford 1–2 s of delays. Spectra works by matching resource pool with service requests to predict if the applications should execute locally or should be offloaded to remote servers, for maximum efficiency. As with MAUI [25], for Spectra, the developers need to specify which modules can be potential offloading candidates or which components can benefit from offloading.

Another framework based on application partitioning techniques, much like MAUI and Spectra, is Chroma [26] that involves offloading individual RPCs to the cloud (or remote servers). Because of coarse-grained execution, less offloading overheads are involved. Chroma is a tactic-based remote execution system where the "tactics" are the useful partitions of the system and these tactics vary in the amount of resources used and quality of apps. Like Spectra, it responds very quickly to the changing resource needs and in Chroma too, the developers have to manually specify the methods for offloading. It is also similar to Spectra in resource monitoring and predictions. However, like Spectra and Alfredo, there isn't much discussion on risk management or fault handling.

The work by Satish et al. [27] has also proposed a client-server architecture based on the Mobile Web Services Mediation Framework (MWSMF) for Mobile Enterprises which consist of Mobile Hosts acting as service providers for client devices. Whenever a mobile device requests some services, these mobile hosts provide seamless integration of requesting nodes with the enterprise. The focus of this framework is to provide proper quality-of-service and discovery mechanisms for successful adoption of mobile web services into enterprise environments. MWSMF uses Enterprise Service Bus (EBS) technology to act as intermediary between the web service clients and the Mobile Hosts within the Mobile Enterprise.

The virtual mobile computing (VMC) framework is presented in Huerta-Canepa and Lee [28]. Like Hyrax, VMC is also based on Hadoop and supports virtual MCC. Mobile device "location' is the basic aspect in this framework and much like conventional MCC, it relies on neighbouring mobile devices. Because of this, the framework requires continuous (and fast) discovery and selection of suitable mobile devices for computation offloading. Distribution of tasks is carried out via Hadoop. VMC framework doesn't currently support risk management.

Another client-server based "Cuckoo" framework is presented by Roelof et al. [29] for computation offloading and targets the Android platform. The Cuckoo Resource Manager is responsible for identifying, selecting, and registering remote resources for offloading. The authors have also presented the evaluation of this framework on real-life applications.

Further frameworks based on application partitioning are proposed in Luo [30] and Zhang et al. [31]. The Cloud-Mobile Convergence for Virtual Reality (CMCVR) framework [30] allows user-friendly convergence of mobile devices to cloud resources. Like a few others, this framework also works by using a mechanism of task partitioning; however, the unique attribute is "scanning tree" which is a data structure for managing cluster nodes at multiple levels. This framework targets mainly media applications and the main focus is dynamic provisioning of context-aware multimedia services and rendering comparable to virtual environments for mobile users. The framework proposed by Zhang et al. [31] works by partitioning a single application into elastic components that can be executed dynamically. The elastic applications consist of one or more "weblets" (application partitions/components) that function independently but communicate with each other. These weblets are platform independent. The application weblets and resource demands are monitored continuously by the "elasticity manager" which then uses this information for decision making for where and how to launch the weblets. However, unlike CMCVR, the authors of Weblets framework have also proposed the authentication and communication mechanisms for elastic applications.

VM Based The frameworks belonging to this category work by offloading the tasks to a server having preinstalled image of VM of the mobile device that initiated the task. Because of virtualization, such offloading is usually seamless but slower.

A cloudlet-based framework concept is proposed by Satyanarayanan et al. [14]. The "cloudlets" approach presented by authors tries to overcome the latency issues by making use of a local cloudlet that is connected to the Internet and comprising nearby mobile devices. This technique reduces the latency and ensures speedy response. Instead of connecting to a distant server for application processing over the Internet, this approach uses the local cloudlet and enables the mobile device to perform all processing at just one-hop latency. If for any reason the local cloudlet is unable to carry out the required processing and computations, the mobile device can go into a safe/failure mode and use the distant cloud's services for the time being. The cloudlets are designed to keep only cached copies of data unlike clouds. The authors have also discussed the various scenarios where such a cloudlet can be deployed in different ways. To enable simple (self-)management without compromising application diversity and potential, "transient cloudlet customization using hardware VM technology" is applied where the guest device's software environment is hidden from that of cloudlet's software infrastructure. The cloudlet customizes each job and later cleans itself up after each operation. Like the cloudlet framework, the CloneCloud framework proposed in Chun and Maniatis [32] also addresses the challenges of mobile devices limitations via an "augmented execution" technique. In this approach, the execution is offloaded to the cloud containing/running the "clone" or replica of mobile devices' (smartphone's) software. Depending on the complexity of tasks, either full or partial execution is offloaded to the cloned cloud. It gives the illusion of increased computational power to the users. The authors have also discussed the categories of CloneClouds for possible augmented executions. However, though they have slightly touched upon the concerns relevant to each framework, but both these frameworks have failed to address the potential risks, and hence, risk avoidance and management.

MAUI framework [25] can be considered as an improvement in CloneCloud and cloudlet frameworks as unlike these, MAUI combines virtual machine migration with code partitioning. This approach allows the required flexibility but within necessary limits of control. The developers annotate which methods to offload while programming; however, the decision to offload the methods is made at run-time on the basis of parameters like profiling information, connectivity, bandwidth, and latency. Unlike many other frameworks, the decision making for offloading is done at the single method level instead of complete modules. This approach can be prone to risks as in some situations as the single method-based decisions cannot represent the whole picture, and an effective decision making should consider more than just one method at a time.

Mobicloud [33] is a mobile cloud framework that treats mobile devices as service nodes to improve ad hoc networking operations, and increases the capability of cloud computing for securing mobile ad hoc networking (MANET) applications. This framework is also based on virtual machine technology where every mobile node is treated as a virtualized component and mirrored in cloud as one on more extended semishadow images (ESSI). These ESSIs are not necessarily the same as virtual images and can be either exact clones or partial clones. MobiCloud provides intermediary services for access management, security isolations, and risk assessment and intrusion detection for MANETs. In short, it works to provide security service architecture in multiple security domains. They have also presented virtual trusted and provisioning domain (VTaPD) to enable isolated information flow and access controls across multiple virtual domains. However, again we see a lack of potential risks discussion.

7.2.2.2 Communication Protocols. Three types of common network protocols in MCC are Wi-Fi, Bluetooth, and 3G/4G. These terms refer to the way the mobile devices can connect to the Internet, and to each other. Each of these communication protocols have their own benefits and drawbacks, and this subcategorization can be useful in understanding whether if/how the use of some specific communication protocol within a framework can possibly contribute toward the risk of using a relevant framework.

WI-FI Most of the frameworks discussed earlier such as Hyrax, Alfredo, Spectra, Chroma, VMC, Cuckoo, Cloudlets, CloneCloud, MAUI, and MobiCloud have suggested Wi-Fi as a communication mechanism. Although Wi-Fi provides improved performance in terms of reliability and increased bandwidth and data rates, the frameworks using Wi-Fi as their embedded communication protocol should be comparatively more prone to risks of security threats and interference due to external objects. Also, the Wi-Fi connectivity is dependent on the availability of hotspots, unlike 3G which can be connected from anywhere. Related to Wi-Fi, is also Wi-Fi Direct for mobile to mobile communication which has higher bandwidth than Bluetooth and a much longer range, tens of metres or more, compared to the often noted 10 metres of classic Bluetooth. Wi-Fi can also be used for tethering purposes.

3G/4G 3G/4G provides users with more consistent networking conditions and better security; nevertheless, it takes its toll in terms of slower data transfers, increased battery drainage, and higher costs. For MCC, it provides users with advantages of connectivity from anywhere, anytime; however, its battery consumption rates and the response times make it a secondary choice for mobile cloud developers. The frameworks currently employing this approach have generally suggested using this in conjunction with Wi-Fi for optimum access and connectivity. None of the frameworks have used 3G/4G as their primary communication medium, but some of the frameworks like MAUI and CloneClouds have used this in their experiments to compare performances of different approaches.

BLUETOOTH Despite the advantages of Bluetooth, like ease-of-use and no-cost wireless access, Bluetooth is not a very popular medium for widespread Internet connectivity because of lower ranges and potentially more susceptible to interference. With an exception of Alfredo, none of the surveyed frameworks have utilized Bluetooth exclusively and the only mention we find is within the Cuckoo framework as it uses Ibis middleware which can be run with any of the above discussed communication protocols. Bluetooth low energy (BLE) is more energy efficient than Classic Bluetooth, and generally more energy efficient than Wi-Fi-based protocols though with lower bandwidth and range than Wi-Fi protocols.

7.2.2.3 Risk Management Strategies. An interesting thing to note is that with an exception of Zhang et al. [31], none of the frameworks have discussed the risks associated with their use. However, we see some mentions of fault tolerance and risks in Hyrax and MobiCloud, respectively. Hyrax framework implies fault tolerance implementation as it's based on Hadoop and uses the same fault tolerance mechanism to recover from failures. However, no explicit mention of relevant risks and risk management is given. Similarly, the MobiCloud framework has proposed using context-aware information for aiding risk assessment and intrusion detection. Despite this, no thorough discussions of risks and risk management have been provided and little attempt has been made to generalise these beyond the scenarios described.

7.2.3 Discussion

As stated earlier, except for the framework proposed by Zhang et al. [31], none of the frameworks have discussed the risks associated with the use of that particular framework.

To analyse each framework for inherent risks, we have tried to identify the risks associated with each aforementioned aspect, as all frameworks belonging to same category share, on most part, similar risks.

7.3 RISKS IN MCC

This section first explores the risk management concept in general and provides an insight into the "risk" and "risk management" definitions and the basic steps in any risk management methodology. Then, we present a survey of risks inherent in cloud and MCC domains. Identification of MCC risk factors is needed to fully understand what types of risks are being faced in the MCC domain and what is the pattern and intensity of these risks, at present. Recognizing the loopholes in current MCC technologies and analyzing the identified risks for their causes and solutions, could be a key starting point to explore the possibilities of a risk management system that would be able to deal with these risks.

7.3.1 Risk Management

Risk can be defined as the "possibility" of something happening that can affect the outcome negatively; it's measured in terms of "probability" and "impact" [34, 35] and usually derived by formula: $\mathrm{RE} = \mathrm{P(O)} \times \mathrm{L(O)}$

where RE is risk exposure, P(O) is probability of negative outcome, and L(O) is the loss or impact of that negative outcome [36].

This formula can be taken as a standard risk calculation device, as the similar formulas are being used in other domains as well for calculating respective risks, for example, finance, insurance, and health. The presentation of these formulas could be slightly different in different domains, but the basic logic is similar, comprising two basic elements: probability and impact [37–39].

Risk management is the process of managing risks in a given system with the aid of formal processes, methods and tools, for example, providing a disciplined environment for continuously analyzing the risk factors, calculating the relative importance of each risk item and designing strategies to deal with these risks. Any risk management system usually comprises these basic activities: (i) risk identification, (ii) risk analysis, (iii) treatment, and (iv) monitoring and control.

Risk identification refers to identifying the risk factors, that is, proactively diagnosing what could potentially go wrong. *Risk analysis* includes calculating the likelihood and impact of identified risk factors and its prioritization, whereas *risk treatment* refers to exploring the possible treatment options for prioritized risk factors and detecting best treatment solutions. *Monitoring and control,* on the other hand, is a continuous activity carried out throughout the risk cycle; it involves overall risk planning and monitoring for any change in status or priority of identified factors and to keep a look-out for any new risks surfacing. Such proactive decision making reduces the system's exposure to risks and minimizes the potential loss from these risks. A formal risk management process provides an auditable system for risk mitigation and contingency [40].

As with any new technology, cloud computing also has some associated risks that need to be managed for successful and efficient utilization of clouds. The complexity of cloud infrastructure poses many serious risks, and there is an ever-increasing need for managing these risks effectively and proactively. Shipley [41] holds the view that at present cloud computing is implemented without any proper risk management. The same is true for MCC. In MCC, the intricacy of the system makes it further risk prone.

In the previous section, we have surveyed and analyzed a sample of mobile cloud frameworks and models. In this section, we will be discussing the multiple risk factors that have been identified so far within cloud and MCC domains. The analysis of frameworks on the basis of their ability to deal with prevalent risks will then be presented in the following sections. We will also be categorizing these risk factors according to the nature of individual risk items. This section has been divided into two parts: the first part will present the risks that are common to both cloud and MCC domains. The second part will present the risks specific to MCC area.

7.3.2 Risks in Cloud Computing (Inherited by Mobile Clouds)

Despite the benefits of cloud computing and its potential to improve efficiency and productivity, there's some reluctance to its usage. It's because of the fact that it's a relatively newer technology and there's no formal mechanism or standards for managing the risks associated with cloud usage. Many of the identified risk factors (Sections 7.3.2.1–7.3.2.3) are common to both cloud computing and MCC systems. This is because the basic attributes of both domains are similar; as MCC itself is based on the fundamentals of cloud computing technology, so in addition to the basic characteristics, MCC inherits the risks of the cloud computing domain as well. This phenomenon makes it important to explore the risks in the cloud computing domain as well to have coverage of possible risk factors that can affect MCC systems.

There are a number of sources in the literature that discuss the different types of cloud computing risks. These risks factors are classified into three categories: *security risks*—including all the risks related to the security aspect of cloud computing networks and data, *performance risks*—including the risks related purely to performance attributes of a cloud infrastructure, and *legal/environmental risks*—including risks related to legislation and the operational environment of cloud providers and users. The risk factors are represented in separate tables according to their categorization. Sections 7.3.2.1–7.3.2.3 represent the risks common to both the cloud computing and MCC systems, whereas Section 7.3.3 discusses the risks specific to MCC.

7.3.2.1 Security Risks. The typical characteristics of cloud environments such as abstraction, virtualization, shared resources, and ad hoc nature make it very difficult to implement proper security and safety mechanisms. Also, with cloud the users usually having no knowledge or control over these mechanisms makes the situation even worse. As with any other network, "security and privacy" is one of the biggest risk factor that needs proper consideration and management. This category comprises the risks related to security and privacy aspects of the infrastructure from both the user and provider's perspective; this includes the risks to communication networks itself and the data.

- *Unauthorized access*: As most of the data storage and processing is done externally, it's difficult to implement physical and logical controls over access rights which bring with it the risk of unauthorized access. This issue is inherent in all remote and distributed systems, but the abstraction and virtualization of resources in cloud environments makes it even more difficult to deal with [1, 33, 41–49].
- *Security defects in technology itself*: Failure to implement security controls that are essential to protect customers' assets is yet another risk factor that needs to be managed properly. At present, the cloud providers tend to keep their functional procedures and policies a secret, and so there's no way of knowing the level of security provided to its users by the vendor [41, 50].
- *Security defects in Web services*: Cloud applications are usually provided as services over the web. However, unlike other Web-based applications, the cloud applications are not user specific, and hence present serious vulnerability issues. Potential loopholes in security of Web services pose a potential risk for cloud computing. Moreover, the attackers can also use the weakness of web application security for gaining access to other users' data as well [50].
- *Leak of customers' information*: The risk of losing customer's private information (i.e., passwords and profile information) to unwanted attackers is yet another risk shared equally by the cloud vendors and users [41, 51].
- *Leak of proprietary information*: This risk of losing confidential proprietary information is very serious, especially if the organisations involved are government or other national agencies. A survey given by Shipley shows that percentage of commercial organisations reluctant to use cloud computing for fear of losing proprietary information is 28%. Understandably, this percentage will be much higher for security critical organizations [41, 45, 50].
- *Data location*: In case of clouds, the cloud users are never sure of the exact location of their data storage. This makes data location an important security issue. Some of the authors perceive data location as major performance and legislative risks as well [1, 42, 43, 47, 48, 50].
- *Physical location of the system*: As with data, the physical location of the system infrastructure (e.g., servers) is not made public by cloud providers. Being unaware of the location of your system can sometimes become a security issue [1].
- *Data segregation/isolation*: Cloud computing is based on the notion of shared resources; multiple users' data can be stored on the same servers, making data segregation one of the biggest risk factor faced by cloud users. It makes the system prone to the risks such as unwanted access to one user's data by another [33, 42, 43, 47, 50–52].
- *Data recovery*: Failure to recover the data and services properly after some disaster can be a problem. It's different from "faulty backup" risk factor as sometimes, despite proper backup mechanisms, data are lost due to other legislative or environmental reasons. For example, consider a scenario where many cloud providers (primary) themselves rely on cloud services of other bigger providers (secondary). In such cases, if the intermediate primary cloud provider goes out of business, it might not be possible for its users to retrieve their data as they were not the direct

customers of the secondary cloud provider and hence have no access rights on the data hosted by the secondary cloud provider, even though they are the owners of that data. Such legislative gaps can result in data losses. Similarly, if the secondary service provider goes down, it can also cause the service to become unavailable, creating risks for data recovery [42, 43, 48].

- *Weaknesses of browsers*: Cloud computing uses Web-based browsers for service provisioning. The weaknesses of browsers is one of the security risks associated with clouds as any attacker can use these weaknesses to get access to confidential data [43].
- *Data security*: Failure to have an effective data security model in place can be disastrous to any cloud vendor or users. Sharing of resources, unawareness of data location, and malicious attacks to the infrastructure from inside or outside the cloud provider organisation all contribute to the risk of data security. In addition to usual network-related security issues, abstraction and virtualization of cloud infrastructure make the data even more vulnerable [13, 50, 51, 53].
- *Network security*: Network security is an important risk factor because if the network is compromised, the whole of the cloud infrastructure will be compromised. Networks can be attacked in many ways (e.g., spoofing and sniffing), and any loophole in network security can bring the whole cloud to risk [50, 54].
- *Data integrity*: This refers to the accuracy of data that is managed by the clouds. Issues of shared storage can affect data integrity adversely, making it more risk prone [45, 49, 50].
- *Faulty backup mechanisms*: Faulty data backups and backup procedures can cause problems. The cloud is typically a pay-as-you-go service which makes backup more risk prone as it has to keep up with ever changing user environment continuously. Any minor loophole can be disastrous to the security of the system data [13, 50].
- *Insecure/incomplete data deletion*: Cloud computing is utility-based service, and every time a user leaves the system, the cloud infrastructure should be able to delete that user's relevant data and reallocate resources to other users. However, the failure to delete the leaving user's data properly or completely is a risk that can give rise to unauthorized access of one user's data to other [51].
- *Natural disasters*: It's a risk as natural disasters can cause interruption to cloud service availability, for example, Amazon EC2 June 2009 incident. It is often uncontrollable, but there should be proper risk management planning to deal with such situations [43].
- *Malicious insider*: Cloud service providers usually place multiple users' data in one place. This can be very risky as any malicious insider can just attack one single location to gain access to thousands of customers/users' data. The situation will be even worse if the customer is an organization as by attacking a single point, a hacker can access complete organizations' information [43, 52].

Most of these risk factors are similar to those in general communication networks and applications risks. This is mainly because the cloud infrastructure is itself based on

Web and wireless network technology and all the inherent risks of those domains, are propagated to cloud infrastructures as well. However, owing to the scalable and flexible nature of clouds, these risks are further magnified in the cloud computing domain. Some of these risk factors may be common to other categories as well; it's because different authors have slightly different perspectives regarding these risks, with some considering a factor as security risk and others may perceive it as performance or legislation risk (e.g., for some "data location" is a security risk while others perceive it as legislative or performance risk). To cover all different perspectives, we have stated these factors sometimes in multiple risk categories, with each representing the related school of thought.

7.3.2.2 Performance Risks. This category represents the risks related purely to performance attributes of cloud infrastructure. This comprises the risks of reliability, usability, and efficiency, which can directly affect the performance (i.e., quality and effective execution) of the system. Application or system performance is perhaps the most important risk, every cloud provider and user should consider. This is the only risk that is mutually dependent on all other risk factors (or most of it). Occurrence of any risk factor is a risk to effective performance of system and application.

- *Features and general maturity of technology*: Although getting increasingly popular, MCC is still in its early stages and needs maturity in terms of processes and functionality. For the same reason, the mobile cloud-based services/applications are also still in their premature stages. This immaturity of MCC technology and services is comparatively more risk prone than other traditional computing platforms. Any weaknesses or loopholes in technology are a major risk to both the providers and customers equally [41, 43, 50].
- *Data location*: In MCC, the data isn't located on the same premises as the organisation itself. Also, the location of data isn't known to the users. In such case, there is always the risk that data access and also the transfer and processing would be more complex and more time consuming, than if it's located locally [43, 50].
- *Data segregation/isolation*: Besides being a security and privacy issue, data isolation is a risk to application performance as well. If the data isn't properly isolated, there is a risk that user application is unable to access required data efficiently, resulting in delayed processing, and so its performance would be degraded [33, 42, 43, 50, 51].
- *Portability*: Application running on one platform may not give the same performance on other platforms. A basic aspect of a mobile cloud system is that it might be made up of or used from multiple heterogeneous mobile devices, and hence various different platforms. Also as MCC is still in its early stages, there aren't any standards for data formats or interfaces, which makes portability a risk to the performance of MCC services and applications [45, 55].
- *Data availability*: It's a real-time performance risk, for example, if required data aren't available when user need them [1, 33, 45, 47, 50].
- *Service availability*: Timely and continuous availability of mobile cloud services is an important risk factor that needs much consideration and proper risk planning.

Any disruption to the infrastructure can cause the cloud to become unavailable [1, 13, 45, 47, 51, 53].

- *Reliability*: Risks of service downtimes, faults, and failures [13, 45].
- *Resource exhaustion*: In the mobile cloud environment, resources are at risk of exhaustion. The reason is that in clouds, there are usually more users per application and more applications per server. The scalability aspect of the cloud further contributes to this risk; any problems in service mechanisms (i.e., inappropriate modelling of resource usage or inadequate resource provisioning) can also lead to resource or memory exhaustion which in turn can substantially degrade system performance [51].
- *Complexity*: Elasticity, abstraction, sharing of resources and its ad hoc nature make cloud computing a complex infrastructure. There would be more users per application, and the servers might be hosting more applications/server. Increased complexity of this infrastructure makes it more susceptible to risks that affect the efficient performance of the system.
- *Network constraints*: Wireless and mobile networks are comparatively more prone to risks of disconnections, limited bandwidth, and high latency. Any attempt to improve these figures will put a strain on power resources [11].

Some of the factors cited here are similar to those of security risks. As mentioned earlier, this is because different authors can perceive a single risk item in multiple ways, according to their own values. Some of the factors like portability and reliability are included in the performance risk category as these factors can directly affect the performance of a cloud computing infrastructure.

7.3.2.3 Legislative/Organizational Risks. These are the risks related to legislation, business organisation, and the operational environment of cloud providers and users.

- *Vendor lock-in*: Due to lack of standards it's difficult to switch from one provider to other; or worse, to move it back in-house (e.g., in case of price increase or poor service quality) [41, 48, 51, 53, 56].
- *Business viability of provider*: If the cloud provider goes out of business or terminates cloud services, it's a big risk for customers as they are exposed to loss of service, loss of investment, and this can lead to loss of their own customers and users [1, 41, 42, 51, 56].
- *Unpredictable costs*: Cloud services are usually pay as you go, and even a slight change in costs by provider can affect the total costs of usage tremendously. The risk of unpredictable costs can be managed via thorough negotiation before getting into contracts [41, 56].
- *Regulatory compliance*: As the users' data are located elsewhere, there is a risk of regulatory compliance between cloud providers and customers. As the users are not managing their own data and services, there's a risk that their own data

won't be compliant with their organizational policies and standards. Moreover, the certifications (current and future) will also be potentially put to risk with migration to the cloud [42, 43, 45, 47, 51].

- *Data location (risks of changing jurisdictions)*: As the users are unaware of locations where their data are stored or handled, they usually have no control over their own data. Loss of governance is a business risk that needs consideration [41–43, 47, 50, 51]. Some other authors have presented the more or less similar concept as "Data-access risks" due to changing jurisdictions. It's very much possible that customers' data are located in some other legal jurisdiction, making it difficult to apply one's own state laws or regulations on data; for example, some states have restriction on certain types of data and if your data are located on servers in that area, it could become a problem [45]. Moreover, the data location can also present the risks of "auditing," due to changing jurisdictions. Data stored in other countries or jurisdictions will be subject to the laws and conventions of the hosting country which can be a risk for data auditing. Also, if someone has claims against the cloud service providers, it could become a risk too with multiple state laws involved regarding data and service. This data storage over multiple jurisdictions and inherent risks of ignorance about other states' laws makes it a very risky legal activity [48, 51].
- *Investigative support*: Investigating inappropriate or illegal activity may be impossible in cloud computing, because logging and data for multiple customers may be co-located and may also be spread across an ever-changing set of hosts and data centers. This makes it very risk susceptible [42, 45, 48, 53].
- *Lack of organizational learning*: Your staff isn't managing your technology/data, and there is a risk that cloud customers won't be able to train their own staff for their own data and services [56].
- *Business risks from co-tenant activities*: Cloud resources are shared between multiple customers and that means there is a risk that some malicious activity by one customer can affect other customers as well who are sharing same resources, which can, in turn, be a huge risk to the reputation of innocent tenant organizations [51].
- *Licensing risks*: The ad hoc nature of the cloud makes it difficult to audit licencing compliances [51]

This category of risks is very important owing to the remote processing and remote storage characteristics of cloud computing. In cloud computing, the location of servers and data is usually unknown, and this means that some or all of the users' data could be in some place that is outside of their own jurisdiction. This poses many risks to already fragile cloud computing environments. Such risks can be as extreme as loss of control over data or in the worst case, loss of data completely in matters of changed jurisdiction or disputes with providers. Moreover, the auditing or any legal disputes could also be complicated in cloud computing systems because of the aforementioned reasons.

7.3.3 Further Risks in MCC

As discussed earlier, MCC poses some additional risks to conventional cloud computing. The following segment presents the risk factors that primarily relate to the MCC environment:

- *Resource limitation of mobile devices*: Mobile devices are at risk of resource exhaustion if mobile cloud application development fails to consider this factor; this risk is even greater with smaller mobile device environments and sensors [10, 11].
- *Application mobility*: An application may have to continuously switch between device and cloud (or different local clouds), every time the mobile environment changes. If not managed properly at development and later levels, it can cause interruptions as well as put tremendous constraint on mobile device resources as well [12].
- *Device mobility*: This presents the risks of connectivity to correct base stations and with other mobile devices in the cloud [10, 11].
- *Portability*: A mobile cloud is made up of varied mobile devices hence posing a risk for portability, that is, an application running on one device platform may not be suitable for other devices [11, 12].
- *Metering risk*: In a mobile cloud environment, a single user will be connecting and disconnecting multiple times, thus bringing the efficient "metering of services" to higher risk levels [57, 58].
- *Context-awareness*: The cloud services need to be aware of a user's current context and adapt to it automatically. Failure of these services to self-adapt is a risk to efficient and effective MCC [11, 12].
- *Physical risks*: Portable devices are more prone to physical risks like damage, and theft than desktop PCs or standalone machines [11].

7.3.4 Discussion

A fact worth noticing here is that most of the literature has covered what we can call "system-level" risks. These mostly include the risk factors prevalent in high-level cloud systems and communications networks. The low-level risks at mobile cloud application development or frameworks level have not been effectively discussed.

Furthermore, irrespective of the huge lists of risks prevailing in MCC, relatively little attention has been given to managing these risks in comparison to other issues like saving on resources or saving on costs of mobile devices. This is in contrary to other matured IT systems and technologies. In addition to identifying the risks, it's crucial to devise strategies to appropriately manage and mitigate these risks. A better way would be to deploy a proper risk management process in place before moving into the cloud so as to proactively identify, monitor, assess and manage the risks in order to avoid them or mitigate them.

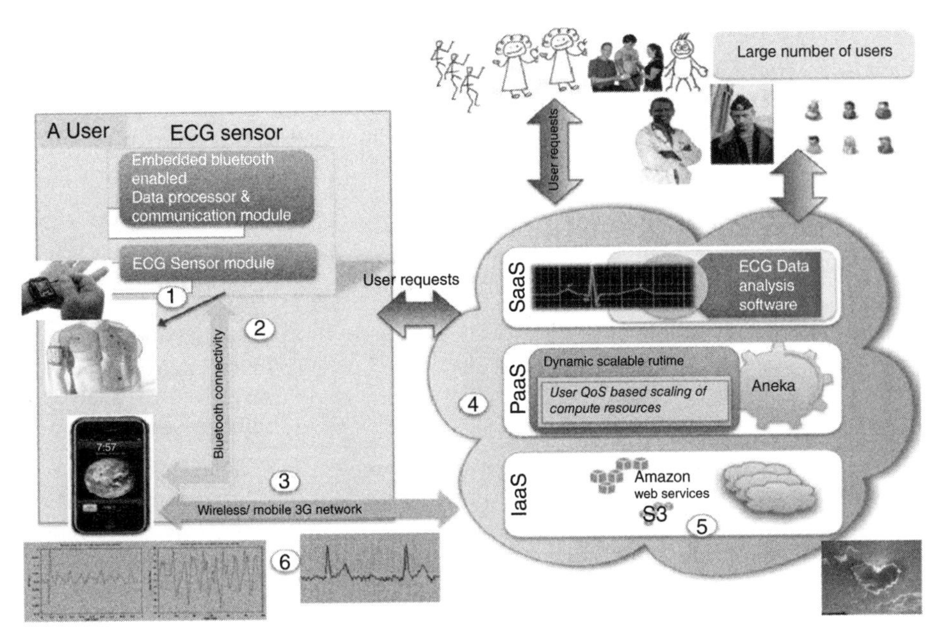

Figure 7.1. ECG data analysis software as service [59].

A detailed analysis of the current situation in MCC from the risk perspective is given in the following sections.

7.3.5 An Illustration of Risks for a Real-Life Application

To see that how these risks can affect any application, we have taken an example of a mobile cloud-based e-health application proposed in Pandey et al. [59].[1] They have designed a scalable real-time health monitoring and analysis system and have used an electrocardiogram (ECG) analysis system prototype as their case study. This prototype system collects patient data (e.g., pulse and heart beat rates) through an ECG sensor device attached to a patient's body. This sensor device transmits the data to the patient's mobile device via Bluetooth without manual intervention. A client software on a mobile device then transfers the data to an ECG analysis Web-service hosted on a cloud computing stack, either using Wi-Fi or mobile device's 3G network (Fig. 7.1). They have used the Aneka cloud computing platform and Amazon's S3 storage services. The software then analyses the patient's data, generates results and appends the latest findings to the patient's medical record. Depending on analysis, the data could be sent to patients, doctors, or emergency services as needed. When we use this application for demonstrating the inherent risks and their consequences, we see that its impacts could be disastrous

[1] Note that this is not a critique on ECG analysis prototype as understandably, risk is not their research focus.

(that can even cost a precious human life) if no proper risk management is implemented. It's a time-critical application that requires extreme levels of reliability for the results that are being generated and needs to be always up to date with the latest findings. An overview of how different risks can impact the application and users is given in Table 7.1.

TABLE 7.1. Overview of risk factors in mobile clouds

Risk category	Risk subcategory	Risk factors	ECG analysis application potential vulnerabilities (data = patient data, service = ECG analysis software service)
Security and privacy risks	Network security risks	• unauthorized access • security defects in technology • security defects in Web services • weakness of browsers • network security • physical location of system infrastructure	Conformity (because of manipulation), sabotage (malware, virus, worm), data integrity, service availability, credibility
	Data security and privacy risks	• leak of customer information • leak of proprietary information • data location • data segregation/isolation • data recovery • data security • data integrity • insecure/incomplete data deletion • faulty backup mechanism	Privacy, conformity, sabotage, identity theft, customer confidence, business loss/competitive edge, espionage, wrong diagnosis, record mix up, data integrity, data loss
	Others	• natural disaster • malicious insider	
Performance risks	Availability and reliability risks	• data location • data segregation/isolation • data availability • service availability • reliability	Faulty data collection, wrong analysis, prolonged response times, correct and efficient data availability, loss of business, life threat

TABLE 7.1. (Continued)

	Service and application usability risks	• features and general maturity of technology • complexity • portability • resource exhaustion	Misdiagnosis, ease of use, time critical, mobile device limitations, compatibility for different devices, compatibility of device to SaaS, robustness, and load balancing
Legal/environmental risks	Organisational/ business risks	• vendor lock in • business viability of provider • unpredictable costs • data location • investigative support • lack of organisational learning • business risks from co-tenant activities	Varied costs due to different usage patterns, lack of organizational expertise (software malfunction can go unnoticed, wrong use of application, cloud dependence), regulatory compliance (less options for legal support)
	Environmental risks	• natural disaster	Data and service availability, life risks
	Different legal jurisdictions	• data access risks due to changing jurisdictions • auditing risks due to changing jurisdictions • licensing risks	Reduced support for legal auditing, ignorance about other countries laws, different laws within different jurisdictions
Mobile cloud computing risks		• Resource limitation risks	Performance, incompatibilities, erroneous or slow response, memory exhaustion, application crash, limited power, Bluetooth connectivity issues, ECG monitoring device and mobile phone compatibility, patient's 3G plan limitations
	Mobility risks (between device and cloud)	• application mobility • device mobility	Patients won't be static, networking conditions different for different geographical locations, networking conditions different for different users within same geographical regions

(Continued)

TABLE 7.1. (Continued)

	• Portability risks (between multiple devices)	Hugely varied mobile devices (of patients)
	• Metering risk	Usage patterns and frequency different for different patients
	• Context-awareness	Patient's location can impact data collection (if s/he at home or gym)
	• Physical risks	Patient can damage/loose the monitoring device or mobile phone

7.3.5.1 Security and Privacy Risks. The authors have tried to reduce the risks of security breaches to the application data by implementing encryption mechanisms which are expected to minimize the risks of unauthorized access to the data. For that purpose, they have suggested using a third-partyp key infrastructure which will make it difficult for attackers to access the confidential data. When it comes to privacy and leak of information, we see that though being an important aspect, it's not as critical as reliability and efficiency in this scenario. However, any loop holes in security can sabotage the overall reliability and efficiency of the whole setup. Risk factors like unauthorized access, security defects in technology, Web-services or networks, or malicious insider might mean attackers can access patient's personal information, misuse it, or even manipulate it. In this particular case, it could mean the integrity of patients record is at risk and any tampering or mishandling can lead to false alarms for emergencies or no alarms when needed. In case of alarming results for a patient's ECG analysis, the patient needs immediate medical assistance; however any meddling in patient's data can lead to his/her death. Also, in case if the patient's ECG is normal, a false alarm can be sent to emergency services that could have huge financial impacts and the availability of emergency services for genuine cases can also be endangered.

In addition to the risks of unauthorized access, there are certain other risk factors that can meddle with the efficient processing as well and can't possibly be handled by encryption mechanisms only. Risk factors like faulty backups, data integrity, and insecure/incomplete data deletion can pose a risk to the availability of patient's medical history and records for doctors. Any weaknesses in backup and data recovery procedures can mean longer times for patient diagnosis or even loss of medical records. The risks to data integrity can be very grave in such time critical applications when a human life is at risk. The privacy risk from insecure/incomplete data deletion can be of less impact in this case unless attacked by malicious attackers.

7.3.5.2 Performance Risks. For a time critical application like this, application and system performance is an aspect that needs to be critically assessed for risks. The only quality-of service-(QoS) parameter the authors have selected is shorter response times. To ensure fast responses, the authors have implemented dynamic scalable runtime (DSR) module that continuously monitors the average response time and if it's above a threshold value, the DSR assigns more resources until response time decreases. However, besides short response times, any degradation in other performance attributes (e.g., reliability, efficiency, and availability) can also lead to consequences ranging from erroneous or unavailability of patient records, to the possible loss of human life. Also, as this application intends to target a huge population of cardiac patients worldwide, any defects in robustness or load-handling will be a risk to efficient performance. Despite being increasingly popular and evaluated, MCC systems and cloud based e-health applications like this still need maturity in terms of features of technology. Any weakness in technology or feature-sets can be a risk to QoS, and hence customer satisfaction, and business objectives as well. For example, in this particular scenario the target population is cardiac patients, which on most part, comprises people above the ages of 55 years,[2] and it will be much more difficult for these people to get accustomed to complex applications and features than younger people. If patients don't find the application easy to use or of good quality or cost effective, they might misuse the device or may be hesitant in using it; a fact that could have serious implications on ROI. Also, each error contributes to increased dissatisfaction in customers (in this case, doctors and patients) and can be a risk to patient health as well.

Data isolation and segregation from the security perspective is a risk to privacy and loss of confidential patient information and records. However within the performance perspective, faulty data deletion or improper data isolation can cause a mix-up of patients' records and can result in wrong diagnosis for the respective patients. Data location on other hand can affect timely availability of patient data but its probability is very low.

As this application targets wide patient population, the mobile devices involved could be hugely varied, which poses a risk of application portability for multiple devices. The analysis software is hosted by cloud services, but the client software resides on mobile devices itself, and so it should be extensively evaluated for processing efficiency for multiple devices and mobile platform. Any problems with portability could mean a percentage of patients being unable to use this application.[3] Also, the compatibility of mobile devices with cloud services should also be risk free for otherwise the use of this application will be limited to a subset of patients only.

Data availability, service availability, and reliability risk factors are extremely critical owing to the nature of this application and for reasons discussed earlier. Any error in these aspects can be life-threatening for some patients with heart risk. Application crashes and downtimes can be very dangerous. For entertainment applications, we can tolerate some risk but in applications like this ECG monitoring and analysis system, such risks are intolerable and should be managed very carefully. Also, resource exhaustion on the mobile can directly affect availability and it's a possibility in this scenario.

[2] Can be less than 55 years as well, but major percentages belong to age category over 55 years http://www.worldheartfailure.org/index.php?item=75, http://www.rosscountyhealth.com/brochures/MensHeart.pdf

[3] Age of patients is already a risk factor with respect to smartphone usage, as elderly find smartphones much too complex for their comfort level.

7.3.5.3 Legal/Environmental Risks. Most of the legislative risks don't have direct noticeable impacts in this application; however, unpredictable costs, lack of organization learning, and licensing risks are applicable. The ECG data analysis is provided as a service to users who can pay on per-analysis basis. However this software service is itself dependent on cloud based storage provided by Amazon. As there's flexibility for patients to use this application with different frequencies, depending on health condition and doctor's recommendations, keeping tracks and auditing licensing compliances will be very complex. Also, because of similar reasons (different frequency and usage patterns), the costs won't be fixed and can be unpredictable making it difficult for both the patients and doctors to keep track of their relevant costs. Similarly, the lack or organisational learning can create cloud dependence for a long time and also it will be difficult for employees to gain expertise on technology usage and analysis.

7.3.5.4 Mobile-Specific Risks. Like any other mobile application, resource limitation of mobile devices is a risk that can affect the performance of the ECG application negatively. Considering the target user population (i.e., cardiac patient and people with heart risks worldwide), it's safe to assume that a large number of different types of mobile devices will be used. In addition to general issues of possible incompatibilities, limited power resources, and limited memory, there could be some devices with comparatively slower response times due to reasons like memory exhaustion (because of large data files stored on a patients mobile OR if a patient owns a mobile phone with lesser memory capacity). Also, some mobile devices face Bluetooth connectivity issues (e.g., iPhone connectivity with Nokia and Sony Ericson mobile devices via Bluetooth might not be reliable, and some Nokia devices might have connectivity issues with other devices like HP notebooks). Asking every patient to buy a compatible smartphone isn't an attractive option; likewise, usage of an application shouldn't be restricted to owners of specific mobile devices. It can be fatal if application developers fail to cater to this problem. Another aspect that needs consideration is device mobility. As the patients won't be static to one geographical location (a single patient can be frequent traveller), so the devices connectivity to cloud and availability of services accordingly is very risk prone. Also, context-awareness needs to be considered in analysis as for this particular application; it can tamper with the results falsely. For example, consider a scenario if the patient decides to check his/her ECG after a workout at gym or elsewhere, his/her heart and pulse rate would be very different from his/her normal rates. If the analysis software fails to recognize this, it can generate false alarms and hence a wrong diagnosis. Cases like this can negatively affect users' confidence in the system and can eventually lead to application failure.

7.3.6 Understanding Risks with Near-to-Life Scenarios

Let's consider an example to see that how ever-changing mobile environments can effect overall risk situation of a mobile cloud system. To understand these risks and their impacts better, let's consider some scenarios briefly. Consider an elderly heart patient John as a user of the above mentioned ECG analysis app. He must wear the heart-rate-monitor all the time (preferably) so that his condition could be monitored by his doctors. Now consider this person John in following scenarios:

Scenario 1: John is working in a well-developed city that has many offerings in terms of security and mobile and network services at better speeds.

Scenario 2: John is travelling to his home village that's not very well developed, and hence offers fewer options for network and mobile services.

Scenario 3: John goes to a third-world country for an official assignment for 2 years. The crime rates in that country are noticeably higher than his own country. The quality of mobile devices isn't very good and the most commonly used mobile devices are different from those in X's home country.

As mentioned earlier, to calculate risk we need to determine the probability and impact of that risk in the given mobile cloud instance. Now let's consider the effect of these seemingly innocent scenarios on a few risk factor values (Table 7.2).

In Samad et al. [60], these risks and calculations are discussed in detail with examples and its highlighted that how the current context is important in assessing risks in mobile cloud systems and how any slightest change in context parameters can influence the whole risk picture.

7.4 RISK MANAGEMENT FOR MCC

This section analyzes the MCC from risk perspective. We have surveyed multiple MCC frameworks and also identified the risk factors currently prevalent in this domain. We have also demonstrated using a real life application that how different risk factors can affect application processing. This section provides an analysis of the surveyed frameworks from a risk perspective and discusses a current situation in the MCC domain regarding risk management.

7.4.1 Analysis of MCC Frameworks from the Risk Perspective

As mentioned previously, four of the frameworks have been selected for in-depth analysis in this research for better understanding of current situation. These are (1) hyrax, (2) Cloudlet, (3) Clonecloud, and (4) Amazon's EC2 for mobiles. This selection is solely on basis of (our assumption of) most widely used commercially and popularity among research community.

The Hyrax framework is designed for computations that need only the data on mobile devices and usually requires no interaction with traditional servers for large scale computations. The system is targeted at multimedia and sensor data that doesn't need to be changed frequently. The authors of this framework claim to provide scalability, fault tolerance, privacy, and hardware interoperability. However, the Hyrax architecture lacks in energy efficiency, reduced efficiency for CPU and memory usage, and is not suitable for slow networking conditions. From the framework's analysis, we can see that where it

TABLE 7.2. Example of effects on risk probabilities and impacts

	Scenario 1	Scenario 2	Scenario 3
Effect on Probabilities	The probability values for risk factors like battery hardware problem, memory leak by apps, and different memory sizes would have similar values as those in scenarios 2 and 3.		
		The risk factors such as poor connection, blind spots, infrastructure problems, etc., will have higher probability in scenarios 2 and 3, than in scenario 1	
			The risk factors such as compatibility issues, and different data formats supported by devices will have higher probability values than scenarios 1 and 2. The probability of the risk "device theft" will be highest in scenario 3.
Effect on impacts		In scenarios 2 and 3, the probability of John facing low bandwidth or noise issues is understandably high, but the *impact* value would be dependent on the level of bandwidth deterioration or level of noise, for example.	
		i. If the bandwidth deteriorates from 79 Mbps to 50Mbps, then its impact would be say medium	ii. But if the bandwidth drops from 79 Mbps to 20 Mps, then its impact would be high or very high

considers the security and privacy risks, it hasn't considered several risks for performance and availability of the system when and where needed.

The Cloudlet framework, on other hand, tends to improve the performance (fast responses) and overcomes slow networking issues. It lacks in mechanisms dealing with security and privacy risks and focuses on most part in improving the performance and dealing with resource limitations of mobile devices. As the mobile devices are not directly connected to a major remote cloud, so it can be assumed that in some situations certain risks won't be a major problem at higher levels. However, at cloudlet level, these privacy and security risks will pose some serious threats and if not dealt with properly, can be very damaging for business owners and users. This is because the cloudlet infrastructure is fundamentally based on a "self-management" concept and the basic criterion of this framework is to make things easier and self-manageable with minimal manual intervention. As a result there is a potentially bigger risk factor as compared to when the infrastructure management was done by domain experts with more technical knowledge of security and safety procedures are required. Moreover, even if the users are not directly connected to remote cloud server, any failures or weaknesses at cloudlet level can jeopardize the whole system and can even pose risks to the whole cloud infrastructure to which that cloudlet is connected by passing on the failure instances to the main infrastructure.

Some cloud service providers offer users with the option for different availability zones with each having a separate infrastructure. Different zones can be insulated from each other and users can host their applications on more than one geographical location to ensure higher availability. In case of failures, one backs up the other. However, there is a possibility that weaknesses of this system as a network glitch can cause the servers to start automatic back up, causing complete server congestion and a possible downtime of several hours. It's a risk that can affect availability and performance of application usage. In case of time critical applications, the risk is manifold. This further raises various questions regarding the loop holes in their risk contingency and management.

One thing worth noting is that with an exception of Hyrax, the rest of the frameworks are more concerned about saving costs and resources related to mobile computing (understandable given their research foci). Not much focus has been put on the risks that could arise by using that particular framework or risks that can occur by processing apps over that framework. For example, some of frameworks have talked about dividing jobs between mobile devices and cloud servers; however, there is no mention of what risks could be associated by such division of jobs, how those risks can be dealt with and what risks users can face if they run their applications based on that framework.

One of the questions we planned to answer via this analysis was what kind of applications will be supported by each framework so that it can be analyzed what type and what level of risk management shall be applied to each of these frameworks. Figure 7.2 discusses the rationale behind this risk hierarchy diagrammatically—sources of risk can be at the application-specific level, framework level, or underlying system level. One assumption that each of these frameworks have made is that the applications will be stable for longer times without any need for frequent/immediate updates and modifications. Apart from that, it can be assumed that these frameworks are targeting a range of applications as there is no mention of any specific set of apps that will be most suitable for each framework. This factor can be positive in a sense that now we can assume that a single

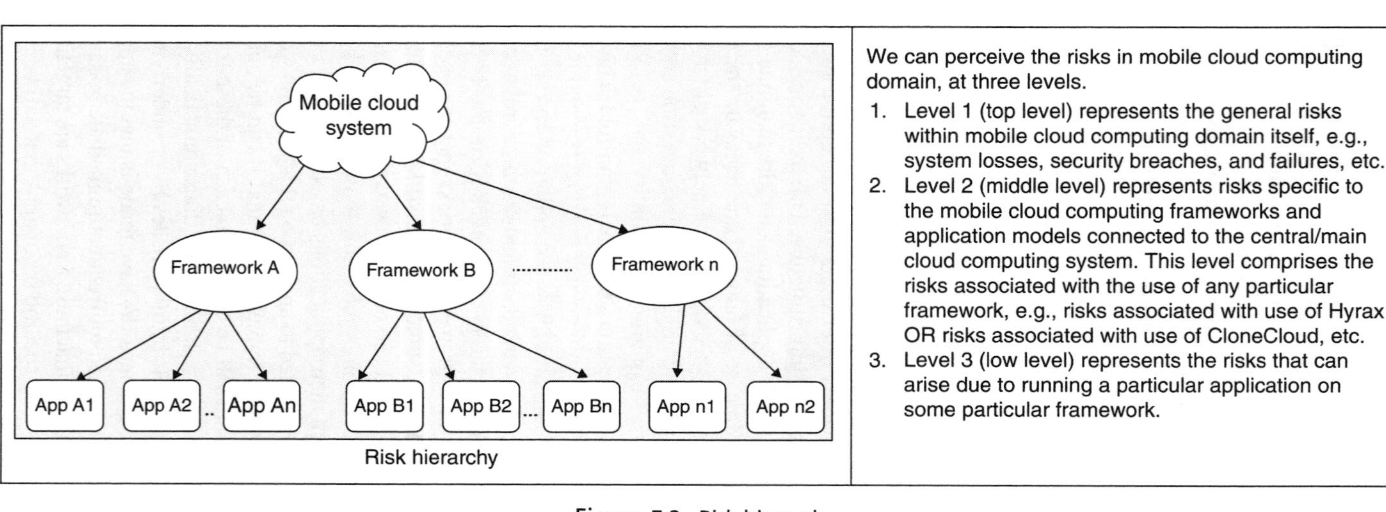

Mobile cloud system
Framework A
Framework B
Framework n
App A1
App A2
App An
App B1
App B2
App Bn
App n1
App n2
Risk hierarchy
We can perceive the risks in mobile cloud computing domain, at three levels.
1. Level 1 (top level) represents the general risks within mobile cloud computing domain itself, e.g., system losses, security breaches, and failures, etc.
2. Level 2 (middle level) represents risks specific to the mobile cloud computing frameworks and application models connected to the central/main cloud computing system. This level comprises the risks associated with the use of any particular framework, e.g., risks associated with use of Hyrax OR risks associated with use of CloneCloud, etc.
3. Level 3 (low level) represents the risks that can arise due to running a particular application on some particular framework.

Figure 7.2. Risk hierarchy.

risk management process can be applied to all of these frameworks; but the major disadvantage of this practice would be that if there are any risks specific to some particular framework, they might be left unattended and can prove to be fatal later on. For example, the Hyrax framework is not suitable for applications requiring immediate responses, and likewise CloneCloud and Cloudlets don't provide much in terms of data security and privacy. So, it can be assumed the time critical applications shouldn't be run on Hyrax without due consideration to risk and contingency measures. Similarly, for security critical applications Cloudlets and CloneCloud may not be the best choice in terms of risks involved.

Considering this situation prevailing in MCC world, there is a need for a formal risk management process that

1. is general enough to be applicable to all frameworks/categories as an extended module, and
2. explicit enough to allow catering to framework specific risks.

Table 7.3 provides a tabular analysis of these frameworks at a glance. The column "applications supported" represents what applications are used as references or examples in these frameworks as there are no mentions of targeted/supported applications by authors of these frameworks.

7.4.2 Effectiveness of Traditional Risk Management Approaches in Managing Mobile Cloud Risks

A number of approaches have been proposed in literature for managing risks of software development at many levels. However, most of these processes focus on traditional development environment only. These approaches may not be customized to respond to the challenges of application development for distributed, remote processing cloud environments. Our survey of the MCC frameworks has also highlighted a fact that there isn't a formal risk management process in place to deal with risks associated with the use of these frameworks or the applications based on these framework.

In order to make the MCC environment more effective and efficient, we need to devise a mechanism to deal with related risks. Such a risk management process can be made to work at two levels: framework level and application level. Also the MCC risk management process should be able to cope with ad hoc nature of MCC environment. We need a risk management methodology that can be appended to any framework and it will work equally effectively in all mobile cloud domains/environments. For this purpose, the most suitable risk management process can be selected and modified to work in dynamic and robust environments to match with cloud computing requirements. In absence of any such process, a specialized risk management process can be designed based on the strengths of existing traditional processes as starting point. For designing such a process, we need to explore mobile computing, grid computing and distributed computing settings individually too see how each of these domains deal with respective development risks. This information, along with the analysis of surveyed mobile

TABLE 7.3. Framework analysis

Framework	Category	Features	Applications supported	Inherent risk management
Hyrax	Application partitioning/Client server	• Used primarily for computations that involve data on mobile devices, not for generic distributed computation • Do not expect to replace or effectively collaborate with traditional servers for generic large-scale computation • Sufficient space to store multimedia data and sensor logs *Pros*: scalability, fault tolerance, privacy, hardware interoperability *Risks*: no energy efficiency, less effective for CPU and memory usage, can't cope with slow n/w conditions	System is targeted at multimedia and sensor data, which can be considered historical records that do not need to be changed	None
Cloudlets	VM-based	• Location-based services, fast interactive response, rapid customization of infrastructure, peak bandwidth load issue solved • Deal with resource poverty via nearby resource rich cloudlet • Cloud solves mobile resource poverty and cloudlets solve latency issues • Assumption that devices making up cloudlet won't be moving during request processing	Local business systems, multimedia apps, human cognition applications (voice/face recognition)	None

(*Continued*)

TABLE 7.3. (Continued)

CloneCloud	VM-based	• Four augmentation categories • Assumption that applications won't change much over time • Assumption that requesting devices would remain within range of clone running server during request processing	Designed to cater for wide range of apps	None
Amazon EC2	Client/server	• Flexibility for users to create as many "instances" (virtual servers) as needed • Four "availability zones' for fault tolerance • Most widely used commercially	Supports all kinds of application via multiple Web service components like S3, EBS, SimpleDB, RDS, CF	Multiple zones to handle failures

cloud frameworks and the identified risk factors, can then be used to design a (preferably automated) risk management framework for MCC.

7.4.3 Related Work for Risk Management in Cloud Computing

When it comes to risk management in cloud computing, we see that most of the contributions are limited to the security and privacy issues in clouds. Many authors have highlighted the security issues in clouds and suggested numerous approaches in dealing with those security issues, each suggesting a solution depending on their cloud usage scope and domain [61–66]. Brender and Markov [48] have also highlighted the risks in cloud computing but their focus is on the adaptability of clouds by different companies, based on their analysis of few Swiss companies of different sizes. NIST has also provided some security and privacy guidelines for clouds in Wayne and Grance [47]. These solutions range from different authorization and access control mechanisms to audits conducted by third parties to ensure security. However, most of these suggestions/proposals are discursive in nature with none providing complete solutions to the problem.

An important contribution has been made by Saripalli and Walters [67] in which they have proposed a model for quantitatively assessing security risks in cloud platforms. In the literature we also see some more advanced intrusion detection and security frameworks like [65, 66] but as mentioned previously, they have not been implemented yet (to the author's knowledge) and they are mostly proposed as a guidelineframework that requires cloud computing users to follow some certain steps to achieve security.

Ko et al. [66] have proposed a framework "TrustCloud" for accountability in clouds. They have defined four components of trust in clouds as security, privacy, accountability and auditability. Another classification of trust is given as preventative and detection security measures taken to enhance confidence. The detection framework proposed by them consists of five layers and usually revolves around maintaining logs for accountability at different layers. Similarly, Zhang et al. [62] have proposed an information security risk framework focusing on data security issues. The framework proposed by them is qualitative in nature and very basic with no actual strategies to deal with the risks or threats identified. Like a process outline, it defines only the steps that need to be taken for mitigating risks. On slightly different lines, the authors in Houmansadr et al. [65] have proposed a cloud based intrusion detection framework for mobile devices. This framework however doesn't deal directly with cloud/mobile cloud security issues.

There is also some work being done in privacy issues in cloud computing explicitly. Authors like Pearson et al. [68] have specifically focused on privacy issues as a separate entity from "security" in the cloud computing domain. The suggestion for privacy issues range from encryption of data to personalization or preference classification, to prevent misuse or theft of private data.

In summary, we see a lot of work in cloud computing related to security and privacy issues, but there is less work in dealing comprehensively with the range of aspects of cloud computing risks like performance, connectivity, and mobility. These issues and their solutions are mentioned implicitly in some of the work but not explicitly, and hence no concrete work to deal with or mitigate these risks. Also, most of the work we see in regards to risk management and mitigation is theoretical or discursive in nature. Another important point highlighted by the literature analysis is that though we see work in cloud computing regarding security and privacy issues, we don't see much work being done in MCC in the same areas. Also, as discussed earlier, being proactive in MCC domain needs the system to be efficiently context-aware. Some authors such as La et al. [69] and Papakos et al. [70] have suggested at using context awareness in cloud domains, but their focus is not on using this information for risk management.

7.5 CONCLUSIONS

As MCC is still in its early stages, it still needs some maturity in terms of processes, technical and nontechnical aspects and auditing. This immaturity along with the intricacy of mobile cloud system introduces many risks. At system level, these include the risks of connectivity, limited resources, security, and limited power supply. Moreover, as the system complexity increases, both the technical and nontechnical risks increase and so is the need to manage these risks. The ad hoc nature and mobility in MCC environments means that the development of these system need more rigorous and specialized risk management to deal with all the risks. This further burdens the developers of MCC frameworks and applications. In addition to the complexity of mobile cloud infrastructure, they also have to deal with the risks at framework/application level including but not limited to efficient job distribution, virtualization, scalability, and so on.

We make the following observations and notes on interesting future trends:

- Our analysis highlighted the fact that most of the current frameworks and applications tend to focus on savings costs and resources, but not much attention is given to managing the risks. We have also highlighted the reasons that why risk management is important and why it should be incorporated in MCC frameworks and applications. No one can deny the importance of identifying, and hence treating the risks at earlier stages instead of when it has already become a problem. Being proactive in managing risks can save mobile cloud providers and users from many negative outcomes like service disruptions, financial losses, data losses, loss of customer satisfaction and confidence and at worst, lost of lives.
- An ideal MCC risk management system should be able to assess continuously (and automatically) what can go wrong, identify risk areas, implement treatment solutions for identified factors, and in case of contingency failure, ensure safe system closure. Such an ideal system can also make use of historical data of previous incidents and risk management records from past incidents. Also, the ideal risk management system shouldn't focus on just the service providers; rather, it should also take into consideration the other stakeholders as well, such as customers and users.
- While we have focused on risk, there is a range of issues surrounding risk so that it is a complex and multifaceted concern, including security, trust and privacy, not just system reliability and performance. We have only provided a broad overview but there remains future work in studying complex issues of risks in specific MCC applications, in those that involve remote cloud servers and those that use a collection of surrounding devices, the so-called *mobile computing crowd*. Emerging mechanisms such as homomorphic encryption[4] provides a partial solution to protect data computed with remotely, but its applicability in the mobile cloud environment needs further exploration.
- There is emerging a much larger range of mobile devices, from the Google Glass to watch computers as well as smart jackets and smart shoes, each of which could participate in a resource pool to provide local cloud-like services or utilise the greater Cloud. An approach that considers the range of devices on multiple users working together in a risk-managed manner could be an interesting avenue of work.
- With the rapid growth in crowdsourcing and crowdsensing, crowd-sourced clouds become interesting together with their human-related risks—while SETI@home has been around for a long time, mobile versions of that and variants of the idea for different applications have emerged (e.g., BOINC[5]).
- Modern mobile computing development is not only resulting in a range of wearable and mobile devices of different forms, but the Internet of Things has emerged with everyday objects forming potential resource providers, participating in future

[4] http://www.infoworld.com/t/encryption/ibms-homomorphic-encryption-could-revolutionize-security-233323

[5] http://boinc.berkeley.edu/

resource clouds. Interesting developments include vehicular clouds (e.g., initiated by Gerla [71]), and the increasing uptake of drones for non-military uses—one can even consider fly-in/fly-out cloud servers where mobile computing infrastructure can be flown into a disaster stricken zone to provide computing services for a time, a situation where perhaps need outweighs the risks.

REFERENCES

1. Livingstone, R. (2011). *Navigating through the Cloud.* Createspace, ASIN: B005ILWCGG.
2. Huang D. (2011). *Mobile Cloud Computing.* IEEE COMSOC MMTC E-Letter.
3. Wang Q. A. (2011). *Mobile Cloud Computing.* Thesis for Master of Science, University of Saskatchewan, Canada.
4. Ragnet F., Conlee R. G. (2010). *Can You Trust Cloud? A Practical Guide to the Oppurtunities and Challenges of Document 3.0 Era.* Cloud Computing-White Paper, Xerox Corporation.
5. Vaquero L. M., Merino L R., Caceres J., Lindner M. (2009). A break in the clouds: Towards a cloud definition. *ACM SIGCOMM Computer Communication Review*, 39: 50–55.
6. Sosinsky B. (2011). *Cloud Computing Bible.* Hoboken, NJ: John Wiley & Sons, Inc.
7. Baun C., Kunzel M. (2012). A taxonomy study on cloud computing systems and technologies. In *Cloud Computing: Methodology, System, and Applications*, Wang L., Ed., Boca Raton, FL: CRC Press, p. 73–90.
8. Huang D., Xing T., Wu H. (2013). Mobile cloud computing service models: A user centric approach. *IEEE Network*, 27: 6–11.
9. Huang T. D., Lee C., Niyato D., Wang P. (2011). A survey of mobile cloud computing: Architecture, applications, and approaches. *Wireless Communications and Mobile Computing.* DOI:10.1002/wcm.1203.
10. Satyanarayanan M. (1996). *Fundamental Challenges in Mobile Computing.* Philttdelphia PA: ACM PODC 96.
11. Gupta A. K. (2008). *Challenges of Mobile Computing.* Proceedings of 2nd National Conference on Challenges & Opportunities in Information Technology (COIT-2008) March 29, Mandi Gobindgarh, India, pp. 86–90.
12. Kovachev D., Cao Y., Klamma R. (2010). *Mobile Cloud Computing: A Comparison of Application Models*, Computing Research Repository, CoRR, vol. abs/1009.3088, 2010.
13. Kumar K., Lu Y. H. (2010). Cloud computing for mobile users: Can offloading computation save energy? *IEEE Computer*, 43(4): 51–56.
14. Satyanarayanan M., Bahl P., Cáceres R., Davies N. (2009). The case for VM-based cloudlets in mobile computing. *IEEE Pervasive Computing*, 8: 14–23.
15. Kovachev D., Renzel D., Cao Y., Klamma R. (2010). *Mobile Community Cloud Computing: Emerges and Evolves.* Proceedings of Eleventh International Conference on Mobile Data Management May 23–26, Kansas City, MO, pp. 393–395.
16. Fernando N., Loke S., Rahayu W. (2012). Mobile cloud computing: A survey. *Future Generation Computer Systems*, 29: 84–106.
17. Rajiv R., Zhao L. (2011). *Peer-to-Peer Service Provisioning in Cloud Computing Environments.* Berlin: Springer, pp. 1–31.

18. Srikumar V., Buyya R., Ramamohanarao K. (2006). A taxonomy of data grids for distributed data sharing, management, and processing. *ACM Computing Surveys*, 38(1): 1–53.
19. Lamia Y., Da Silva D., Butrico M., Appavoo J. (2010). *Understanding the Cloud Computing Landscape.* Cloud Computing and Software Services, pp. 1–6.
20. Hadoop, A. http://hadoop.apache.org/. Accessed Novemer 21, 2014.
21. Dean J., Ghemawat S. (2008). Mapreduce: simplified data processing on large clusters. *Communications of the ACM*, 51(1): 107–113.
22. Marinelli E. (2009). *Hyrax: Cloud Computing on Mobile Devices Using Mapreduce.* Masters Thesis, Carnegie Mellon University, Pittsburgh, PA.
23. Jan R., Riva O., Alonso G. (2008). *Alfredo: An Architecture for Flexible Interaction with Electronic Devices.* Proceedings of the 9th ACM/IFIP/USENIX International Conference on Middleware (Middleware 2008), December 1–4, Leuven, Belgium, pp. 22–41.
24. Flinn J., Satyanarayanan M., Young P. (2002). *Balancing Performance, Energy and Quality in Pervasive Computing.* IEEE 22nd International Conference on Distributed Computing Systems, July 2–5, Vienna, Austria, pp. 217–226.
25. Cuervo E., Balasubramanian A., Cho D., Wolman A., Saroiu S., Chandra R., Bahl P. (2010). *MAUI: Making Smartphones Last Longer with Code Of?oad.* Proceedings of 8th ACM MobiSys'10, June 15–18, San Francisco, CA, pp. 49–62.
26. Balan R. K., Satyanarayanan M., Park S. Y., Tadashi O. (2003). *Tactics-Based Remote Execution for Mobile Computing.* Proceedings of the 1st International Conference on Mobile Systems, Applications and Services, May 5–8, 2003, San Francisco, CA, pp. 273–286.
27. Satish S., Vainikko E., Šor V., Jarke M. (2010). *Scalable Mobile Web Services Mediation Framework.* IEEE 5th International Internet and Web Applications and Services Conference, May 9–15, Spain, pp. 315–320.
28. Huerta-Canepa G., Lee D. (2010). *A Virtual Cloud Computing Provider for Mobile Devices.* ACM Workshop on Mobile Cloud Computing & Services: Social Networks and Beyond. MCS'10, June 15, San Francisco, CA.
29. Roelof K., Palmer N., Kielmann T., Bal H. (2010). *Cuckoo: A Computation Offloading Framework for Smartphones.* IEEE 2nd International Conference on Mobile Computing, Applications and Services MobiCase'10, October 25–28, San Francisco, CA.
30. Luo X. (2009). *From Augmented Reality to Augmented Computing: A Look at Cloud-Mobile Convergence.* IEEE International Symposium on Ubiquitous Virtual Reality (ISUVR'09), July 8–11, South Korea, pp. 29–32.
31. Zhang X., Schiffman J., Gibbs S., Kunjithapatham A., Jeong S. (2009). *Securing Elastic Applications on Mobile Devices for Cloud Computing.* ACM Cloud Computing Security Workshop (CCSW'09), November 13, Chicago, IL, pp. 127–134.
32. Chun B. G., Maniatis P. (2009). *Augmented Smartphone Applications through Cole Cloud Execution.* Proceedings of the 12th conference on Hot topics in operating systems, May 18–20, 2009, Monte Verità, Switzerland, p. 8.
33. Huang D., Zhang X., Kang M., Luol J. (2010). *Mobicloud: Building Secure Cloud Framework for Mobile Computing and Communication.* Proceedings of the Fifth IEEE International Symposium on Service Oriented System Engineering, SOSE, June 4–5, Nanjing, China, pp. 27–34.
34. ACTIA (2004). *Guide to Risk Management: Insurance and Risk Management Strategies.* Australian Capital Territory: ACT Insurance Authority.

35. OB-007. (2009). *Risk Management-Principles and Guidelines.* AS/NZS ISO 31000:2009 Standards Australia and Standards Newzealand, ACT Australia.
36. Boehm B. W. (1997). Software risk management: Principles and practices. *IEEE Transactions*, January: 32–41.
37. AIRMIC, et al. (2002). *A Risk Management Standard.* Technical Report. Institute of Risk Management (IRM), Association of insurance and risk managers (AIRMIC) and National forum for risk management in public sector (ALARM).
38. Stoneburner G., Goguen A., Feringa A. (2002). *Risk Management Guide for Information Technology Systems.* Nationtal Institute of Standards and Technology (NIST), Department of Commerce.
39. Shimonski R. J. (2004) *Risk Assessment and Threat Identification.* Security Plus Study Guide http://www.windowsecurity.com/articles-tutorials/misc_network_security/Risk_Assessment_and_Threat_Identification.html. Accessed February 14, 2015.
40. Samad J., Ikram N., Usman M. (2007). *Managing Risks: An Evaluation of Risk Management Processes.* IEEE International Multi-topic Conference, December 23–24, Islamabad, Pakistan, pp. 281–287.
41. Shipley G. (2010). *Cloud Computing Risks.* Information Week-Cover Story, pp. 22–24. UBM LLC. http://www.informationweek.com/. Accessed February 14, 2015.
42. Brodkin J. (2008). *Seven Cloud Computing Risks.* InfoWorld Canada. Downsview: July 2.
43. Mansfield-Devine S. (2008). Danger in clouds. *Network Security*, 2008: 9–11.
44. Ovadia S. (2010). Navigating the challenges of the cloud. *Behavioral & Social Sciences Librarian*, 29(3): 233–236.
45. Scott P., Jaeger P. T., Wilson S. C. (2010). Identifying the security risks associated with governmental use of cloud computing. *Elsevier: Government Information Quarterly*, 27: 245–253.
46. Glimmer B. (2011). Navigating the cloud. *Broadcast Engineering; ProQuest Telecommunications*, 53: 24–26.
47. Wayne J., Grance T. (2011). *Guidelines on Security and Privacy in Public Cloud Computing.* NIST Special Publication 800-144, Computer Security Division- National Insititue of Standards & Technology (NIST)/U.S. Department of Commerce.
48. Brender N., Markov L. (2013). Risk perception and risk management in cloud computing: Results from a case study of Swiss companies. *International journal of Information Management*, 33: 726–733.
49. Khan A. N., Mat K., Khan S., Madanic S. (2013). Towards secure mobile cloud computing: A survey. *Future Generation Computer Systems*, 29, p. 1278–1299.
50. Subashini S., Kavitha V. (2010). A survey on security issues in service delivery models of cloud computing. *Elsevier Journal of Network and Computer Applications*, 34: 1–11.
51. Catteddu D., Hogben G. (2009). *Cloud Computing: Benefits, Risks and Recommendations for Information Security.* European Network and Information Security Agency (ENISA). Web Application Security Communications in Computer and Information Science, vol. 72, p. 17.
52. Glott R., Husmann E., Sadeghi A., Schunter M. (2011). *Trust Worthy Clouds Underpinning the Future Internet.* Berlin: Springer, pp. 209–221.
53. Armbrust M., Fox A., Griffith R., Joseph A., Katz R., Konwinski A., Lee G., Patterson D., Rabkin A., Stoica I., Zaharia M. (2009). *Above the Clouds: A Berkeley View of Cloud*

Computing, Research Report. Berkeley, CA: UC Berkeley Reliable Adaptive Distributed Systems Laboratory.

54. Yan Y., Hao X. (2014). *Privacy Security Issues under Mobile Cloud Computing Mode.* Proceedings of International Conference on Computer, Communications and Information Technology (CCIT 2014), January 2014, Beijing, China, p. 49–52.
55. Wasserman, A. I. (2010). *Software Engineering Issues for Mobile Application Development.* ACM FoSER 2010, November 7–8, 2010, Santa Fe, NM.
56. Sakthivel S. (2007). Managing risk in offshore systems development. *European Journal of Operational Research*, 174: 245–264.
57. Kaliski B. S., Pauley W. (2010a). *Towards Risk Assessment as Service in Cloud Environments.* Hopkinton, MA: EMC Corporation.
58. Kaliski B. S., Pauley W. (2010b). *Towards Risk Assessment as Service in Cloud Environments.* 2nd Unisex Workshop on Hot Topics in Cloud Computing (HotCloud'10). Hopkinton, MA: EMC Corporation.
59. Pandey S., Voorsluys W., Niu S., Khandoker A., Buyya R. (2012). An autonomic cloud environment for hosting Ecg data analysis services. *Elsevier Future Generation Computer Systems*, 28: 147–154.
60. Samad J., Loke S. W., Reed K. (2013). *Quantitative Risk Analysis for Mobile Cloud Computing: A Preliminary Approach and a Health Application Case Study.* Proceedings of 12th IEEE International Conference on Trust, Security and Privacy in Computing and Communications (TrustCom), July 16–18, Melbourne, Australia, pp. 1378–1385.
61. Ramgovind S., Eloff M., Smith E. (2010). *The Management of Security in Cloud Computing.* IEEE Information Security for South Africa (ISSA) August 2–4, Johannesburg, South Africa.
62. Zhang X., Wuwong N., Li H., Xuejie Z. (2010). *Information Security Risk Management Framework for the Cloud Computing Environments.* 10th IEEE International Conference on Computer and Information Technology (CIT 2010), June 29–July 1, Bradford, United Kingdom, pp. 1328–1334.
63. Bisong A., Rehman S. (2011). An overview of the security concerns in enterprise cloud computing. *International Journal of Network Security & Its Applications*, 3(1): 99–110.
64. Carroll M., Merwe A. V., Kotzé P. (2011). *Secure Cloud Computing: Benefits, Risks and Controls.* Information Security South Africa (ISSA), August 15–17, Johannesburg, pp. 1–9.
65. Houmansadr A., Zonouz S. A., Berthier R. (2011). *A Cloud-Based Intrusion Detection and Response System for Mobile Phones.* IEEE/IFIP 41st International Conference on Dependable Systems and Networks Workshops (DSN-W), June 28–30, Hong Kong.
66. Ko R. K. L., Mowbray M., Pearson S., Kirchberg M., Liang Q., Lee B. S. (2011). *Trustcloud: A Framework for Accountability and Trust in Cloud Computing.* IEEE World Congress on Services, Cloud & Security Lab, Hewlett-Packard Laboratories.
67. Saripalli P., Walters B. (2010). *QUIRC: A Quantitative Impact and Risk Assessment Framework for Cloud Security.* IEEE 3rd International Conference on Cloud Computing, July 5–10, Florida, pp. 280–288.
68. Pearson S., Shen Y., Mowbray M. (2009). *Privacy Manager for Cloud Computing.* Berlin: Springer LNCS 5931, pp. 90–106.
69. La H. J., Kim S. D. (2010). *A Conceptual Framework for Provisioning Context-Aware Mobile Cloud Services.* IEEE 3rd International Conference on Cloud Computing (CLOUD), July 5–10, Florida, pp. 466–473.

70. Papakos P., Capra L., Rosenblum D. S. (2010). *Volare: Context-Aware Adaptive Cloud Service Discovery for Mobile Systems*. 9th International Workshop on Adaptive and Reflective Middleware, ARM'10, November 29–December 3, India, pp. 32–38.
71. Gerla, M. (2012). *Vehicular Cloud Computing*. Ad Hoc Networking Workshop (Med-Hoc-Net), 2012 The 11th Annual Mediterranean, June 19–22, pp. 152–155.

PART III

CLOUD MANAGEMENT

8

ENERGY CONSUMPTION OPTIMIZATION IN CLOUD DATA CENTERS

Dzmitry Kliazovich[1], Pascal Bouvry[4], Fabrizio Granelli[2], and Nelson L. S. da Fonseca[3]

[1]*Interdisciplinary Centre for Security, Reliability and Trust, University of Luxembourg, Luxembourg City, Luxembourg*
[2]*Department of Information Engineering and Computer Science, University of Trento, Trento, Trentino, Italy*
[3]*Institute of Computing, State University of Campinas, Campinas, São Paulo, Brazil*
[4]*Faculty of Science, Technology and Communications, University of Luxembourg, Luxembourg City, Luxembourg*

8.1 INTRODUCTION

Cloud computing has entered our lives and is dramatically changing the way people consume information. It provides platforms enabling the operation of a large variety of individually owned terminal devices. There are about 1.5 billion computers [1] and 6 billion mobile phones [2] in the world today. Next-generation user devices, such as Google glasses [3], offer not only constant readiness for operation, but also constant information consumption. In such an environment, computing, information storage, and communication become a utility, and cloud computing is one effective way of offering easier manageability, improved security, and a significant reduction in operational costs [4].

Cloud Services, Networking, and Management, First Edition.
Edited by Nelson L. S. da Fonseca and Raouf Boutaba.

Cloud computing relies on the data center industry, with over 500 thousand data centers deployed worldwide [5]. The operation of such widely distributed data centers, however, requires a considerable amount of energy, which accounts for a large slice of the total operational costs [6, 7]. Interactive Data Corporation (IDC) [8] reported that, in 2000, on average the power required by a single rack was 1 kW, although in 2008, this had soared to 7.4 kW. The Gartner group has estimated that energy consumption accounts for up to 10% of the current data center operational expenses (OPEX), and with this estimate possibly rising to 50% in the next few years [9]. The cost of energy for running servers may already be greater than the cost of the hardware itself [10, 11]. In 2010, data centers consumed about 1.5% of the world's electricity [12], with this percentage rising to 2% for the United States of America. This consumption accounts for more than 50 million metric of tons of CO_2 emissions annually.

Energy efficiency has never been a goal in the information technology (IT) industry. Since the 1980s, the only target has been to deliver more and faster; this has been traditionally achieved by packing more into a smaller space, and running processors at a higher frequency. This consumes more power, which generates more heat, and then requires an accompanying cooling system that costs in the range of $2–$5 million per year for corporate data centers [9]. These cooling systems may even require more power than that consumed by the IT equipment itself [13, 14].

Moreover, in order to ensure reliability, computing, storage, power distribution and cooling infrastructures tends to be over provisioned. To measure this inefficiency, the Green Grid Consortium [15] has developed two metrics: the power usage effectiveness (PUE) and data center infrastructure efficiency (DCIE) [16], which measures the proportion of power delivered to the IT equipment relative to the total power consumed by the data center facility. PUE is the ratio of total amount of energy used by a computer *data center* facility to the *energy* delivered to computing equipment while DCIE is the percentage value derived, by dividing *information technology* equipment power by total facility power. Currently, roughly 40% of the total energy consumed is related to that consumed by IT equipment [17]. The consumption accounts approximately, while the power distribution system accounts the other 15%.

There are two main alternatives for reducing the energy consumption of data centers: (1) shutting down devices or (2) scaling down performance. The former alternative, commonly referred to as dynamic Power Management (DPM) results in greatest savings, since the average workload often remains below 30% in cloud computing systems [18]. The latter corresponds to dynamic voltage and frequency scaling (DVFS) technology, which can adjust the performance of the hardware and consumption of power to match the corresponding characteristics of the workload.

In summary, energy efficiency is one of the most important parameters in modern cloud computing data centers in determining operational costs and capital investment, along with the performance and carbon footprint of the industry. The rest of the chapter is organized as follows: Section 8.2 discusses the role of communication systems in cloud computing. Section 8.3 presents energy efficient resource allocation and scheduling solutions. Finally, Section 8.4 concludes the chapter.

8.2 ENERGY CONSUMPTION IN DATA CENTERS: COMPONENTS AND MODELS

This section introduces the energy consumption of computing and communication devices, emphasizing how efficient energy consumption can be achieved, especially in communication networks.

8.2.1 Energy Consumption of Computing Servers and Switches

Computing servers account for the major portion of energy consumption of data centers. The power consumption of a computing server is proportional to the utilization of the CPU utilization. Although an idle server still consumes around two-thirds of the peak-load consumption just to keep memory, disks, and I/O resources running [19, 20]. The remaining one-third increases almost linearly with an increase in the load of the CPU [6, 20]:

$$P_s(l) = P_{\text{fixed}} + \frac{(P_{\text{peak}} - P_{\text{fixed}})}{2}(1 + l - e^{-\frac{l}{a}}), \tag{8.1}$$

where P_{fixed} is idle power consumption, P_{peak} is the power consumed at peak load, l is a server load, and a is the level of utilization at which the server attains power consumption which varies linearly [0.2, 0.5].

There are two main approaches for reducing energy consumption in computing servers: (1) DVFS [21] and (2) DPM [22]. The former scheme adjusts the CPU power (consequently the level of performance) according to the load offered. The power in a chip decreases proportionally to V^2f, where V is a voltage, and f is the operating frequency. The scope of this DVFS optimization is limited to the CPUs, so that the computing server components, such as buses, memory, and disks, continue to function at the original operating frequency. On the other hand, the DPM scheme can power down computing servers but including all of their components, which makes it much more efficient; but if a power up (or down) is required, considerably more energy must be consumed in comparison to the DVFS scheme. Frequency downshifts can be expressed as follows (Eq. 8.1):

$$P_s(l) = P_{\text{fixed}} + \frac{(P_{\text{peak}} - P_{\text{fixed}})}{2}(1 + l^3 - e^{-\frac{l^3}{a}}). \tag{8.2}$$

Figure 8.1 plots the power consumption of computing server.

Network switches form the basis of the interconnection fabric used to deliver job requests to the computing servers for execution. The energy consumption of a switch depends on various factor: (i) type of switch, (ii) number of ports, (iii) port transmission rates, and (iv) employed cabling solutions; these can be expressed as follows [23]:

$$P_{\text{switch}} = P_{\text{chassis}} + n_c \times P_{\text{linecard}} + \sum_{r=1}^{R} n_p^r \times p_p^r \times u_p^r, \tag{8.3}$$

where P_{chassis} is the power related to the switch chassis, P_{linecard} is the power consumed by a single line card, n_c is the number of line cards plugged into the switch, P_p^r is the

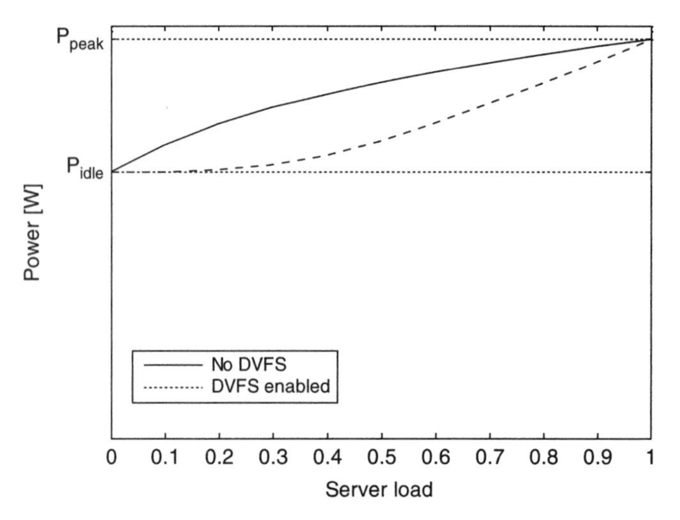

Figure 8.1. Computing server power consumption.

power consumed by a port running at rate r, n_p^r is the number of ports operating at rate r and $u_p^r \in [0, 1]$ is a port utilization, which can be defined as follows:

$$u_p = \frac{1}{T} \int_{t}^{t+T} \frac{B_p(t)}{C_p} dt = \frac{1}{T * C_p} \int_{t}^{t+T} B_p(t) dt \tag{8.4}$$

where $B_p(t)$ is an instantaneous throughput at the port's link at the time t, C_p, is the link capacity, and T is the time interval between measurements.

8.2.2 Energy Efficiency

In an ideal data center, all the power would be delivered to the IT equipment executing user requests. This energy would then be divided between the communication and the computing hardware. Several studies have mistakenly considered the communication network as overhead, required only to deliver the tasks to the computing servers. However, as will be discussed later in this section, communications is at the heart of task execution, and the characteristics of the communication network such as bandwidth capacity, transmission delay, delay jitter, buffering, loss ratio, and performance of communication protocols, all greatly influence the quality of task execution.

Mahadevan et el. [23] present power benchmarking of the most common networking switches. With current network switch technology, the difference in power consumption between peak consumption and idle state is less than 8%; turning off an unused port saves only 1–2 W [24]. The power consumption of a switch comprises three components: (1) power consumed by the switch base hardware (the chassis), (2) power consumed by active line cards, and (3) power consumed by active transceivers. Only the last component scales with the transmission rate, or the presence of the forwarded traffic, while the former two components remain constant, even when the switch is idle. This phenomenon is known as energy proportionality, and describes how energy consumption increases with an increase in workload [24].

Making network equipment energy proportional is one of the main challenges faced by the research community. Depending on the data center load level, the communication network can consume between 30 and 50% of the total power used by the IT equipment [21, 26] with 30% being typical for highly loaded data centers, whereas 50% is common for average load levels of 10–50% [27]. As with computing servers, most solutions for energy-efficient communication equipment depend on downgrading the operating frequency (or transmission rate) or powering down the entire device or its components in order to conserve energy. One solution, first studied by Shang et al. [25] and Benini et al. [28] in 2003, proposed a power-aware interconnection network utilized dynamic voltage scaling (DVS) links [25], and this, DVS technology was later combined with dynamic network shutdown (DNS) to further optimize energy consumption [29]. Refs. [30–34] review the challenges and some of the most important solutions for optimization of energy consumption and the use of resources.

The design of these power-aware networks when on/off links are employed is challenging. There are issues with connectivity, adaptive routing, and potential network deadlocks [35]. Because a network always remains connected, such challenges are not faced when using DVS links. Some recent proposals combined traffic engineering with link shutdown functionality [36], but most of these approaches are reactive and may perform poorly in the event of unfavorable traffic patterns. A proactive approach is necessary for on/off procedures. A number of studies have demonstrated that simple optimization of the data center architecture and energy-aware scheduling can lead to significant energy savings of up to 75% based on traffic management and workload consolidation techniques [21].

8.2.3 Communication Networks

Communication systems have rarely been extensively considered in cloud computing research. Most of the cloud computing techniques evolved from the fields of cluster and grid computing which are both designed to execute large computationally intensive jobs, commonly referred as high-performance computing (HPC) [37]. However, cloud computing is fundamentally different: Clouds satisfy the computing and storage of millions of users at the same time, yet each individual user request is relatively small. These users commonly need merely to read an email, retrieve an HTML page, or watch an online video. Such tasks require only limited computation to be performed, yet their performance is determined by the successful completion of the communication requests but communications involves more than just the data center network; the data path from the data center to the user also constitute an integral part for satisfying a communication request. Typical delays for processing users' requests, such as search, social networks, and video streaming, are less than a few milliseconds, and we sometimes even measured on the level of microsecond. Depending on the user location, these delays are as large as 100 milliseconds for intercontinental links and up to 200 milliseconds if satellite links are involved [38]. As a result, a failure to consider the communication characteristics on an end-to-end basis can mislead the design and operational optimization of modern cloud computing systems.

Optimization of cloud computing systems and cloud applications will not only significantly reduce energy consumption inside data centers, but also globally, in the wide-area network. The world hosts around 1.5 billion Internet users [1] and 6 billion mobile phone users [2], and all of them are potential customers for cloud computing applications. On an average, there are 14 hops between a cloud provider and end users on the Internet [39, 40]. This means that there are 13 routers involved in forwarding the user traffic, each consuming from tens of watts to kilowatts [23]. According to Nordman [41], Internet-connected equipment accounts for almost 10% of the total energy consumed in the United States. Obviously, optimization of the flow of communication between the data center providers and end users can make a significant difference. For example, a widespread adoption of the new Energy-Efficient Ethernet standard IEEE 802.3az [42] can result in savings of €1 billion [43].

At the cloud user end, energy is becoming an even greater concern: More and more cloud users use mobile equipment (smart phones, laptops, tablet PCs) to access cloud services. The only efficient way for these battery-powered devices to save power is to power off most of the main components, including the central processor, transceivers and memory, while also configuring sleeping cycles appropriately [44]. The aim is to decrease request processing time so that user terminals will consume less battery power. Smaller volumes of traffic arranged in bursts will permit longer sleeping times for the transceivers, and faster replies to the cloud service requests will reduce the drain on batteries.

8.3 ENERGY EFFICIENT SYSTEM-LEVEL OPTIMIZATION OF DATA CENTERS

8.3.1 Scheduling

This section addresses issues related to scheduling, load balancing, data replication, virtual machine placement, and networking that can be capitalized on to reduce the energy consumption in data centers.

Job scheduling is at the heart of the successful power management in data centers. Most of the existing approaches focus exclusively on the distribution of jobs between computing servers [45], the targeting of energy efficiency [46], or thermal awareness [47]. Only a few approaches consider the characteristics of the data center network [48–50], such as DPM-like power management [18].

Since energy savings result from such DPM-like power management procedures [18], job schedulers tend to adopt a policy of workload consolidation maximizing the load on the operational computing servers and increasing the number of idle servers that can be put into the "sleep" mode. Such a scheduling policy works well in systems that can be treated as a homogenous pool of computing servers, but data center network topologies require special policies. For example, the most widely used data center architecture [51], fat-tree architecture presented in Figure 8.2, blindly concentrates scheduling and may end up grouping all of the highly loaded computing servers on a few racks, yet this creates a bottleneck for network traffic at a rack or aggregation switch.

Moreover, on a rack level, all servers are usually connected using Gigabit Ethernet (GE) interfaces. A typical rack hosts up to 48 servers, but has only 2 links of 10GE connecting them to the aggregation network. This corresponds to a mismatch of 48GE/20GE = 2.4 between the incoming and the outgoing bandwidth capacities.

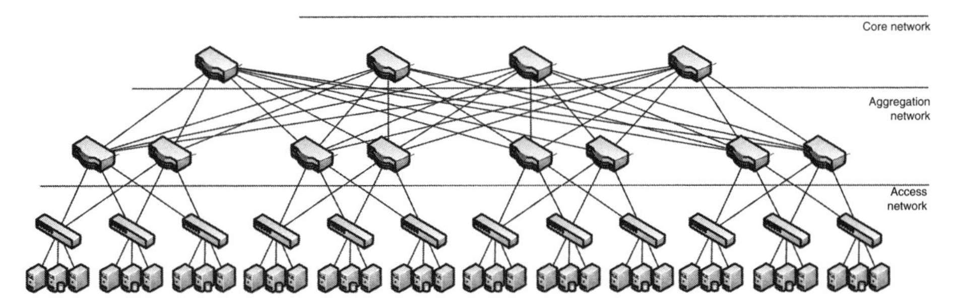

Figure 8.2. Three-tier data center architecture.

Implementation in a data center with cloud applications requiring communication means that the scheduler should tradeoff workload concentration with the load balancing of network traffic.

Any of the data center switches may become congested in either the uplink or downlink direction or both. In the downlink direction, congestion occurs when the capacity of individual ingress links surpasses that of egress links. In the uplink direction, the mismatch in bandwidth is primarily due to the bandwidth oversubscription ratio, which occurs when the combined capacity of server ports overcomes a switch aggregate uplink capacity.

Congestion (or hotspots) may severely affect the ability of a data center network to transport data. The Data Center Bridging Task Group (IEEE 802.1) [52] specifies layer-2 solutions for congestion control in IEEE 802.1Qau standard. This standard introduces a feedback loop between data center switches to signal the presence of congestion. Such feedback allows overloaded switches to backpressure heavy senders by notifying them when congestion occurs. Such technique can avoid some of the congestion-related losses and keep the data center network utilization high. However, it does not address the problem adequately since as it is more efficient to assign data-intensive jobs to different computing servers so that those jobs can avoid sharing common communication paths. To benefit from such spatial separation in the three-tiered architecture (Fig. 8.2), these jobs must be distributed among the computing servers in proportion to job communication requirements. However, such approach contradict the objectives of energy-efficient scheduling, which tries to concentrate all of the active workloads on a minimum set of servers and involve a minimum number of communication resources.

Another energy-efficient approach would be the DENS methodology, which takes the potential communication needs of the components of the data center into consideration along with the load level to minimize the total energy consumption when selecting the best-fit computing resource for job execution. Communicational potential is defined as the amount of end-to-end bandwidth provided to individual servers or group of servers by the data center architecture. Contrary to traditional scheduling solutions that model data centers as a homogeneous pool of computing servers [45], the DENS methodology develops a hierarchical model consistent with the state of the art of topology of data centers. For a three-tier data center (see Fig. 8.2), DENS metric M is defined as a weighted combination of server-level (f_s), rack-level (f_r), and module-level (f_m) functions:

$$M = \alpha \cdot f_s + \beta \cdot f_r + \gamma \cdot f_m \tag{8.5}$$

where α, β, and γ are weighted coefficients that define the impact of the corresponding components (servers, racks, and/or modules) on the metric behavior. Higher α values favor the selection of highly loaded servers in lightly loaded racks. Higher β values will give priority to computationally loaded racks with low network traffic activity. Higher γ values favor the selection of loaded modules.

The selection of computing servers combines the server load $L_S(l)$ and the communication potential $Q_r(q)$ corresponding to the fair share of the uplink resources on the top of the rack ToR switch. This relationship is given as follows:

$$f_S(l,q) = L_S(l) \cdot \frac{Q_r(q)^{\varphi}}{\delta_r} \tag{8.6}$$

where $L_S(l)$ is a factor depending on the load of the individual servers l, $Q_r(q)$ defines the load at the rack uplink by analyzing the congestion level in the switch's outgoing queue q, δ_r is a bandwidth over provisioning factor at the rack switch, and φ is a coefficient defining the proportion between $L_S(l)$ and $Q_r(q)$ in the metric. Given that both $L_S(l)$ and $Q_r(q)$ must be within the range [0, 1] higher φ values will decrease the importance of the traffic-related component $Q_r(q)$.

The fact that the energy consumption of an idle server consumes merely two-third of that at peak consumption [19] suggests that an energy-efficient scheduler must consolidate data center jobs on the minimum possible set of computing servers. On the other hand, keeping servers constantly running at peak loads may decrease hardware reliability and consequently affect job execution deadlines [53]. These issues are addressed with DENS load factor, the sum of two sigmoid functions:

$$L_S(l) = \frac{1}{1+e^{-10\left(l-\frac{1}{2}\right)}} - \frac{1}{1+e^{-\frac{10}{\varepsilon}\left(l-\left(1-\frac{1}{\varepsilon}\right)\right)}} \tag{8.7}$$

The first component in Equation (8.8) defines the shape of the main sigmoid, while the second serves to encourage convergence toward the maximum server load value (see Fig. 8.3). The parameter ε defines the size and the inclination of this falling slope and he server load l is within the range [0,1].

Figure 8.4 presents the combined server load and queue-size-related components. The bell-shaped function obtained favors the selection of servers with a load level above average located in racks with little or no congestion.

8.3.2 Load Balancing

Enabling the sleep mode in idle computing servers and network hardware is the most efficient method of avoiding unnecessary power consumption. Consequently, load balancing becomes the key enabler for saving energy.

However, changes in the power mode introduce considerable delays. Moreover, the inability of instantaneous wake up of a sleeping server means that a pool of idle servers must be available to be able to accommodate incoming loads in the short term and prevent quality-of-service (QoS) degradation. It should be remembered that data centers are required to provide a specific level of quality of service, defined as service-level agreements (SLAs), even at peak loads. Therefore, they tend to over provision computing and communication resources. In fact, on average, data center are functioning

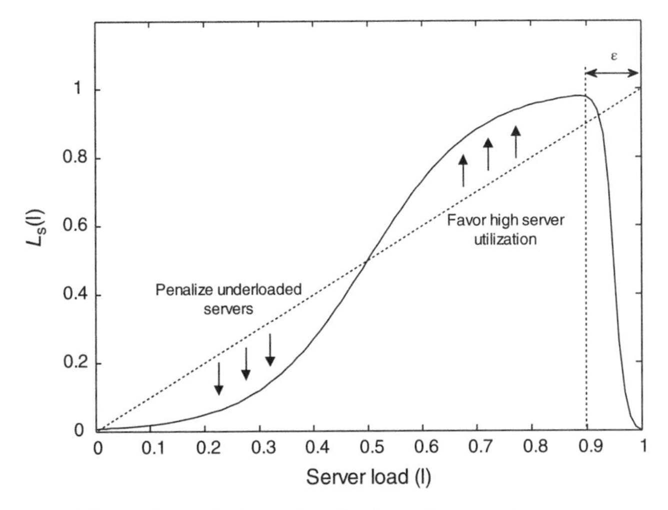

Figure 8.3. DENS metric selection of computing server.

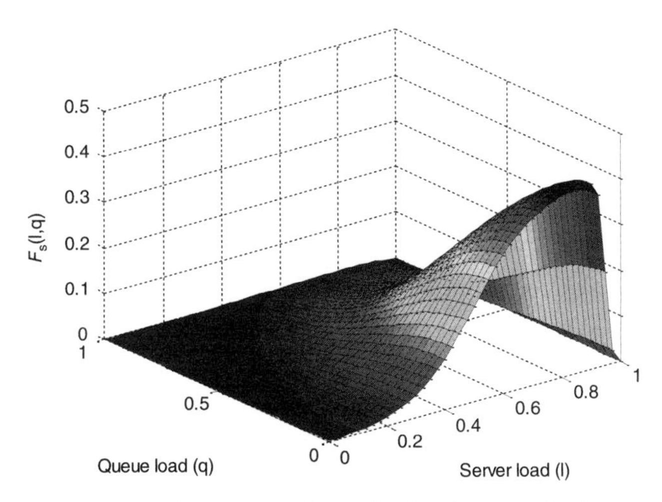

Figure 8.4. Server selection according to load and communication potential.

at only 30% of their capacity. The load in data centers is highly correlated with region and time of the day since more users are active during the daytime hours; the number of users during the day is almost double that at night. Moreover, user arrival rate is not constant, but can spike due to the crowd effect. Most of the time almost 70% of data center servers, switches, and links remain idle, although during peak periods, this usage can reach 90%. However, idle servers still need to run OS software, maintain virtual machines, and power on both peripheral devices and memory. As a result, even when being idle, servers still consume around two-thirds of the peak power consumption. In switches, this ratio is even higher with the energy consumed being shared by the switch chassis, the line cards, and the transceiver ports. Moreover, various Ethernet standards require the uninterrupted transmission of synchronization symbols in the physical layer to guarantee the synchronization required prevents the downscaling of the consumption of energy, even when no user traffic is transmitted.

An energy-efficient scheduler for cloud computing applications with traffic load balancing can be designed to optimize energy consumption of cloud computing data centers, like e-STAB proposed in Ref. [54]. One of these is the e-STAB scheduler, which gives equal treatment to communicational demands and computing requirements of jobs. Specifically, e-STAB aims at (i) balancing the communication flows produced by jobs and (ii) consolidating jobs using a minimum of computing servers. Since network traffic can be highly dynamic and often difficult to predict [55], the e-STAB scheduler analyzes both load on the network links and occupancy of outgoing queues at the network switches. This queuing analysis helps prevent a buildup of network congestion. This scheduler is already involved in various transport-layer protocols [56] estimating buffer occupancy of the network switches and can react before congestion-related losses occur.

The e-STAB scheduling policy involves the execution of the following two steps for each incoming cloud computing data center job:

Step 1: Select a group of servers *S* connected to the data center network with the highest available bandwidth, if at least one of the servers in *S* can accommodate the computational demands of the scheduled job. The available bandwidth is defined as the unused capacity of the link or a set of links connecting the group of servers *S* to the rest of the data center network.

Step 2: Within the selected group of servers, *S*, select a computing server with the least available computing capacity, but sufficient to satisfy the computational demands of the scheduled task.

One of the main goals of the e-STAB scheduler is to achieve load-balanced network traffic as well as to prevent network congestion. A helpful measure is the available bandwidth per computing node within the data center. However, such a measure does not capture the dynamics of the system, such as sudden increase in the transmission rate of cloud applications.

To provide a more precise measure of network congestion, e-STAB adjusts scales the available bandwidth to the component related to the size of the bottleneck queue (see Fig. 8.5). This favors empty queues or queues with minimum occupancy and penalizes highly loaded queues that are on the threshold of buffer overflow (or on the threshold of losing packets).

By using the available bandwidth with the component $Q(t)$ metric, the available per-server bandwidth can be computed for modules and individual racks as follows:

$$Fr_j(t) = \frac{1}{T}\int_{t}^{t+T}\left(\frac{(Cr_j - \lambda r_j(t)) \cdot e^{-(\rho \cdot qr_j(t)/Qr_j \cdot \max)^{\varphi}}}{Sr_j}\right) dt \tag{8.8}$$

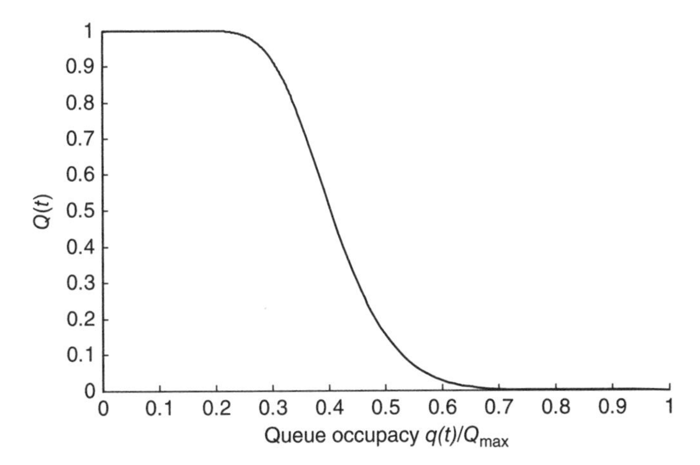

Figure 8.5. Queue-size related component of the STAB scheduler.

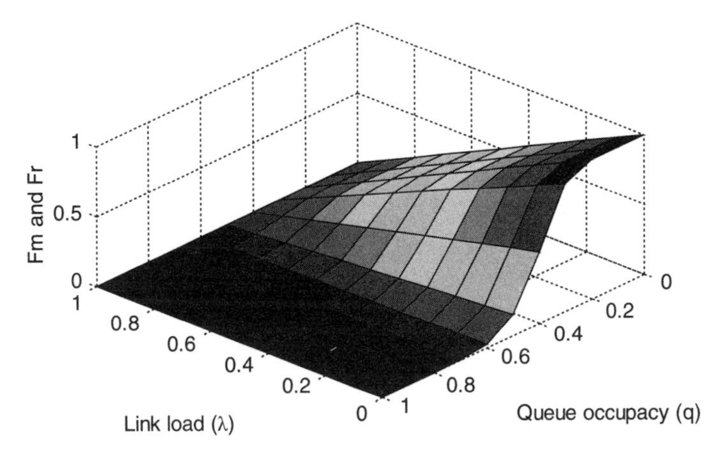

Figure 8.6. Selection of racks and modules by the STAB scheduler.

where $Qr_j(t)$ is the weight associated with occupancy levels of the queues, $qr_j(t)$ is the size of the queue at time t, and $Qr_j \cdot \max$ is the maximum size of the queues allowed at the rack j.

Figure 8.6 presents the evolution of $Fr_j(t)$ with respect to different values of the network traffic and buffer occupancy. The function is insensitive to the level of utilization of the network links for highly loaded queues, while for lightly loaded queues, the links with the lighter load are preferred to the heavily used ones.

Having selected a proper module and a rack based on their traffic load and congestion state indicated by the queue occupancy, we must select a computing server for the job execution. To do so, we must analyze energy consumption profile of the servers.

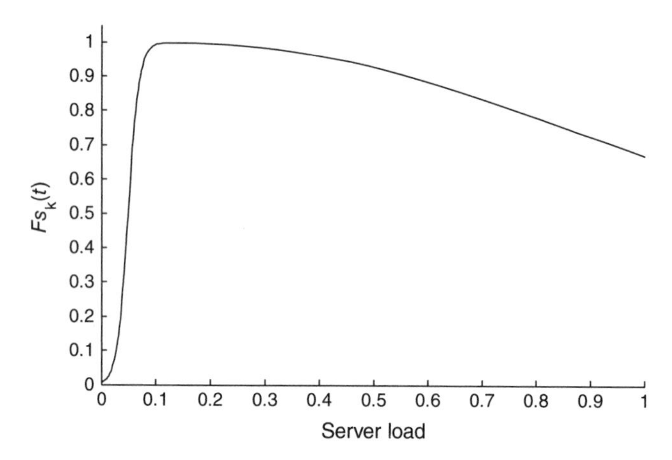

Figure 8.7. Selection of computing servers by the STAB scheduler.

Once the energy consumption of a server is known, it is possible to derive a metric to be used by the e-STAB scheduler for server selection, as follows:

$$FS_k(t) = \frac{1}{T}\int_t^{t+T}\left(\frac{1}{1+e^{-\frac{10}{\varepsilon}}\left(l_k(t)-\frac{\varepsilon}{2}\right)}\right) - \frac{1}{2}\left(1-\frac{P_{idle}}{P_{peak}}\right)\left(1+l_k(t)^3 - e^{-\left(\frac{l_k(t)}{T}\right)^3}\right)dt, \tag{8.9}$$

where $l_k(t)$ is the instantaneous load of server k at time t and T is an averaging interval. While the second summand under the integral in Equation (8.9) is a reverse normalized version of Equation (8.2), the first summand is a sigmoid designed to penalize selection of idle servers for job execution. The parameter ε corresponds to the CPU load of an idle server required to keep the operating system and virtual machines running. Figure 8.7 presents a chart for $FS_k(t)$.

Balancing the load in federated data centers has also been proposed to reduce energy consumption. In Ref. [57], the authors propose an algorithm to migrate virtual machines to data centers, which use renewable energy. They consider a cloud computing scenario with many data centers, some of them powered by brown (non-renewable) energy sources and others with access to green (renewable) energy. They propose an algorithm to decide on a set of long-lived VM's to be migrated to data centers with access to green energy, taking into consideration both the data center network topology and the energy consumption. They also consider the impact of migration and the use renewable energy availability to enhance their strategy. They show that the brown energy can be replaced by green energy with a low increase in overall the consumption.

A comprehensive work on load balancing in distributed data centers is presented in Ref. [58]. Using an optimization modeling, the authors present distributed algorithms for achieving optimal geographical load balancing. They also present a study about the effect of green and brown energy, highlighting its potential benefits.

In Ref. [59], the authors propose an approach to relocate workload in distributed data centers. The algorithm is based on electricity prices in cities where the centers are located and in the cost of migration. The solution was modelled as an optimization problem. Using traces of social network applications and real cost of electricity in different regions of the United States, the authors manage to reduce the average electricity cost. This work does not take into account the energy consumption of data center network, modelling only the overall load of the data center.

The authors of Ref. [60] propose a framework called stochastic power reduction scheme (SAVE) for geographically distributed data centers. The approach was designed for delay tolerant workloads, such as MapReduce jobs, and two different techniques are used for achieving energy savings: switching off the unused hosts, and DVSF. The proposed solution was modeled as an optimization problem in which each data center is represented as a set of physical hosts. Jobs can be executed either in the data center to which they were submitted or in another data center in case the cost of the energy consumed will be lower.

8.3.3 Data Replication

The performance of cloud computing applications, such as gaming, voice and video conferencing, online office, storage, backup, and social networking, depends largely on the availability and efficiency of high-performance communication resources. For better reliability and low latency service provisioning, data resources can be brought closer (replicated) to the physical infrastructure, where the cloud applications are running. A large number of replication strategies for data centers have been proposed in the literature [61–65]. These strategies optimize system bandwidth and data availability between geographically distributed data centers. However, none of them focuses on energy efficiency and replication techniques inside data centers.

In Ref. [61], an energy efficient data replication scheme have been proposed for data center storage. Under utilized storage servers can be turned off to minimize energy consumption, although one of the replica servers must be kept for each data object to guarantee availability. In Ref. [62], dynamic data replication in a cluster of data grids is proposed. This approach creates a policy maker, which is responsible for the replica management. It periodically collects information from the cluster heads,with significance determined by a set of weights selected according to the age of the reading. The policy maker further determines the popularity of a file based on the access frequency. To achieve load balancing, the number of replicas for a file is computed in relation to the access frequency of all other files in the system. This solution follows a centralized design approach, however, leaving it vulnerable to a single point of failure.

Other proposals have concentrated on replication strategies between multiple data centers. In Ref. [63], power consumption in the backbone network is minimized by linear programming to determine the optimal points of replication on the basis of data center traffic demands and the popularity of data objects. This relation of the traffic load to power consumption at aggregation ports is linear and, consequently, optimization approaches that consider the traffic demand can bring significant power savings.

Another proposal for replication is designed to conserve energy by replicating data closer to consumers to minimize delays. The optimal location for replicas of each data object is determined by periodically processing a log of recent data accesses. The replica site is then determined by employing a weighted k-means clustering of user locations and deploying the replica closer to the centroid of each cluster. Migration will take place from one site to another if the gain in quality of service from migration is higher than a predefined threshold.

Another approach is cost-based data replication [65]. This approach analyzes failures in data storage and the probability of data loss probability, which are directly related to each other, and builds a reliability model. Time points for replica creation are then determined from the data storage reliability function.

The approach presented in Ref. [66] is different from all the others replication approaches discussed earlier due to (i) the scope of the data replication, which is implemented both within a single data center and between geographically distributed data centers and (ii) the optimization target, which takes into account system energy consumption, network bandwidth and communication delay to define the replication strategy to be employed.

Large-scale cloud computing systems comprise data centers geographically distributed around the globe (see Fig. 8.8). The central database (Central DB) is located in the wide-area network and hosts all the data required by the cloud applications. To speed

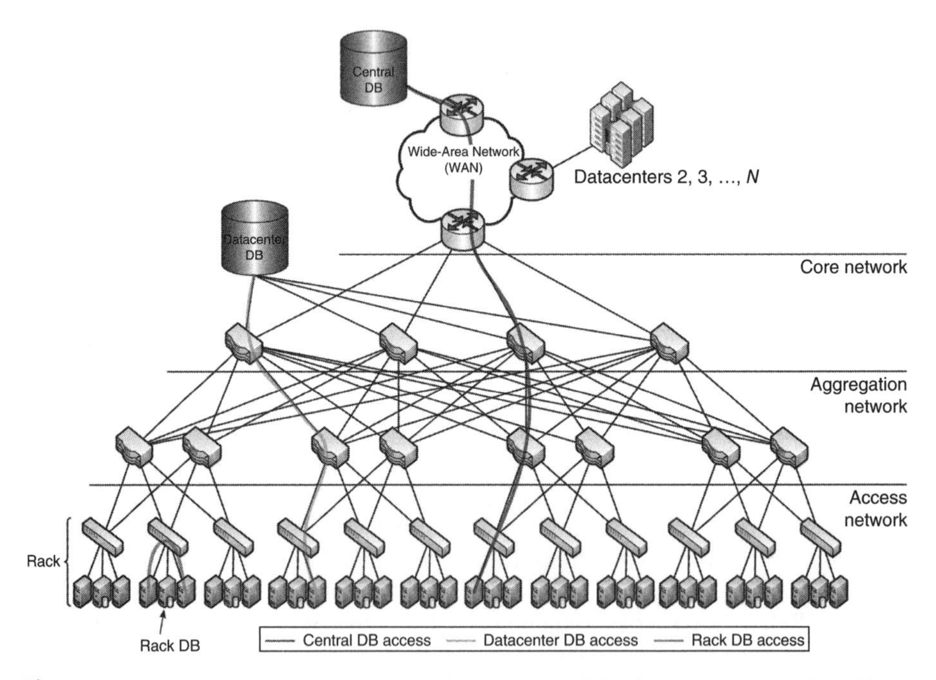

Figure 8.8. Replication in cloud computing data centers. All database requests produced by the cloud applications running on computing servers are first directed to the rack-level database server. Rack DB either replies with the requested data or forwards the request to the Data center DB. In a similar fashion, the Data center DB either satisfies the request or forwards it up to the Central DB.

up database access and reduce access latency, each data center hosts a local database, called a data center database (Data center DB), which is used to replicate the most frequently used data items from the central database. Moreover, each rack hosts at least one server capable of running a local rack-level database (Rack DB), which is used for subsequent replication from the data center database.

When data are requested, the information about requesting server, rack, and data center is stored. Moreover, the statistics showing the number of accesses and updates are maintained for each data item. The access rate (or popularity) is measured as the number of access events per period of time. While accessing data items, cloud applications can also modify them. Such modifications must be sent back to the database so that all replica sites will be updated.

A module located at the central database, the replica manager, periodically analyzes data access statistics to identify what items are the most suitable for replication and at which replication sites. The availability of these access and update statistics makes it possible to project data center bandwidth usage and energy consumption.

Figure 8.9 presents the requirements of downlink bandwidth. Since it is proportional to both the size of a data item and the rate of update, the bandwidth consumption grows rapidly and easily overtakes the corresponding capacities of the core, aggregation and access segments of the data center network requiring replication.

Figure 8.10 reports the trade-off between data center energy consumption, including the consumption of both the servers and network switches, and the downlink residual bandwidth. For all replication scenarios, the core layer reaches saturation first since it is the smallest of the data center network segments and has capacity of only 320 GB/s. The residual bandwidth for all network segments generally decreases with increase in load, except for the gateway link, for which the available bandwidth remains constant for both Data center DB and Rack DB replication scenarios, since data queries are processed at the

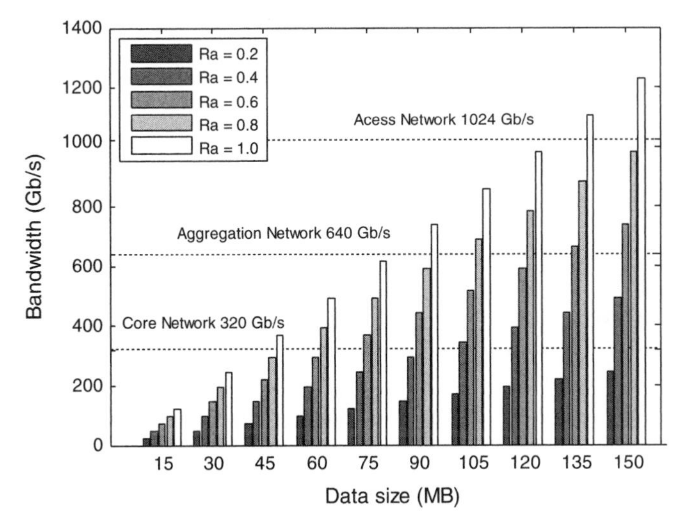

Figure 8.9. Downlink bandwidth requirements.

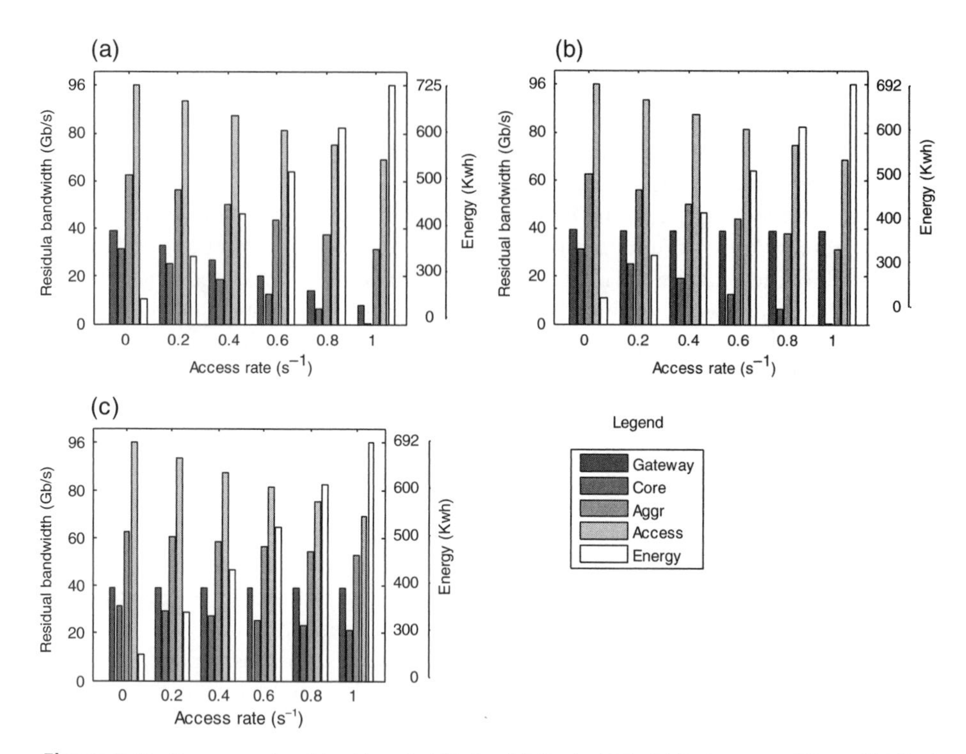

Figure 8.10. Energy and residual bandwidth for (a) Central DB, (b) Data center DB, and (c) Rack DB replication scenarios.

replica databases and only data updates are routed from the Central DB to the Data center DB. The benefit of Rack DB replication is two-fold: on the one hand network, traffic can be restricted to the access network, which has lower nominal power consumption and higher network capacity, while on the other hand, data access becomes localized, thus improving performance of cloud applications.

8.3.4 Placement of Virtual Machines

Virtualization represents a key technology for efficient operation of cloud data centers. Energy consumption virtualized data centers can be reduced by appropriate decision on which physical server a virtual machines should be placed. Virtual machine consolidation strategies try to use the lowest possible number of physical machines to host a certain number of virtual machines. Some proposed strategies are described next.

In Ref. [67], the authors developed a strategy for traditional three-tier data center architectures which takes into consideration the energy consumption of both servers and network switches. The proposed strategy analyzes the load of each network switch to avoid overloading them. It tries to compromise load balancing of data center network traffic and consolidation of virtual machines. Such compromise is important to the operation of data centers running jobs that impose low computational load but produce heavy traffic streams.

The problem of virtual machine placement has been addressed by different formulations of the bin-packing problem. The proposal in Ref. [46] employs a variation of the best fit decreasing algorithm. Although, in this case, only the energy consumption of servers is considered, results showed potential energy savings without a significant number of violation of service level agreements. In Ref. [68], a heuristic is proposed to achieve server utilization close to an optimal level determined by the computation of the Euclidean distance of the allocation state. A first fit decreasing strategy was employed in Ref. [69] for data centers processing Web search and MapReduce applications. The consolidation approach is based on the analysis of CPU usage, and favors the placement of correlated virtual machines in distinct physical servers, to avoid overloading the servers.

The formulation of virtual machine problem presented in [70] includes active cooling control besides the traditional approaches such as DPM and DVFS. This work also does not take into account the contribution of network switches to the energy consumption of a data center and it shown that active cooling control result in small, but relevant, gains.

The work in Ref. [71] promotes energy reduction by consolidating network flows instead of virtual machines; only the consumption of network switches are considered. Correlated flows are analyzed and assigned to network paths in a greedy way. This approach employs link rate adaptation and shutting down of switches with low utilization. Results derived using simulations based on real traces of Wikipedia traffic demonstrated that this approach can in fact reduce energy consumption.

8.3.5 Communications Infrastructure

The energy efficiency of a data center also depends on the underlying communication infrastructure. Indeed, at the average load level of a data center, the communication network consumes between 30 and 50% of the total power used by the IT equipment; this in turn represents roughly 40% of the total energy budget.

Moreover, an analysis of the distribution of data traffic in clouds suggests that the majority of the traffic is transferred within the data center itself (around 75%), with rest being split between communication with users (18%) and data center to data center exchanges (7%) [72].

Based on these facts, it is clear the need to develop energy efficient solutions for communication technologies and architectures to interconnect the servers in data centers. Since high speed and high capacity are required, the most suitable communication technology for cloud data centers is optical. In the remainder of this section, some possible architectures addressing energy efficient solutions for internal communications in data centers are presented.

Optical interconnection networks are a novel alternative technology to provide high bandwidth, low latency and reduced power consumption. Up until recently, such optical technology has been used only for point-to-point links to connect the electrical switches (fiber optics) thus reducing noise and leaving smaller footprints. However, since the switches operate in the electrical domain, power hungry electrical-to-optical (E/O) and optical-to-electrical (O/E) transceivers are required.

New modules connecting the silicon chip directly with optical fibers have been developed, thus enabling switching to be performed in the optical domain.

Optical interconnections can be based on circuit switching or packet switching, each generating different trade-off in terms of energy versus performance. Solely in terms of energy efficiency, optical circuit switching represents the most efficient solution, but it leads to high reconfiguration times due to the nature of circuit switching. On the other side, packet switching, although less energy efficient, potentially the source of greater latency, achieves better performance, since its reconfiguration time is lower and its scalability higher.

One recent alternative is the usage of optical OFDM. Optical OFDM distributes the data on a large number of low data rate subcarriers, and can thus provide fine-granularity capacity to connections by the elastic allocation of subcarriers according to connection demands.

The use of optical OFDM as a bandwidth-variable and highly spectrum-efficient modulation format can provide scalable and flexible sub- and super-wavelength granularity, compared to the conventional, fixed-bandwidth fixed-grid WDM network. However, this new concept poses new challenges for the routing and wavelength assignment algorithms. Indeed, traditional algorithms for routing and wavelength assignment will no longer be directly applicable for such new kinds of communication infrastructure.

8.4 CONCLUSIONS AND OPEN CHALLENGES

Costs and operating expenses have become a growing concern in the cloud computing industry, with energy consumption accounting for a large percentage of the operational expenses in the data centers used as backend computing infrastructure. This chapter emphasizes the role of communications and network awareness of this consumption and presents suggested solutions for energy efficient resource allocation in clouds.

The challenge of energy efficiency will largely determine the future of cloud computing systems, at present experiencing unprecedented growth. Most of the existing energy-efficient and performance optimization solutions in the IT domain focus on computing, with communications-related processes relegated to a secondary role or unaccounted for. In reality, however, communications are at the heart of cloud systems, and network characteristics, such as bandwidth capacity, transmission delay, delay jitter, buffering, loss rate and performance of communication protocols, often determine the quality of task execution. However, most current research is restricted to processes inside data centers, yet the models must also account for communication dynamics in the wide-area network, and at the user end.

Open research challenges are essentially related to improving the energy scalability of cloud computing. The previous sections have underlined the need for the joint optimization of computing and communication while maintaining an appropriate balance between performance and energy consumption for the overall architecture.

The following specific research challenges have been identified:

- Integration of novel and more efficient energy consumption models for the different components of the cloud computing architecture. As the concept of energy-proportional computing is emerging in the design of computing hardware and software infrastructures, it is also becoming relevant in the design of communication equipment. These emerging models will drive the need for improved and innovative approaches for the joint optimization and balancing of performance and energy consumption in cloud computing.
- The concept of Mobile Cloud, deriving from the clear trend toward user mobility (and the "always on" paradigm) and the availability of ever more powerful devices in the hands of the cloud services' users are shaping the possibility of even more pervasive usage of the cloud computing infrastructure. Users' request for 24/7 availability of cloud services even in sparsely "covered" areas, will lead to a redefinition or least an evolution, of the cloud architecture, which will involve the need for efficient dissemination of both information and services across the Internet, whether in data centers, on users devices, or somewhere in between. This is sure to have an impact on the way data are replicated and services are provided.

REFERENCES

1. Internet World Statistics, available at http://www.internetworldstats.com. Accessed November 18, 2013.
2. J. Ekholm and S. Fabre "Forecast: Mobile Data Traffic and Revenue, Worldwide, 2010–2015," Market report, Gartner Inc., 2011.
3. Google Glass project, available at https://plus.google.com/111626127367496192147. Accessed November 18, 2013.
4. A. Weiss, "Computing in the clouds," netWorker, vol. 11, no. 4, pp. 16–25, 2007.
5. "State of the Data Center 2011," Emerson Network Power, Columbus, OH, 2011.
6. X. Fan, W.-D. Weber, and L. A. Barroso, "Power provisioning for a warehouse-sized computer," Proceedings of the ACM International Symposium on Computer Architecture, San Diego, CA, June 2007.
7. R. Raghavendra, P. Ranganathan, V. Talwar, Z. Wang, and X. Zhu, "No "Power" Struggles: Coordinated Multi-level Power Management for the Data Center," APLOS, ACM, New York, 2008.
8. Interactive Data Corporation, available at: http://www.interactivedata.com/. Accessed November 18, 2013.
9. Gartner Group, available at: http://www.gartner.com/. Accessed November 18, 2013.
10. A. Vasan and A. Sivasubramaniam "Worth their watts?-an empirical study of data center servers," IEEE 16th International Symposium on High Performance Computer Architecture (HPCA), Bangalore, India, January 9–14, 2010.
11. IDC, "Worldwide Server Power and Cooling Expense 2006–2010," Market Analysis, 2006.
12. J. G. Koomey, "Growth in Data Center Electricity Use 2005 to 2010," Analytics Press, Oakland, CA, 2011.

13. "Reducing Data Center Cost with an Air Economizer," Intel IT@Intel Brief, August 2008.
14. N. Rasmussen, "Calculating Total Cooling Requirements for Data Centers," White Paper #25, American Power Conversion, 2007.
15. The Green Grid Consortium, available at http://www.thegreengrid.org/. Accessed November 18, 2013.
16. A. Rawson, J. Pfleuger, and T. Cader, "Green Grid Data Center Power Efficiency Metrics: PUE and DCIE," edited by C. Belady, White Paper #6, The Grid Grid, 2008.
17. R. Brown et al. "Report to congress on server and data center energy efficiency: public law 109-431," Lawrence Berkeley National Laboratory, Berkeley, 2008.
18. J. Liu, F. Zhao, X. Liu, and W. He, "Challenges Towards Elastic Power Management in Internet Data Centers", Proceedings of the 2nd International Workshop on Cyber-Physical Systems (WCPS 2009), in conjunction with ICDCS 2009, Montreal, Quebec, June 2009.
19. G. Chen, W. He, J. Liu, S. Nath, L. Rigas, L. Xiao, and F. Zhao, "Energy-aware server provisioning and load dispatching for connection-intensive internet services," 5th USENIX Symposium on Networked Systems Design and Implementation, Berkeley, CA, April 16–18, 2008.
20. Server Power and Performance characteristics, available at: http://www.spec.org/power_ssj2008/. Accessecd November 18, 2014.
21. J. Pouwelse, K. Langendoen, and H. Sips, "Energy priority scheduling for variable voltage processors," International Symposium on Low Power and Design, ACM, New York, pp. 28–33, Huntington Beach, CA, August 6–7, 2001.
22. L. Benini, A. Bogliolo, and G. De Micheli, "A survey of design techniques for system-level dynamic power management," IEEE Transactions on Very Large Scale Integration (VLSI) Systems, vol. 8, no. 3, pp. 299–316, June 2000.
23. P. Mahadevan, P. Sharma, S. Banerjee, and P. Ranganathan, "A power benchmarking framework for network devices," IFIP Networking, May 2009.
24. B. Heller, S. Seetharaman, P. Mahadevan, Y. Yiakoumis, P. Sharma, S. Banerjee, and N. McKeown, "ElasticTree: saving energy in data center networks," 7th USENIX Conference on Networked Systems Design and Implementation (NSDI). USENIX Association, Berkeley, CA, 2010.
25. L. Shang, L.-S. Peh, and K. N. Jha, "Dynamic voltage scaling with links for power optimisation of interconnection networks," International Symposium on High Performance Computer Architecture, Anaheim, CA, February 12, 2003.
26. D. Kliazovich, S. T. Arzo, F. Granelli, P. Bouvry, and S. U. Khan, "Accounting for load variation in energy-efficient data centers," IEEE International Conference on Communications (ICC), Budapest, Hungary, June 9–13, 2013.
27. D. Abts, M. Marty, P. Wells, P. Klausler, and H. Liu, "Energy Proportional Datacenter Networks," Proceedings of the International Symposium on Computer Architecture, pp. 338–347, Saint-Malo, France, June 19–23, 2010.
28. L. Benini and G. D. Micheli, "Powering networks on chips: Energy-efficient and reliable interconnect design for SoCs," International Symposium on Systems Synthesis, ACM, New York, pp. 33–38, Montreal, Canada, October 1–3, 2001.
29. J. S. Kim, M. B. Taylor, J. Miller, and D. Wentzlaff, "Energy characterization of a tiled architecture processor with on-chip networks," International Symposium on Low Power Electronics and Design, ACM, New York, pp. 424–427, Seoul, Korea, August 2003.

30. M. A. Sharkh, M. Jammal, A. Shami, and A. Ouda, "Resource allocation in a network-based cloud computing environment: design challenges," IEEE Communications Magazine, vol. 51, no. 11, pp. 46–52, November 2013.
31. X. Leon and L. Navarro, "Limits of energy saving for the allocation of data center resources to networked applications," IEEE INFOCOM, pp. 216–220, Shanghai, China, April 10–15, 2011.
32. B. Guenter, N. Jain, and C. Williams, "Managing cost, performance, and reliability tradeoffs for energy-aware server provisioning," IEEE INFOCOM, pp. 1332–1340, Shanghai, China, April 10–15, 2011.
33. J. Doyle, R. Shorten, and D. O'Mahony, "Stratus: load balancing the cloud for carbon emissions control," IEEE Transactions on Cloud Computing, vol. 1, no. 1, pp. 1, January–June 2013.
34. L. Hongyou, W. Jiangyong, P. Jian, W. Junfeng, and L. Tang, "Energy-aware scheduling scheme using workload-aware consolidation technique in cloud data centres," China Communications, vol. 10, no. 12, pp. 114–124, December 2013.
35. J. Duato, "A theory of fault-tolerant routing in wormhole networks," IEEE Transactions on Parallel and Distributed Systems, vol. 8, no. 8, pp. 790–802, August 1997.
36. G. Wei, J. Kim, D. Liu, S. Sidiropoulos, and M. Horowitz, "A variable frequency parallel I/O interface with adaptive power-supply regulation," Journal of Solid-State Circuits, vol. 35, no. 11, pp. 1600–1610, 2000.
37. S. K. Garg, Chee Shin Yeo, A. Anandasivam, and R. Buyya, "Energy-efficient scheduling of HPC applications in cloud computing environments," CoRR, abs/0909.1146, 2009.
38. B. Huffaker, D. Plummer, D. Moore, and K. Claffy, "Topology discovery by active probing," Symposium on Applications and the Internet (SAINT), IEEE Computer Society, Washington, DC, pp. 90–96, 2002.
39. M. E. Crovella and R. L. Carter, "Dynamic server selection in the Internet," Third IEEE Workshop on the Architecture and Implementation of High Performance Communication Subsystems (HPCS), pp. 158–162, August 23–25, 1995.
40. X. Chen, L. Xing, and Q. Ma, "A distributed measurement method and analysis on Internet hop counts," 2011 International Conference on Computer Science and Network Technology (ICCSNT), IEEE, Dates, pp. 1732–1735, Harbin, China, December 24–26, 2011.
41. B. Nordman, "What the real world tells us about saving energy in electronics," Proceedings of 1st Berkeley Symposium on Energy Efficient Electronic Systems (E3S), Berkeley, CA, May 2009.
42. IEEE Std 802.3az-2010, "Media access control parameters, physical layers, and management parameters for energy-efficient ethernet," pp. 1–302, October 27, 2010.
43. K. Christensen, P. Reviriego, B. Nordman, M. Bennett, M. Mostowfi, and J. A. Maestro, "IEEE 802.3az: the road to energy efficient ethernet," IEEE Communications Magazine, vol. 48, no. 11, pp. 50–56, November 2010.
44. G. Y. Li, Z. Xu, C. Xiong, C. Yang, S. Zhang, Y. Chen, and S. Xu, "Energy-efficient wireless communications: tutorial, survey, and open issues," IEEE Wireless Communications, vol. 18, no. 6, pp. 28–35, December 2011.
45. Y. Song, H. Wang, Y. Li, B. Feng, and Y. Sun, "Multi-tiered on-demand resource scheduling for VM-based data center," IEEE/ACM International Symposium on Cluster Computing and the Grid (CCGRID), pp. 148–155, Shanghai, China, May 18–21, 2009.

46. A. Beloglazov and R. Buyya, "Energy Efficient Resource Management in Virtualized Cloud Data Centers," IEEE/ACM International Conference on Cluster, Cloud and Grid Computing (CCGrid), pp. 826–831, Melbourne, Australia, May 18–21, 2010.
47. Q. Tang, S. K. S. Gupta, and G. Varsamopoulos, "Energy-efficient thermal-aware task scheduling for homogeneous high-performance computing data centers: a cyber-physical approach," IEEE Transactions on Parallel and Distributed Systems, vol. 19, no. 11, pp. 1458–1472, November 2008.
48. M. Al-Fares, S. Radhakrishnan, B. Raghavan, N. Huang, and A. Vahdat, "Hedera: dynamic flow scheduling for data center networks," Proceedings of the 7th USENIX Symposium on Networked Systems Design and Implementation (NSDI'10), San Jose, CA, April 2010.
49. A. Stage and T. Setzer, "Network-aware migration control and scheduling of differentiated virtual machine workloads," Proceedings of the 2009 ICSE Workshop on Software Engineering Challenges of Cloud Computing, International Conference on Software Engineering. IEEE Computer Society, Washington, DC, May 2009.
50. X. Meng, V. Pappas, and L. Zhang, "Improving the scalability of data center networks with traffic-aware virtual machine placement," IEEE INFOCOM, San Diego, CA, March 2010.
51. Cisco, "Cisco Data Center Infrastructure 2.5 Design Guide," Cisco Press, March 2010.
52. IEEE 802.1 Data Center Bridging Task Group, available at: http://www.ieee802.org/1/pages/dcbridges.html. Accessecd November 18, 2014.
53. C. Kopparapu, "Load Balancing Servers, Firewalls, and Caches," John Wiley & Sons Inc., New York, 2002.
54. D. Kliazovich, S. T. Arzo, F. Granelli, P. Bouvry, and S. U. Khan, "e-STAB: Energy-efficient scheduling for cloud computing applications with traffic load balancing," IEEE International Conference on Green Computing and Communications (GreenCom), Beijing, China, pp. 7–13, Beijing, China, August 20–23, 2013.
55. A. Sang and S.-q Li, "A predictability analysis of network traffic," Nineteenth Annual Joint Conference of the IEEE Computer and Communications Societies (INFOCOM), vol. 1, pp. 342–351, 2000.
56. C. Barakat, E. Altman, and W. Dabbous, "On TCP performance in a heterogeneous network: a survey," IEEE Communications Magazine, vol. 38, no. 1, pp. 40–46, January 2000.
57. U. Mandal, M. Habib, S. Zhang, B. Mukherjee, and M. Tornatore, "Greening the cloud using renewable-energy-aware service migration," IEEE Network, vol. 27, no. 6, pp. 36–43, November 2013.
58. Z. Liu, M. Lin, A. Wierman, S. Low, and L. Andrew, "Greening geographical loadbalancing," IEEE/ACM Transactions on Networking, vol. PP, no. 99, pp. 1, 2014.
59. M. Ilyas, S. Raza, C.-C. Chen, Z. Uzmi, and C.-N. Chuah, "Red-bl: energy solution for loading data centers," INFOCOM, 2012 Proceedings IEEE, pp. 2866–2870, Orlando, FL, March 25–30, 2012.
60. Y. Yao, L. Huang, A. Sharma, L. Golubchik, and M. Neely, "Data centers powerreduction: a two time scale approach for delay tolerant workloads," INFOCOM, 2012 Proceedings IEEE, pp. 1431–1439, Orlando, FL, March 25–30, 2012.
61. B. Lin, S. Li, X. Liao, Q. Wu, and S. Yang, "eStor: energy efficient and resilient data center storage," 2011 International Conference on Cloud and Service Computing (CSC), pp. 366–371, Hong Kong, China, December 12–14, 2011.

62. R.-S. Chang, H.-P. Chang, and Y.-T. Wang, "A dynamic weighted data replication strategy in data grids," IEEE/ACS International Conference on Computer Systems and Applications, pp. 414–421, West Bay Lagoon, Doha, April 2008.
63. X. Dong, T. El-Gorashi, and J. M. H. Elmirghani, "Green IP over WDM networks with data centers," Journal of Lightwave Technology, vol. 29, no. 12, pp. 1861–1880, June 2011.
64. F. Ping, X. Li, C. McConnell, R. Vabbalareddy, and J.-H. Hwang, "Towards optimal data replication across data centers," International Conference on Distributed Computing Systems Workshops (ICDCSW), pp. 66–71, Minneapolis, MN, June 20–24, 2011.
65. W. Li, Y. Yang, and D. Yuan, "A novel cost-effective dynamic data replication strategy for reliability in cloud data centres," International Conference on Dependable, Autonomic and Secure Computing (DASC), pp. 496–502, Sydney, Australia, December 12–14, 2011.
66. D. Boru, D. Kliazovich, F. Granelli, P. Bouvry, and A. Y. Zomaya, "Energy-Efficient Data Replication in Cloud Computing Datacenters," Springer Cluster Computing, pp. 1–18, 2015.
67. D. Kliazovich, P. Bouvry, and S. U. Khan, "DENS: data center energy-efficient network-aware scheduling," Cluster Computing, vol. 16, no. 1, pp. 65–75, 2013.
68. S. Srikantaiah, A. Kansal, and F. Zhao, "Energy aware consolidation for cloud computing," Proceedings of the 2008 Conference on Power Aware Computing and Systems, ser. HotPower'08, USENIX Association, Berkeley, CA, pp. 10, 2008.
69. J. Kim, M. Ruggiero, D. Atienza, and M. Lederberger, "Correlation-aware virtual machine allocation for energy-efficient data centers," Design, Automation Test in Europe Conference Exhibition (DATE), pp. 1345–1350, Grenoble, France, March 18–22, 2013.
70. D. G. d. Lago, E. R. M. Madeira, and L. F. Bittencourt, "Power-aware virtual machine scheduling on clouds using active cooling control and dvfs," Proceedings of the 9th International Workshop on Middleware for Grids, Clouds and e-Science, ser. MGC'11. ACM, New York, pp. 2:1–2:6, Lisbon, Portugal, December 12, 2011.
71. X. Wang, Y. Yao, X. Wang, K. Lu, and Q. Cao, "Carpo: correlation-aware power optimization in data center networks," INFOCOM, 2012 Proceedings IEEE, pp. 1125–1133, Orlando, FL, March 25–30, 2012.
72. Cisco, "Global cloud index: forecast and methodology, 2011–2016," White Paper, Cisco, 2011.

9

PERFORMANCE MANAGEMENT AND MONITORING

Mark Shtern[1], Bradley Simmons[2], Michael Smit[3], Hongbin Lu[1], and Marin Litoiu[2]

[1]*Department of Computer Science and Engineering, York University, Toronto, Ontario, Canada*
[2]*School of Information Technology, York University, Toronto, Ontario, Canada*
[3]*School of Information Management, Dalhousie University, Halifax, Nova Scotia, Canada*

9.1 INTRODUCTION

Organizations are transitioning from private data centers to infrastructure-as-a-service (IaaS)-style resource management where resources are acquired on-demand from a large pool, managed internally (i.e., a private cloud), or by a third-party supplier (i.e., a public cloud). Interest is growing in creating a single computational fabric across a set of cloud providers, a *multicloud* [1–4]. Multiclouds are a natural evolution of cloud computing; also called the intercloud [5, 6], or clouds-of-clouds, in which multiple cloud systems (typically IaaS) are composed together to add value to users. For example, a private and public cloud can be combined to address data privacy concerns while still enjoying some public cloud benefits (i.e., *hybrid clouds*, or public/private cloud overlays [7]). Multiple public clouds can be federated to improve availability [1], reduce lock-in, and optimize costs [8] beyond what can be achieved with a single cloud provider.

Cloud Services, Networking, and Management, First Edition.
Edited by Nelson L. S. da Fonseca and Raouf Boutaba.

As this transition to multicloud progresses, there are several critical differences that affect application management: (i) Every application is sandboxed from every other application. (ii) For an individual application, the potential for resource contention is effectively zero. (iii) Resources can be acquired on-demand according to a pay-as-you-go pricing model. These differences permit applications to be more easily managed on a per-application basis, rather than managing the entire IT infrastructure of an organization as a whole. This results in a shift of responsibility from established practices. A set of cloud providers (or, for private clouds, the IT operations teams) manages the physical infrastructure and provides virtualized containers (for IaaS, virtual machines or VMs) to clients who wish to deploy applications. The client assumes responsibility for both the functional and nonfunctional quality of a deployed application; increasingly, the client is the development team, a scenario referred to as *devops*[1] and/or *noops*.[2,3,4] Devops relies on automation for cost-efficient management of software systems. We provide more details about this transition in Section 9.2.

These changes in operational context (i.e., private datacenter versus multicloud) motivate an evolution in the approach to management of applications. If developers are expected to manage nonfunctional aspects of their applications, there is value in supporting best-practices with regard to the design and implementation of management logic and infrastructural support, while simultaneously incorporating established management best-practices into the overall approach. Additionally, the developer should be shielded from the complexity of acquiring and releasing resources in the context of the multicloud. Finally, they should be able to harness their own domain-specific languages (DSLs) and intimate knowledge of the application in support of management objectives instead of being prescribed a particular approach.

We introduce the X-Cloud[5] Application Management Platform (XCAMP), a platform to enable whomever has assumed responsibility for automating management—application developers, researchers, operations teams, and so on—to focus on how best to manage their application's runtime behavior (i.e., its management logic) rather than focusing on the minutiae of running on a multicloud. In the classic MAPE-k model of autonomic systems [9], XCAMP implements and integrates both the Monitoring and Execution stages while placing the onus for Analysing and Planning on the user. The management logic (i.e., operational policies guiding the runtime behaviour of the managed application) is specified in the language of the user's choice using their preferred environment and according to the methodology of their choice. In Section 9.3 we position our work in relation to the state of the art. We then describe the architecture of this platform and the challenges in managing the complexity of the multicloud in Section 9.4.

[1] http://devopsdays.org

[2] http://blogs.forrester.com/mike_gualtieri/11-02-07-i_dont_want_devops_i_want_noops

[3] In devops, developers collaborate with the operations team to build and manage services while in noops it is only the developers who do this.

[4] In situations where there is no operations team devops is equivalent to noops. For the remainder of this chapter, we will simply refer to devops.

[5] The X is pronounced "cross."

The main contribution of this chapter is the creation, definition, implementation, and evaluation of a novel approach to application management on multi-clouds that confers autonomic properties on applications at runtime and that embraces devops-style management and facilitates experimentation with diverse autonomic management approaches (e.g., model-based, rules/threshold driven, classic control, etc.) while abstracting away many of the low-level cloud programming details and nuisances. An important use case for XCAMP will be as the management framework for the SAVI[6] testbed[7] to streamline the life-cycle management of applications on a novel cloud architecture and to simplify the process of deploying runtime management, facilitating research on this two-tier cloud system by noncloud experts and students. XCAMP has already been presented in a hands-on tutorial at the SAVI Annual General Meeting (2013) in Toronto, Canada to a group of approximately 75 project members (i.e., students, researchers, and industrial participants).

To demonstrate the effectiveness of our framework we have implemented a prototype. We use this implementation to demonstrate the feasibility of our approach with an experiment demonstrating the autonomic cloud bursting of a legacy application.[8] Additionally, we have run XCAMP on a two-tier cloud architecture and we present a case study in which we diagnosed the root cause of a performance bottleneck observed on the SAVI testbed. Finally, an experiment measuring the throughput of our implementation, ensuring it is practical for managing large systems, is presented. The experiments (described in Section 9.5) effectively demonstrate the capabilities of our approach. Based on our implementation experience, we describe (Section 9.7) several ongoing challenges for management in the multicloud.

We close the chapter (Section 9.8) by offering concluding remarks.

9.2 BACKGROUND CONCEPTS

Historically, a company owned a set of dedicated resources (e.g., a private data center) upon which their business applications were run. Typically, there were many such applications and *how* these applications behaved in relation to each other was of paramount importance. Specifically, issues of ownership and access were critical (i.e., could application A run on machine Z between 5 and 8 PM EST). Further, an IT operations team was responsible for ensuring both the security and operations of the physical infrastructure and also with ensuring the effective functioning and security of all applications, including those developed in-house. Most applications ran on bare metal (i.e., servers) and a ceiling existed on total available resources that was relatively constant (unless

[6]Smart applications on virtual infrastructure (SAVI) is a national research project in Canada: http://savinetwork.ca.

[7]The proposed architecture, implemented by the testbed, introduces a novel architecture (i.e., two tier cloud) where virtualized resources exist close to end-users (i.e., the smart edge), allowing applications to access either low-latency resources near end-users, or standard public cloud data centers (i.e., the core).

[8]That is, acquiring additional resources from a public cloud when a private cloud does not have sufficient resources to handle its workload [10].

machine upgrades were performed or new resources were added to the data center's footprint). Extensive work on management frameworks and methodologies supports these processes.

As described in the introduction, cloud computing is fast removing many of the standard management barriers that once defined the IT landscape. For example, the requirements to carefully plan for capacity is being eclipsed by the ability to programmatically launch VM using a pay-as-you-go model (i.e., the IaaS cloud) as required. The responsibility of managing the physical infrastructure has been separated from the responsibility for managing applications. This affords significant flexibility, allowing for the fine-tuned management of resource acquisition and release, and dynamic configuration of managed applications. This new-found infrastructural flexibility has allowed developers (or, has allowed managers to push developers) to focus on business-level considerations and effective operational strategies (e.g., devops) as an alternative to focusing on highly optimized and tuned code design. The adoption of cloud computing by medium and large enterprises is expected to accelerate as a growing number of suppliers build additional datacenters, as virtualization technologies continue to improve, and as faster networking links provide high-speed connectivity.

While this development of technologies supporting the cloud continues to accelerate, the challenge of how best to manage applications deployed to the cloud remains unresolved. For example, in 2013 the Amazon.com website went down for longer than 20 minutes.[9] One popular approach to the management of applications (including those on clouds) is referred to as *autonomic computing* [9]. Autonomic computing was introduced as a way of dealing with the increasing complexity of systems. It is based on the concept of the autonomic nervous system, which in humans is responsible for the constant beating of the heart among other things. The outwardly observable behavior of an autonomic application (i.e., one managed using this approach) is that of self-optimization, self-configuration, self-healing, and self-protection (i.e., self-*) behavior. This approach involves a key management component: the *autonomic manager*.

The autonomic manager is responsible for adjusting the behaviour of an application in response to both runtime and management policy constraints. More precisely, an autonomic manager's function can be decomposed into a loop composed of four main phases: monitoring, analysing, planning, and execution (i.e., the MAPE-k loop). In the monitoring phase, the autonomic manager monitors the performance of the application (and possibly the environment, etc.). In the analysis phase, the autonomic manager analyses this data to build up an understanding about what possible strategies to apply to improve the application's state. In the planning phase, the autonomic manager selects a strategy from among the possible strategies. In the execution phase, the chosen strategy is implemented.

A key characteristic of autonomic computing is automation. This is also true for devops. Devops and related approaches are used by major industry trend-setters

[9]http://venturebeat.com/2013/08/19/amazon-website-down

(e.g., Amazon[10] and Netflix[11]). Complementary to devops is the process referred to as *continuous deployment* in which the release cycle is shortened from months to days (or even less). For example, Amazon.com deploys a release every 11.6 s [11]. Although traditional applications are faster to develop and deploy and easier to manage in clouds, autonomic applications are still difficult to design, implement, and deploy, and still require substantial knowledge and resources. A goal of this work is to make development and deployment of autonomic applications easier. XCAMP mechanisms for automating the life-cycle (i.e., deploy, manage, and undeploy) of not only the application, but also the management logic responsible for autonomically managing it.

The first step in cloud adoption is often transitioning an on-site datacenter into a private cloud. However, private clouds, while providing many of the benefits of a general cloud (i.e., on-demand resources) lack many of the economies of scale inherent in public clouds such as massive scale and freedom from equipment storage/maintenance/personnel costs, and so on. As a result, both hybrid public–private clouds and cross-provider deployments are becoming more common. It is well known that one of the biggest challenges of constructing both hybrid clouds and/or the multicloud is the bridging together of multiple infrastructures. Difficulties include but are not limited to abstracting away the details of the various provider-specific syntaxes [12], unifying/normalizing the various pricing models [13], providing seamless monitoring across potentially quite disparate provider domains [14], ensuring data ownership, privacy, locality, security, and so on. This motivates the need for abstraction of the low-level operations on the multicloud, a need XCAMP is designed to meet.

9.3 RELATED WORK

The notion of on-demand systems existed well before the advent of cloud computing [15, 16]. Noticing the scale and increasing complexity of systems, IBM introduced the notion of autonomic computing [9] that popularized the notion of a MAPE-k loop and self-* functionality. The concept of autonomics has also been considered by [17, 18]. These concepts can be understood as precursors and/or progenitors in one way or another of the current notion of the cloud.

Managing resources in this emerging cloud environment is a significant challenge; Jennings and Stadler enumerate key aspects of this challenge, including: "the scale of modern data centers; the heterogeneity of resource types and their interdependencies; the variability and unpredictability of the load; as well as the range of objectives of the different actors in a cloud ecosystem" [19]. As the cloud has begun to take shape, several tool-kits and frameworks have been introduced as possible approaches to addressing the challenge of managing resources while extending the cloud's capabilities. Some well-known examples of these include Reservoir [20], OPTIMIS [21], Aneka [22], and VDC

[10] http://aws.amazon.com

[11] http://www.netflix.com

Planner [23]. Often, these approaches include notions of federation, multicloud, hybrid cloud, and so on. However, they are all devised from a more traditional perspective in which a deployment must be carefully designed and optimized in advance so that it may negotiate a correct SLA to ensure its requirements are met. Where these frameworks are forward-looking and require complex architectural components we chose instead to focus on the cloud as it is presently available. This design choice allows us to help bring new users to the cloud and facilitates experimentation with various approaches to the design and implementation of management logic (e.g., model-based, rules and/or threshold driven, and classic control). Our focus was on facilitating management of applications by the developers not on how to manage the cloud from the perspective of an infrastructure provider.

An important aspect of facilitating a multicloud involves the notion of a broker [8, 24]. A broker acts to facilitate resource acquisition and release on behalf of a client application in response to their dynamic requirements at runtime. While in some cases, as demonstrated in this chapter, the management logic suffices to determine from where to obtain and/or release resources to; in other scenarios in which multiple potential competing providers exist, a broker provides a logical component to obtain/release the best selection of resources as required. Therefore, we envision future integration between managers and brokers. The broker will be responsible for resource acquisition/release, while the manager will be responsible for application management tasks.

9.4 X-CLOUD APPLICATION MANAGEMENT PLATFORM

The design of X-Cloud Application Management Platform (XCAMP) is based on the MAPE-k [9] loop, with framework components and developer-specified components working in collaboration to perform MAPE-k-based management of an application. The monitoring and executing portions of the loop are performed by framework components, while the analysis and planning portions are done by developer-specified management logic. This separation of concerns guides runtime operations and is presented graphically in Figure 9.1a. XCAMP was designed to work at multicloud scale (i.e., massive application deployments of thousands of nodes) and is able to support multiple application deployments simultaneously. XCAMP leverages a stream processor paradigm to achieve scalability, fault tolerance, and reliability, and to provide a useful abstraction of streams (long sequences of records) to transfer metrics, key performance indicators (KPI)s, and in general *knowledge* among the components, with each new tuple processed in transit by the various components. The following sections will provide an overview of the XCAMP architecture in terms of the MAPE-k loop and then delve more deeply into its components and the abstraction features of the platform. First, two usage scenarios will illustrate the use of the XCAMP platform, one focusing on the impact on a single application, and the other from the perspective of a service provider.

9.4.1 Usage Scenarios

In this section we introduce two illustrative usage scenarios for XCAMP.

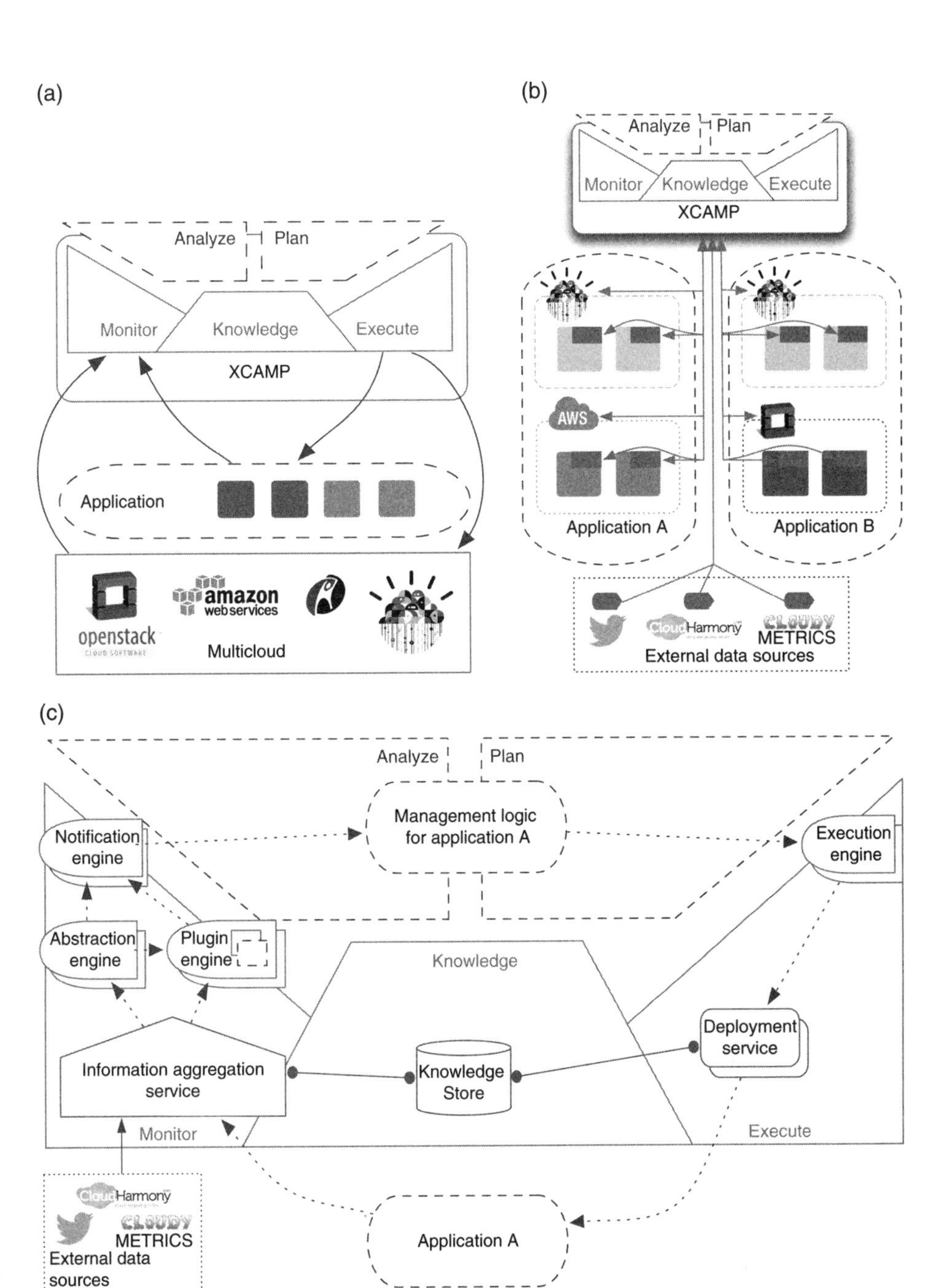

Figure 9.1. Conceptual overviews of XCAMP: the components with red, dashed borders are provided by the developer/deployer, components with solid lines are provided by the platform. Dashed lines with arrows represent the flows of data; solid lines represent relationships. Solid squares indicate VMIs and their color indicates the provider from which they have been acquired. Inner blue rectangles indicate the management agent on the VMI. Arrows denote monitoring data and execution command flows. To simplify the presentation we focus on a single application deployment. However, as was described in the Usage Scenarios, XCAMP is highly scalable and designed to handle multiple application deployments at the same time. (a) High-level architecture view. (b) High-level deployed application view. (c) A detailed look at the components implementing MAPE-k.

9.4.1.1 Scenario 1, Hybrid Clouds. Company A would like to deploy an application to their private cloud. However, they are constrained by a lack of resources to support it during peak periods of demand. They wish to create a hybrid cloud, using public cloud resources when private resources are exhausted, with resources being added and removed autonomically based on demand.

- *Preparation*: After deploying XCAMP to their private cloud[12], they would register both their private cloud and Amazon Elastic Compute Cloud (EC2)[13] with the platform. Next, they would create a deployment document that describes the layout of their application on cloud resources (i.e., this includes describing nodes, images, services, and communication links between services). They would capture their desired autonomic behavior in rules (e.g., one rule might be *when resource utilization in the web-tier exceeds 60%, add a node to the private cloud, unless resources are exhausted, then add a node to the public cloud*). The rules use the terminology defined in their deployment document (e.g., *web-tier*), and can reference any metric captured by XCAMP. This set of rules is called *Management Logic* throughout this chapter, and represents the management policies to be enforced, as implemented by an application capable of accepting monitored metric values at a specified URL, making management decisions based on this stream of metrics, and returning actions to effect change in the deployed application as needed. This Web-based Application is implemented in whatever language the developer prefers, and is deployed automatically by XCAMP into an appropriate container (e.g. Apache Tomcat).
- *Deployment*: The administrator then submits their deployment documents together with application and Management Logic to the system along with any additional automation scripts (i.e., to setup a database). XCAMP automatically instantiate cloud resources and dynamically builds the application according to the given descriptions. Upon instantiation, the platform automatically begins capturing metrics from all configured resources, and feeding this stream of metrics to the Management Logic's defined URL.
- *Runtime Management*: As the Management Logic receives metrics from XCAMP, it returns (as-needed) actions that are realized by XCAMP. In this scenario, the company would author their Management Logic application to detect increases in demand (as reflected by increases in utilization) and in response add application servers first on the private cloud, then on Amazon EC2 when private resources are exhausted. XCAMP handles the process of adding resources, including creating instances (of the specific image) from the correct cloud provider (private cloud, EC2), dynamically installing the correct packages, instantiating the correct services, and connecting these new nodes within the application environment topology (i.e., adding them to the front end load balancer and pointing them to

[12]An automated installation using from 1 to 5 VMs

[13]http://aws.amazon.com/ec2/

the database). Similarly, as demand recedes, these resources can be automatically released and decommissioned.

9.4.1.2 Scenario 2, Edge-Core Clouds. The SAVI two-tier cloud is made of edge nodes, close to the end user, and core nodes, located in a big data center. The architecture is meant to support low latency and high bandwidth applications. SAVI administrators want to provide a management service to the users of a testbed implementing this SAVI cloud architecture. The XCAMP platform must enable users to deploy and manage their applications while accommodating a broad range of practical experience (from novice to expert) with regard to deploying and/or managing applications on the SAVI cloud.

- *Preparation*: The administrators must deploy XCAMP to their two-tier cloud architecture, then provide the XCAMP front-end URL to their users. Administrators can decide where to place the initial deployment, on the edge or core nodes, and then author a deployment descriptor.
- *Deployment*: SAVI researchers submit their jobs (i.e., the application and Management Logic) through the Web interface or RESTful API. XCAMP deploys the application on the edge and core nodes provisioning at the same time the connectivity among components.
- *Runtime Management*: XCAMP ensures all monitored details about a user's deployed application environment is collected and routed to the user's Management Logic, and accepts all commands issued in response, translating them to low-level actions and executing these actions. The management logic can act on application components on edge or core nodes. Typical actions include scaling out/in on edge and core nodes, live migration of VMs for load balancing, and so on. A key design pattern of XCAMP is its utilization of stream-processing that facilitates its ability to scale to massive size and support large number of nodes while collecting massive numbers of measurements about runtime performance and external monitored details. The entire SAVI testbed shares this single platform, avoiding duplication which results in high utilization efficiency.

9.4.2 MAPE-k Loop View

The key XCAMP components and their position in the MAPE-k loop are presented in Figure 9.1c. In this section, we will describe these various components and their contributions to the management of a deployed application on the multicloud.

The *Information Aggregation Service* is the main interface for gathering information about the deployed applications, environment, and from external sources (e.g., Twitter, CloudyMetrics [13], CloudHarmony,[14] and others). Collected information is streamed to the *Notification Engine* which is used to forward an augmented stream of metrics to

[14] http://cloudharmony.com/

the various *Management Logic* components. In one path, data traverse the *Abstraction Engine*. Given information from the knowledge store that describes all existing deployed applications, each metric is translated to a more abstract form based on the terminology defined in the deployment document (e.g., an IP address is translated to a unique identifier that is marked as belonging to *web-tier*).

In a second path, data traverses the **Plugin Engine** (optionally first passing through the abstraction engine) where additional processing is applied to the data stream. For example, aggregation may be applied to individual server metrics constructing tier-specific information (e.g., mean CPU utilization per cluster) or the archive of metrics hosted by the Information Aggregation Service may be queried to produce metric trends. The platform provides several plugins (e.g., calculating the cost of a deployment based on information from CloudyMetrics); the user may add their own. Information that leaves the plugin engine is specific to a given application, either using information from the abstraction engine or from the user-supplied plugins.

The *Management Logic* represents developer-specified management directives (e.g., management policies [25]) designed to guide the runtime behavior of the application. The XCAMP framework does not place any restriction on how the Management Logic is expressed/implemented; it is run within a sandboxed container, on its own VM. The

```
1  public class ManagerServletEx extends HttpServlet{
2  ...
3   public void doGet(HttpServletRequest request, HttpServletResponse res) {
4    ...
5    if (LOAD_ONE.equals(request.getParameter(METRIC_NAME)))
6      updateMetricValue(request.getParameter(SOURCE), LOAD_ONE,
           toDouble(request.getParameter(VALUE)));
7    ...
8    if ((caculatedMeanLoadForAppTier < appTierScaleDownTheshold) &&
         (getSizeOfPublicFootprint()> MIN_PUBLIC_SIZE)) {
9              elasticScaleFootprint(PUBLIC_CONTAINER, size_public - SCALING_INCREMENT);
10             return;
11   ...
12 }
13 ...
14 public class ActionGenerator {
15     ...
16     public void elasticScaleFootprint(String tierName,int finalFootprintSize) {
17      .....
18        JSONMessage msg = new generateJSONMessage(SCALE_FOOTPRINT, tierName,
               finalFootprintSize);
19        sendJSONResponse(msg);
20     }
21     ...
22 }
```

Figure 9.2. Sample code (with exception handling omitted to simplify readability) for the cloud bursting Management Logic implementation is presented. This Management Logic is implemented as a Java servlet. The doGet method is called from the monitoring components of the XCAMP platform with updates about all relevant monitored metrics and the response is either empty or carries an action to be performed by the XCAMP Execution Engine. On line five an update about load_one is processed for a particular node in which the metric METRIC_NAME for the node SOURCE of the application topology is updated with the value VALUE. The management rule presented on line eight is one of the four rules used to implement the elastic bursting strategy and can be stated informally as follows: *IF the mean load for the application server tier is less than a threshold and the size of the application server tier on the public cloud is greater than MIN_PUBLIC_SIZE THEN scale down the public footprint of the application server tier by SCALING_INCREMENT*. The ActionGenerator class on line 14 is used to generate and send JSON messages to indicate what action the execution action should take (e.g., line 18).

Management Logic represents a combination of both the Analyze and Plan components of the MAPE-K loop. It offers a URL to which metrics are submitted, and responds with JSON-formatted actions that are passed to the Execution phase. Each developer uses best practices for filtering requests for their chosen platform (e.g., Java EE Filters) to decide which metrics reach the Web application logic. A Management Logic component is responsible for managing its own data store if required. A partial excerpt from a Java-based implementation of Management Logic is presented in Figure 9.2.

The *Execution Engine* and the *Deployment Service* implement the Execution component of the MAPE-k loop. The Execution Engine accepts requests for changes from the Management Logic and converts this into high-level workflow statements. These statements are forwarded to the Deployment Service, which executes a set of lower level workflows to implement the requested changes to the application's deployment and/or configuration. For example, the Management Logic might request an additional resource be added to its *web-tier*; the Execution Engine translates this request to a parameterized call to the deployment service which creates and provisions the node and re-configures the load balancer. The components collaboratively maintain a knowledge base of system state through the *Knowledge Store* component that stores data about the historical, current, and predicted future state of the system.

9.4.3 Deployment View

The process of application deployment requires that the developer submit a declarative deployment document [26] that describes the application pattern to deploy [27], a deployable version of their application (e.g., a WAR file), and a Management Logic Web application (e.g., a WAR file). The submission component (not shown) passes this information to the deployment service, which deploys the application in accordance with the deployment document (using user-supplied credentials) and registers the deployment in the Knowledge Store. The Management Logic application is deployed, and is automatically registered with XCAMP to receive pertinent information for its associated application. The deployer may specify external data sources from which information should also be retrieved. Application-level metrics are submitted to the Information Aggregation Service and will be available to the Management Logic.

The result is the automatic collection and pushing of well-formatted, high-level, abstract, consistent metrics to the Management Logic. Using whatever approach and methodology the developer prefers—ad hoc Java code, a Web interface wrapper to an existing management system, and so on the analysis and planning steps are completed. If actions are required, a JSON message is passed to the Execution Engine, where it is de-abstracted and passed to the deployment service to modify the running deployment.

Once an application is deployed using XCAMP, its structure will be similar to those of the applications presented in Figure 9.1b. Functionally, an application deployment represents a complex graph of an application in which nodes are VMIs running on the various cloud providers and edges represent communication channels between these nodes. XCAMP deploys a management agent to each VMI in an application. This agent is responsible for transmitting collected monitored data to the Information Aggregation

Service and for modifying configuration settings of the installed application stack, the operating system, and/or altering the set of installed applications on the VMI in response to commands from the deployment service.

To facilitate operations, XCAMP communicates directly with the various cloud provider APIs (e.g., AWS and Openstack), which allows it to perform operations like adding and removing instances, and to collect metrics from the provider when available. XCAMP also monitors data from sources other than cloud resources and further passes it to the Management Logic; for example, data from Twitter, CloudyMetrics, or CloudHarmony can be passed to Management Logic to assist in decision making. For example, should there be a failure of a region in AWS, XCAMP will receive this status update. This data can be utilized by the Management Logic in order to make decisions about where to deploy nodes of the application. This might include transposing [1] application servers to alternative cloud providers on the fly or simply to avoid launching new VMIs in affected regions.

9.4.4 Information Abstraction in XCAMP

A key contribution of XCAMP is the abstraction of low-level details that differ among various cloud providers, and a common metrics format. To illustrate how we hide complexity in XCAMP from the management logic, consider the following example of adding an instance to a deployed application.

After the Management Logic determines that an instance must be added to the application server tier of a deployed application, it sends a JSON-formatted message to the Execution Engine saying *There should be five servers similar to Web server A in cluster my-web-tier*. The Execution Engine translates this declarative request for resources into a high-level workflow, which is passed to the Deployment Service with associated information (e.g., which application deployment to modify). The Deployment Service translates this into a low-level workflow as follows. First, the Deployment Service determines upon which cloud provider Cluster my-web-tier is running. This allows it to connect to the correct cloud provider. It then requests two instances of the same configuration as web server A (determined by the deployment document, e.g., m1.large) and deploys the management agents on both instances. The management agents are then instructed to deploy an identical software stack with the same configuration as Web server A. After the instances are fully configured and ready, the agent will begin streaming monitored data to the Information Aggregation Service, which will ultimately inform the Management Logic of the successful addition of resources to the deployed application.

9.4.5 Management Logic

Much work has been done in the domain of distributed systems management (see, e.g., the proceedings of IEEE/IFIP NOMS, IFIP/IEEE IM, and CNSM). Policy-based management [25, 28, 29] represents an approach to management in which management actions are decoupled from management logic and in which the management logic is interpreted at runtime. This affords a flexible management paradigm in which as management logic changes, policies may be altered thus facilitating the design of elegant autonomic systems. Often, a policy specification language [30–32] is used to encapsulate the rules that govern the behavior of the system. While we embrace the need for specification languages, especially in the context of large distributed systems and network management, we feel that their utility is tightly coupled to the actor who is tasked with using them and the specific environment in which they are used. In devops, we feel freedom should be given to the developers to do it their way and to take advantage of all the intimate details that they possess with regards to the inner workings of the application that they are managing. Unlike in traditional management situations where system administrators, operations teams and/or business people require a mechanism for automating the control of their systems that is semantically clear to them and does not place much emphasis on programming capabilities. Developers have an entire arsenal of tools, libraries, and methodologies for ensuring that a system is functionally correct. These can be embraced to ensure the nonfunctional requirements of a system are being met as well. Due to their experience with programming languages (and likely lack of experience with DSLs like policy specification languages), use of programming languages may be a preferred approach for specifying management logic. Further, our proposed approach does not in any way preclude the use of an existing policy specification language/PBM solution, which could be readily employed as the Management Logic component.

9.5 IMPLEMENTATION

To demonstrate the feasibility of our approach, we authored a proof-of-concept implementation of XCAMP. We leveraged existing libraries and frameworks where possible to allow us to focus on the abstraction task.

Monitoring components are built on the Misure [14] extensible, distributed, and scalable monitoring system. Due to its central importance in XCAMP we provide a brief overview of it in Section 9.5.1. The Information Aggregation Service, Abstraction Engine, Plugin Engine, and Notification Engine were written as elements that used the stream-processing paradigm central to Misure to communicate and scale horizontally.

Execution is provided by the Execution Engine, which like the other engines is built on Misure; and by the deployment service, for which we used a customized version of the pattern-based deployment service[15] (PDS) [26] developed by our team. Similar to Misure the PDS plays an important role in XCAMP and so we elaborate on it in Section 9.5.2. The Execution Engine connects to the deployment service via a RESTful API.

Analysis and Planning is provided by Management Logic applications, one per deployed application, running in a Java EE container (Tomcat) on a dedicated VMI. The responsibility for authoring the Management Logic application rests with the application developer/deployer; they submit WAR files for deployment. Any container is adequate

[15] https://github.com/ceraslabs/pattern-deployer; the customizations have been pulled into the master version.

for this purpose; others could be added with a straightforward extension to the current implementation.

9.5.1 Misure

In previous work, we defined a set of requirements for monitoring in heterogeneous federated clouds [14], defined a suitable architecture built on stream-processing, and implemented a prototype solution. Based on an enhanced publish–subscribe pattern, the design and implementation allow for the gathering of metrics at any level (system, application, etc.) from disparate sources like Ganglia [33], SNMP sources, Amazon Cloudwatch, and various Web APIs. These streams of metrics are transformed (aggregated, annotated, split, etc.) in transit, and published to interested users as live streams (push-type). Metrics are also persisted to long-term storage which can be queried via an API (poll type). The prototype was evaluated on pubic clouds and found effective at handling metrics at scale with low infrastructure cost.

The core abstraction underlying Misure is stream processing, in the family of complex event processing [34]; as an abstract concept, it refers to the generation, manipulation, aggregation, splitting, and transformation of data organized in a long sequence of records. One example is Storm,[16] a Twitter open-source project, on which Misure is built. Storm is billed as a "distributed, scalable, reliable, and fault-tolerant stream processing system," and can be used for stream processing, continuous computation, and distributed RPC.[17]

One of the key features of Storm is the effort to manage the complexity of distributed computation on realtime data entirely behind the scenes. This includes guaranteed message processing; aggressive resource management (garbage collecting defunct processes); fault detection and task reassignment after failure, efficient, and scalable message transmission; streams that consist of any data (serialization occurs behind the scenes); and local development environments for debugging. Storm also allows components to be implemented using many programming languages. Storm is parallel and distributed; there is no central router and no intermediate queue. It is designed to scale horizontally, and has been deployed at scale processing large Twitter data sets.

9.5.2 Pattern-Based Deployment Service

The pattern-based deployment service, PDS [26], emerged out of the need to simplify the process of deploying complex multitier applications to cloud environments and further to adapt them at runtime (i.e., dynamically add/remove nodes to an existing application topology). Specifically, the notion of describing declaratively what you want versus how to achieve it appealed to us. Further, the use of patterns is quite powerful in they allow for simplification, re-use and sharing. The PDS hides the low-level

[16]https://github.com/nathanmarz/storm

[17]https://blog.twitter.com/2011/storm-coming-more-details-and-plans-release

```
<topology id="scale">
   <instance_templates>
      <template id="openstack_small_vm">
         <cloud>OpenStack</cloud>
         <instance_type>2</instance_type>
         <key_pair_id>mark</key_pair_id>
         <image_id>50</image_id>
         <ssh_user>root</ssh_user>
      </template>
   </instance_templates>
   <container num_of_copies="4" id="web_host_container">
      <node id="web_host">
      <use_template name="openstack_small_vm"/>
      <service name="web_server">
          <database_connection node="data_host"/>
          <war_file>
             <file_name>petstore.war</file_name>
             <datasource>jdbc/pet</datasource>
          </war_file>
      </service>
      </node>
      </container>
      <node id="data_host">
          <use_template name="openstack_small_vm"/>
          <service name="database_server">
             <script>petstore.sql</script>
          <service/>
      </node>
      <node id="web_balancer">
          <use_template name="openstack_small_vm"/>
          <service name="web_balancer">
             <member node="web_host"/>
      </service>
   </node>
</topology>
```

Figure 9.3. A sample deployment document written in an XML-based DSL. This particular document describes a deployment on six nodes in total (i.e., one Web balancer, four Web hosts, and one database server.

details about how to deploy services to any cloud provider, making it quite useful in the context of multicloud. Specifically, the PDS facilitates application topologies to be described, deployed, and adapted across multiple cloud providers. The PDS has been used successfully on EC2 and Openstack and has support for all Fog[18] compliant cloud providers. Further, the PDS has been open-sourced and is available for download and contributions.[19]

With PDS, a user describes the "pattern" of their application in an XML-based domain-specific language (DSL). This DSL is quite easy to understand and comprises several key elements that can be grouped as functional elements (e.g., Topology, Node, and Service) and syntactic sugar (e.g., instance_templates, Container, and num_of_copies). A sample deployment document written in this XML-based DSL is presented in Figure 9.3.

[18]http://fog.io/about/provider_documentation.html

[19]https://github.com/ceraslabs/pattern-deployer

9.6 EXPERIMENTS AND A CASE STUDY

Using our implementation, we performed two experiments (Sections 9.6.1–9.6.3). In the first, we demonstrated the process of implementing Management Logic and using it to manage an application facing changing workloads; in the second, we showed early performance results demonstrating that a Management Logic component can handle millions of metrics. We also used this implementation in a case study demonstrating how researchers running on an experimental testbed can more easily perform complex experiments repeatedly to obtain meaningful results (Section 9.6.4).

9.6.1 Experimental Setup

Experiment 1: For the managed application, we used a sample Java EE web application that accepts requests; connects to a database (mysql) to perform selects, inserts, or updates; and returns a response. We defined a declarative deployment document (see Section 9.4.1) with a database server, a load balancer, and a cluster of web application servers initialized to a single instance running in the private cloud and no instances in the public cloud.

We authored the application's Management Logic, see Figure 9.2 for an excerpt of this code, as a second Java EE Web application, implementing the RESTful interface defined by the platform to receive new metrics and information about resources deployed. The Management Logic collected the 1-min load averages[20] for each web application server, calculated an average, and requested additional resources when a configurable threshold was surpassed. Resources were released when the average fell beneath a second configurable threshold. Given the limited capacity of private clouds, after two instances are running on the private cloud, the manager requests resources from the public cloud. To limit churn, a refractory period was introduced (as a configurable parameter that could be changed on the fly through the RESTful interface): 10 min between adding nodes, and 5 min between removing nodes. Aside from features to allow run-time configuration of various parameters, the Management Logic consisted of 65 source lines of code.[21]

The PDS was deployed to an Amazon EC2 m1.small instance. We created a deployment package including the Web application WAR, its Management Logic WAR, and the various keys and credentials required to provision instances on the clouds in our topology, and submitted this package to the PDS. The Management Logic ran on a t1.micro instance; Web application servers were deployed to a local Openstack installation running on a dedicated IBM Bladecenter in a university data center, with a 100 Mbps uplink. An openstack.small instance was defined with 2 GB of RAM and one virtual CPU. The public cloud deployment; if necessary; was to m1.small instances on Amazon EC2.

[20]Load average refers to the number of processes ready for CPU time on average over some period of time, 1 min in this case.

[21]Clearly, there is substantial room to improve this algorithm; the focus is on the enabling platform and not the adaptive scaling algorithm employed.

The XCAMP implementation ran on a four-core cluster in Amazon EC2. Ganglia[22] monitors acquired the metrics from each machine and passed them to the Information Aggregation Service.

Finally, an Apache JMeter[23] test plan was used to generate load. The workload was two-thirds read requests (e.g., browse catalog), one-sixth write requests (e.g., checkout), and one-sixth update requests (e.g., modify user profile). The design was to peak workload at 120 simultaneous threads sending requests as quickly as possible, launching in 4 groups of 30 threads, each ramping up over a 5-min period. The first group launched at start time t minutes; the second at $t+10$, the third at $t+25$, and the fourth at $t+33$. Peak workload was maintained until $t+90$, when the fourth group was terminated, followed by the second group at $t+100$, and the final two groups at $t+120$. This plan was executed by an m3.xlarge (quad-core) instance in Amazon EC2.

Experiment 2: Using the same implementation and components, we examined the performance of the Management Logic when run on three different instance sizes in Amazon EC2 (t1.micro, m1.small, and m1.medium) to assess the scalability of this approach. We created simulated metrics and submitted them via the RESTful API as quickly as the service could handle them for a 1-h period. The mix of metrics was proportional to reality, with many requests being irrelevant to the actual decision-making. Our primary question was whether it would be necessary to autonomically scale the Management Logic Container as a cluster for large topologies.

9.6.2 Results

Experiment 1: Figure 9.4 illustrates what happened during the experiment, showing the addition and removal of instances, the size of the workload, the average load over all deployed resources, the average response time, and the total throughput. The deployment began with a single private instance. This was sufficient for the first workload group;[24] but shortly after adding the second workload group, the autonomic manager detected load average greater than 1.0 (Fig. 9.4a). A private cloud instance was requested (light red band). There are brief spikes in response time (up to 4 seconds) when the node is added to the balancer manager and when the node is first enabled and receives its first requests which are not shown due to the smoothing (for readability). The two private instances are sufficient to handle 60 workload threads, but not 90 where a third instance is required. This instance is requested from Amazon EC2 (light orange band).

As the experiment continues, workload continues to increase and the load average remains high. Amazon m1.small instances are substantially smaller than openstack.small instances; a total of five are required (added as soon as possible given the refractory period) to meet the generated workload. After the activation of the final public instance, load average settles at around 1 and remains there, providing stable response time and

[22]http://ganglia.sourceforge.net

[23]http://jmeter.apache.org

[24]Note that instances running on this Openstack installation using KVM use a virtual CPU and report load averages differently, counting processes waiting in a queue and NOT running processes.

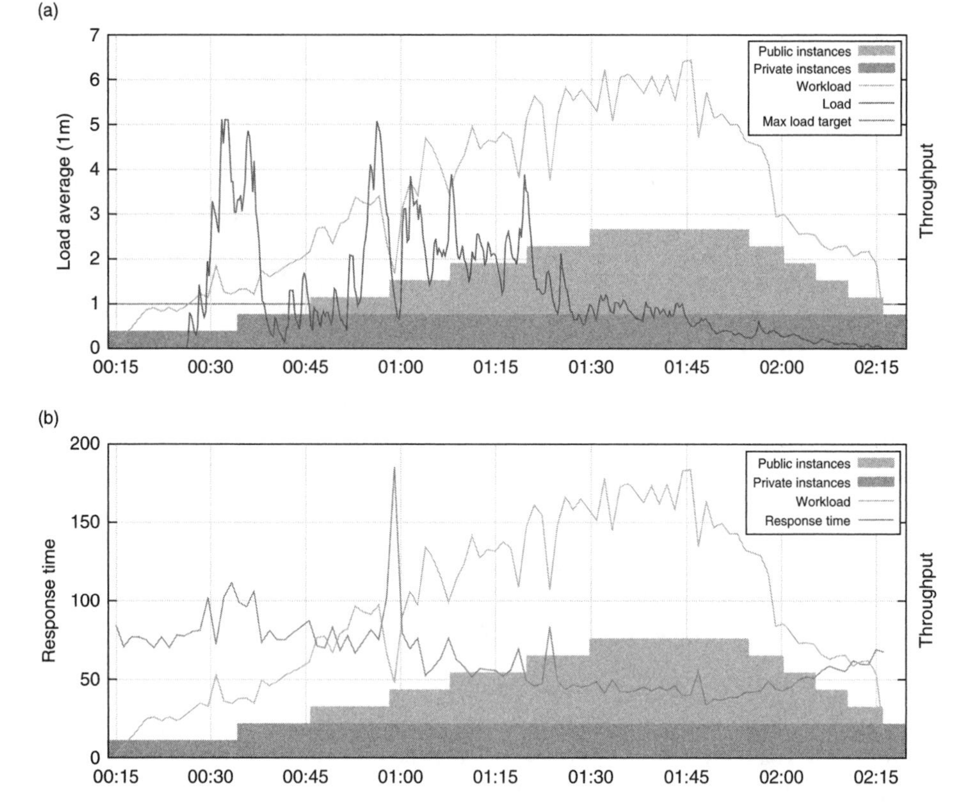

Figure 9.4. Measurements from the scaling experiment, first adding private instances then bursting to the public cloud. The stacked graphs show the instance counts, with the blue line representing throughput (smoothed to be more readable). (a) One-minute load average, averaged over all active instances. (b) Average response time over a 1 seconds window, smoothed using splines to improve readability. Sharp spikes (peaking at 4 seconds, not shown) are due to load balancer restarts when adding a new node, and an initial period of slow response times for new nodes.

maximum throughput. Once the workload decreases, the additional instances are gradually removed—first the public instances, then the private instance (at the end of the experiment).

Experiment 2: Figure 9.5 presents the results of the scaling, showing the total throughput and average response time achieved by the three instances running the straightforward bursting adaptation policy. A gradual ramp-up was included in the load generation. The t1.micro instance is specified for only periodic or bursting workloads, not for sustained load; this is evident in the results as performance varies dramatically. There are several drops in throughput, due to either other running tasks competing for resources or the variation inherent in the public cloud [35] which is most noticeable with smaller single instance sizes. The t1.micro response time results make it difficult to see corresponding degradation in response time.

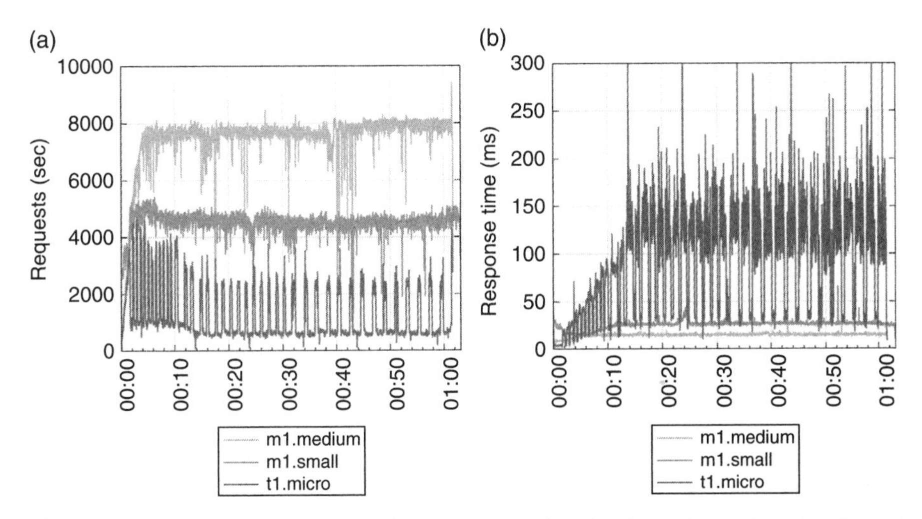

Figure 9.5. Measurements from a scaling test measuring the throughput of our implemented Management Logic on three Amazon EC2 instance sizes. (a) Throughput (b) Average response time; peaks (not shown) at 800 milliseconds.

9.6.3 Discussion

Experiment 1: We demonstrated the ability to use a standard Java EE Web application with a simple adaptation policy written in a programming language familiar to the original application developer using the common RESTful API pattern. There is room to improve the Management Logic; for example, to better handle the time required for a new instance to become active and handle requests.

While designing the Management Logic, we considered several metrics as the basis for adaptation. We have been aware of the limitations of existing monitoring tools on public clouds for some time, but our trials with CPU utilization metrics and load averages demonstrated the unreliability of these numbers. The individual metrics for each of the seven instances launched in Experiment 1 varied per cloud. The load averages from OpenStack were zero even when the machine was clearly loaded; it was only when overloaded that they would produce higher load averages, which resulted in slower reactions from the Management Logic. The data from EC2 had higher peaks, particularly during bootstrapping. More notably, despite high load averages, they rarely exceeded 20% CPU utilization (Fig. 9.6b).

In contrast with 1-minute load averages (Fig. 9.6a), the 15-min load average offers a better understanding of the overall trend of the system. Figure 9.6c shows this load average for each instance. All of the managed instances trended toward a load average of 1.0, the target set by our Management Logic. Much of the difficulty in achieving this load average on EC2 instances hinged on load incurred during bootstrapping. This indicates that launching from machine images with the required software pre-installed is important to effective adaptive management.

Experiment 2: The performance numbers measured indicate an m1.small instance running our Management Logic could process over 270,000 metrics per minute;

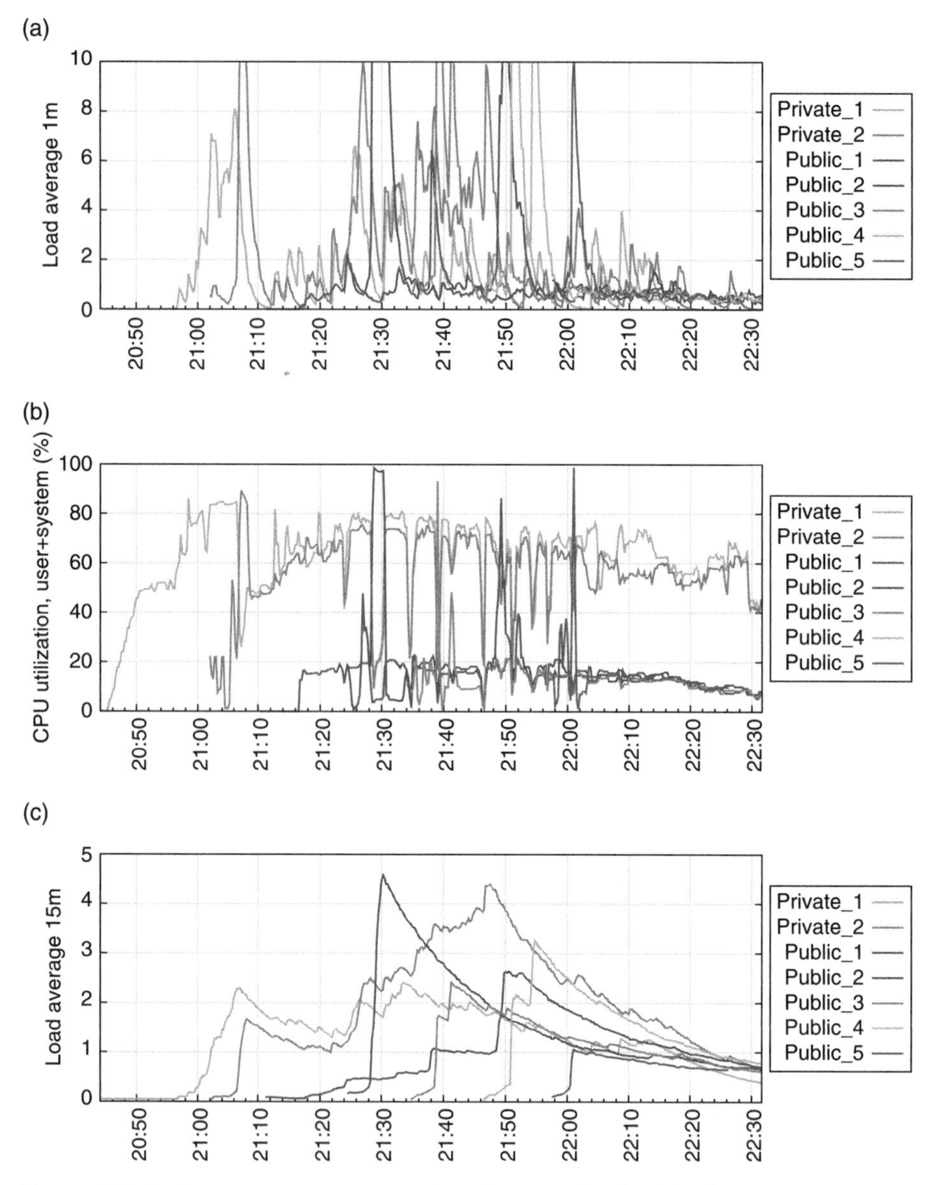

Figure 9.6. Performance measurements from instances involved in the autonomic scaling experiment. (a) One-minute load average (peaks reaching 15–20, 30 for public_2, not shown). (b) CPU Utilization, for user and system processes. (c) 15-minute load average.

collecting the standard 18 Ganglia core metrics once per minute suggests an ability to manage 15,000 active instances. This indicates there is currently no need to autonomically scale a cluster of containers. There is no strict bound on complexity for alternative Management Logic applications, and so this need may arise in the future if computation-intensive planning and analysis is performed. The scalability of Misure has been discussed previously [14]; it is similarly capable of handling thousands and even millions of metrics.

9.6.4 Case Study

One of the goals of XCAMP is to make systems management painless for developers. This case study illustrates this ability in action. It was noted that in practice, when several SAVI users deployed applications simultaneously using XCAMP, they observed a major degradation in performance of the SAVI testbed.[25] We used XCAMP as the platform for a series of experiments to explore this phenomenon to contribute to the ongoing improvement of the testbed.

Initial exploratory runs: Using XCAMP's deployment service, we deployed a three-node Java EE application to the SAVI testbed (note that all three nodes are deployed simultaneously by default). Once deployed, we dynamically added an additional node to the application topology. Once the scale out operation had completed we removed this additional node by scaling down. Finally, we undeployed the application. We kept measurements of how long each stage of this process took. We conducted variations of this experiment with various numbers of simultaneous application deployments (1, 2, 5, and 10), each deploying three-node Java EE applications (for between 3–30 VM instantiations). Each experiment configuration was run three times. The mean timing results (with standard deviations) are presented in Figure 9.7a. We observed that the Deploy stage is the slowest of the four and that it was most impacted by the number of concurrent users; scale-out, which is like deploy but with one instance instead of three, was also impacted.

Examining the deployment stage: We decided to explore the Deploy stage more carefully by examining the two phases: downloading required files from the PDS on the internal network, versus downloading and installing software packages from an external package repository. We also performed the same experiments on Amazon EC2 in order to use the results as a comparator. Figure 9.7c shows a linear increase in download time as the number of concurrent users increases for both EC2 and the SAVI testbed. However, software installation on EC2 appears to be constant no matter how many concurrent users there are while on the SAVI testbed a dramatic increase is observed as the number of concurrent users increases. We hypothesized it might be the network and/or a disk IO-related problem. The network-bottleneck hypothesis is that additional concurrent users are creating traffic on the network, causing congestion or bandwidth-cap related issues. The IO-bottleneck hypothesis is the increased number of VMs running on a single physical host and performing random read-writes overextends disk resources. We know from observation that CPU and memory utilization are not excessive, so we did not create additional resource-contention hypotheses.

Evaluating hypotheses: To confirm or reject each hypothesis, we designed a final experiment. We created a full image containing all required packages for the application. We compared the time required to deploy this image, versus the time required to

[25] The testbed implements a two-tier cloud extending OpenStack, with a single core and seven edges distributed across Canada.

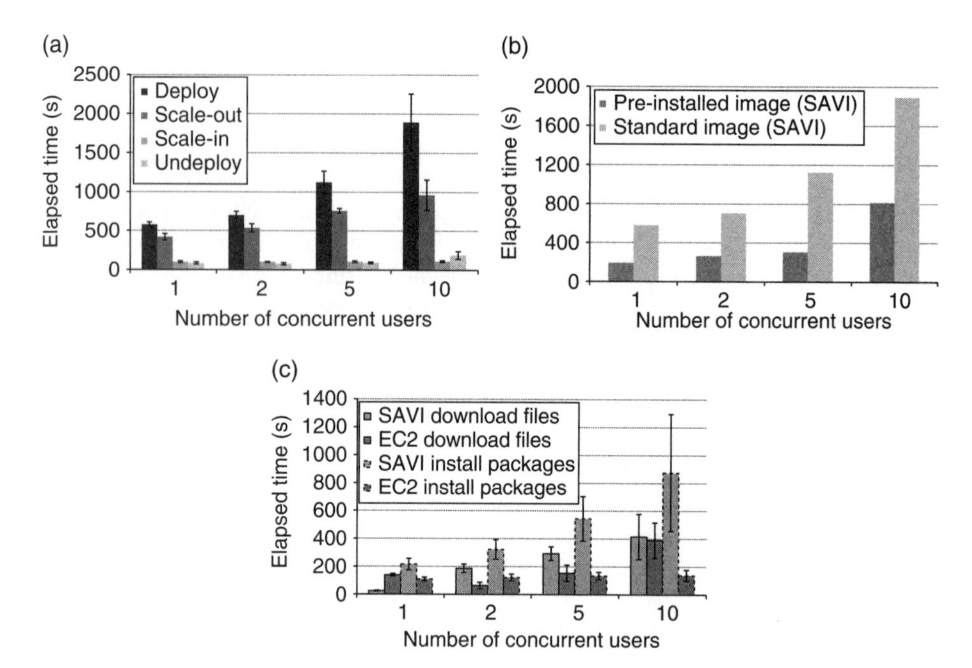

Figure 9.7. Various experiment results for the case study exploring performance of the SAVI two-tier cloud testbed. (a) Temporal breakdown of deploying and scaling an application on the SAVI testbed. (b) Comparing the performance of bootstrapping a node versus using a node with all software pre-installed on the SAVI testbed. (c) A breakdown of performance for two phases of the Deployment stage on the SAVI testbed and Amazon EC2: downloading files from the PDS, and installing software packages.

deploy a standard Ubuntu image and bootstrap it (i.e., download and install all required software packages from a central repository). A complete image has similar bandwidth requirements, but can be written to disk with sequential writes (versus random read-write) thus reducing IO load. If deploying the full image is faster than bootstrapping, we would regard the IO-bottleneck hypothesis as confirmed. The results are presented in Figure 9.7b. Notice that the full image outperforms the standard image, suggesting the presence of an IO-bottleneck.

Using XCAMP in this case study allowed us to easily run and monitor a variety of experiment configurations, systematically and repeatedly, to collect and present evidence of system performance issues in the SAVI testbed.

9.7 CHALLENGES IN MANAGEMENT ON HETEROGENEOUS CLOUDS

Based on our experience designing, implementing, and testing a multicloud adaptive system, we offer the following reflections on the particular challenges that apply to adaptively managing heterogeneous clouds.

Heterogeneous monitoring systems: Many cloud providers offer monitoring services to provide information about the performance of provisioned resources; these systems vary significantly, and typically require relatively detailed information to query (e.g., instance IDs for Amazon EC2). The state of monitoring in private clouds is even more varied, with various solutions deployed based on each organization's whim. Existing cloud abstraction layers largely disregard monitoring, focusing instead on acquiring resources.

Rapid reaction: Automated management requires current and accurate monitoring data. The ability of an aggregating monitoring service to meet this requirement depends on the timeliness of metrics received from third-party monitoring systems being aggregated. The automated manager can be run on the public cloud, which introduces more delays outside of the control of the monitoring system. It remains an open question how to best ensure timely decisions are made.

Inaccuracy in traditional monitoring techniques: It is generally understood that virtualized resources offer more variable performance than bare-metal resources, and the variance in the performance of Amazon EC2 instances has been benchmarked [35]. However, it is less understood that standard monitoring techniques will report inaccurate information that can mislead adaptive managers.

In a public cloud environment, the desire to sell fractions of a CPU's processing power have removed the meaning of many of these standard metrics. For example, Amazon EC2 configures instances using a measure called elastic compute units (ECUs), which they document as a 1.0–1.2 GHz 2007 Opteron or 2007 Xeon processor. One ECU is approximately 40% of a single core of a Intel Xeon CPU E5430 @ 2.66 GHz, a common processor in the first-generation Amazon infrastructure. The default instance size, small, is 1 ECU; the hypervisor enforces this 1 ECU limit. However, Xen's paravirtual mode offers limited ability to abstract the processor for performance reasons, so instances and Linux kernels running on instances perceive a full core available to them. Xen enforces the limits by refusing access to the CPU if the allotted quota has been used, which the Linux kernel reports as steal. The exact time spent in steal may vary over time, and in any case can only be measured when the system is operating at capacity. An idle machine will report 100% idle time, giving no indication of the actual limits on the CPU. It is not trivial to calculate the actual load on a machine reporting 20% user and 80% idle.

9.8 CONCLUSION

This chapter introduced a framework for managing the life-cycle of applications in multicloud environments and for conferring autonomic properties on them at runtime. The framework allows the specification of the Management Logic, its deployment and instantiation, and its execution alongside the managed application. The Management Logic runs in a container that is seamlessly connected to XCAMP's monitoring and execution engines. XCAMP provides application developers with effortless access to monitoring sensors, third-party data sources and actuators (i.e., the monitoring and executing stages of the MAPE-k loop) from across the multicloud, while placing control of both the analysis and planning stages in their hands and allowing them to express their management policies in their own vernacular and harnessing all their personal expertise.

We validated XCAMP in multiple ways. First, we demonstrated the ability to elastically cloud burst a legacy Java EE application from a private cloud to a public cloud and reported our findings and our experience in automating applications in multicloud environments. Additionally, we demonstrated a capability of the XCAMP framework to facilitate the diagnosis of a bottleneck on the SAVI (i.e., two-tier cloud) testbed. We demonstrated, through experimentation, the capabilities of our design to scale to large size and for our autonomic manager (i.e., Management Logic) to process massive numbers of metric updates per minute. Finally, we presented a hands-on tutorial to a group of approximately 75 SAVI members at the 2013 Annual General Meeting in Toronto, Canada and received positive feedback from the participants.

The XCAMP platform will provide a useful middleware upon which to base much future research for both the SAVI project and other areas of multicloud research. By allowing developers to harness their vast skill sets, different approaches to management can be considered with ease and this is a critical benefit we provided through the introduction of XCAMP.

ACKNOWLEDGMENTS

This research was supported by IBM Centres for Advanced Studies (CAS), Natural Sciences and Engineering Council of Canada (NSERC) under the Smart Applications on Virtual Infrastructure (SAVI) Research Network, and Ontario Research Fund under the Connected Vehicles and Smart Transportation (CVST) project.

REFERENCES

1. M. Shtern, B. Simmons, M. Smit, and M. Litoiu, "Navigating the cloud with a MAP," in *13th IFIP/IEEE International Symposium on Integrated Network Management (IM)*, 2013, pp. 464–470.
2. G. Baryannis, P. Garefalakis, K. Kritikos, K. Magoutis, A. Papaioannou, D. Plexousakis, and C. Zeginis, "Lifecycle management of service-based applications on multi-clouds: a research roadmap," in *Proceedings of the International Workshop on Multi-Cloud Applications and Federated clouds*, 2013, pp. 13–20.
3. N. Loutas, V. Peristeras, T. Bouras, E. Kamateri, D. Zeginis, and K. Tarabanis, "Towards a reference architecture for semantically interoperable clouds," in *Cloud Computing Technology and Science (CloudCom)*, 2010, pp. 143–150.
4. D. Petcu, C. Craciun, and M. Rak, "Towards a cross platform cloud API," in *CLOSER*, 2011, pp. 166–169.
5. D. Bernstein, E. Ludvigson, K. Sankar, S. Diamond, and M. Morrow, "Blueprint for the intercloud—protocols and formats for cloud computing interoperability," in *Proceedings of the 2009 Fourth International Conference on Internet and Web Applications and Services*, 2009, pp. 328–336.

6. R. Buyya, R. Ranjan, and R. N. Calheiros, "Intercloud: Utility-oriented federation of cloud computing environments for scaling of application services," in *ICA3PP (1)*, 2010, pp. 13–31.
7. M. Shtern, B. Simmons, M. Smit, and M. Litoiu, "An architecture for overlaying private clouds on public providers," in *8th International Conference on Network and Service Management, CNSM 2012, Las Vegas, USA*, 2012.
8. P. Pawluk, B. Simmons, M. Smit, M. Litoiu, and S. Mankovski, "Introducing STRATOS: A cloud broker service," in *IEEE 5th International Conference on Cloud Computing*, 2012, pp. 891–898.
9. J. Kephart and D. Chess, "The vision of autonomic computing," *Computer*, vol. 36, no. 1, pp. 41–50, 2003.
10. M. Smit, M. Shtern, B. Simmons, and M. Litoiu, "Partitioning applications for hybrid and federated clouds," in *CASCON '12: Proceedings of the 2012 Conference of the Center for Advanced Studies on Collaborative Research*, 2012, pp. 27–41.
11. J. Jenkins, 2011, Presentation from O'Reilly Velocity Conference, http://assets.en.oreilly.com/1/event/60/Velocity%20Culture%20Presentation.pdf. Accessed November 20, 2014.
12. M. Smit, M. Shtern, B. Simmons, and M. Litoiu, "Supporting application development with structured queries in the cloud," in *New Ideas and Emerging Results (NIER) track, Proceedings of the 2013 International. Conference on Software Engineering (ICSE)*, 2013.
13. M. Smit, P. Pawluk, B. Simmons, and M. Litoiu, "A web service for cloud metadata," in *IEEE Congress on Services*, 2012, pp. 24–29.
14. M. Smit, B. Simmons, and M. Litoiu, "Distributed, application-level monitoring of heterogeneous clouds using stream processing," *Future Generation Computer Systems*, vol. 29, no. 8, pp. 2103–2114, 2013.
15. I. Foster, C. Kesselman, and S. Tuecke, "The anatomy of the grid: Enabling scalable virtual organizations," *International Journal of High Performance Computing Applications*, vol. 15, no. 3, pp. 200–222, 2001.
16. M. A. Rappa, "The utility business model and the future of computing services," *IBM Systems Journal*, vol. 43, no. 1, pp. 32–42, 2004.
17. J. Strassner, N. Agoulmine, and E. Lehtihet, "Focale: A novel autonomic networking architecture," *International Transactions on Systems Science and Applications (ITSSA)*, vol. 3, no. 1, pp. 64–79, 2007.
18. L. Baresi, A. Ferdinando, A. Manzalini, and F. Zambonelli, "The cascadas framework for autonomic communications," in *Autonomic Communication*, A. V. Vasilakos, M. Parashar, S. Karnouskos, and W. Pedrycz, Eds. New York: Springer US, 2009, pp. 147–168.
19. B. Jennings and R. Stadler, "Resource management in clouds: Survey and research challenges," *Journal of Network and Systems Management*, pp. 1–53, 2014.
20. B. Rochwerger, D. Breitgand, E. Levy, A. Galis, K. Nagin, I. M. Llorente, R. Montero, Y. Wolfsthal, E. Elmroth, J. Cáceres, *et al.*, "The reservoir model and architecture for open federated cloud computing," *IBM Journal of Research and Development*, vol. 53, no. 4, pp. 4–11, 2009.
21. A. J. Ferrer, F. Hernández, J. Tordsson, E. Elmroth, A. Ali-Eldin, C. Zsigri, R. Sirvent, J. Guitart, R. M. Badia, K. Djemame *et al.*, "Optimis: A holistic approach to cloud service provisioning," *Future Generation Computer Systems*, vol. 28, no. 1, pp. 66–77, 2012.

22. R. Buyya, C. Yeo, S. Venugopal, J. Broberg, and I. Brandic, "Cloud computing and emerging it platforms: Vision, hype, and reality for delivering computing as the 5th utility," *Future Generation Computer Systems*, vol. 25, no. 6, pp. 599–616, 2009.
23. M. F. Zhani, Q. Zhang, G. Simona, and R. Boutaba, "Vdc planner: Dynamic migration-aware virtual data center embedding for clouds," in *2013 IFIP/IEEE International Symposium on Integrated Network Management (IM 2013)*, 2013, pp. 18–25.
24. N. Grozev and R. Buyya, "Inter-cloud architectures and application brokering: Taxonomy and survey," *Software: Practice and Experience*, Vol. 24: 369–390, 2012.
25. M. Sloman, "Policy driven management for distributed systems," *Jounal of Network Systems Management*, vol. 2, no. 4, pp. 333–360, 1994.
26. H. Lu, M. Shtern, B. Simmons, M. Smit, and M. Litoiu, "Pattern-based deployment service for next generation clouds," in *IEEE Congress on Services*. IEEE Computer Society, 2013.
27. T. Eilam, M. Elder, A. Konstantinou, and E. Snible, "Pattern-based composite application deployment," in *2011 IFIP/IEEE International Symposium on Integrated Network Management (IM)*. IEEE, 2011, pp. 217–224.
28. D. C. Verma, *Policy-based Networking: Architecture and Algorithms*. Indianapolis, In: New Riders Publishing, 2000.
29. J. Strassner, *Policy-based Network Management: Solutions for the Next Generation*. Boston, MA: Morgan Kaufmann, 2003.
30. N. Damianou, N. Dulay, E. Lupu, and M. Sloman, "The ponder policy specification language," in *Policies for Distributed Systems and Networks*. M. Sloman, E. Lapu, and J. Lobo, Eds. Berlin: Springer, 2001, pp. 18–38.
31. L. Kagal, T. Finin, and A. Joshi, "A policy language for a pervasive computing environment," in *IEEE 4th International Workshop on Policies for Distributed Systems and Networks, 2003. Proceedings. POLICY 2003*. IEEE, 2003, pp. 63–74.
32. R. Boutaba and I. Aib, "Policy-based management: A historical perspective," *Journal of Network and Systems Management*, vol. 15, no. 4, pp. 447–480, 2007.
33. M. L. Massie, B. N. Chun, and D. E. Culler, "The ganglia distributed monitoring system: design, implementation, and experience," *Parallel Computing*, vol. 30, no. 7, pp. 817–840, 2004.
34. D. Luckham, *The Power of Events: An Introduction to Complex Event Processing in Distributed Enterprise Systems*. Boston, MA: Addison-Wesley, 2002.
35. J. Schad, J. Dittrich, and J.-A. Quiane-Ruiz, "Runtime measurements in the cloud: Observing, analyzing, and reducing variance," *Proceedings of the VLDB Endowment*, vol. 3, no. 1, 2010.

10

RESOURCE MANAGEMENT AND SCHEDULING

Luiz F. Bittencourt, Edmundo R. M. Madeira, and Nelson L. S. da Fonseca

Institute of Computing, State University of Campinas, Campinas, São Paulo, Brazil

10.1 INTRODUCTION

As computer networks have evolved, processing demands have migrated from local computing devices to distributed computing environments. In this context, the capacity of distributed processing has also progressed from job-based computing to the more user-friendly service-oriented computing. This paradigm shift has been accompanied by an evolution of distributed system architectures: job-oriented cluster computing gives rise to through job- and service-oriented grid computing, and then to service-oriented utility computing, now known as *cloud computing* [1].

Cloud computing is currently being offered and used by many companies [2]. For example, the Amazon Web Services (AWS – http://aws.amazon.com/) offers various services for database, e-Commerce, storage, and processing power. Google Applications[1] also offers a variety of applications as services, including Google Application Engine (GAE),[2] which allows application development to be performed directly in the cloud

[1] http://www.google.com/apps/
[2] http://code.google.com/appengine/

Cloud Services, Networking, and Management, First Edition.
Edited by Nelson L. S. da Fonseca and Raouf Boutaba.

through Google's application programming interfaces (APIs). Other examples are Microsoft Azure, Salesforce.com, Globus Nimbus, and Eucalyptus [3].

It has been estimated[3] that spending on public cloud services represented a market of US$132 billion in 2013, and that it would exceed US$244 billion by 2017. In such a competitive market, resource management is crucial for seizing a significant market share. A service-oriented distributed environment demands quality of service (QoS), which must be accompanied by cost reduction for both service provider and users. This raises new challenges which must be addressed [4, 5].

In the next section, basic concepts related to the management of cloud computing are introduced, followed by a discussion on types of application that can be allocated to cloud resources, as well as formalization of the cloud system and a description of the problems of both application scheduling and virtual machine (VM) allocation. After that, resource management and resource allocation in clouds are discussed, with a focus on infrastructure providers, followed by an overview of techniques for the scheduling of tasks of applications and the allocation of VMs. Challenges and future perspectives are presented at the end of the chapter.

10.2 BASIC CONCEPTS

Clouds are capable of offering usually virtualized computing resources as dynamically scalable services to users over the Internet, without any need to worry about the technical aspects of resource management. According to the *National Institute of Standards and Technology* (NIST), cloud computing can be defined as follows [6]:

> Cloud computing is a model for enabling convenient, on-demand network access to a shared pool of configurable computing resources (e.g., networks, servers, storage, applications, and services) that can be rapidly provisioned and released with minimal management effort or service provider interaction.

Cloud providers add resource management layers over computing clusters and grids to make their infrastructure available as computing services to users, who ideally require minimal management effort and knowledge to use such infrastructure [7]. These management layers conceal the physical infrastructure from the user through the adoption of a series of automatic resource management actions. In this section, we introduce an overview of the terminology of cloud computing as utilized in this chapter as well as of some problems in cloud management and resource allocation.

10.2.1 Cloud Service Models

Resource management in clouds must consider the type of computing resources to be offered as a service. Clouds are usually classified by the service offered; the most common are *software as a service* (SaaS), *platform as a service* (PaaS), and *infrastructure*

[3]https://www.gartner.com/doc/2598217

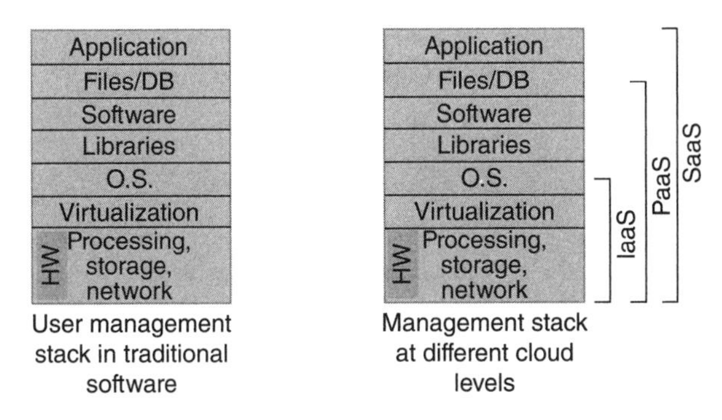

Figure 10.1. Management stack for different cloud service levels.

as a service (IaaS). In the SaaS model, the consumer simply utilizes an application provided by the cloud provider, having control neither over the application development nor the host on which the application is run. Popular examples of this model include Google Apps and Salesforce.com. The PaaS model makes available a framework in which consumers can develop and deploy their own applications in the cloud. Examples of clouds that offer this model are Google App Engine and Amazon Web Services. In the Infrastructure as a Service model (IaaS), a cloud provider offers computing resources and administrative privileges for users, usually as VMs running on the provider infrastructure, so that the users can control their computing environment, including software development and deployment. The Amazon Elastic Compute Cloud (Amazon EC2), Globus Nimbus, and the Eucalyptus are examples of this model. Other models also exist, such as *network as a service* (NaaS)[4] and *database as a service* (DbaaS) [8].

The levels of the management stack involved in the three types of service can be seen in Figure 10.1. In traditional systems, the client is responsible for managing every layer in the stack, from hardware configuration, including operating system management and application deployment. For the IaaS service model, the provider is responsible for managing only the lower layers in the stack, including hardware and software for networking, storage, and processing, as well as virtualization technologies to share these resources among cloud clients. However, the clients are responsible for managing the operating system and its softwares (libraries, middleware), as well as data/databases and applications.

A PaaS provider must perform all the management performed by IaaS providers, as well as managing the operating system, libraries, software, and middleware, thus offering to the client a development platform over a self-managed execution environment. In SaaS, however, users have no responsability in managing any layer in the hardware/software stack. They can utilize the software provided and change the software configuration, but cannot change the software nor manage the infrastructure. This makes SaaS easy to use, but harder to customize than PaaS- or IaaS-based cloud services.

A cloud provider itself can also be a client of another provider offering different type of services. Figure 10.2 illustrates a scenario in which an SaaS (or PaaS) provider relies

[4]http://www.scaledb.com/

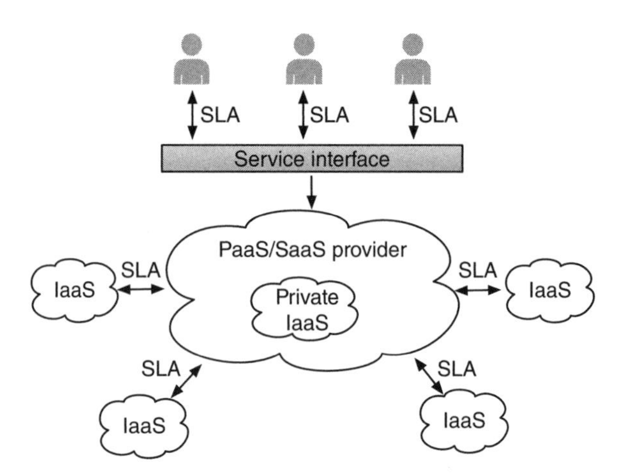

Figure 10.2. Scenario with SLAs at two levels.

on other IaaS providers to serve its customers. In this case, more than one level of service-level agreements (SLAs) are necessary: one between clients and the SaaS provider, and another between the SaaS provider and IaaS providers. Clearly, arrangements between providers on more than two service levels are possible, as in the case of various types of business.

Management problems involving different levels of SLAs must be dealt with, including the dependencies of the upper layers on the services of lower layers. In order to offer a service level guarantee to the user, the cloud provider must receive guarantees from its lower level providers. Moreover, the cloud provider (the SaaS in Figure 10.2) must consider its margin of profit when pricing its services, which depends on the agreement it has with other providers.

10.2.2 Cloud Types

Cloud computing can be classified according to access policy. The classification presented here is generally associated with IaaS, although it can be extended to SaaS and PaaS. Depending on the access to cloud resources, an IaaS provider is classified as public, private, or hybrid:

- Public cloud offers virtualized computational resources as services to any user who can access that service through the Internet; resources are provided in a pay-per-use basis. Public IaaS cloud offers certain advantages because it has computational capacity on demand, while avoiding upfront investment in processing/storage pools for handling eventual peaks in demand. On the other hand, it does not offer controlled access to physical machines and communication channels, which can results in a compromise of the security of critical applications or sensitive data.

- Private cloud is more of a virtualized cluster or computational grid which offers a more transparent interface to the user. Usually restricted to a single organization, it can provide fine-tuned performance as well as flexibility. Although it does not completely avoid upfront investment, it can be implemented over an existing computational infrastructure and prevent further capital investment. Moreover, it offers better access control to computer resources, thus improving security, which could be critical for the data security of an organization.
- Hybrid cloud combines public and private clouds. This type of cloud allows users and organizations to keep using their private resources, yet provides access to extended computational capacity when necessary using public cloud resources on a pay-per-use basis. The flexibility in meeting demands for computational capacity known as elasticity is fundamental in reducing costs during increased demands in comparison to fully in-house computing infrastructures.

10.2.3 SLAs and Charging Models

The offering of cloud services is based on SLAs on a pay-per-use basis. In PaaS and SaaS, the charge to users is often based on a variety of criteria such as predefined quantity of hours of use of storage space and number of I/O requests. The IaaS model, however, is often more flexible, allowing users to choose the types of resources as well as the model of charging.

Depending on the SLA established between users and cloud providers, different management systems will be necessary. In SaaS and PaaS, management should be able to automatically increase or decrease computing power for a user according to his/her demands. Monitoring entities, however, are essential to achieve automatic elasticity without compromising QoS. Moreover, the ability to increase/decrease computing power involves deploying and/or resizing VMs to cope with demands.

In IaaS, on the other hand, elasticity involves client choice: It is the client who decides the capacity and number of VMs to be leased. Commonly leased examples of VMs features include: CPU cores/speed, amount of RAM, amount of storage and access speed, and network bandwidth. In this case, management entities are necessary to handle VM allocation. They must act according to client demands for different VM types, allocating these requests to physical machines on the basis of the policies defined by the provider.

Actual VM allocation may depend on the model of charging selected by the client. On-demand VM leasing establishes a price per hour or minute of use, and the user is charged for each VM from its deployment to release. A reserved model provides a predefined price for access to VMs on demand. The spot model, on the other hand, works as a market, with VM prices varying with demand. In this case, the user offers a price he/she is willing to pay for a type of VM, and it can be used as long as the actual price remains lower than the initial offer. If demands on the provider increase, however, the price can increase, and the provider has the right to deny access to VMs which, at least momentarily, cost more on the market.

10.2.4 Resource Allocation and Scheduling

The perspective of cloud providers and their clients are potentially conflicting in regard to computing resources. On the one hand, the clients want to receive a high QoS for the minimum payment, whereas cloud providers want to charge as much as possible.

Cloud clients who have their own computing resources are interested in making the best utilization of their "free" resources, utilizing the public cloud resources only when necessary. They want to maximize the utilization of local resources while minimizing the monetary costs for the use of the public cloud, although they are unwilling to sacrifice QoS. In hybrid clouds, the users must schedule applications, with the scheduler choosing how to schedule the application jobs on both private and public cloud resources on the basis of information of various kinds of processing capacity of the private and public cloud resources, job computing requirements, public cloud costs, and data transfer costs.

Cloud providers, on the other hand, must allocate VMs in a way that respects SLAs, yet reduces costs so they can make a profit. This cost reduction comes from sharing resources among users and maximizing resource utilization. Thus, fewer physical machines are needed; moreover, power consumption can be reduced by turning off idle physical machines. When the demand increases, the cloud management system must decide if new physical machines will need to be turned on to cope with the new VM requests.

Resource allocation and scheduling are vital to both cloud users and providers, but each has its own specifics. Different management entities and allocation algorithms are necessary to make the best use of the cloud from both perspectives.

10.3 APPLICATIONS

Large hardware and networking capacities have leveraged a whole new set of applications, many of which can benefit from cloud computing. To explore private and public cloud resources to the full extent when running applications, computing resources must be efficiently used and resource allocation and scheduling play a fundamental role in this efficiency. In particular, application scheduling should be able to decide on which resource each application (or part) should be run, given the demands of the applications and resource capacities [9–12].

The decisions made by the scheduler have a direct impact on the QoS of the application. Taking application characteristics into consideration when allocating resources can lead to a variety of approaches to the problem. These characteristics include cost of the computation of jobs, data transfer between jobs, and data source localization, all which can be accounted in the scheduler objective function and have important influence on the decision-making process.

One conceptual difference arising from service-oriented computing is the invocation of services instead of job dispatching. Service invocation assumes an already deployed code with an interface to be called, and which will remain running after results are delivered so that service can be called again with different parameters. Job dispatching, on the other hand, involves code that is to be transferred, run, and finished with the results delivered to the user or to another application. This conceptual difference leads to certain distinct management needs, such as a service repository to control where each service is already running and service deployment to transfer and deploy services across the resources. In this chapter, we adopt the term *task* to refer to a job or a service invocation, regardless of whether the service is already running or must be transferred and deployed.

A user can submit different applications to run in the cloud. The simplest application is a single task that must be run and results returned to the user. Such a single task may or may not have parallel codes; but if it does, they can run on separate processors or cores in the same machine. The user can also submit parallel tasks that can be run in separate machines, commonly called bag-of-tasks (BoTs), because they can be run with no communication with the other tasks. For example, parameter sweep applications are independent tasks that can often be parallelized with no constraints, and job running several times but with different input parameters.

Data transfers can have a strong impact on the running time of applications. These can take place between different tasks, between a task and a data source, and between a task and a user. When the tasks of an application are dependent data transfers between these tasks are precedence-constrained by such transfers, forming a workflow. The topological ordering of workflows with dependent tasks that can be represented by a directed acyclic graph (DAG) $G = (V, E)$ with n nodes (or tasks), where $t_a \in V$ is a *workflow* task with an associated computation cost (weight) $w_{t_a} \in \mathbb{R}^+$, $e_{a,b} \in E$ represents a dependency between t_a and t_b with an associated communication cost of $c_{a,b} \in \mathbb{R}^+$.

Various scientific workflows can be represented by DAGs, these include Montage (Fig. 10.3a; from Ref. [13]); AIRSN (Fig. 10.3b; from Ref. [14]); CSTEM (Fig. 10.3d; from Ref. [15]); LIGO (Fig. 10.3c; from Ref. [16]); and Chimera [17].

Variations in application types can also be found in the literature, including mixes of independent and dependent tasks as in campaign scheduling [18], where one level of independent tasks must finish running before the next level can start, similarly to a concatenation of fork-join DAGs. In this type of application, the join task may not even be a computer task, but rather a human-dependent one, such as setting up a new experiment based on the results of the previous campaign. Other DAG-related variations include applications where the DAG can change itself during execution, due to the presence of conditional tasks or loops in the DAG specifications, which can generate a different number of tasks as a function of input parameters.

Different applications demand different scheduling algorithms and management approaches in the cloud. Various such approaches to scheduling applications in clouds are detailed in this chapter.

10.4 PROBLEM DEFINITION

This section contains a formalization of the system model, the scheduling problem, and the VM allocation problem in cloud computing are formalized. The two problems must be dealt in the resource management of clouds.

Figure 10.4 illustrates how entities that solve these problems act in resource allocation. Scheduling output can direct applications to VMs in four different states: (1) VMs that are already allocated and running in the private cloud, (2) unallocated VMs in the

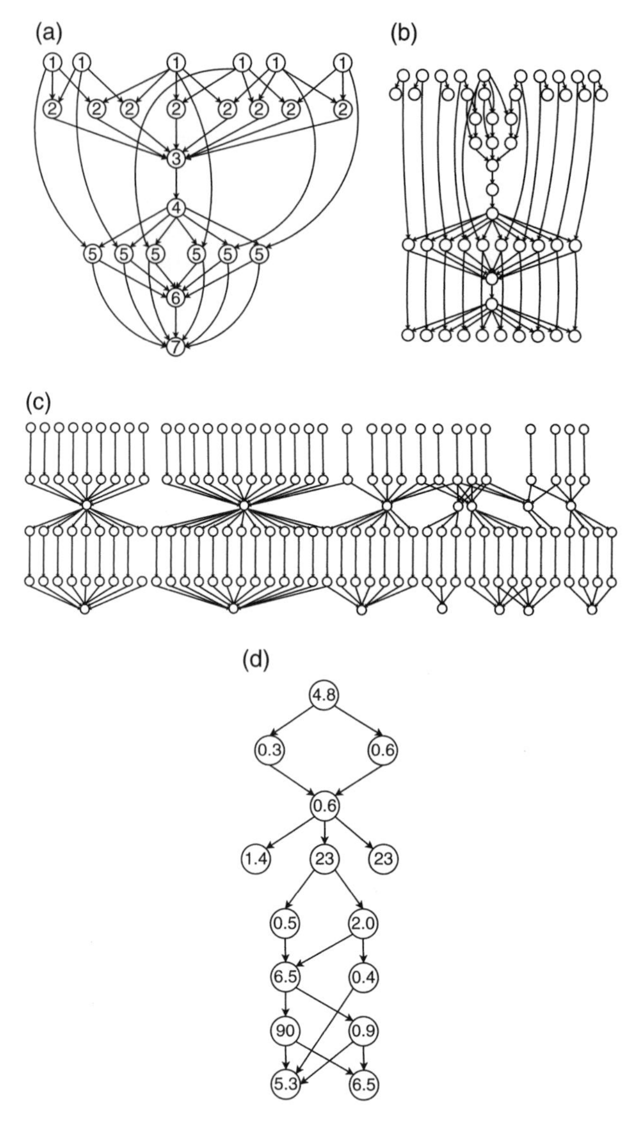

Figure 10.3. Examples of DAGs representing *workflow* applications. (a) Montage; (b) AIRSN; (c) LIGO; (d) CSTEM.

private cloud, (3) allocated VMs in the public cloud, and (4) unallocated VMs in the public cloud. Tasks scheduled to run on already deployed VMs (either in private or public clouds) do not have to interact with the VM allocator, as their VMs are already running on a given server. If, on the other hand, the scheduler decides that new, unallocated VMs are necessary, it must determine which physical machines can be used for these VMs in order to guarantee QoS. Thus, each submission is scheduled independently, and the

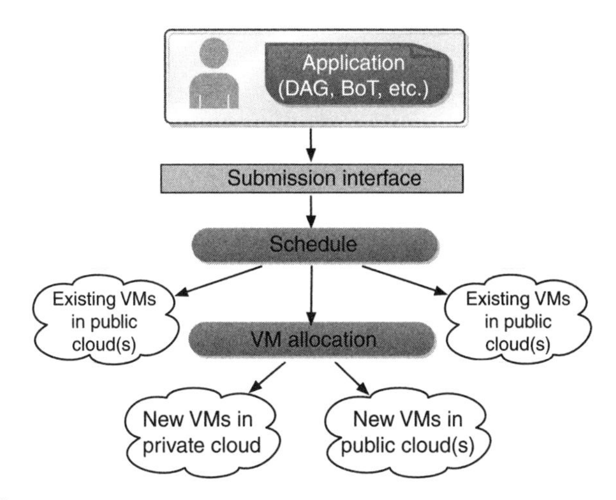

Figure 10.4. Scheduler and VM allocation in user tasks submission.

scheduler sends the resource allocator a list of VMs to be created. The resource allocator then maps these requests to the cloud infrastructure.

10.4.1 Infrastructure

Let $\mathcal{I} = \{i_1, \ldots, i_m\}$ be the set of m IaaS providers available to the users. Each cloud provider i has a set $\mathcal{M}_i = \{m_1^i, m_2^i, \ldots, m_{n_i}^i\}$ of physical machines in the system. Each $m_k^i \in \mathcal{M}_i$ is a 4-tuple $m_k^i = \{c_k^i, p_k^i, q_k^i, d_k^i\}$, where $c_k^i \in \mathbb{N}^+$ is the number of processing cores, $p_k^i \in \mathbb{M}^+$ the processing capacity of each core, $q_k^i \in \mathbb{N}^+$ is the amount of memory, and the 2-tuple $d_k^i = \{d_{k,a}^i, d_{k,s}^i\}$ represents the amount of disk storage, $d_{k,a}^i$, and data access speed, $d_{k,s}^i$. Physical machines in the IaaS i are connected by a set of links $\mathcal{L}_i$, where $l_{h,j}^i \in \mathcal{L}_i$, with $1 \leq h \leq n_i$; $1 \leq j \leq n_i$, is the bandwidth in the link between the resources m_h^i and m_j^i.

10.4.2 Service-Level Agreements

In IaaS, VMs are offered in accordance to SLA, which provides features of the VM to be leased. The cloud provider usually defines a set of VM types and SLAs, and the user chooses from these options the number of VMs and their types. The definition of an SLA depends on the underlying hardware in the data center. In other words, the VM types offered by provider i to cloud users rely on $\mathcal{M}_i$ and $\mathcal{L}_i$. Let $\mathcal{S}_i = \{s_1^i, \ldots, s_o^i\}$ be the set of SLAs offered by IaaS provider i. Each $s_j^i \in \mathcal{S}_i$ is a 7-tuple $s_j^i = \{c_j^i, p_j^i, q_j^i, d_j^i, l_j^i, p_j^i, o_j^i\}$ offering different VM QoS for the number of processing cores c_j^i, the processing capacity of each core (p_j^i), the amount of memory (q_j^i), disk storage space and data access speed ($d_j^i = \{d_{j,a}^i, d_{j,s}^i\}$), the bandwidth of the link (l_j^i), as well as price p_j^i, and model for charging o_j^i. The lefthand side of Figure 10.5 illustrates the relation between provider infrastructure and SLA in the context of scheduling.

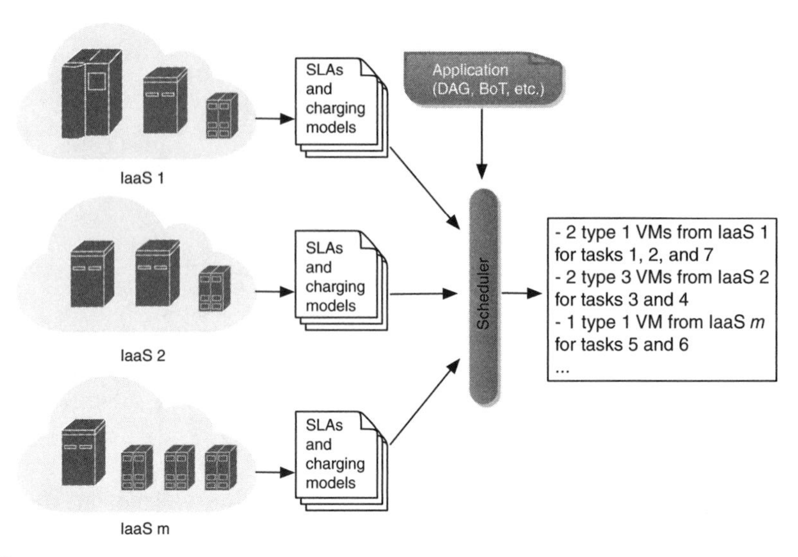

Figure 10.5. SLAs offered to the client scheduler rely on the provider infrastructure.

10.4.3 Scheduling

The scheduling problem is commonly defined as a 3-tuple $\alpha \mid \beta \mid \gamma$; α describes the execution environment, and has a single entry; β provides details about the processor characteristics and constraints; and γ describes the objective to be minimized; it frequently has a single entry [19]. In this chapter, we are interested in the following problem [19]: $\alpha = R_m$ are nonrelated machines running in parallel. There are m machines in parallel, and machine i can process task t at a speed of $v_{i,t}$. Note that given two tasks t_1 and t_2 and two resources r_1^i and r_2^i, $v_{r_1^i,t_1} > v_{r_2^i,t_2} \nRightarrow v_{r_1^i,t_2} > v_{r_2^i,t_1}$, that is, the running speeds of different tasks on different resources are unrelated.

Let $\mathcal{T} = \{t_1, t_2, \ldots, t_n\}$ be the set of tasks submitted by the users for execution. The scheduler receives as input the application, $\mathcal{T}$, and the set of SLAs available for each cloud provider, $\mathcal{S} = \cup_{i \in \mathcal{I}} \mathcal{S}_i$. The scheduler is a noninjective and nonsurjective function $\mathcal{F}_s : \mathcal{T} \to \mathcal{S}$. The scheduler defines a multiset $\mathcal{R}_{vm}$ over $\mathcal{S}$, that is, $\mathcal{R}_{vm} = (S, \mu)$ where μ: $\mathcal{S} \to \mathbb{N}_{>0}$. The multiset $\mathcal{R}_{vm}$ establishes the number of SLAs needed for each type (i.e., the number and type of VMs to be used in the application schedule). Moreover, the scheduler output includes information about the task sequencing on each VM type. This information as a whole defines both the number and type of VMs needed as well as the queue of tasks for each VM. The righthand side of Figure 10.5 provides an example of scheduler output.

10.4.4 VM Allocation

VM allocation is of paramount importance for the best utilization of physical machines. This allocation must comply with the provider objectives as well as fulfilling QoS requirements specified in the SLAs.

A VM allocation algorithm of a provider i receives as input the set of VMs, $\mathcal{R}_{vm}$, to be instantiated and the set of physical machines $\mathcal{M}_i$ available in the data center. Let $\mathcal{U}_i = \{u_1^i, u_2^i, \ldots, u_{n_i}^i\}$ represent the utilization in the datacenter, where the 3-tuple $u_k^i = \{u_k^{c^i}, u_k^{q^i}, u_{k,s}^{d^i}\}$ contains the number of resources currently allocated to existing VMs in resource i_k, namely number of cores ($u_k^{c^i}$), amount of memory ($u_k^{q^i}$), and disk storage space ($u_{k,s}^{d^i}$). The VM allocation algorithm produces a mapping of each VM from $\mathcal{R}_{vm}$ to a physical machine from $\mathcal{M}_i$.

10.4.5 Optimization Techniques

To optimize the desired objective function, the scheduler and the VM allocator utilize information about the current state of the system to guide the decision on tasks and VMs should run. Information includes processing power, amount of volatile or nonvolatile memory, and bandwidth. The weight of each of these bits of information depends on the application or the objective function, as well as on the types of resources available in the system.

Scheduling in general is an NP-Complete problem [19]; therefore, no algorithm that optimally and deterministically solves the problem in polynomial time is known. Some techniques attempting to approximate the optimal solution with polynomial time complexity have been proposed. A few of the more common techniques and the type of solutions provided are listed below:

- *Heuristics*: can produce solutions with low complexity and fast execution time; however, they ocassionally produce solutions that differ significantly from the optimal one.
- *Metaheuristics*: can obtain good quality solutions, but they take longer to run. Execution time depends on the stopping condition (e.g., number of iterations) imposed by the programmer/configuration. Moreover, they do not guarantee bounds on the quality of the solution, and local optima are commonly taken as the final solution.
- In *Linear programming*, the execution time and solution quality depend on the relaxation of constraints and a reduction in the number of variables in the problem. Heuristics can be adopted to reduce the search space, thus reducing the problem size so that solutions can be found more rapidly.
- *Approximation algorithms* with low complexity and reasonable approximation for generic problems are hard to obtain and involve tools for obtaining tight bounds [20] to the exact solution. Approximation algorithms provide solutions that guarantee quality bounds at some distance from the optimum. The more generic the problem specification is, the harder it is to obtain a satisfactory approximation.

Numerous heuristics have been proposed in the literature for scheduling and resource allocation. An overview of some of these approaches for resource allocation and scheduling in clouds is presented next.

10.5 RESOURCE MANAGEMENT AND SCHEDULING IN CLOUDS

In this section, we describe general solutions for scheduling and resource allocation in clouds. The aim is not to present a complete survey of these problems, but to provide an overview of existing approaches that can be extended to the cloud context.

10.5.1 Scheduling

As described previously, a scheduler maps tasks from the application submitted by the user to computational resources in the system. Scheduling in clouds has one important difference from scheduling in physical machines: the algorithm must consider the ability of the system to "create" computer resources as needed, given the elasticity provided by the cloud. In this section, we examine some well-known algorithms, and discuss how they could be adapted to function in the cloud.

10.5.1.1 Independent Tasks. There are a handful of algorithms to schedule independent tasks in distributed computing systems. Two traditional straightforward approaches for the scheduling a set of tasks $\mathcal{T} = \{t_1, t_2, \ldots\}$ are available: the *random* and *round-robin* algorithms. Both work in a *first-come first-served* (FCFS) basis, with the first task arriving, t_1, being the first to be scheduled. They are more suitable for homogeneous systems. Let R_{vm} be the set of VMs already rented (i.e., SLAs already established). For each task in $t_i \in \mathcal{T}, i = 1, 2, \ldots$, a scheduler *random* $(\mathcal{T}, \mathcal{R}_{vm})$ randomly takes a resource $r_j \in \mathcal{R}_{vm}$ and sends t_i to r_j's queue. A scheduler *round-robin* $(\mathcal{T}, \mathcal{R}_{vm})$, on the other hand, first transforms the $\mathcal{R}_{vm}$ into a circular queue, it then takes one task from the incoming queue and sends it to execution on the next r_j from the circular queue. No information about the duration of tasks or the capacity of resources is needed for this type of scheduling. Moreover, knowledge about the tasks queue length is not necessary, since the number of tasks existing in the queue is not taken into account in scheduling decisions.

From the client's point of view, both random and round-robin algorithms are directly applicable to scheduling in IaaS clouds over a set of already instantiated VMs. In order to take advantage of the elasticity of the cloud, both would need support from an elasticity management entity that decides when to lease new VMs (or release existing ones), dynamically changing $\mathcal{R}_{vm}$ without the interference of the schedulers. Such an entity could, for example, be invoked when the length of queues for resources execution surpasses a certain threshold. After that, one possible action would be to reschedule queued tasks using the same algorithm, or to fill new VM queues up to the size of previously existing VMs before resuming the original scheduling algorithm.

Algorithm 1 illustrates the utilization of a cloud management entity along with random or round-robin schedulers. Figure 10.6 shows an example scenario: five tasks to be scheduled using a random or round-robin algorithm, where three VMs are already rented through SLAs established with IaaS providers. Assume that tasks 1, 2, and 3 are scheduled to the first, second, and third VMs available, respectively. After that, an attempt is made to allocate task 4 on VM1, but the maximum threshold for the queue (i.e., turnaround time for task 4) is exceeded. Therefore, the elasticity manager is called

Algorithm 1 Random and round-robin adaptation for clouds

1: *scheduler* $\mathcal{F}_s = random$ OR *round-robin*
2: **while** $\mathcal{T} \neq \emptyset$ **do**
3: $\quad t =$ first task in $\mathcal{T}$
4: $\quad$ select r from $\mathcal{R}_{vm}$ following *scheduler* policy
5: $\quad$ **if** $queue(r) > threshold_{max}$ OR $queue(r) < threshold_{min}$ **then**
6: $\quad\quad$ Call elasticity management entity to determine new $\mathcal{R}_{vm}$ from $\mathcal{S}$
7: $\quad\quad$ Reschedule queued tasks
8: $\quad$ **end if**
9: **end while**

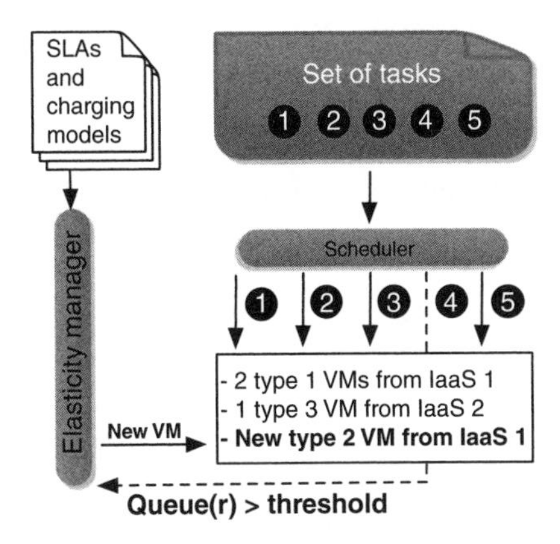

Figure 10.6. Example scenario for Algorithm 1. Task 4 exceeds the resource queue threshold, triggering a call to the elasticity manager to acquire a new VM.

to add a new VM to the pool, establishing a new SLA with an IaaS provider. As a consequence, task 4 can be scheduled to this new VM and will finish before the threshold is reached, and task 5 can also benefit from this new VM set. From this point on, a new VM will be added if new tasks arrive and the threshold is again exceeded.

The round-robin algorithm can be also adapted for use in heterogeneous systems. If relative performance among machines can be established, round-robin algorithm can assign a performance value to each machine, and assign a number of tasks proportional to this value, and advance in the circular queue. Moreover, both random and round-robin algorithms can process BoTs as well as sequences of incoming tasks over time (i.e., in an "online scheduling", in which the whole set of tasks is not known beforehand). For online scheduling, another common approach is to schedule the incoming task to the machine that currently has the shortest queue, in accordance with a specific load-balancing policy.

When tasks running time can be estimated, heuristics can utilize this information jointly with an estimation of machine performance to improve the decision taken. Two of the most well-known scheduling heuristics for BoT are the *min–min* and *max–min* algorithms [21]. Min–min first selects the task which can be completed in the shortest time and schedules it to the machine that can finish it at this earliest time. This is repeated until all tasks have been scheduled. Similarly, max–min selects first the task that takes longest to run, and schedules it on the machine that can finish it most quickly. Another well-known algorithm is *suffrage* [22], which establishes a value for a task that is the difference between its minimum completion time and its possible second minimum completion time. Tasks with the maximum suffrage are scheduled first. As with random and round-robin algorithms, these heuristics can also be adapted following an approach similar to that adopted in Algorithm 1. Several other heuristics exist and can be adopted by the elastic management entity in clouds. A comparison of eleven heuristics for the scheduling independent tasks can be found in Ref. [23].

10.5.1.2 Elasticity Management Entity. Elasticity provided by the use of cloud resources increases and decreases the capacity available to cope with current demands, as shown in Figure 10.7. The available computing power must be as close as possible to the demand, considering the QoS requirements involved in providing the resource. In this way, the elasticity management entity avoids both over provisioning in low-demand periods and under provisioning upon peak demand, consequently reducing costs.

The elasticity entity must be oriented by an objective function, which can either be incorporated into a monitoring system that invokes the elasticity entity, or invoked by the scheduler itself whenever more resources are necessary. In the first case, the monitoring system must detect when the current system load is above the desired capacity for handling the incoming workload and invoke a decision-maker in the elasticity entity to decide which is the best type of VM to be leased at that time. In the second case, the scheduler output computes the number and types of VM needed to cope with the current workload submitted. In both cases, the elasticity entity is responsible for instantiating the necessary VMs in the corresponding IaaS provider and preparing them to run tasks.

Figure 10.8 illustrates the interactions and components of the elasticity management entity. Such a decision to request more resources will be supported by information

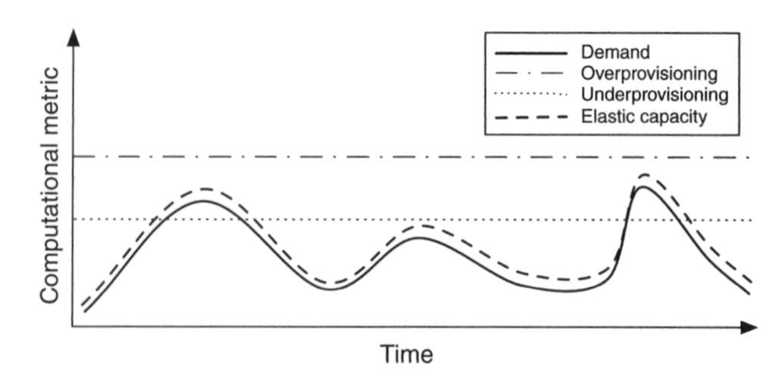

Figure 10.7. Desired elasticity versus demand in comparison to underprovisioning and overprovisioning.

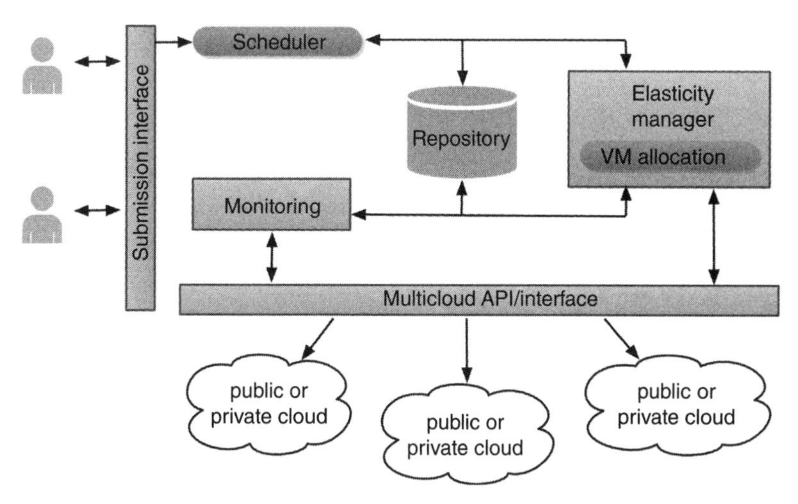

Figure 10.8. Elasticity manager and its interactions in the management system.

maintained in a repository containing data about all resources currently available (leased), including load and performance, as well as those which can be made available (SLAs offered by IaaS providers). Objective functions commonly found in the literature include the minimization of the application makespan and minimization of monetary costs [5, 10, 24].

The elasticity manager must also be able to act over multiple IaaS providers in order to manage VMs. Different providers have different management interfaces, which can lead to a cloud lock-in problem if a single interface is utilized by the elasticity manager. To avoid this problem, standardization efforts, such as the multicloud toolkit of Apache JClouds[5] can be utilized.

10.5.1.3 Dependent Tasks. Dependent tasks, as in directed acyclic graphs (DAGs) and workflows, present a topological order which must be respected by the schedule. Moreover, data transfer costs between tasks must be considered when scattering them to the available resources. As a consequence, algorithms for independent task scheduling are not directly applicable for the scheduling of dependent tasks, unless some prior selection of tasks is carried out.

Schedulers for independent tasks can be utilized for DAGs if a ready task selection is performed. A ready task has all predecessors already scheduled. Therefore, by construction, a set of ready tasks in a DAG is composed of independent tasks. At any moment, an independent task scheduler can take a task from the ready set and schedule it without violating precedences. By doing so, the scheduling of dependent tasks is transformed into a sequence of the scheduling of independent tasks, as illustrated in Algorithm 2.

[5]http://jclouds.apache.org/

Algorithm 2 Dependent task scheduling using algorithms for independent task scheduling

1: $scheduler\mathcal{F}_s$ = independent task scheduler (e.g., Algorithm 1 with random or round-robin)
2: G = DAG to be scheduled
3: **while** there exist a not scheduled task $t_a \in G$ **do**
4: $\mathcal{T}$ = set of tasks in G with all predecessors already scheduled
5: Call scheduler $(\mathcal{T}, \mathcal{R}_{vm})$
6: **end while**

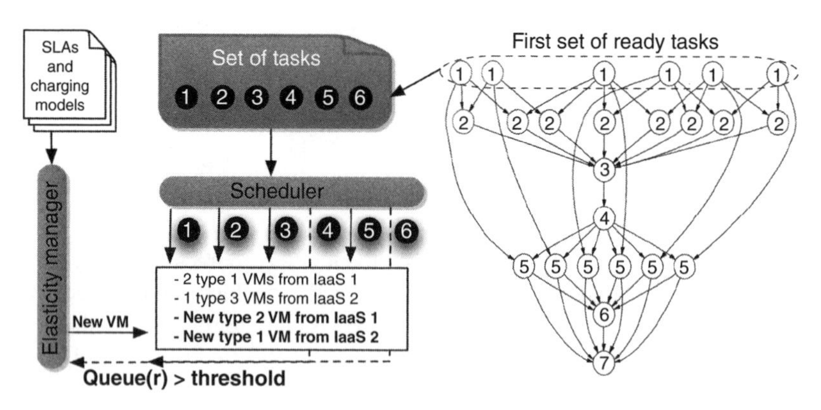

Figure 10.9. Example scenario for Algorithm 2. Ready tasks are selected to be scheduled independently using the same technique from Algorithm 1. Tasks 4 and 6 trigger the elasticity manager in this example.

Figure 10.9 shows an example of a scenario for Algorithm 2. The DAG is broken into sets of independent tasks with the first comprising all the tasks at the first level of the DAG. These tasks are sent to the scheduler as independent tasks, and then scheduled. The next set of independent tasks is not, however, necessarily composed of all tasks at the second level of the DAG, since a subset of the tasks at the first level can finish earlier, and if so a new set of independent ready tasks can be immediately computed and scheduled. The DAG is scheduled as a set of independent tasks, thus the elasticity manager can act in the same way as in the example in Figure 10.6.

The approach presented in Algorithm 2 is used by HTCondor DAGMan.[6] The main drawback of this approach is that tasks are scheduled regardless of their dependencies, since these are not considered during resource selection, even if the scheduler considers resources performance and task duration (e.g., min–min, max–min, and suffrage). In systems such as hybrid clouds and multiple IaaS providers, which resources can be geographically distant, application characteristics (e.g., high edge density and high

[6]http://research.cs.wisc.edu/htcondor/dagman/dagman.html

communication-to-computation ratio (CCR) can make data transfer times a dominant part of the application makespan. If data transfer times are disregarded during scheduling, the results can slow down the application rather than speeding it up.

Special DAG scheduling algorithms have been proposed to consider communication delays during the execution of an application. One technique commonly used for the scheduling of dependent tasks is *list scheduling*, in which tasks are first prioritized in a list according to an objective function, and then taken in order of priority for scheduling. This prioritization often takes dependencies and running times of tasks into account, resulting in higher priorities for tasks on longer paths in the DAG. One well-known example of a scheduling list for DAGs is the Heterogeneous earliest finish time (HEFT) [25]. HEFT considers heterogeneous tasks in heterogeneous nonrelated systems, and has been reported to provide good results [26–28]. Moreover, HEFT has been modified by other authors [29, 30].

HEFT and similar algorithms are usually focused on the single objective of makespan minimization. With the emergence of utility computing and clouds, a variety of budget-oriented scheduling algorithms have been proposed, including heuristics [31] and meta-heuristics [12]. Makespan minimization may not always be the main issue in scheduling, such as when an application must run within a certain timeframe, although speed is not necessary an issue. By setting a maximum finish time for the application, that is a *deadline*, the user can control his needs for results coming from applications being run. In clouds, there is usually a trade-off between cost and makespan: the user pays more for a faster resource to run applications.

Running applications with deadline constraints in clouds (e.g., hybrid clouds and/or multiple IaaS providers) requires a scheduler that is both makespan- and cost-aware, transforming the scheduling problem into a *cost minimization within a maximum makespan*. The hybrid cloud optimization cost (HCOC) algorithm [24] approaches this problem by scheduling the DAG on the private cloud resources ("costless" resources) by use of a DAG scheduling algorithm (e.g., HEFT [25] or PCH [32]), and then iteratively selecting tasks to be run in public clouds while the deadline is obeyed. The DAG scheduling algorithm utilized is typically a list scheduling and tasks to be sent to public clouds are iteratively selected based on their priorities. A generalization of this approach is presented in Algorithm 3.

Figure 10.10 presents an example of scenario for Algorithm 3. First, the DAG is scheduled on the costless resources, and a deadline violation is detected. Task 1 is the highest priority task, and it is rescheduled during the first iteration of the algorithm. Then, assuming the deadline is not satisfied with this first iteration, in the second iteration task 3, which is the task with the second highest priority in this hypothetical list scheduling, is also added to the set $\mathcal{T}$. The final schedule is achieved with tasks 1 and 3 on VM3 from a public cloud provider. The elasticity manager is invoked after the deadline is satisfied in the scheduling, and only after that the VM3 leased through the IaaS provider interface. Moreover, after task 3 finishes at VM3, it transfers the necessary data to its successor (task 6), and then VM3 can be released.

Lines 8 and 10 in Algorithm 3 can vary depending on the scheduler and policies utilized. Different algorithms use different prioritization schemes in line 3, and the sequence

---

Algorithm 3 Dependent task scheduling in clouds with deadline constraints

---

1: *scheduler* $\mathcal{F}_s$ = dependent task scheduler (e.g., HEFT, PCH)
2: G = DAG to be scheduled
3: $\mathcal{R}_c$ = costless VMs available (e.g., in private cloud or grid)
4: Schedule G in $\mathcal{R}_c$ using *scheduler*
5: $T = \emptyset$ //stores tasks to reschedule; initially empty
6: **while** $makespan_G \geq deadline_G$ **do**
7: $\quad\mathcal{R}_{vm} = \mathcal{R}_c$
8: $\quad t_p$ = task from G with highest priority in the list scheduling
9: $\quad T = T \cup \{t_p\}$
10: $\quad R_{pub}$ = select VMs from public cloud according to a resource selection policy considering the set T of tasks to be rescheduled
11: $\quad\mathcal{R}_{vm} = \mathcal{R}_c \cup R_{pub}$
12: $\quad$reschedule tasks in T to $\mathcal{R}_{vm}$ using *scheduler*
13: **end while**
14: Call elasticity management entity to allocate/setup the necessary VMs from $\mathcal{R}_{vm}$ to compose the hybrid cloud

---

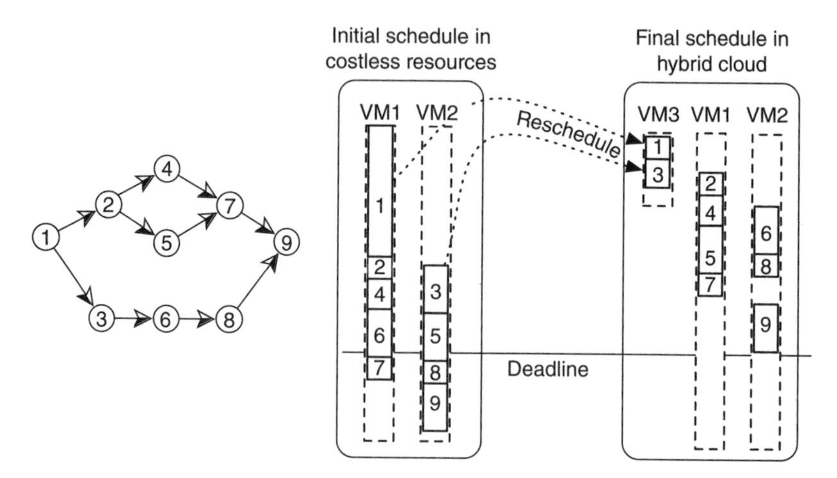

Figure 10.10. Example scenario for Algorithm 3. Tasks 1 and 3 are rescheduled to the VM3 in a public cloud provider.

of tasks rescheduled can be different. In line 8, the resource selection policy can be adapted according to the application and/or system characteristics. For example, HCOC, utilizes a multicore-aware policy to select new VMs to be leased according to the number of parallel tasks being rescheduled. This multicore-awareness also takes into account the processor performance and prices to reduce application running costs.

Yu et al. [33] have also proposed a deadline-driven cost-minimization algorithm. The Deadline-Markov decision process (MDP) algorithm breaks the DAG into partitions, assigning a maximum finish time for each partition (subdeadlines) according to the deadline given by the user. Based on this, each partition is scheduled to the resource which results in both lowest cost and lowest estimated finishing time.

10.5.2 VM Allocation

While cloud clients focus on scheduling applications, providers focus on client requests for VM allocation. Thus, IaaS cloud providers (both public and private) are concerned with the characteristics of the VMs requested by the client (expressed by an SLA), and then allocating such a VM to the available physical machines (e.g., a datacenter), according to a pre-defined objective. Allocation is important in orchestrating VMs in the computational infrastructure [34].

Objective functions for VM allocation commonly include maximization of the utilization of physical machines [35, 36], reducing power consumption [36, 37], and minimization of network traffic [38], as well as increasing security [39]. Allocation decisions should also consider the QoS requirements in accordance with the SLAs established with the cloud clients [40]. To achieve this, the cloud provider can allocate a VM to a selected physical machine, but it can also try to migrate VMs already allocated if this would significantly improve the achievement of objectives. Therefore, the cloud management system must be able to detect when VM migrations are necessary. Such a necessity can arise when users deallocate VMs and leave physical resources partially allocated. Reallocating VMs can improve resource utilization and allow physical machines to be turned off to reduce power consumption.

A general view of VM allocation is presented in Algorithm 4. The algorithm receives a set of VMs to be allocated ($\mathcal{R}_{vm}$), a set of physical machines available ($\mathcal{M}$), and the current utilization of physical machines ($\mathcal{U}$). The algorithm allocates all VMs to physical machines by first selecting a VM to be allocated, and then selecting a physical machine that can run this VM. This selection is based on the VM characteristics (according to the SLA), and the characteristics of the physical machine, as well as on the resources currently unallocated in physical machines (i.e., utilization $\mathcal{U}$). This algorithm also serves as a basis for resource reallocation when a need is detected by monitoring the amount of unallocated resources of physical machines. In this case, the set of VMs to be allocated, $\mathcal{R}_{vm}$, would comprise all VMs currently allocated.

Other algorithms that follow the reasoning of Algorithm 4 can work for different objective functions. Beloglazov and Buyya focus on energy efficient allocation [37]; they

---

Algorithm 4 Virtual machine allocation overview

1: $\mathcal{R}_{vm}$ = set of VMs to be allocated to the physical machines $\mathcal{M}$
2: $\mathcal{U}$ = current utilization of physical machines in $\mathcal{M}$
3: **while** there are not allocated VMs in $\mathcal{R}_{vm}$ **do**
4: $\quad VM$ = virtual machine from $\mathcal{R}_{vm}$ //selected according to a heuristic
5: $\quad r$ = select_machine ($VM, \mathcal{M}, \mathcal{U}$) //selected according to an objective function
6: $\quad$ Update u_r: information about resources already allocated to VMs in machine r
7: **end while**
8: Turn off idle physical machines

---

propose heuristics that consolidate virtual machines by constantly calling a VM reallocation algorithm, using live migration to switch off underutilized hosts. On the other hand, physical machines with high utilization can also trigger VM migration in order to avoid SLA violations.

VM migration can be triggered as a result of the monitoring of three situations: (1) users switching VM off, (2) users switching VMs on, and (3) monitoring physical machine utilization. The first two cases are covered, respectively, by periodically calling the VM reallocation algorithm when the number of unallocated resources in physical machines surpasses a threshold and there are a sufficiently large number of unallocated resources distributed over the physical machines to handle the new VM requests. In the first situation, reallocating VMs will probably lead to switching off physical machines, while in the second, the reallocation can avoid the turning on of new physical machines to cope with demands. The third situation requires the use of a more sophisticated mechanism to detect hotspots of utilization in VMs to decide whether it is possible to avoid *overconsolidation* [41]. Overconsolidation means that the amount of resources allocated to VMs on a host exceeds the physical resources available in that host. Thus, if the applications running in all VMs use all the resources available in the VMs, performance bottlenecks can be created and SLAs will potentially be violated.

10.6 CHALLENGES AND PERSPECTIVES

Cloud computing resource management can be partially handled by adapting techniques developed for other distributed systems such as grids and clusters. Some new management problems arise from handling any type of data coming from the omnipresent computing devices connected to the Internet. Here, we discuss some of the challenges and perspectives during cloud computing management for the next few years.

10.6.1 Scheduler and VM Allocation Cooperation

Currently, resource allocation involves in two separate phases: application scheduling and VM allocation. These two phases are often treated as independent of each other since a client is uninformed about the underlying physical system to be used for an application, and the cloud provider has no knowledge about the application requirements when allocating VMs. One way to improve QoS, as well as resource utilization is to connect application scheduling (client) with VM allocation (provider). By feeding the VM allocator with information about application requirements, VM allocation algorithms could consider the computational/networking demands of applications in each VM, thus being able to allocate VMs better to physical machines. This need for cooperation may result in privacy issues and challenges, although these could be resolved in different ways, depending on the relationship between client and provider.

10.6.2 Big Data

In the era of Big Data, large datasets are constantly generated and often accessed and processed to summarize information. One challenge in this scenario is when/where to move these large datasets to achieve faster application execution/response time and reduce costs. This involves making decisions about where application (or parts of applications) will be executed, depending on input/output data size and frequency of use. Some tasks may involve large datasets that are often utilized; in this case, it may be better to leave this data ready for use in the cloud to prevent incurring costs for transferring the data, to potentially even higher than those of running the task. In other cases, it may be worthwhile to remove the task output from the public cloud and regenerate it. This depends on the data size, computational demand for its generation and how often this data is utilized, as well as the cost of storage of the data in the cloud. Scheduling algorithms that consider data transfer times, data transfer costs, and storage costs in this context will be challenging yet necessary.

10.6.3 Greeness

A current concern in cloud computing is energy consumption. This can be reduced in cloud data centers by energy-efficient hardware design, but data center management efforts can also play a role. Two main aspects of energy-aware data center management are VM consolidation and green networking. VM consolidation allows physical machines to be turned off while not in use, while green networking techniques allow the switching off of network equipment, at least or partially, or reduction of power consumption by reducing port operating speeds. Both VM consolidation and green networking involve decisions to improve utilization and reduce network usage during allocation. Both profiling or cooperation between VM allocation and schedulers can help to achieve a greener usage of the cloud infrastructure.

10.6.4 Scheduling Multiple Workflows

The problem of scheduling a *single* workflow on clouds has been studied extensively [12, 42–45]. Nevertheless, the scheduler must also handle the concurrent execution of *multiple* workflows this issue has yet been barely considered [46–48]. When multiple workflows share the same execution environment, they compete for the same set of computational resources. In such a situation, there may be conflicts which must be dealt with to guarantee the efficiency of the workflow management system as a whole. For example, it is important that the execution of a workflow fulfil the objective function of that, but also that the agreed upon QoS be guaranteed. Therefore, besides coping with dataset management for each workflow, a scheduling algorithm should consider fairness in the sharing of resources as equally as possible among workflows. Moreover, datasets utilized/generated by one workflow may be reutilized by other workflows to be run within a limited timeframe. The decision on when to maintain or remove such datasets from the cloud will have an impact on workflow execution time and costs.

10.6.5 Hybrid Clouds and Uncertainty

Hybrid cloud management by itself is a challenge, as discussed in this chapter. One complicating factor is the uncertainty present in the public communication channels traversing the Internet and interconnecting the hybrid cloud components. A decision to compose a hybrid cloud is based on current computing demands, while the distribution of applications in that cloud must take into account data transfers to/from public clouds. Estimation of data transfer time is strongly related to the available bandwidth, which cannot be precisely predicted for the application execution horizon. Often, bandwidth is inaccurate, as is the bandwidth availability predicted. The uncontrollable and unpredictable of bandwidth variation in public Internet channels makes the application execution times in hybrid clouds prone to variation. These uncertainties in the estimation of bandwidth availability in communication channels must be considered by an application scheduler when deciding whether to lease VMs from public clouds to run applications, specially dependent tasks and applications involving large data sets.

10.7 CONCLUSION

Resource management and scheduling in cloud computing can be seen from two perspectives: viewpoint of the client and that of the provider. While the client is focused on running his/her applications with the best possible QoS and lower costs, the provider is willing to provide these services for the client with the agreed on QoS. Application scheduling and management by the cloud user involve resource management, with the user responsible for keeping track of the resources leased and current demands to determine if new machines must be leased to maintain QoS or if the currently leased machines can be released to reduce costs. VM management also encourages cost reduction by maximizing the utilization of physical infrastructure. The intrinsic conflict in these two objectives brings independent challenges to both entities. Cooperation between the two parties can, however, result in a gain–gain scenario, where application information could be explored by VM allocation to improve QoS while reducing costs.

In this chapter, we have provided an overview of aspects and requirements in resource management and application scheduling. We have discussed how these two frameworks can be handled in the cloud computing context and presented a summarizing promising research topics. Both a cloud computing model and a resource management model were presented along with VM allocation and scheduling issues. Moreover, we have discussed how existing resource allocation approaches can be extended to incorporate the elasticity intrinsic to cloud computing. A brief discussion of challenging aspects facing further developments in cloud computing research is also presented.

REFERENCES

1. M. Armbrust, A. Fox, R. Griffith, A. D. Joseph, R. Katz, A. Konwinski, G. Lee, D. Patterson, A. Rabkin, I. Stoica, and M. Zaharia. A view of cloud computing. *Communications of the ACM*, 53:50–58, 2010.
2. A. Khajeh-Hosseini, I. Sommerville, and I. Sriram. Research challenges for enterprise cloud computing. *CoRR*, abs/1001.3257, 2010.

3. D. Nurmi, R. Wolski, C. Grzegorczyk, G. Obertelli, S. Soman, L. Youseff, and D. Zagorodnov. The Eucalyptus open-source cloud-computing system. In *9th IEEE/ACM International Symposium on Cluster Computing and the Grid (CCGRID)*, pages 124–131, 2009.
4. Q. Zhang, L. Cheng, and R. Boutaba. Cloud computing: state-of-the-art and research challenges. *Journal of Internet Services and Applications*, 1(1):7–18, 2010.
5. L. F. Bittencourt, E. R. Madeira, and N. L. Da Fonseca. Scheduling in hybrid clouds. *Communications Magazine, IEEE*, 50(9):42–47, 2012.
6. P. Mell and T. Grance. The NIST Definition of Cloud Computing. Technical Report, National Institute of Standards and Technology (NIST), 2009.
7. I. Foster, Y. Zhao, I. Raicu, and S. Lu. Cloud computing and grid computing 360-degree compared. In *Grid Computing Environments Workshop, 2008. GCE '08*, pages 1 –10, 2008.
8. P. Costa, M. Migliavacca, P. Pietzuch, and A. L. Wolf. NaaS: Network-as-a-service in the cloud. In *Proceedings of the 2nd USENIX conference on Hot Topics in Management of Internet, Cloud, and Enterprise Networks and Services, Hot-ICE*, 2012.
9. M. de Assunção, A. di Costanzo, and R. Buyya. A cost-benefit analysis of using cloud computing to extend theÂcapacity of clusters. *Cluster Computing*, 13:335–347, 2010. 10.1007/s10586-010-0131-x.
10. R. Van den Bossche, K. Vanmechelen, and J. Broeckhove. Cost-optimal scheduling in hybrid IaaS clouds for deadline constrained workloads. In *2010 IEEE 3rd International Conference on Cloud Computing (CLOUD)*, pages 228–235, 2010.
11. C. Vecchiola, R. N. Calheiros, D. Karunamoorthy, and R. Buyya. Deadline-driven provisioning of resources for scientific applications in hybrid clouds with aneka. *Future Generation Computer Systems*, 2011.
12. S. Pandey, L. Wu, S. Guru, and R. Buyya. A particle swarm optimization-based heuristic for scheduling workflow applications in cloud computing environments. In *2010 24th IEEE International Conference on Advanced Information Networking and Applications (AINA)*, pages 400 –407, 2010.
13. E. Deelman, G. Singh, M.-H. Su, J. Blythe, Y. Gil, C. Kesselman, G. Mehta, K. Vahi, G. B. Berriman, J. Good, A. Laity, J. C. Jacob, and D. S. Katz. Pegasus: A framework for mapping complex scientific workflows onto distributed systems. *Scientific Programming Journal*, 13(3):219–237, 2005.
14. Y. Zhao, J. Dobson, I. Foster, L. Moreau, and M. Wilde. A notation and system for expressing and executing cleanly typed workflows on messy scientific data. *SIGMOD Records*, 34(3):37–43, 2005.
15. A. Dogan and F. Özgüner. Biobjective scheduling algorithms for execution time-reliability trade-off in heterogeneous computing systems. *Computer Journal*, 48(3):300–314, 2005.
16. A. Ramakrishnan, G. Singh, H. Zhao, E. Deelman, R. Sakellariou, K. Vahi, K. Blackburn, D. Meyers, and M. Samidi. Scheduling data-intensive workflows onto storage-constrained distributed resources. In *CCGRID '07: Proceedings of the Seventh IEEE International Symposium on Cluster Computing and the Grid*, pages 401–409, 2007.
17. J. Annis, Y. Zhao, J. Voeckler, M. Wilde, S. Kent, and I. Foster. Applying Chimera virtual data concepts to cluster finding in the Sloan Sky Survey. In *Supercomputing '02: Proceedings of the 2002 ACM/IEEE Conference on Supercomputing*, pages 1–14, 2002.
18. V. Pinheiro, K. Rzadca, and D. Trystram. Campaign scheduling. *2012 19th International Conference on High Performance Computing*, pages 1–10, 2012.

19. M. L. Pinedo. *Scheduling: Theory, Algorithms, and Systems*. Springer, New York, 2008.
20. V. V. Vazirani. *Approximation Algorithms*. Berlin: Springer, 2004.
21. O. H. Ibarra and C. E. Kim. Heuristic algorithms for scheduling independent tasks on nonidentical processors. *Journal of ACM*, 24(2):280–289, 1977.
22. M. Maheswaran, S. Ali, H. Siegal, D. Hensgen, and R. Freund. Dynamic matching and scheduling of a class of independent tasks onto heterogeneous computing systems. In *Heterogeneous Computing Workshop, 1999. (HCW '99) Proceedings. Eighth*, pages 30–44, 1999.
23. T. D. Braun, H. J. Siegel, N. Beck, L. L. Bölöni, M. Maheswaran, A. I. Reuther, J. P. Robertson, M. D. Theys, B. Yao, D. Hensgen, et al. A comparison of eleven static heuristics for mapping a class of independent tasks onto heterogeneous distributed computing systems. *Journal of Parallel and Distributed Computing*, 61(6):810–837, 2001.
24. L. F. Bittencourt and E. R. M. Madeira. HCOC: A cost optimization algorithm for workflow scheduling in hybrid clouds. *Journal of Internet Services and Applications*, 2(3):207–227, 2011.
25. H. Topcuoglu, S. Hariri, and M.-Y. Wu. Performance-effective and low-complexity task scheduling for heterogeneous computing. *IEEE Transactions on Parallel and Distributed Systems*, 13(3):260–274, 2002.
26. H. Zhao and R. Sakellariou. An experimental investigation into the rank function of the heterogeneous earliest finish time scheduling algorithm. In *Euro-Par 2003 Parallel Processing*, pages 189–194, 2003.
27. M. Wieczorek, R. Prodan, and T. Fahringer. Scheduling of scientific workflows in the askalon grid environment. *ACM SIGMOD Record*, 34(3):56–62, 2005.
28. H. Arabnejad and J. Barbosa. List scheduling algorithm for heterogeneous systems by an optimistic cost table. *IEEE Transactions on Parallel and Distributed Systems*, 25(3):682–694, 2014.
29. F. Suter, F. Desprez, and H. Casanova. From heterogeneous task scheduling to heterogeneous mixed parallel scheduling. In *Euro-Par 2004 Parallel Processing*, pages 230–237, 2004.
30. L. F. Bittencourt, R. Sakellariou, and E. R. M. Madeira. DAG scheduling using a lookahead variant of the heterogeneous earliest finish time algorithm. In *2010 18th Euromicro International Conference on Parallel, Distributed and Network-Based Processing (PDP)*, pages 27–34, 2010.
31. R. Sakellariou, H. Zhao, E. Tsiakkouri, and M. D. Dikaiakos. Scheduling workflows with budget constraints. In *Integrated Research in GRID Computing*, pages 189–202, 2007.
32. L. Bittencourt and E. Madeira. Hcoc: A cost optimization algorithm for workflow scheduling in hybrid clouds. *Journal of Internet Services and Applications*, 1–21, 2011. 10.1007/s13174-011-0032-0.
33. J. Yu, R. Buyya, and C. K. Tham. Cost-based scheduling of scientific workflow applications on utility grids. In *e-Science and Grid Computing*, pages 140–147, 2005.
34. B. Sotomayor, R. S. Montero, I. M. Llorente, and I. Foster. Virtual infrastructure management in private and hybrid clouds. *Internet Computing, IEEE*, 13(5):14–22, 2009.
35. J. Xu and J. A. B. Fortes. Multi-objective virtual machine placement in virtualized data center environments. In *Green Computing and Communications (GreenCom), 2010 IEEE/ACM Int'l Conference on Int'l Conference on Cyber, Physical and Social Computing (CPSCom)*, pages 179–188, 2010.

36. D. G. d. Lago, E. R. M. Madeira, and L. F. Bittencourt. Power-aware virtual machine scheduling on clouds using active cooling control and dvfs. In *Proceedings of the 9th International Workshop on Middleware for Grids, Clouds and e-Science*, MGC '11, pages 2:1–2:6, 2011.
37. A. Beloglazov and R. Buyya. Energy efficient allocation of virtual machines in cloud data centers. In *2010 10th IEEE/ACM International Conference on Cluster, Cloud and Grid Computing (CCGrid)*, pages 577–578, May 2010.
38. X. Meng, V. Pappas, and L. Zhang. Improving the scalability of data center networks with traffic-aware virtual machine placement. In *INFOCOM, 2010 Proceedings IEEE*, pages 1–9, 2010.
39. D. Stefani Marcon, R. Ruas Oliveira, M. Cardoso Neves, L. Salete Buriol, L. Gaspary, and M. Pilla Barcellos. Trust-based grouping for cloud datacenters: Improving security in shared infrastructures. In *IFIP Networking Conference, 2013*, pages 1–9, 2013.
40. T. Wood, P. J. Shenoy, A. Venkataramani, and M. S. Yousif. Black-box and gray-box strategies for virtual machine migration. In *USENIX Symposium on Networked Systems Design and Implementation (NSDI)*, 7, 17–17, 2007.
41. C. Hyser, B. Mckee, R. Gardner, and B. J. Watson. Autonomic virtual machine placement in the data center. Hewlett Packard Laboratories, Technical Report HPL-2007-189, 2007.
42. T. A. L. Genez, L. F. Bittencourt, and E. R. M. Madeira. Workflow scheduling for SaaS / PaaS cloud providers considering two SLA levels. In *IEEE/IFIP Network Operations and Management Symposium - NOMS 2012*. IEEE, 2012.
43. C. Lin and S. Lu. Scheduling scientific workflows elastically for cloud computing. In *2011 IEEE International Conference on Cloud Computing (CLOUD)*, pages 746–747, 2011.
44. K. Bessai, S. Youcef, A. Oulamara, C. Godart, and S. Nurcan. Bi-criteria workflow tasks allocation and scheduling in cloud computing environments. In *IEEE 5th International Conference on Cloud Computing (CLOUD)*, pages 638–645, 2012.
45. W.-N. Chen and J. Zhang. A set-based discrete pso for cloud workflow scheduling with user-defined qos constraints. In *IEEE International Conference on Systems, Man, and Cybernetics (SMC)*, pages 773–778, 2012.
46. H. Zhao and R. Sakellariou. Scheduling multiple DAGs onto heterogeneous systems. In *20th International Conference on Parallel and Distributed Processing (IPDPS)*, 2006.
47. L. F. Bittencourt and E. R. M. Madeira. Towards the scheduling of multiple workflows on computational grids. *Journal of Grid Computing*, 8(3):419–441, 2010.
48. R. F. da Silva, T. Glatard, and F. Desprez. Workflow fairness control on online and non-clairvoyant distributed computing platforms. In *19th International Conference on Parallel Processing (Euro-Par)*, pages 102–113, Springer-Verlag, 2013.

11

CLOUD SECURITY

Tianyi Xing[1], Zhengyang Xiong[1], Haiyang Qian[2], Deep Medhi[3], and Dijiang Huang[4]

[1]*School of Computing, Informatics, and Decision systems Engineering, Arizona State University, Tempe, AZ, USA*
[2]*China Mobile Technology, Milpitas, CA, USA*
[3]*Computer Science and Electrical Engineering Department, University of Missouri-Kansas City, Kansas City, MO, USA*
[4]*School of Information Technology and Engineering, Arizona State University, Tempe, AZ, USA*

Security is one of the highest concerns with cloud-based services. Intrusion detection and prevention systems (IDPS) have been widely deployed to enhance cloud security. The use of software-defined networking (SDN) approaches to enhance system security in a virtualized cloud networking environment has been recently presented [1, 2]. These approaches incorporate IDS/IPS agents in cloud servers by reconfiguring the cloud networking environment on-the-fly to counter malicious attacks. However, the performance and feasibility studies have not been well investigated. In this chapter, we provide a comprehensive study on the existing cloud security solutions and analyze its challenges and trend. Then we present an OpenFlow-based IDPS solution, called FlowIPS, that focuses on the intrusion prevention in the cloud virtual networking environment. FlowIPS is a software-based approach that implements SDN-based control functions based on Open vSwitch (OVS). FlowIPS provides network reconfiguration (NR) features

Cloud Services, Networking, and Management, First Edition.
Edited by Nelson L. S. da Fonseca and Raouf Boutaba.

by programming POX controllers to enable the FlowIPS mitigation approaches. Finally, the performance evaluation of FlowIPS demonstrates the feasibility of the proposed solution, which is more efficient compared to traditional IPS approaches.

11.1 INTRODUCTION

11.1.1 Cloud Network Security Issues

Cloud computing technologies have been widely adopted today due to its resource provisioning capabilities, such as scalability, high availability, efficiency, and so on. However, security is one of the critical issues [3] that have not been fully addressed. Attackers may compromise vulnerable virtual machines (VMs) to form botnets, and then deploy distributed denial-of-service (DDoS) attacks or send spams, which have become a major security concern of using cloud services. We highlight four critical cloud network security issues:

1. *Abuse and Nefarious Use of Cloud Computing*: IaaS providers offer their customers the illusion of unlimited compute, network, and storage capacity. By abusing the relative anonymity behind these registration and usage models, spammers, malicious code authors, and other criminals have been able to conduct their activities with relative impunity. Platform-as-a service (PaaS) providers have traditionally suffered most from this type of attacks; however, recent evidence shows that hackers have begun to target infrastructure-as-a service (IaaS) vendors as well [4]. Future areas of concern include password and key cracking, launching dynamic attack points, hosting malicious data, botnet command and control, and so on.
2. *Malicious Insiders*: The threat of a malicious insider is well known to most organizations. In traditional computer networking systems, security protection is usually deployed at the edge of the system, for example, the firewall system. However, an attacker can break the firewall or DMZ and get access into the internal network, these attack consequences can be very servere. Since all resources in the same domain is trusted among each other by default, insider attacks can cause more damage than outsider attacks.
3. *Data Integrate*: Storage is one of the most important and common scenarios in clouds. Therefore, compromising stored data, for example, deletion or alteration of records without a backup of the original content, becomes another critical security issue in clouds. The authentication and authorization of the data must securely guarantee that unauthorized or unauthenticated parties must be prevented from gaining access to privacy data. The threat of data compromise increases in the cloud, due to the number of and interactions between risks and challenges that are either unique to cloud, or more dangerous because of the architectural or operational characteristics of the cloud environment.
4. *Virtualization Hijacking*: One of the significant characteristics of the cloud computing is the virutalization, which enables better resources utilization and fine-grained resource isolation. IaaS vendors provide their services by sharing

the physical infrastructure in a scalable fashion. However, the underlying components building up the infrastructure (e.g., CPU and GPU) were not dedicated designed to deliver strong isolation capability in a multi-tenant environment. To address this issues, hypervisor is designed and introduced to fill the gap between the physical infrastructure and guest operating system. However, the existing hypervisor is not flawless and can still be compromised in that it enables users to gain access to inappropriate level of control to guest OS. A defense in depth strategy is recommended, and should include compute, storage, and network security enforcement and monitoring. Strong compartmentalization should also be employed to guarantee that individual customers do not impact the operations of other tenants running on the same cloud service provider. Customers should not have access to any other tenant's actual or residual data, network traffic, and so on.

11.1.2 Cloud Security Approach Design Challenges

Here, we describe two major design requirements that should be considered to establish a secure cloud networking environment:

1. *Robust Network Architecture Design*: Before building the cloud system, a robust security system design is highly desired. The following criteria should be followed when designing the cloud network architecture:

    - Network isolation should be provided for multiple purposes. For example, data networks should be separated from the management network because it is not secure to make users have privilege to access the cloud management network. Moreover, different networks needs to be separated from each other physically by using different network devices, e.g., switch, or virtually by deploying the network virtualization technology, such as VLAN, GRE tunnel, and so on.
    - The system should allocate sufficient resources based on the usage of system components. For example, storage network usually should be allocated with more bandwidth than management network because only control messages are sent over management network while Gigabit-sized VM images may be transmitted over the storage network. Besides network resources, host resources should be also considered. For example, the rabbitMQ, that is, the message queuing system, should be allocated with better resources due to its higher processing workload than other servers, otherwise, it will become the bottleneck and introduce vulnerability for the whole system.
    - They system should be enabled with high availability (HA), for example, redundant or backup, to avoid the single point (link) failure. The HA can be reflected in either network or host perspective. It is recommended to enable the HA for the services that especially can be directly accessed by users.

    The challenge of building a robust network architecture is that there cannot be a perfect system design, which means, system architect or administrator can only design a near perfect system and always have other security solutions to prevent the system from being impact by any possible malicious behavior.

2. *Intrusion Detection and Prevention System (IDS/IPS)*: An intuitive solution to address the cloud security issues is to deploy an IDPS (IDS/IPS), for example, Snort [5], Suricata [6], and so on. Detecting and alerting natures of IDS solutions demand the human-in-the-loop to inspect the generated security alerts, which cannot respond to attacks in a prompt fashion. Recently, the SDN technologies provide a programmable networking environment, which enables the IPS to become a key research area in the cloud automated defensive mechanism. In general, the IPS can be constructed based on IDS. For instance, Snort can be configured as inline mode and work with a common firewall system, for example, Iptables, to implement the IPS in the cloud networking environment [7]. However, there are several issues in the Snort+Iptables based IPS system, and our presented solutions target at addressing these issues:

   - *Latency*: The IPS detection engine usually uses a buffer to queue incoming packets for inspection purpose, and a packet will be dropped when the incoming packets exceeds the buffer's capacity. This mechanism ensures the IPS for packets inspection and possible blocking actions on each network packet. IPS usually consumes more cloud system resources compared to IDS, and it also increases the packet delivery delay due to the packets inspection procedure.
   - *Resource Consumption*: Enabling new services in the system will consume more resources and downgrade the system performance. For the service that is highly interactive with all the network traffic generated in the cloud virtual networking system, resources utilization becomes very critical since the security services availability depends on it. Under the same hardware resources, the one with better processing capability, for example, detection rate, has better resources consumption performance.
   - *Network Reconfigurations*: Programmable virtual networking system in the cloud environment provides the IPS a flexible way to reconfigure the virtual networking system and provide a secure traffic inspection and control. How to incorporate the deep-packet inspection (DPI) with fine-grained traffic control in the cloud virtual networking environment to reduce the intrusiveness to normal traffic is a key research challenge.

11.1.3 Arrangement of the Book Chapter

In the rest of this book, chapter is organized as follows. Section 11.2 discusses the technical background of the SDN (i.e., OpenFlow) and intrusion detection system. Section 11.3 presents the existing solutions of the cloud security. Section 11.4 disscusses the transformation from the existing cloud security solutions to the next-generation SDN-based solutions. The FlowIPS design and process flow are presented in Section 11.5. The FlowIPS is compared with traditional Snort/Iptables IPS from principle perspective in Section 11.6. NR is proposed based on the proposed architecture in Section 11.7. The thorough evaluation is conducted in Section 11.8. Finally, the future work is discussed in Section 11.9 and this book chapter is concluded in Section 11.10.

11.2 TECHNICAL BACKGROUND

In this part, we will discuss the technical background of the SDN and IDPS that will be utilized in the proposed SDN-based cloud security solutions.

11.2.1 Software Defined Networking

SDN is a new concept to evolve traditional networking technologies by separating the control plane and data plane. *OpenFlow* is the most representative protocol implementing the SDN concept to manage SDN enabled devices, and defines standard control interfaces implement packet-forwarding rules in OpenFlow switch (OFS)'s flow tables, which can handle data packets in line rate. As shown in Figure 11.1, OpenFlow introduces a centralized and separate controller and defines standard interfaces to the controller for installing the packet-forwarding rules in the flow table, which can rapidly handle incoming packets. In the OpenFlow architecture, a controller executes all control tasks of the switches and is also used for deploying new networking frameworks, such as new packet forwarding protocols or optimized cross-layer packet-switching algorithms. When a packet arrives at an OFS, the switch processes the packet in the following three steps:

1. It checks the header fields of the packets and attempts to match any entry in the local flow table. If there is no any matching entry in the flow table, the packet will be sent to the controller for further processing, for example, installing a flow

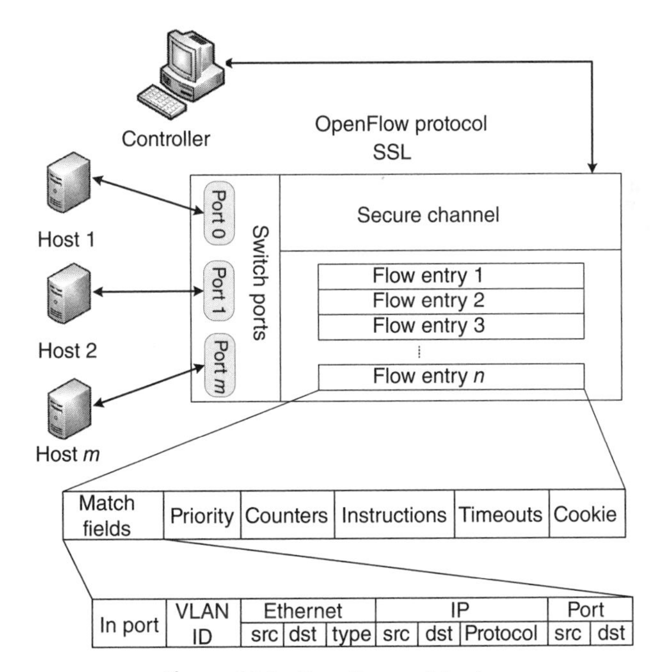

Figure 11.1. OpenFlow architecture.

table rule for forward this traffic flow in the future. There can be multiple flow control rules in the controller. It follows a best matching procedure to pick the best rule.

2. It then updates the byte and packet counting information associated with the rules for statistic logging purposes.
3. Once a matching rule is decided, the OFS takes the action based on the corresponding flow table entry, for example, forward to a specific port, or drop.

OFS separates the control plane and data plane by virtualizing the network control as a network OS layer. The network controller is considered as the software engine to deploy the control functions that can be implemented through automatic control algorithms. With these features, dynamic NR can be implemented in the cloud virtual networking environment. There are several OFS controllers available following the OpenFlow standard, such as NOX/POX [8]. Both OVS and OFS are OpenFlow protocol enabled switches. OVS is implemented as a software OFS in the cloud environment, where OVS is usually implemented in the privilege domain of a cloud server, for example, Domain 0 of XenServer [9] and the host domain of KVM [10]; while we use OFS to represent physical OFS.

11.2.2 Intrusion Detection and Prevention System

Snort is a multi-mode packet analysis IDS/IPS tool, which basically consists of sniffer, packet logger, and data analysis tools. In its detection engine, rules form signature to judge if the detected behavior is a malicious behavior or not. It has both host and network-based detection engines, and it has a wide range of detection capabilities including stealth scans, OS fingerprinting, buffer overflows, back doors, and so on. Network Intrusion Detection System (NIDS) mode has been widely used and focuses only on detection; thus, the action to be taken when the rules are matched are usually log or alert, without disabling the ongoing attacks. The combination of Snort and Iptables is the most common way to implement the Snort IPS mode that is also known as inline mode [11]. The IPS mode is different from IDS mode in that the IPS can prevent the attacks from happening in addition to intrusion detection. As mentioned in Section 11.1, one of the main challenges of IPS is the performance issue since Snort serves as a traffic proxy and every packet needs to wait until Snort tells if it is safe to pass or block. In Section 11.8, we will be discussing the performance issues of snort under IDS mode and IPS mode.

11.3 EXISTING SOLUTIONS

The related and representative security solutions are discussed in the following way: we first discuss the traditional security solution for general attacks in cloud environment, and then investigate the general SDN-based security solutions. We then investigate the solutions addressing the security issues of SDN itself. At last, we discuss current SDN-enabled IDS/IPS solutions that is highly related to the proposed SDN-based IPS in this book chapter.

11.3.1 Traditional Non-SDN Solutions

IDS and IPS are traditional efforts to monitor and secure cloud computing system. As an efficient security appliance, IDS can utilize signature-based, statistical-based, and stateful protocol analysis method to achieve its detection capability. Furthermore, IPS can respond detected threats by triggering variety prevention actions to tackle malicious activities. In Ref. [12], the authors introduce an effective management model to optimally distribute different NIDS and NIPS across the whole network. This work differs from single-vantage viewpoint NIDS/NIPS placement, it is a scalable solution for network-wide deployment.

A stateful intrusion detection system is introduced in Ref. [13]. This paper applies a slicing mechanism to divide overall traffic into subsets of manageable size and each subset contains enough evidences to detect a specific attack. As a distributed network detection system, after the system is configured, it is not easy to reconfigure and migrate the detection sensors for users on demand.

In Ref. [14], the authors propose a Host-based intrusion detection system that deploys IDS to each host in the cloud computing environment. This design enables behavior-based technique to detect unknown attacks and knowledge-based one to identify known attacks. However the data are captured from system logs, services, and node messages, this system cannot detect any intrusion which running on the VM. To apply VM compatible IDS, another architecture is provided in Ref. [15]. It extends the capability of typical IDS to incorporate VM-based detection approach to the system. The provided IDS management can detect, configure, recover the VMs, and prevent VMs from visualization layer threats. However, this is not a lightweight solution, and multiple IDSs instances are needed to build this system.

FireCol [16] is a dedicated flooding DDoS attacks detection solution implemented in traditional network system. In this design, IPSs are distributed in the network to form a virtual protection rings with selected traffic exchange to detect and defend DDoS attacks. This collaborative system addresses the hardly detection problem and single IDS/IPS crashing problem under overwhelming traffic. However, this method is not a lightweight solution such as [17], and the flexibility and dynamism is limited in this system and the deployment and management is complicated.

In Ref. [18], the authors propose a dynamic resource allocation strategy to counter DDoS attacks against individual cloud customers. When a DDoS attack occurs, they employ the idle resources of the cloud to clone sufficient intrusion prevention servers for the victim in order to quickly filter out attack packets and guarantee the quality of the service for benign users simultaneously. However, this paper focused on how to allocated idle resource for IPS but did not discuss how the IPS prevent the DDoS attack.

Similar to the FireCol, [19] presents a multiple layers game-theoretic framework for DDoS attack and defense evaluation. An innovative point in this work is the strategic thinking of attacker's perspective benefit the defense decision maker in this interaction between attacks and defenses. However, this framework is not suitable for deploying dynamic network threats countermeasure and has no real-time security solution for real-time attacks.

Packet marking technique is widely used for IP traceback even tracing back the source of attacks is extremely difficult. In Ref. [20], the authors present a marking mechanisms for DDoS traceback, which injects a unique mark to each packet for traffic identification. As a probabilistic packet marking (PPM) method, it has a potential that leads attackers to inject marked Packet and spoofed the traffic. Reference [21] is another important traceback method by using deterministic packet marking (DPM). The victim could track the packets from the router which splits the IP address into two segments. Differ from previous methods, in Ref. [22], the authors present an independent method to traceback attacker based on entropy variations. However, most of these works do not handle the IP spoofing very well and packet modification is needed to implement these methods.

11.3.2 SDN-Based Security Solutions

OpenFlow-enabled solutions provide the programming capabilities with high flexibility and scalability, which have been largely deployed to enable new services or enhance the agility for networking systems [23–25]. Combining the OpenFlow with other opensource packages creates new networking service opportunities. QuagFlow [26] integrates the Quagga opensource routing suite with OpenFlow to provide a centralized control over the physical OFS and Quagga router in VM. However, using OpenFlow as a way for security purpose, especially in cloud environment, is still in an early stage. SDN has been researched to establish monitoring system [27–29] due to its centralized abstract architecture and its statistics capability. OpenSafe [28] is a network monitoring system that allows administrators to easily collect usage statistics of networking and detect malicious activities by leveraging programmable network fabric. It uses OpenFlow technique to enable some manipulations of traffic, such as selective rules matching and arbitrary flows directing, to achieve its goal. Furthermore, ALARMS is designed as a policy language to articulate paths of switches for easily network management. OpenNetMon [29] is another approach for network monitoring application based on OpenFlow platform. This work is implemented to monitor per-flow metrics to deliver fine-grained input for traffic engineering. Benefiting from the OpenFlow interfaces that enable statistic query from controller, the authors proposed an accurate way to measure per-flow throughput, delay and packet loss metrics. In Ref. [27], the authors proposed a new framework to address the detection problem by manipulating network flows to security nodes for investigation . This flow-based detection mechanism guarantees all necessary traffic packets are inspected by security nodes. For dynamism purpose, provided services could be easily deployed by users through a simple script language. However, all three aforementioned studies [27–29] do not further discuss the countermeasure for intrusion malicious activities but only provide the monitoring service.

FortNox [30] is a Security Enforcement Kernel to address the conflict of rule to secure OpenFlow network. Different rules are inserted by various OpenFlow applications can generate rule conflict, which has potential to allow malicious packets bypass the strict system security policy. FortNox applies a rule reduction algorithm to detect conflicts and resolves a conflict by assigning authorization roles with different privilege for the candidate flow rule. This kernel overcomes the potential vulnerability of OpenFlow rules installment and enables an enforceable flow constraint to enhance SDN security. The authors another research, Fresco, [31] implements an OpenFlow Security

Application Development Framework based on Security Enforcement Kernel. It encapsulates the network security mitigation in the framework and provide an APIs to enable legacy application to trigger FRESCO module. However, this work does not have the capability to defend and protect network assets independently because predefined policies are needed to drive this system.

CONA [32] is a content-oriented networking architecture build on NetFPGA-OpenFlow platform. In this design, hosts request contents and agents deliver the requested contents while the hosts can not. Under the content-aware supervision, system can perform prevention by: (1) collecting suspect flows information from others agents for analysis, and (2) applying rate limit to each of relevant agents to slow down the overwhelming malicious traffic.

11.3.3 Security Issues of SDN

With the emerging of SDN, researchers start to concern the security of SDN itself. In Ref. [33], a replication mechanism is brought up to handle the weakness of centralized controlled network architecture, which is that one single point of failure could lead a downgrade of network resilient for the whole system. CPRecovery component is able to update the flow entry in secondary controller dynamically and secondary controller can take control the switch automatically when primary controller is down due to overwhelming traffic or DDoS attack. This work could be considered as a solution for DDoS attack; however, simple replication mechanism hardly promise that all the secondary controllers are able to tolerate high pressure attack even more backups could be deployed in this system.

AvantGuard [34] is an SDN extension, which enhances the security and resilience of OpenFlow itself. To address the two bottlenecks, scalability, and responsiveness challenge, in OpenFlow, this paper introduces two new modules: connection migration module and actuating trigger module. The former component is efficient to filter incomplete TCP connection by establishing a handshake session before packets arriving the controller. TCP connections are maintained by migration connection module to avoid the threats of TCP saturation attack. Actuating trigger module enables the data plane report network status and active a specific flow rule based on predefined traffic conditions. This research improved the robustness of SDN system and provided additional data plane information to control plane to acquire higher security performance.

11.3.4 SDN-Based Intrusion Detection and Prevention System

As we discussed before, IDS and IPS are critical security appliance to protect cloud computing network. When we apply SDN to the cloud system, the decoupled switch with separated control plane and data plane, creates a network OS layer to allow programmable interface and open network control. This feature leads a flexibility and dynamic NR, which can efficiently and effectively control the network and enable security manipulation for higher level guards. However, only a small number of works are done to implement SDN-based IDS and even fewer works on SDN-based IPS.

L-IDS [35] is a learning intrusion detection system to provide a network service for mobile devices protection. It is able to detect and respond to malicious attack with the deployment of existing security system. It is more like a network service that can transparently configured for end-host mobility and enable already known countermeasures to mitigate detected threats. The authors do not provide a comprehensive solution for detected attack and more evaluations are needed to figure the most efficient response action for threats.

In a recent work [2], the authors present an SDN-based IDS/IPS solution to deploy attack graph to dynamically generate appropriate countermeasures to enable the IDS/IPS in the cloud environment. The originality and contribution of this work mainly comes from using the attack graph theory to generate a vulnerability graph and achieve the optimal decision result on selecting the countermeasure. SnortFlow [1] is another recent work that focuses on the design of OpenFlow-based IPS with preliminary results.

Improving the accuracy and efficiency of NIDS is another important research that has attracted many researchers in this area. For example, selective packet discarding is proposed in Ref. [36]. They built up the prototype by using Snort to improve the accuracy of NIDS. In Ref. [37], the authors show good throughput performance of IDS/IPS by proposing a string matching architecture.

In Ref. [38], the authors propose a mechanism called OpenFlow Random Host Mutation (OFRHM) in which the OpenFlow controller frequently assigns each host a random virtual IP that is translated to/from the real IP of the host. This mechanism can effectively defend against stealthy scanning, worm propagation, and other scanning-based attack.

We believe the dynamic and adaptive capability of the SDN framework could benefit the development of IDPS. This area is worth to be well explored for SDN-enabled cloud system to build suitable and on demand IDS/IPS system. Thus, we have been setting our research target on establishing the SDN-based IDPS in cloud environment. This research outcome includes design and implementation of a full lifecycle SDN-based intrusion detection and prevention system in cloud virtual networking environment.

11.4 TRANSFORMING TO THE NEW IDPS CLOUD SECURITY SOLUTIONS

11.4.1 Limitations of Existing Solutions

After investigating the traditional cloud security solutions and SDN-based one, we find the existing solutions still have limitations in the following aspects:

- The detection solutions cannot efficiently detect and monitor the traffic. The most common way for detecting the traffic is to configure the SPAN port mirror, which means that all the traffic need to be duplicated and forwarded to a port in which an IDS is directly attached. Doubling the ongoing traffic definitely downgrades the performance such as delay, available bandwidth, and so on.
- The prevention solutions are not sufficiently flexible. The most common way to prevent the attack traffic is to drop it. However, all detection engine has false

positive and false negative (FN), which means drooping actions on all suspect traffic may kill the good traffic. Other prevention solutions, for example, OFRHM [38], is not performed in a reactive way. It proactively performs the moving target defense to prevent the malicious traffic, which does not work for the malicious insider case.

- The comprehensive cloud security solutions including both detection and prevention can be hardly found.

11.4.2 New IDS/IPS in Cloud Virtual Networking Environments

A straightforward approach to implement the IDS/IPS is to deploy existing solutions such as Snort-based IDS/IPS solutions without changes in clouds. In Ref. [11], Rafeeq discusses an approach on how to implement the Snort and Iptables-based IPS in clouds. In [7], the authors classify the types of traditional IPS based on desktop, host, and network, where the network-based IPS usually involves security inspections such as DPI.

SDN-based security approaches in a cloud virtual networking environment have been considered as the trend for future virtual networking security solutions [39]. In our recent work [2, 40], we present an SDN-based IDS/IPS solution using attack graph techniques to guide the cloud security management system to dynamically generate appropriate countermeasures to enable the IDS/IPS services. SnortFlow [1] is another recent work focusing on the design and evaluation of OpenFlow-[41] based IPS in the cloud environment. These existing solutions demonstrated that Snort can be used to detect intrusions in clouds; however, there are still a few important issues that current work has not addressed and can be regarded as the guidance for designing future IDPS solutions:

- Will SDN-based IDS/IPS has better performance than traditional snort-based IPS?
- How to establish an efficient software-based SDN solution in the cloud virtual networking?
- How to design the SDN-based IDS/IPS networking architecture that provides a dynamic defensive mechanism for clouds?

To address the aforementioned enumerated issues, we proposed a high-level architecture to realize the IPS by integrating Snort and OpenFlow [41] components in Ref. [1]. By utilizing the power of SDN OpenFlow, the cloud networking environment can be dynamically reconfigured based on the detected attacks in realtime. Our prototyping is established based on the Open Virtual Switch (OVS) and Xen-based [42] cloud environment. The evaluation results show that the proposed system is feasible in the cloud environment and provides valuable guidance for re-designing FlowIPS and further conducting thorough evaluations.

11.5 FLOWIPS: DESIGN AND IMPLEMENTATION

FlowIPS provides several salient features to advance the security research and development for cloud computing. It presents a new design of IDS/IPS based on SDN

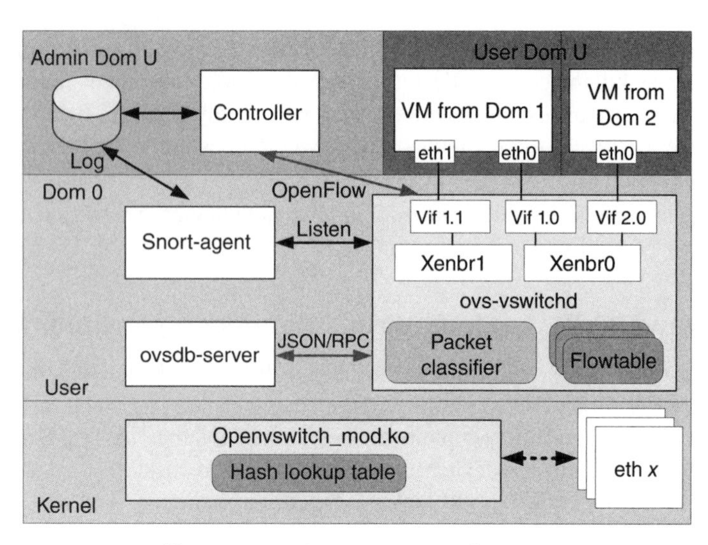

Figure 11.2. FlowIPS system architecture.

approaches, that is, using programmable OVS. It supports a dynamic defensive mechanism supporting programmable NR.

In the rest of this section, we present the designed architecture and components of the proposed FlowIPS, and the processing flow. The architecture and components are presented in Figure 11.2.

11.5.1 System Components

Cloud Cluster hosts cloud resources and the proposed FlowIPS. A cloud cluster contains one or multiple cloud servers with major cloud-based OS installed. All major cloud-based OS with SDN feature enabled, such as OpenStack, CloudStack, Xenserver, KVM, and so on, can be compatible with our proposed system. In this work, we demonstrate and establish the system based on Xenserver that is an efficient parallel virtualization solution. There are two types of domains in Xen-based cloud: Dom 0 and Dom U. Dom 0 is the management domain that belongs to the cloud administrative domain. We introduce one Dom U dedicated for administrative purpose to place controller and log component; while all other Dom Us are for hosting VMs for users. Dom U resources are managed by Dom 0 and must go through Dom 0 to access the hardware.

Open vSwitch (OVS) is a software implementation of the OFS. OVS is usually implemented in the management domain or privilege domain of cloud servers. In our established prototype, OVS is natively implemented in the Dom 0 of XenServer cloud system. Inter-VM communication within the same physical server is controlled by the OVS without exposing the traffic out of the physical box. Each Dom 0 in Xenserver runs a userspace daemon (flow path) as well as a kernel space module (fast path). In userspace, there are two modules; they are *ovsdb-server* and *ovs-switchd*. The module *ovsdb-server* is the log-based database that holds switch-level configuration; while the module ovs-switchd is the core OVS component that supports multiple independent datapaths (bridges). In Figure 11.2, the *ovs-switchd* module is able to communicate with the *ovsdb-server* module through the management protocol. They communicate with the controller through OpenFlow protocol, and with the kernel module through netlink protocol. In the kernel space, the kernel module handles packet switching, lookup and forwarding, tunnel encapsulation and decapsulation. Every virtual interface (VIF) on each VM has a corresponding virtual interface/port on OVS, and different VIFs connecting to the same bridge can be regarded on the same switch. For example, VIF 1.0 (the virtual port of eth0 on VM from Dom 1) has the layer 2 connection with VIF 2.0 (the virtual port of eth0 on VM from Dom 2). OVS forwards packets based on the entries in flow table.

Snort is a multi-mode packet analysis IDS/IPS tool, and it has better performance compared to many other products [43]. It has several components such as sniffer, packet logger, and data analysis tools. In its detection engine, rules form attack signatures to judge if the detected behavior is a malicious behavior or not. It has both host- and network-based detection engines; and it also has a wide range of detection capabilities including stealth scans, OS fingerprinting, buffer overflows, back doors, and so on. To establish the IPS in the cloud environment, the first step is to interface the detection engine Snort to the cloud networks management component, that is, OVS. In a cloud server, Snort can be implemented in Dom 0 (privilege domain) or Dom U (unprivileged domain) based on Xen virtualization techniques. In this architecture, we deploy the Snort in Dom 0, which makes it easily sniff the traffic through the software bridge in OVS. All the logging information generated from the Snort is output into a CSV file so that the controller can access in real time. The Snort component can be simply replaced with other IDS solutions, for example, Suricata, because the mitigation and detection is decoupled, which is different from the traditional IPS solution (e.g., Snort+Iptables). The performance evaluation of Snort and Suricata in the cloud is discussed in Ref. [43], the overall performance of Snort is better than Suricata, which is also the reason why we choose Snort as the candidate for the detection engine in this implementation.

Controller is the component providing a centralized view and control over the cloud virtual networks. The controller contains three major components, FlowIPS daemon, alert interpreter, and rules generator. FlowIPS daemon is mainly for collecting alerts generated from Snort agents deployed in Dom 0. Alert interpreter takes care of parsing the alert and targets the suspect traffic. Then, the parsed and filtered information is passed to rules generator who is in charge of the rules to be configured on OpenFlow-enabled software or hardware switches. A database is used to store the generated rules and switches' original states for future operations like resuming functions, and so on.

11.5.2 FlowIPS Processing Flow

The processing flow of the FlowIPS is illustrated in Figure 11.3. The network traffic is generated from the cloud resources, that is, VMs. All network traffic must be generated from the VIFs that are attached to virtual bridges in OVS. The virtual bridge can be regarded as the virtual switch, which means all VIFs connecting to the same bridge are on the same network. The Snort agent in Dom 0 has the advantage of directly detecting through the bridge. When any traffic matching the Snort rules is alerted into the log file,

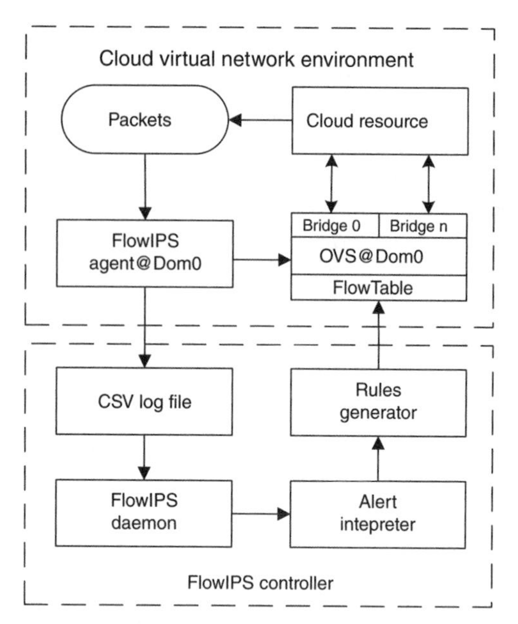

Figure 11.3. FlowIPS processing flow.

the FlowIPS daemon will fetch the alert information from the CSV log file in real time. Then, the alert interpreter will parse the alert information to get the current network security situation. Finally, the rules generator will generate the OpenFlow flow table rule entries and push them to the OVS to update its flow table. Therefore, the following suspect traffic with respect to flow table entries in the OVS can be swiftly handled with a deployed countermeasure.

11.6 FLOWIPS VS SNORT/IPTABLES IPS

Motivated by the limitation of the representative traditional IPS, for example, Snort/Iptables IPS, FlowIPS is designed to take advantages of SDN to provide security countermeasures to increase the flexibility and efficiency. This section mainly discusses the comparison between the proposed FlowIPS and the Snort/Iptables IPS focusing on the FlowIPS working mechanism and new capabilities.

Tractional IPS system is not specially designed for cloud virtual networking environment, but for a general network environment. The major difference between the general network environment and cloud virtual networking environment is that the latter one usually has difference network domains, that is, management network domain and user network domain. Those two domains are at difference layers, and therefore have difference efficiency. User network is on top of the management network, which means the lower layer can be expected with better efficiency. Thus, we design the FlowIPS especially for cloud virtual networking environment and take advantages of OVS in management domain in order to achieve better performance.

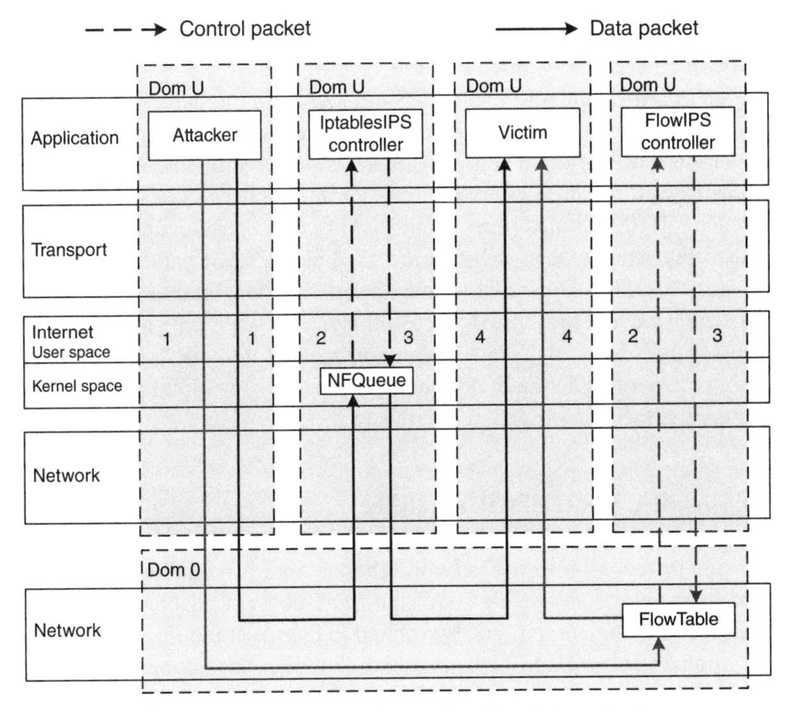

Figure 11.4. FlowIPS and Snort/Iptables IPS mechanism.

The compared two IPS solutions are different in terms of the working mechanism and operation levels. Figure 11.4 indicates the scenario on how the Iptables IPS (light lines) and FlowIPS (dark lines) detect and prevent attacks. The number beside each line represents the sequence of the packet flow. Solids lines and dotted lines represent the data traffic and control traffic, respectively. For Snort/Iptables IPS, Snort needs to be configured as inline mode and recompiled with Iptables. Besides detection engines, one of most important components of the Iptables is the NFQUEUE, which is an Iptables and Ip6tables target delegating the decision on packets to a userspace software. It issues a verdict on a detected packet. Snort/Iptables IPS can not be placed in the Dom 0 of Xenserver because Snort/Iptables IPS is a higher level proxy-based solution comparing with the OVS-based IPS solution, which means that it needs to be placed in the middle of two or more communication end virtual hosts. Thus, Snort/Iptables IPS needs to be placed at the same level with all VMs at Dom U. Moreover, it is noted that OVS in Dom 0 is the same as network stack in OS kernel level, which means that OpenFlow feature is not enabled when the flow table is empty. As shown in Figure 11.4, when attacking packets generated from attacker's virtual interface, all the packets need to be passed through Dom 0 before being forwarded to the destination (line 1). When Snort detects any suspect traffic, it needs to inform the NFQUEUE to take the actions defined in the rules. The Iptables IPS needs to consult the controller who sends out control messages to issue command (lines 2 and 3). Finally, the suspect packet is handled at the kernel space at Dom U and will be either forwarded to victim or dropped (line 4).

Unlike the Snort/Iptables IPS, FlowIPS deploys both the detection engine and the packet processing module in Dom 0. This is an efficient approach especially when handling large amount of traffic. When packets arrive at Dom 0 (line 1), Snort detection engine is able to sniff the bridges, and only few traffic between OVS at Dom 0 and the controller at Dom U is generated (lines 2 and 3). After the controller update the flow table, all traffic with the same pattern will be processed at OVS fast path in Dom 0 (line 4). From the Figure 11.4, it is also obvious that packets in Snort/Iptables IPS scenario need to be forwarded in and out the Dom 0 twice, while the FlowIPS only needs once to fulfill the same task. Thus, due to the IPS working mechanism, FlowIPS should significantly outperformance any other Dom U IPS solution especially in cloud virtual networking environment, which will be proved in Section 11.8.

11.7 NETWORK RECONFIGURATION

NR is a means to reconfigure the network characteristics including topology, packet header, QoS parameters, and so on. With the SDN concept enabled in the cloud virtual networking environment, NR can be applied to construct the IPS system. Major NR actions are summarized in Table 11.1, including the following actions:

- *Traffic Redirection (TR)* can redirect the traffic to a secure appliance (e.g., DPI unit, Honeypot) by rewriting the packet header. TR is usually implemented by using MAC/IP address rewriting. Controller can push entry to flow table which can take packet header rewriting action on matching packets.
- *QoS Adjustment (QA)* is a very efficient way to handle flood type of attacks. OVS is able to adjust the QoS parameters of any attached VIF. After lower the TX/RX rate, suspect attack traffic will generate less impact on the network and hosts nearby. Sometimes, QA can be configured to work with other NR like traffic isolation.
- *Traffic Isolation (TI)* is different from the TR in that TI provides an isolated virtual networking channel separated from others, for example, separated virtual bridges,

TABLE 11.1. Network reconfiguration actions

No.	Countermeasure
1	Traffic redirection
2	QoS adjustment
3	Traffic isolation
4	Filtering
5	Block switch port
6	Quarantine

isolated ports or GRE tunnel. Malicious traffic will be only impact any host on its isolated virtual channel and will not impact other normal traffic.

- *Filtering* is similar with the filter in Iptables, but they are different in that filtering in NR will handled packets at OVS kernel space and will not forwarded to a remote controller. MAC/IP address change is a very straightforward way to prevent the victim from being attacked by the malicious traffic. The default IPS action, that is, drop, can be also regarded as a filtering rule that drop the matching packets.
- *Blocking switch port* actions can be set up in the flow table as filtering rules. Some attacks are performed by exploring a certain port, especially a public service port. By blocking those ports, the attack can be prevented as the attacking path is disconnected.
- *Quarantine* is a comprehensive approach to do the isolation in cloud virtual networking environment. It works similarly with TI, but it isolates the suspect network resources (not just the suspect traffic). Another difference between the normal traffic isolation and quarantine is that more flexible self-defined policies can be applied in quarantine mode. Quarantine can be also regarded as the superset of many NR set. For example, the system can quarantine suspect network targets (VMs) with only ingree permission and without egree permission. Thus, such VM can only receive traffic but can not generate traffic to the network.

11.7.1 Representative NR Actions

Before introducing the NR actions, the default action taken by the IPS is blocking or dropping the malicious traffic, which can be regarded as filtering function we described in Table 11.1. Since NIDS incurs FP and FN when judging the network packets, decisions such as dropping packets may incurs high FP or FN. In this section, we present two representative NR actions besides the default IPS action, TR and QoS adjustment (QA). The reason why we exclude the traffic isolation, block switch port, is because they can be performed with similar ways, for example, rewrite packet header. And quarantine is a comprehensive solution that can be performed by combining several individual action. Thus, we only discuss the default NR, that is, dropping, and other two NR actions, which are displayed in Figure 11.5.

11.7.1.1 Traffic Redirection. There are three ways to implement the TR based on OVS in a cloud environment: MAC address rewriting, IP address rewriting, and OVS port rewriting. When detection engine detects any suspect packet, the controller firstly pushes the OpenFlow entry (i.e., matching packet header fields and corresponding actions) to OVS to update the flow table. IP and MAC addresses changes are done by flow table when certain packets are matching specific entries. Then corresponding actions will be taken for matching packets. Actions can be set as changing on any header field of the flow table, for example, source IP, destination IP, source MAC, destination MAC. TR mostly depends on the destination address (DA) field. When destination IP or MAC address is changed, the OVS will forward the packet to the changed destination address in the packet header. This NR function is especially useful when dealing

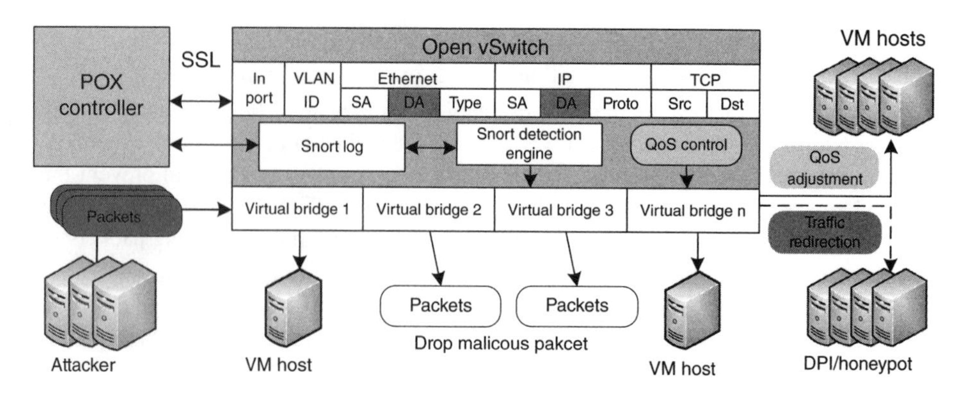

Figure 11.5. Network reconfiguration mechanism.

with the suspect packets, which cannot be determined as malicious. Suspicious packets are expected not to forward to a possible victim but a detection site for further checking, for example, applying DPI or sending to a honeypot. As shown in Figure 11.5, the block represents the corresponding flow table fields that TR may change. Moreover, IP and MAC field rewriting can be combined with other NR function to implement many network function, for example, NAT.

Beside the MAC and IP address change, there is another way to realize the TR, port rewriting. This method is also natively enabled by the OVS architecture. As shown in Figure 11.2, each bridge created in an OVS can be regarded as a virtual switch. All VM VIFs are connected to virtual bridges through virtual port (i.e., virtual interface). Thus, by forwarding any packet to the virtual port, the VM VIF that connects to that virtual port will be able to receive the forwarding packets. Through this mechanism, FlowIPS is also able to set any virtual port as the output port of any packet to implement the TR function without changing the packet header. One of benefits of using port rewriting is that any packet header will not be changed while TR is being realized, which is efficient and useful to some components (e.g., security appliance) when collecting original network data for further learning.

11.7.1.2 QoS Adjustment. QA is a desired feature when dealing with flood type of attack, for example, DoS and DDoS. When one or multiple victims are under stress from receiving a huge amount of traffic from one or multiple sources that can not be confidently determined as attackers, it is always expected to slow down the current extremely fast flow and to determine if the traffic is malicious or not after further inspection. In general, there are two ways to implement the QA, reset the QoS parameters on either VIF or port on OVS. Setting the QoS limitation on VIF or OVS port has different applied scenarios. When setting the QoS limitation on VIF, it is necessary to first locate the packet source, for example, the attacker. Thus, the number of suspect attackers would better not be large. On the other hand, when there is a DDoS attack on the network, the attack source may be a large set, which means it is infeasible and impractical to locate all zombie attackers and adjust their VIFs. To solve this issue, we introduce a smart way to implement the QoS for such situation, by limiting the incoming port on OVS. When any packet arrived at OVS, it must have an event port also called in port, as shown on the flow table in Figure 11.5. We can limit the QoS of that incoming port so that any arriving packet exceeding the QoS limit will be dropped without further process, which significantly enhances the performance of QA. Also, attacker may deploy IP spoofing technology to modify the source IP address of attacking packets to avoid being traced back. Thus, an intuitive approach is to modify QoS parameters. It is also possible to integrate the QA with TI, for example, forward the packet into a VLAN with specified QoS limitation. Here, we do not discuss the the QoS model used by FlowIPS, and we only investigate in the capability provided by FlowIPS.

11.7.2 NR Selection Policy

In Figure 11.3, there is a component in FlowIPS controller called rules generator. The rules generator is designed to choose NR and generate the corresponding OpenFlow rules based on the detection engine's alerts. The rules can be generated based on different algorithms that are not our focus here. Based on two representative NRs we mentioned above and one default drop action, we summarize the IPS action selection policy in Table 11.2.

Degree of confidence (DoC) represents the degree of how confident the detection engine believes the traffic is a malicious one. Since one of the biggest challenges of NIDS is to reduce the FN and FP, it is impossible for a detection engine to detect attacks with 100% accuracy. When the traffic is suspected and the detection engine can not draw the conclusion that it is definitely the malicious traffic, it is wise to choose the appropriate NR to mitigate the potential attack consequence. In this article, the cost means the resources consumption in the system when taking countermeasure actions. Various countermeasure actions consume different amount of resources due to the frequent OpenFlow operations, for example, updating flow table, taking OpenFlow actions. In practice, certain types of NR can be establish automatically to respond to a particular attack scenario.

In general, a packet dropping action is the default NR for whitelist-based countermeasure approaches, and it has the lowest cost since the packet match the entry in flow table can be established easily. This countermeasure usually implies high DoC to prevent attackers from introducing malicious traffic into the cloud system.

TR is appropriate for the traffic with medium DoC. When FlowIPS detects possible attacking traffic with a medium DoC, the traffic can be redirected to a secure appliance, for example, DPI proxy or Honeypot for further inspection and learning. After the FlowIPS inspects the traffic, the traffic can be possibly forwarded to the original

TABLE 11.2. FlowIPS actions selection guidance

Major actions	DoC	Cost	Preferred scenario
Drop	High	Low	Any determined malicious traffic
TR	Medium	Medium	Attacking traffic requiring further inspection
QA	Low	Medium	Attacks with overwhelmed traffic, e.g., DoS

destination or take other actions, for example, drop. Using TR, the suspect traffic will not be forwarded to the original destination until a further process is done. Since TR needs to use packet header or OVS port rewriting technology for every single suspect packet matching the flow table entry, it costs more resource than simply *drop* action.

QA has two ways as we mentioned earlier, VIF-based QA and virtual port-based QA. We are mainly focusing on the virtual port-based QA since it can be applied to a broad range of scenarios. QA is preferred to be taken for traffic with lower DoC than TR since the malicious packets with lower DoC can be sent to the original destination. Packets with low DoC can not be determined as malicious. Thus, such traffic does not need to be dropped or redirected. QA incurs similar overhead compared to the TR approach since all the packets need to be handled by the OVS. QA is a good approach to mitigate resource consumption attacks, such as DDoS attacks.

11.8 PERFORMANCE COMPARISON

After implementing the FlowIPS described in Figure 11.2, we present a comparison of the performance of proposed FlowIPS and one traditional IPS candidate, that is, Iptables-based IPS. For fair comparison, FlowIPS deploys just the default NR action (drop) since Snort/Iptables IPS does not have additional NR capabilities besides drop. All the implementation and evaluation is conducted on a Dell R510 Server with two Intel Xeon Quad-core processors and 32 GB Memory total.

Figure 11.6 evaluates the IPS forwarding capability under overwhelmed workload. The IPS itself is set as the proxy of two virtual end hosts. We use hacking tools [44] to initiate the DoS attack toward the IPS target at a fixed rate of 150,000 packets per second as the interference source. For demonstration purpose, we choose two major DoS attacks as candidates, which are ping of death (PoD) and SYN flood attack. To measure the IPS health traffic forwarding capability, a VM sends packets to another one via IPS at various rate, i.e., packets per second. In traditional IPS solution, DoS packets are first captured by the IPS detection engine that further matches the rules and takes drop action on the packets. In the FlowIPS approach, the OVS fulfills the same task as Iptables does but also handles packets in a different and more efficient mechanism. After Snort finds packets matching an attack's signature, the controller is able to be aware of the current threats in real time by parsing CSV log file and then pushes corresponding flow entries into the flow table. After flow table is updated, the malicious traffic can be handled by the OVS fast path that can dramatically increase the system performance. From Figure 11.6, FlowIPS under both type of attacks has almost 100% forwarding rate, which means that all normal traffic can be properly forwarded even the FlowIPS is under the significant stress. For Snort/Iptables IPS it has about 70% and 40% success forwarding rate under SYN flood and PoD attacks, respectively. The reason why IPS SYNFlood has better performance over IPS PoD is because the PoD attack packets are averagely bigger than SYNFlood ones and therefore consume more resource from the IPS than the SYNFlood scenario.

In Figure 11.7, we evaluate the alert generation capacity of both IPS and FlowIPS under flood interference. This metrics also states how IPS can process the attacking

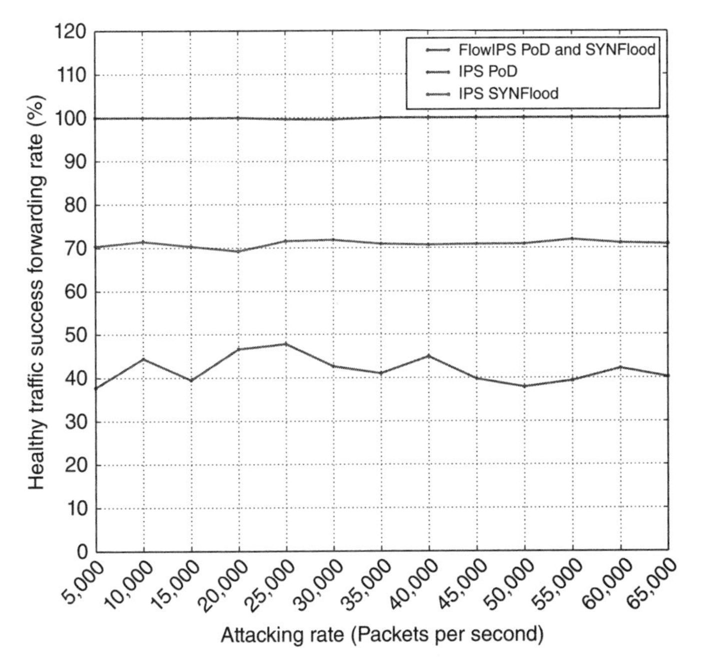

Figure 11.6. Health traffic impact.

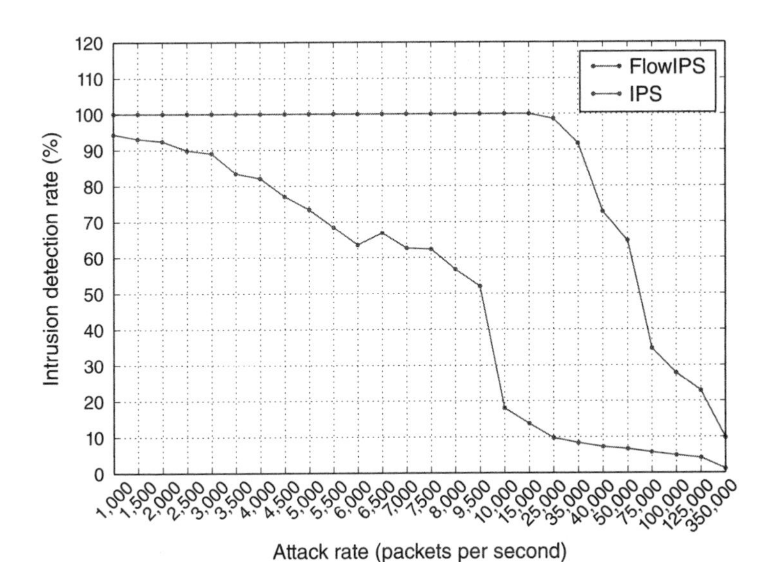

Figure 11.7. Intrusion detection rate.

packets from security perspective. To evaluate this performance, we generate two different types of attacks, which are DoS flooding attack acting as the interference source and ICMP flood attack acting as an potential threat to be tested. This evaluation mainly indicates whether IPS and FlowIPS can generate alert under high workload stress or interference. The figure shows the successful alert generation rate of ICMP attack under DoS attack interference. It suggests that alert generation of traditional IPS is impacted by DoS interference and most resources of IPS system are used to handle DoS attack therefore the performance of alert generation rate decreases as the ICMP attack speed increases. When the speed of the ICMP attack reaches to 15,000 packets per second, IPS can only generate 13.72% alerts of total ICMP attack. On the other hand, FlowIPS is able to efficiently avoid interference from DoS flooding attack due to OVS capability, so it can successfully alert all the threats which are sent at the speed of 15,000 packets per second. When the speed of the ICMP attack reaches to 30,000 packets per second, the performance of FlowIPS start decreasing, and when the speed of ICMP attack increases to 300,000 packets per second, Snort agent in FlowIPS is not able to capture packets and launch alerts because the snort detection engine itself almost reached its threshold.

Thus, the evaluation of the proposed IPS validates the analysis mentioned in Section 11.6. The FlowIPS has better network and security performance, especially in cloud virtual networking environment.

11.9 OPEN ISSUES AND FUTURE WORK

Even though the evaluation results in previous section show expected results. There are still several issues that FlowIPS cannot address. First of all, the intrusion detection system deployed in this system is a signature-based detection engine, which means that the detection capability is limited. Some attack behaviors that do not fall into any signature pattern, for example, DDoS attack, are not able to be efficiently detected by using the signature based introduction detection system. Secondly, the current IDS is planted in the management domain of the cloud virtual networking environment and different detection engines are logically disconnected from each other. This is also an critical issue because some distributed and collaborative attacks, e.g., DDoS attack, will also escape from the detection engines without appropriate sychronization among distributed detection engines. Last, how to interpret the alert and generate the prevention strategy efficiently is also one of the top concerns to improve the proposed system.

Therefore, based on the previous open issues, the future work of FlowIPS involves the following three aspects: (1) Signature- and anomaly-based detection: Beside the signature based detection engine, it is expected to incorporate the anomaly-based detection as well. Thus, the majority of malicious behaviors will be efficiently captured. (2) Synchronization: Currently, there is only one Snort detection agent in the system. We are going to introduce more detection engine placed in different servers and collect alerts from all of them, which further help to generate the NR rules by correlating some alerts. (3) Algorithms: optimized algorithms are required in alert interpreter module, rules generator, and snort agent partition to increase the efficiency of proposed FlowIPS without breaking down the detected vulnerable service.

11.10 CONCLUSION

In this chapter, we summarize the issues and design challenges of cloud network security and comprehensively study the existing work of cloud security. Then, we propose an OpenFlow-based IPS called FlowIPS in the cloud virtual networking environment. It inherits the intrusion detection capability from Snort and flexible NR from OpenFlow. FlowIPS is firstly compared with traditional IPS from principle perspective and then through real-world evaluation. NR actions are also designed and developed based on OVS and POX controller in cloud virtual networking environment. The evaluation results show the performance difference between the proposed FlowIPS and Iptable/Snort IPS, and therefore validate the superior of proposed solution.

ACKNOWLEDGMENTS

The presented work is sponsored by ONR YIP award, NSF grants CNS-1029546 and CNS-1217736, and China Mobile research grant.

REFERENCES

1. T. Xing, D. Huang, L. Xu, C.-J. Chung, and P. Khatkar, "Snortflow: A openflow-based intrusion prevention system in cloud environment," in *GENI Research and Educational Experiment Workshop, GREE*, Salt Lake City, UT, 2013.
2. C.-J. Chung, P. Khatkar, T. Xing, J. Lee, and D. Huang, "Nice: Network intrusion detection and countermeasure selection in virtual network systems," in *IEEE Transactions on Dependable and Secure Computing (TDSC), Special Issue on Cloud Computing Assessment*, vol. 10, no. 4, pp. 198–211, 2013.
3. C. C. S. Alliance, "Top threats to cloud computing v1.0," in *Cloud Security Alliance*, 2010.
4. K. Kell, "Ec2 security revisited," in *Online Blog*, 2013.
5. "SourceFire Inc." [Online]. Available: http://www.snort.org.
6. "Suricata Inc." [Online]. Available: http://suricata-ids.org.
7. W. Morton, "Intrusion prevention straitegies for cloud computing," 2011.
8. N. Gude, T. Koponen, J. Pettit, B. Pfaff, M. Casado, N. McKeown, and S. Shenkes, "Nox: Towards an operating system for networks," in *ACM SIGCOMM Computer Communication Review*, Vol. 38, no. 3, pp. 105–110, July 2008.
9. "Citrix Systems, Inc." [Online]. Available: http://www.citrix.com/products/xenserver.
10. "KVM." [Online]. Available: http://www.linux-kvm.org.
11. R. U. Rehman, *Intrusion Detection Systems with Snort: Advanced IDS Techniques Using Snort, Apache, MySQL, PHP, and ACID*. Prentice Hall Professional, Upper Saddle River, NJ, 2003.
12. V. Sekar, R. Krishnaswamy, A. Gupta, and M. K. Reiter, "Network-wide deployment of intrusion detection and prevention systems," in *Proceedings of the 6th International Conference, ser. Co-NEXT '10*. New York, NY, USA: ACM, 2010, pp. 18:1–18:12. [Online]. Available: http://doi.acm.org/10.1145/1921168.1921192.

13. C. Kruegel, F. Valeur, G. Vigna, and R. Kemmerer, "Stateful intrusion detection for high-speed network's," in *Proceedings. 2002 IEEE Symposium on Security and Privacy, 2002.* Oakland, CA, 2002, pp. 285–293.
14. K. Vieira, A. Schulter, C. Westphall, and C. Westphall, "Intrusion detection for grid and cloud computing," *IT Professional*, vol. 12, no. 4, pp. 38–43, July 2010.
15. S. Roschke, F. Cheng, and C. Meinel, "An extensible and virtualization-compatible IDs management architecture," in *Fifth International Conference on Information Assurance and Security, 2009 (IAS '09).* vol. 2, Xi'An, China, 2009, pp. 130–134.
16. J. Francois, I. Aib, and R. Boutaba, "Firecol: A collaborative protection network for the detection of flooding ddos attacks," *IEEE/ACM Transactions on Networking*, vol. 20, no. 6, pp. 1828–1841, Dec 2012.
17. R. Braga, E. Mota, and A. Passito, "Lightweight DDoS flooding attack detection using nox/openflow," in *2010 IEEE 35th Conference on Local Computer Networks (LCN)*, Denever, CO, Oct 2010, pp. 408–415.
18. S. Yu, Y. Tian, S. Guo, and D. Wu, "Can we beat DDoS attacks in clouds?," *IEEE Transactions on Parallel and Distributed Systems*, vol. 25, no. 9, pp. 2245–2254, September 2014.
19. G. Yan, R. Lee, A. Kent, and D. Wolpert, "Towards a Bayesian network game framework for evaluating DDoS attacks and defense," in *Proceedings of the 2012 ACM Conference on Computer and Communications Security, ser. CCS '12.* New York: ACM, 2012, pp. 553–566. [Online]. Available: http://doi.acm.org/10.1145/2382196.2382255.
20. M. T. Goodrich, "Probabilistic packet marking for large-scale ip traceback," *IEEE/ACM Transactions on Networking*, vol. 16, no. 1, pp. 15–24, Feb. 2008. [Online]. Available: http://dx.doi.org/10.1109/TNET.2007.910594.
21. A. Belenky and N. Ansari, "IP traceback with deterministic packet marking," *IEEE Communications Letters*, vol. 7, no. 4, pp. 162–164, 2003.
22. S. Yu, W. Zhou, R. Doss, and W. Jia, "Traceback of ddos attacks using entropy variations," *IEEE Transactions on Parallel and Distributed Systems*, vol. 22, no. 3, pp. 412–425, 2011.
23. B. Koldehofe, F. Durr, M. A. Tariq, and K. Rothermel, "The power of software-defined networking: Line-rate content-based routing using openflow," in *Proceedings of the 7th Workshop on Middleware for Next Generation Internet Computing, MW4NG '12.* New York: ACM, 2012, pp. 3:1–3:6.
24. R. Kappor, G. Porter, M. Tewari, G. M. Voelker, and A. Vahdat, "Chronos: Predictable low latency for data center applications," in *Proceedings of the Third ACM Symposium on Cloud Computing, SoCC '12.* New York: ACM, 2012, pp. 9:1–9:14.
25. M. Suchara, D. Xu, R. Doverspike, D. Johnson, and J. Rexford, "Network architecture for joint failure recovery and traffic engineering," in *Proceedings of the ACM SIGMETRICS Joint International Conference on Measurement and Modeling of Computer Systems, SIGMETRICS '11*, New York: ACM, 2011, pp. 97–108.
26. M. R. Nascimento, C. E. Rothenberg, M. R. Salvador, and M. F. Magalhaes, "QuagFlow: Partnering quagga with openflow," in *Proceedings of the ACM SIGCOMM 2010 Conference, SIGCOMM '10.* New York: ACM, 2010, pp. 441–442.
27. S. Shin and G. Gu, "Cloudwatcher: Network security monitoring using openflow in dynamic cloud networks (or: How to provide security monitoring as a service in clouds?)," in *Proceedings of the 2012 20th IEEE International Conference on Network Protocols (ICNP), ser. ICNP '12.* Washington, DC: IEEE Computer Society, 2012, pp. 1–6. [Online]. Available: http://dx.doi.org/10.1109/ICNP.2012.6459946.

28. J. R. Ballard, I. Rae, and A. Akella, "Extensible and scalable network monitoring using openSAFE," in *Proceedings of the 2010 Internet Network Management Conference on Research on Enterprise Networking, INM/WREN '10*. Berkeley, CA: USENIX Association, 2010, p. 8.
29. N. L. van Adrichem, C. Doerr, and F. A. Kuipers, "Opennetmon: Network monitoring in openflow software-defined networks," in *Network Operations and Management Symposium (NOMS)*. IEEE, 2014.
30. P. Porras, S. Shin, V. Yegneswaran, M. Fong, M. Tyson, and G. Gu, "A security enforcement kernel for openflow networks," in *Proceedings of the First Workshop on Hot Topics in Software Defined Networks, ser. HotSDN '12*. New York: ACM, 2012, pp. 121–126. [Online]. Available: http://doi.acm.org/10.1145/2342441.2342466.
31. S. Shin, P. A. Porras, V. Yegneswaran, M. W. Fong, G. Gu, and M. Tyson, "Fresco: Modular composable security services for software-defined networks," in *Proceedings of the 20th Annual Network and Distributed System Security Symposium (NDSS '13)*, February 2013.
32. J. Suh, H.-g. Choi, W. Yoon, T. You, T. Kwon, and Y. Choi, "Implementation of content-oriented networking architecture (cona): A focus on DDoS countermeasure." September 2010.
33. P. Fonseca, R. Bennesby, E. Mota, and A. Passito, "A replication component for resilient openflow-based networking," in *Network Operations and Management Symposium (NOMS), 2012 IEEE*, April 2012, pp. 933–939.
34. S. Shin, V. Yegneswaran, P. Porras, and G. Gu, "Avant-guard: Scalable and vigilant switch flow management in software-defined networks," in *Proceedings of the 2013 ACM SIGSAC Conference on Computer & Communications Security, ser. CCS '13*, 2013, pp. 413–424.
35. R. Skowyra, S. Bahargam, and A. Bestavros, "Software-defined IDs for securing embedded mobile devices," in *High Performance Extreme Computing Conference (HPEC), 2013 IEEE*, Waltham, MA, Sept 2013, pp. 1–7.
36. A. Papadogiannakis, M. Polychronakis, and E. O. Markatos, "Improving the accuracy of network intrusion detection system under load using selective packet discarding," in *Proceedings of the Third European Workshop on System Security, EUROSEC '10*. New York: ACM, 2010, pp. 15–21.
37. L. Tan and T. Sherwood, "A high throughput string matching architecture for intrusion detection and prevention," in *the 32nd Annual International Symposium on Computer Architecture (ISCA)*, Madison, WI, 2005.
38. J. H. Jafarian, E. Al-Shaer, and Q. Duan, "Openflow random host mutation: Transparent moving target defense using software defined networking," in *Proceedings of the First Workshop on Hot Topics in Software Defined Networks, HotSDN '12*. New York: ACM, 2012, pp. 127–132.
39. V. Inc., "Vmware vcloud networking and security overview," in *white paper*, 2012.
40. C.-J. Chung, J. Cui, P. Khatkar, and D. Huang, "Non-intrusive process-based monitoring system to mitigate and prevent vm vulnerability explorations," in *2013 9th International Conference on Collaborative Computing: Networking, Applications and Worksharing (Collaboratecom)*, Austin, TX. IEEE, Washington, DC, 2013, pp. 21–30.
41. N. McKeown, T. Anderson, H. Balakrishnan, G. Parulkar, L. Peterson, J. Rexford, S. Shenker, and J. Turner, "Openflow: Enabling innovation in campus networks," in *ACM SIGCOMM Computer Communication Review*, vol. 38, no. 2, pp. 69–74. Apr, 2008.

42. P. Barham, B. Dragovic, K. Fraser, S. Hand, T. Harris, A. Ho, R. Neugebauer, I. Pratt, and A. Warfieldh, "Xen and the art of virtualization," in *Proceedings of the Nineteenth ACM Symposium on Operating Systems Principles (SOSP)*, New York 2003.
43. A. Alhomoud, R. Munir, J. P. Disso, I. Awan, and A. Al-Dhelaan, "Performance evaluation study of intrusion detection systems," in *The 2nd International Conference on Ambient System, Networks and Technologies*, Niagara Falls, Canada, 2011.
44. "Back Track Linux." [Online]. Available: http://www.backtrack-linux.org.

12

SURVIVABILITY AND FAULT TOLERANCE IN THE CLOUD

Mohamed Faten Zhani[1] and Raouf Boutaba[2]

[1]*Department of Software and IT Engineering, École de technologie supérieure, University of Quebec Montreal, Canada*
[2]*D.R. Cheriton School of Computer Science, University of Waterloo, Waterloo, Ontario, Canada*

12.1 INTRODUCTION

In recent years, cloud computing has emerged as a successful model to offer computing resources in an on-demand manner for large-scale Internet services and applications. However, despite the success of cloud computing, many companies are still reluctant to embrace the cloud, mainly because of the lack of hard guarantees on the survivability and reliability of the offered services. As a result, cloud providers are urged to put in place strategies to deal with failures, mitigate their impact, and improve the fault tolerance of their infrastructures in order to ensure high availability of services.

Recent reports and studies have highlighted the devastating impact of failures and service outages on any enterprise in terms of profitability, reputation, and even viability. During the last couple of years, service outages have affected millions of online customers around the world [1]. Although the root causes of such outages may differ (e.g., software bugs, hardware failures, unexpected demand, human mistakes, denial of service attacks, and misconfiguration), the consequences can be disastrous for many businesses.

Cloud Services, Networking, and Management, First Edition.
Edited by Nelson L. S. da Fonseca and Raouf Boutaba.

Obviously, the impact of service downtime varies considerably with the application and the business [2]. For some critical applications, the cost of downtime can run between $84,000 and $108,000 per hour [3].

Even for less-critical services, recurrent outages can damage the company's reputation. Although it may not be possible to directly assess monetary loss due to reputation impairment, there is no doubt that it impacts customer's loyalty, which in the long term affects the revenue and even viability of an enterprise. In North America alone, IT downtime cost businesses more than 26 billion dollars in revenue in 2010 [2]. As more and more critical services run on the cloud, ensuring high availability and reliability of the cloud resources has become a vital challenge in cloud computing environments.

This chapter provides a comprehensive study of fundamental concepts and techniques related to survivability and reliability in cloud computing environments. It first lays out key concepts of the cloud computing model and concepts related to survivability, and then it presents an overview of the outcomes of recent analyses of failures in the cloud. Finally, it reviews and discusses existing techniques aimed at improving fault tolerance and availability of cloud services. The ultimate goal is to develop a comprehensive understanding of state-of-the-art solutions for improving cloud survivability and reliability, and to provide insights into the critical challenges to be addressed in the future.

12.2 BACKGROUND

In this section, we first provide a brief overview of the cloud computing paradigm. We then review the fundamental concepts related to fault tolerance and survivability. These concepts have been well studied in the computer industry, and they can be easily applied to the systems that make up cloud computing environments.

12.2.1 Cloud Computing Fundamentals

In recent years, cloud computing has arisen as a cost-effective platform for hosting large-scale Internet services and applications. In typical cloud environments, the main stakeholders are: the cloud provider (CP), service providers (SPs), and end users. The cloud provider owns the physical infrastructure and leverages virtualization technology to partition the available resources and lease them to multiple service providers [4]. Each SP then uses the leased resources to deploy its services and applications and offer them to end users through the Internet.

Currently, CPs like Google Compute Engine [5] and Amazon EC2 [6] only offer computing and storage resources (i.e., virtual machines) without any guarantees on network performance. The lack of such guarantees results in variable and unpredictable performance and also several potential security risks as applications can impact each other [7].

To address these issues, recent research proposals have advocated offering both computing and networking resources in the form of virtual data centers (VDCs) (also known as virtual infrastructures) [7–9]. Basically, a VDC is made up of virtual machines (VMs) and virtual switches connected through virtual links. Virtual machines and switches are characterized by their capacity in terms of processing, memory, and disk size, whereas virtual links provide guaranteed bandwidth and eventually bounded propagation delay. From the cloud provider's perspective, VDCs are a means to ensure better performance isolation between different user services. In addition, as the resource requirement of each VDC is provided by SPs, CPs are able to take more informed management decisions and develop fine-grained resource allocation schemes. At the same time, SPs' benefit from using VDCs by taking full advantage of the cloud computing model (particularly, in terms of costs) with assured guarantees in terms of the computing and networking resources allocated for their applications and services, as well as greater security, thanks to better isolations between VDCs.

One of the key challenges faced by cloud providers is the VDC embedding (also known as mapping) problem, which aims at allocating computing and networking resources to the VMs and virtual links with the goal of achieving several objectives:

1. Maximize the revenue generated from the embedding of VDCs.
2. Minimize VDC request queuing delay, which is the time an SP has to wait before its requested VDC is allocated.
3. Minimize the energy consumed by the physical infrastructure, which is usually achieved by consolidating VMs in a minimal number of servers.
4. Provide guarantees on (or at least maximize) the availability of the resources allocated to VDCs.

Existing VDC embedding schemes typically attempt to achieve simultaneously more than one of these objectives, which may sometimes conflict with each other. In this chapter, we focus mainly on the schemes that have targeted at least the fourth objective, which is related to VDC fault tolerance and availability.

12.2.2 Survivability-Related Concepts

In the following, we provide the definition of the basic terms related to survivability and fault-tolerance.

- *Fault tolerance*: Fault tolerance is the property of a system that is able to operate correctly despite the presence of hardware or software failures. Generally speaking, fault tolerance is achieved through creating backups that take the place of failed components, and thereby ensure the continuity of the service.
- *Reliability*: Reliability is the conditional probability that a system remains operational for a stated interval of time given that the system was operating flawlessly. Generally speaking, there are two widely used metrics that can capture the reliability of a system: namely, the mean time between failures (MTBFs) and the failure rate. The MTBF is the mean up time between failures of a system, and the failure rate is the expected number of failures per a given time period
- *Availability*: Availability of a system is defined as the percentage of time for which the system in question is operational. It can also be seen as the probability that the system is up at any given time. Specifically, the availability $A_{\bar{n}} \in [0, 1]$ of a physical device $\bar{n}$ is given by

TABLE 12.1. Availability vs. daily and monthly downtimes

System availability (%)	Tolerable daily downtime	Tolerable monthly downtime
95	1 h:12 min	1 day 12 h 31 min
99	14 min:2 s	7 h:18 min:17 s
99.9	1 min:26 s	43 min 49 s
99.99	8.6 s	4 min:23 s
99.999	0.9 s	26.3 s
99.9999	0.1 s	2.6 s

$$A_{\bar{n}} = \frac{MTBF_{\bar{n}}}{MTBF_{\bar{n}} + MTTR_{\bar{n}}} \tag{12.1}$$

where $MTBF_{\bar{n}}$ and $MTTR_{\bar{n}}$ represent the MTBFs and the mean time to repair for the device $\bar{n}$, respectively. Both $MTBF_{\bar{n}}$ and $MTTR_{\bar{n}}$ can be computed based on the historical records of failure events. Table 12.1 provides the tolerable daily downtime associated with some availability value. It is worth noting that the availability is usually expressed in 9s. For instance, five 9s means an availability of 99.999%. The 9s are a logarithmic measures; that is, a system with five 9s availability is 10s times more available than another one with four 9s.

- *Fault domain*: A fault domain is a set of devices that share a single point of failure [10]. For instance, servers connected to the same top-of-rack switch belong to the same fault domain. It is also worth noting that a device may belong simultaneously to multiple fault domains.

12.3 FAILURE CHARACTERIZATION IN CLOUD ENVIRONMENTS

In this section, we briefly review recent works on failure characterization in cloud environments, and then summarize the main outcomes of these studies.

Wu et al. [11] proposed *NetPilot*, an automated failure mitigation system. NetPilot deals automatically with failures in large-scale data centers without human intervention. The system is built based on the analysis and characterization of failures reported in production data centers over a period of 6 months. The authors identified three main causes of failures: software failures, which account for 21% of the total number of failures; hardware failures, which represent 18% of the total failures; and misconfiguration, which is the main cause of failure with 38% of the total number of failures. The study reported also that simple failure mitigation operations are very effective in cutting down failure repair times. However, some failures may require a high repair time, and thus may lead to significant service downtimes. These results are in line with the ones reported in Ref. [12] that revealed that 95% of network failures can be repaired within 10 min, and only 0.09% of the failures may need more than 10 days to be repaired. This shows that repair times and failure impact on services may vary significantly depending on the type of failure.

Vishwanath et al. [13] analyzed the failures of more than 100,000 servers running in multiple data centers owned by Microsoft. The authors analyzed the logs collected over 14 months. Their main finding was that server failures are mainly caused by hard disk, memory, and raid controller failures. The study revealed that hard disk failures account for 78% of the total failures. It also reported that there is a high correlation between the number of disk drives in the server and the number of server failures. Finally, they found that failures are recurrent: devices that have experienced failures are more likely to fail again in the near future. This shows that device failure rates have a skewed distribution.

Gill et al. [14] performed a characterization of failures which occurred in several Microsoft data centers. In their study, they first identified networking devices that were more prone to failure. Then, they assessed the impact of failures on application performance and evaluated the effectiveness of network redundancy. They reported that 75% of networking equipment was top-of-rack switches, 15% were core and aggregation switches, and 10% were load balancers (LBs). They noticed that the failure rates of the equipment varied significantly with the type and the model (LBs, servers, top-of-rack switches, aggregation switches, routers). In particular, LBs had the highest failure probability (20%) during a 1-year period. Switches have much lower failure probability at less than 5%. Furthermore, the failure rates of different devices were unevenly distributed; for instance, the number of failures across LBs was highly variable. Some LB devices experienced more than 400 failures during 1 year. Finally, failure traces show that correlated failures were extremely rare.

Based on the observations of the aforementioned studies, we can summarize the main characteristics of failures in data centers as follows. The duration of failures is extremely variable [12, 15]. Indeed, some failures can last for seconds, whereas others last for days. Data center equipment exhibit high heterogeneity in terms of failure rates and availability. This suggests that such heterogeneity should be taken into account when designing and deploying fault-tolerance mechanisms.

12.4 AVAILABILITY-AWARE RESOURCE ALLOCATION SCHEMES

In the following, we provide a survey of the most representative proposals that have addressed the VDC embedding problem while taking into consideration VDC requirements in terms of resources and availability.

12.4.1 Survivable Mapping

Xu et al. [16] proposed a survivable VDC (termed "virtual infrastructure" in the paper) embedding scheme that allocates resources not only to VDCs but also to backup VMs and virtual links with the goal of minimizing total consumed resources. The authors defined basic requirements to ensure that backup VMs can take over in case of failures. These requirements are translated into placement constraints that have to be considered while mapping the resources. For instance, the first constraint is that each VM and its backup should be placed in two different machines. The second constraint is that there must be sufficient available bandwidth for each backup VM to communicate with other VMs so that it can replace the failed VM without any impact on service performance. However, the proposed scheme does not consider the availability of the physical machines and assumes that the number of backups is known beforehand (a backup VM and a backup virtual link for each VM and virtual link, respectively). Finally, this approach does not consider cases where switches fail resulting in the disconnection of a set of servers at the same time. Hence, it only allows mitigating failures that occur at the servers.

12.4.2 ORP

Yeow et al. [17] studied the problem of allocating resources to VDCs (termed virtual infrastructures in the paper) in a physical infrastructure such that the desired availability for each VDC is guaranteed. Figure 12.1 provides an example of a virtual data center and shows how this VDC is mapped to a physical data center. It also shows backup nodes and links that are provisioned in order to achieve a desired availability.

The first challenge addressed by the authors was how to estimate the availability of a VDC. Hence, they developed a formula to compute the availability of a VDC based on the provisioned number of backups and the availability of physical machines hosting the VDC. The study makes two main assumptions: (1) the data center is homogenous, that is, all physical devices have same availability and failure probability; and (2) node failures are independent. Hence, the availability of a VDC that includes K backup virtual nodes is given by

$$A_{VDC} = \sum_{i=0}^{K} \binom{N+K}{i} A^i (1-A)^{N+K-i} \tag{12.2}$$

where N is the number of virtual nodes comprising the VDC, and A is the availability of the physical nodes hosting the VDC components. This formula allows the estimation of the number of backup nodes required to achieve a desired availability. The authors then proposed an opportunistic redundancy pooling (ORP) mechanism allowing multiple VDCs to share backups. The idea is to estimate the number of shared backups based on Equation 12.2 simply by setting the variable N to the sum of the number of nodes of

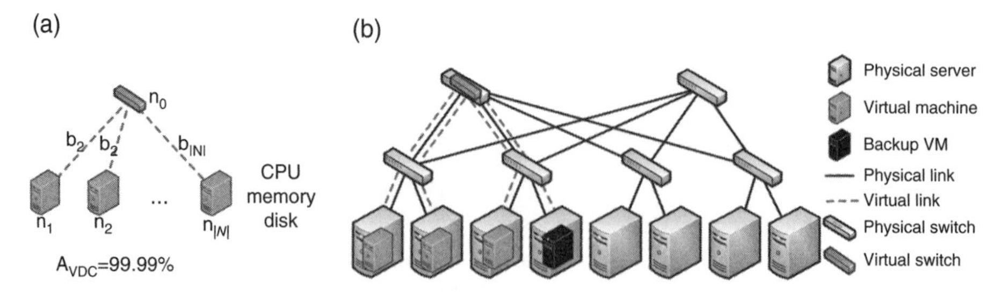

Figure 12.1. VDC embedding [18]. (a) Example of a virtual data center (star topology). (b) Example of embedding a virtual data center into a physical infrastructure.

multiple VDCs. This approach can reduce the amount of allocated backups by up to 31.25%. Finally, the authors adapted the multicommodity flow technique used in Ref. [19] to formulate the node and link joint resource allocation problem while considering sharing backup resources among multiple VDCs.

Although this work presents many advantages, its application to real-world environments is limited. Indeed, machine failure rates and availability are highly heterogeneous in production data centers [20]. Furthermore, the approach does not consider the availability of networking elements (e.g., switches and routers) in the computation of VDC availability.

12.4.3 VENICE

Zhang et al. [20] put forward an aVailability-aware EmbeddiNg framework In Cloud Environments (VENICE). They first presented a technique to compute the availability of an embedded VDC that considers the heterogeneity of the physical devices in terms of failure rate and availability. They then proposed an embedding that uses the availability computation technique to achieve the desired availability for the hosted VDCs.

In their work, the authors considered that each VDC is hosting a multi-tier service application (e.g., a three-tier Web application, as shown in Fig. 12.2a). Each tier contains a set of VM replicas that communicate with the VMs of the following tier. The authors presented a technique to compute the availability of a particular VDC mapping based on the availability of the underlying physical devices. The key idea behind this technique is to compute the VDC availability by considering all possible failure scenarios. Specifically, a failure scenario is a specific configuration in which some physical components have failed. Figure 12.2b shows an example of 3 failure scenarios (s_1, s_2, and s_3) that could affect the VDC drawn in Figure 12.2b. Hence, VENICE first identifies all possible failure scenarios that could impact the operation of the VDC, computes the availability

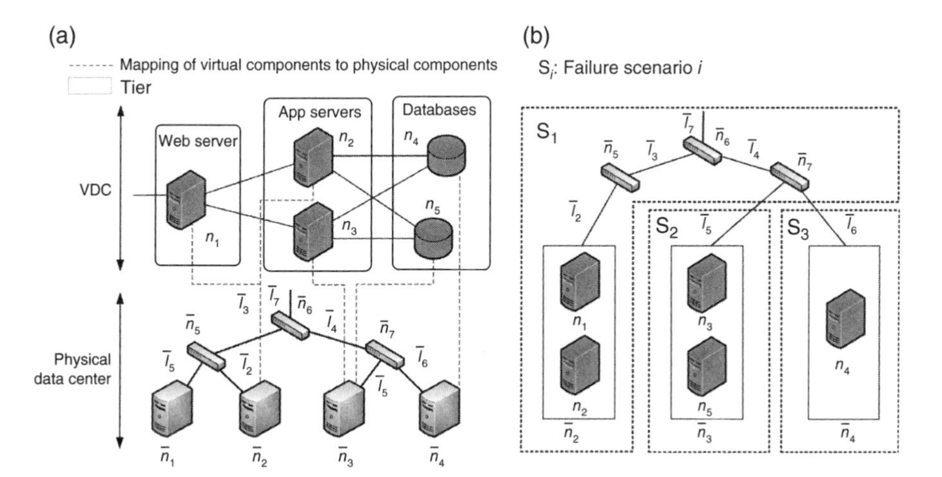

Figure 12.2. VDC embedding [20]. (a) Embedding of a VDC running a three-tier web application. (b) Example of three failure scenarios.

of the VDC under each of them and finally estimates the overall VDC availability using conditional probability. More formally, define $\bar{N}$ as the set of physical components and $s_i(\bar{n}) \in \{0, 1\}$ as a boolean variable indicating whether or not the device $\bar{n}$ is down. Let $\mathbf{s}_i = (s_i(\bar{n}))_{\bar{n}\in\bar{N}}$ denote a failure scenario that involves k simultaneous physical failures, and $S = (\mathbf{s}_i)_{i\in|S|}$ denote the set of all possible failure scenarios. The availability $A^{s_i}_{VDC}$ of a particular VDC under failure scenario $\mathbf{s}_i \in S$ is computed as follows:

$$A^{\mathbf{s}_i}_{VDC} = \prod_{\bar{n}\in\bar{N}:s_i(\bar{n})=1} (1 - A_{\bar{n}}) \prod_{\bar{n}\in\bar{N}:s_i(\bar{n})=0} A_{\bar{n}} \tag{12.3}$$

The overall availability of the VDC is then computed as follows:

$$A_{VDC} = \sum_{i=1}^{|S|} P(\mathbf{s}_i) A^{\mathbf{s}_i}_{VDC} \tag{12.4}$$

where $P(s_i)$ is the probability that failure scenario s_i occurs.

Using this technique, the authors addressed the availability-aware VDC embedding problem where each service provider specifies not only the resource requirements of the VDC but also its desired availability. VENICE tries then to achieve the desired VDC availability by carefully placing the VDC components. The goal of the VDC allocation algorithm is to maximize the total revenue of the cloud provider while minimizing the penalty incurred due to service unavailability. Unfortunately, neither the proposed availability computation technique nor the embedding scheme consider the case where backup nodes and links are provisioned.

12.4.4 Hi-VI

Rabbani et al. [18] proposed a high-availability virtual infrastructure management framework (Hi-VI). The Hi-VI framework dynamically provisions backup resources (i.e., virtual nodes and virtual links) for each VDC in order to achieve the desired availability for the VDCs. The originality of this approach is that it takes into consideration the heterogeneity of data center computing and networking equipments. Hence, it considers the case where equipments have different failure rates and availabilities.

The authors derived a formula to compute the availability of a particular VDC. The formula uses the availability of the physical equipment hosting the VDC components and also considers the number of provisioned backup nodes and links. However, two simplifying assumptions were made: (1) VDCs have a star topology, that is, each VDC comprises a set of VMs connected to a single virtual switch; and (2) equipment does not fail simultaneously, that is, only a single physical failure may occur at a time. The intuition behind the proposed formula is that since there are K backups, it is possible to replace up to K failed VMs. In other words, the availability of a VDC with K backups is given by the probability of having fewer than K failures. Mathematically, the availability of a particular VDC as denoted by A_{VDC} is written as:

$$A_{VDC} = \left(\prod_{\bar{n}:y_{\bar{n}}=1} y_{\bar{n}} A_{\bar{n}} \right) + \sum_{k=1}^{K} \left(\sum_{\bar{n}:g_{\bar{n}}=k} ((1 - A_{\bar{n}}) \prod_{\bar{l}\in\bar{N}\setminus\{\bar{n}\}:y_{\bar{l}}=1} y_{\bar{l}} A_{\bar{l}}) \right) \tag{12.5}$$

where $A_{\bar{n}}$ is the availability of a physical component $\bar{n} \in \bar{N}$, K is the number of backup VMs (which is equal to the number of backup virtual links), and $y_{\bar{n}}$ is a boolean variable that takes 1 if the physical node $\bar{n} \in \bar{N}$ either hosts one of the VMs of the VDC or is used as an intermediate node to embed a virtual link. The variable $g_{\bar{n}}$ is the number of VMs mapped to physical machine $\bar{n}$. The first term of the equation is the probability that all virtual nodes are operational (that is, all physical nodes hosting the VDC are available), whereas the second term represents the probability that a single physical node failure occurs, incurring less or equal than K virtual node failures.

Using the availability formula (Eq. 12.5), the authors proposed a VDC embedding algorithm that jointly allocates computing and networking resources for VDCs and the virtual backup nodes and links. The idea is to first embed the VDC and then gradually add new backups in high availability nodes until the desired availability is achieved. If it is not possible to meet the desired availability, the VDC is simply rejected. The ultimate goal of the algorithm is to ensure all embedded VDCs satisfy the desired availability while minimizing the number of backups and the number of servers used to host the virtual resources.

12.4.5 WCS

Bodik et al. [10] studied resource allocation in data centers that achieve the best tradeoff between fault tolerance and bandwidth usage. Indeed, when VMs of the same VDC (termed "service" in the paper) are spread across the data center, they are less likely to be affected by the same failure (e.g., top-of-rack failures) but they consume significant bandwidth in the data center network, as they are far from each other (Fig. 12.3). Conversely, when these VMs are placed close to each other, they consume less bandwidth, but a single failure (e.g., at the top-of-rack level) may simultaneously affect many VMs.

Based on this observation, the authors proposed an allocation scheme that mitigates failure impact on the virtual data center while minimizing bandwidth usage in the data center network. They hence put forward a new metric named the *worst-case survival* (WCS) to measure the fault-tolerance of a particular VDC. The WCS of a VDC is defined as the number of its VMs that remain available during a single worst-case failure divided by the total number of its VMs.

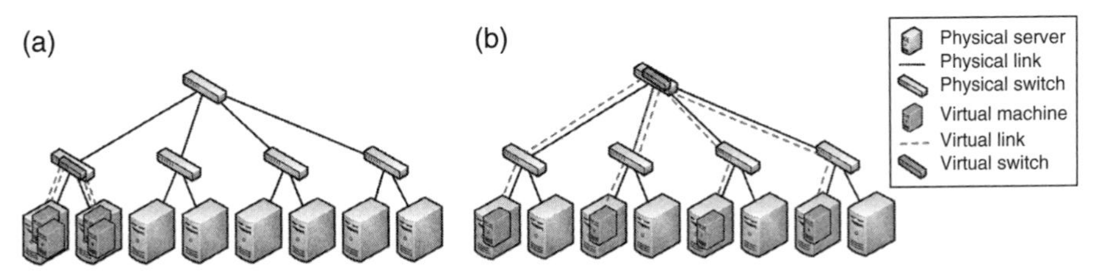

Figure 12.3. Tradeoff between fault tolerance and bandwidth usage [10]. (a) Allocation optimized for bandwidth but with low fault-tolerance. (b) Allocation optimized for fault-tolerance: more bandwidth is consumed.

The proposed resource allocation scheme includes two basic operations: (1) bandwidth minimization and (2) fault-tolerance optimization. The bandwidth minimization is performed only once, at the initial allocation of the VDC. It consists of applying the K-way min-cut to split VMs of the same VDC into partitions such that their inter communication bandwidth is minimized. These partitions are initially placed into different racks in the data center. The fault-tolerance optimization is then accomplished by gradually spreading out the VMs one by one across multiple fault domains in order to maximize the VDC worst-case survival while ensuring that bandwidth consumption remains low.

This solution has several limitations. First, it does not consider the availability of the underlying physical components. Also, it considers only the worst-case failure, which occurs mainly in aggregation switches. Hence, it ignores other types of failures that may happen, for example in top-of-rack switches or servers. Finally, the paper assumes that a physical machine only hosts one VM of the same VDC. Consequently, large VDCs will be mapped onto a large number of distinct servers, and hence will lead not only to a high number of used physical machines but also to high bandwidth usage in the data center.

12.4.6 SVNE

Rahman et al. [21] were the first to introduce and study the problem of survivable virtual network embedding (SVNE). The problem relates to protecting a virtual network (equivalent to a VDC but deployed over an ISP network rather than a data center), and more specifically virtual links, against physical link failures. As a virtual link between two nodes maps onto a path (set of connected physical links), the authors considered two types of virtual link protection and restoration mechanisms. The first mechanism is called *link protection and restoration* and basically aims at protecting the virtual link by protecting each physical link comprised in its associated path. Hence, a backup detour is provided for each physical link in the path. The second mechanism is the *path protection and restoration mechanism*, which requires the provision of a backup path for each primary path associated to a virtual link. Of course, it is mandatory that the primary and backup paths have disjoint links.

Figure 12.4 shows an example of a virtual link embedded into a physical network. The continuous line shows the primary path to which the virtual link is mapped. The dashed lines represent the protection of each physical link when the link protection and restoration mechanism is used. Finally, when the path protection and restoration mechanism is adopted, the dotted line represents the allocated protection path, which is disjoint from the primary path (i.e., initial embedding).

The authors mathematically formulated the SVNE problem as an integer linear program and proposed two heuristic solutions: a proactive solution and a reactive one. The first one addresses failures by proactively provisioning backup resources for potential failures in the future. The main drawback of this approach is that it may result in wasting up to 50% of physical network resources. The second solution addresses this drawback by reactively handling failures. Hence, it determines the backup path when a failure occurs. The advantage of such approach is the use of fewer resources for backups and hence

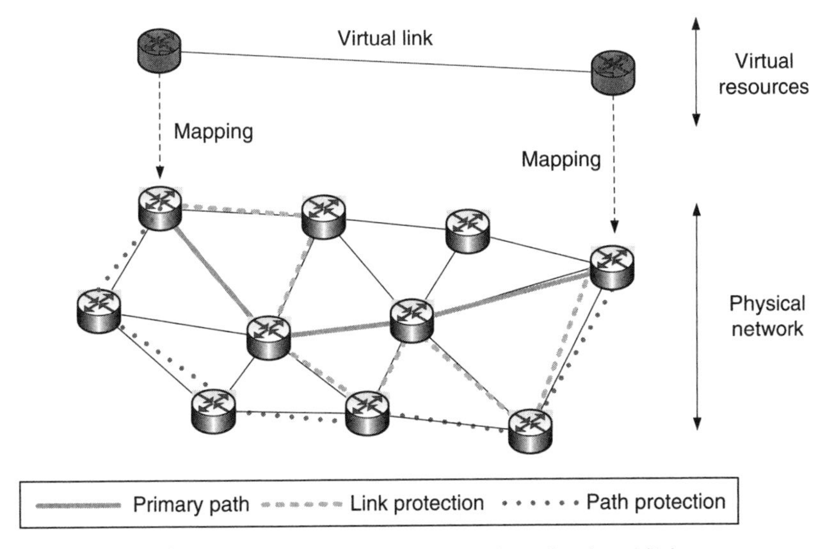

Figure 12.4. Link and path protection of a virtual link.

more virtual networks can be embedded in the physical infrastructure. The main limitation of the proposed solutions is that they assume that only a single failure may occur at a time.

12.4.7 Discussion

Generally speaking, these proposals can be classified into (1) availability-aware resource placement schemes, which attempt to improve VDC availability by carefully selecting the physical nodes hosting the virtual data center; and (2) redundancy provisioning techniques that allocate backup resources in order to achieve the desired availability.

Table 12.2 compares different features of the surveyed schemes. For each, the table provides the following information:

1. What type of backup resources are provisioned (i.e., virtual nodes, virtual links, or both);
2. Whether or not the backup resources are shared among different VDCs;
3. Whether or not the scheme provides a technique to estimate the number of virtual links/nodes provisioned as backups;
4. Whether or not the scheme provides a technique to compute the availability of a VDC;
5. Whether the technique to compute the VDC availability takes into consideration the heterogeneity of the equipment in terms of availability and failure rates;
6. Whether the scheme makes the assumption that one physical server can only host one VM from the same VDC.

TABLE 12.2. Comparison of survivable embedding schemes

Proposal	Backup resources	Shared or not	Estimate the # of backups	Computing avail.	Heterogeneity	VM col-location
Sur. Map. [16]	Nodes	No	No	No	N/A	Yes
ORP [17]	Nodes & Links	Yes	Yes	Yes	No	No
SVNE [21]	Links	No	N/A	No	N/A	N/A
WCS [10]	Nodes	No	No	No	N/A	No
Hi-VI [18]	Nodes & Links	No	Yes	Yes	Yes	Yes
VENICE [20]	No	No	N/A	Yes	Yes	Yes

Based on the surveyed works presented earlier, we can make the following observations:

- Most of the existing proposals did not take into account the availability of the physical devices (e.g., Refs. [10, 16]). Furthermore, many of them have assumed that the cluster is homogenous, that is, all devices have the same availabilities and failure rates (e.g., Ref. [17]), whereas in practice, cloud computing environments are extremely heterogeneous. Indeed, physical devices have different types, capacities, failure rates, MTBFs, availability, and reliability. As a result, proposals overlooking this heterogeneity have limited applicability in real-world cloud environments.
- Some proposals (e.g., Refs. [17, 22]) make the assumption that one physical node is able to host only one virtual component from the same VDC. This assumption does not hold in practice, since the main benefit of virtualization is the possibility that multiple virtual components share the same physical device. In addition, such an assumption has an impact on the way resources are allocated. For instance, in order to satisfy this assumption, for a VDC containing 1000 VMs, the VMs should be placed into 1000 physical nodes even when it is possible to consolidate them into 500 machines. The resulting allocation is hence suboptimal, as it leads to the usage of a higher number of servers in addition to the consumption of more bandwidth between the VMs (as they are scattered across 1000 machines rather than 500). Consequently, an effective solution should allow multiple VMs from the same VDC to share a single host whenever possible.
- Some other proposals (e.g., Ref. [17]) do not consider the availability of the networking devices (e.g., physical routers and switches) and middleboxes. However, recent studies [14] have revealed that these devices, and in particular top-of-rack switches and LBs, are more prone to failure than other types of equipment. Consequently, it is necessary to take into consideration the availability of such components in the VDC availability computation.
- Finally, all the proposals, without exception, assumed that only a single failure could occur at a given time. Although according to some studies [14], it might be reasonable to make such an assumption, it is still not realistic as simultaneous and correlated failures may occur in practice. In this case, many challenges remain

unsolved like computing the availability of a virtual data center, estimating the number of backup nodes and links required to achieve a desired availability, and also deciding the placement of these backups.

12.5 CONCLUSION

Despite the widespread adoption of cloud computing, failures and service outages loom as major concerns, pressing cloud providers to put in place strategies to deal with failures, mitigate their impact and improve the fault tolerance of their infrastructures in order to provide more stronger guarantees on the availability of their resources.

This chapter provided a comprehensive study of cloud survivability and reliability concepts and solutions, including an overview of recent studies of failures in production environments and a survey of the relevant solutions proposed to improve the availability of virtual data centers in the cloud. We discussed the main features of each of the solutions and highlighted their advantages and limitations.

We believe that there are still a lot of challenges to overcome in order to offer highly reliable and available cloud services. Specifically, more work should be dedicated to studying failure characteristics, and particularly the correlation between failures. There is also a pressing need to develop more sophisticated solutions for improving the survivability and fault tolerance of cloud services, taking into account the heterogeneity of the cloud infrastructures as well as scenarios where multiple failures occur simultaneously.

REFERENCES

1. Costs and scope of unplanned outages. http://www.evolven.com/blog/2011-devastating-outages-major-brands.html. Accessed on November 17, 2014.
2. Downtime outages and failures, understanding their true costs. http://www.evolven.com/blog/downtime-outages-and-failures-understanding-their-true-costs.html. Accessed on November 17, 2014.
3. Costs and scope of unplanned outages. http://www.evolven.com/blog/costs-and-scope-of-unplanned-outages.html. Accessed on November 17, 2014.
4. Md. Faizul Bari, Raouf Boutaba, Rafael Esteves, Lisandro Zambenedetti Granville, Maxim Podlesny, Md Golam Rabbani, Qi Zhang, and Mohamed Faten Zhani. Data center network virtualization: A survey. *IEEE Communications Surveys & Tutorials*, 15(2):909–928, 2013.
5. Google compute engine. https://cloud.google.com/. Accessed on November 17, 2014.
6. Amazon elastic compute cloud (Amazon EC2). http://aws.amazon.com/ec2/. Accessed on November 17, 2014.
7. Hitesh Ballani, Paolo Costa, Thomas Karagiannis, and Ant Rowstron. Towards predictable datacenter networks. In *ACM SIGCOMM*, Toronto, Ontario, Canada, August 2011.
8. Mohamed Faten Zhani, Qi Zhang, Gwendal Simon, and Raouf Boutaba. VDC Planner: Dynamic migration-aware virtual data center embedding for clouds. In *IEEE/IFIP International Symposium on Integrated Network Management (IM)*, Ghent, Belgium, May 27–31, 2013.

9. Chuanxiong Guo, Guohan Lu, Helen J. Wang, Shuang Yang, Chao Kong, and Peng Sun. Secondnet: A data center network virtualization architecture with bandwidth guarantees. In *ACM CoNEXT*, Philadelphia, PA, 2010.

10. Peter Bodík, Ishai Menache, Mosharaf Chowdhury, Pradeepkumar Mani, David A. Maltz, and Ion Stoica. Surviving failures in bandwidth-constrained datacenters. In *ACM SIGCOMM*, Helsinki, Finland, 2012.

11. Xin Wu, Daniel Turner, Chao-Chih Chen, David A. Maltz, Xiaowei Yang, Lihua Yuan, and Ming Zhang. Netpilot: Automating datacenter network failure mitigation. *SIGCOMM Computer Communication Review*, 42(4): 443–454, August 2012.

12. Albert Greenberg, James R. Hamilton, Navendu Jain, Srikanth Kandula, Changhoon Kim, Parantap Lahiri, David A. Maltz, Parveen Patel, and Sudipta Sengupta. VL2: A scalable and flexible data center network. In *ACM SIGCOMM*, Barcelona, Spain, 2009.

13. Kashi Venkatesh Vishwanath and Nachiappan Nagappan. Characterizing cloud computing hardware reliability. In *ACM Symposium on Cloud Computing (SOCC)*, Indianapolis, IN, 2010.

14. Phillipa Gill, Navendu Jain, and Nachiappan Nagappan. Understanding network failures in data centers: Measurement, analysis, and implications. In *ACM SIGCOMM*, Toronto, Ontario, Canada, 2011.

15. Ming Zhang, Chi Zhang, Vivek Pai, Larry Peterson, and Randy Wang. Planetseer: Internet path failure monitoring and characterization in wide-area services. In *Symposium on Opearting Systems Design & Implementation (OSDI)*, San Francisco, CA, 2004. USENIX Association, Seattle, WA, pages 12–12. http://dl.acm.org/citation.cfm?id=1251266. Accessed on December 9, 2014.

16. Jielong Xu, Jian Tang, K. Kwiat, Weiyi Zhang, and Guoliang Xue. Survivable virtual infrastructure mapping in virtualized data centers. In *IEEE International Conference on Cloud Computing (CLOUD)*, Honolulu, HI, 2012. http://www.thecloudcomputing.org/2012/. Accessed on December 9, 2014.

17. Wai-Leong Yeow, Cedric Westphal, and Ulas C. Kozat. Designing and embedding reliable virtual infrastructures. *SIGCOMM Computer Communication Review*, 41(2): 53–56, April 2011.

18. Md Golam Rabbani, Mohamed Faten Zhani, and Raouf Boutaba. On achieving high survivability in virtualized data centers. *IEICE Transactions on Communications*, E97-B(1): 10–18, January 2014.

19. N. M. Mosharaf Kabir Chowdhury, Muntasir Raihan Rahman, and Raouf Boutaba. Virtual network embedding with coordinated node and link mapping. In *IEEE INFOCOM*, Rio De Janerio, Brazil, 2009, pages 783–791.

20. Qi Zhang, Mohamed Faten Zhani, Meyssa Jabri, and Raouf Boutaba. Venice: Reliable virtual data center embedding in clouds. In *IEEE International Conference on Computer Communications (INFOCOM)*, Toronto, Ontario, Canada, April 27–May 2, 2014.

21. Muntasir Raihan Rahman and Raouf Boutaba. SVNE: Survivable virtual network embedding algorithms for network virtualization. *IEEE Transactions on Network and Service Management*, 10(2):105–118, 2013.

22. Hongfang Yu, Vishal Anand, Chunming Qiao, and Gang Sun. Cost efficient design of survivable virtual infrastructure to recover from facility node failures. In *IEEE International Conference on Communications (ICC)*, Kyoto, Japan, June 2011.

PART IV

CLOUD APPLICATIONS AND SERVICES

13

SCIENTIFIC APPLICATIONS ON CLOUDS

Simon Ostermann, Matthias Janetschek, Radu Prodan, and Thomas Fahringer

Institute for Computer Science, University of Innsbruck, Innsbruck, Austria

13.1 INTRODUCTION

Today, scientific applications require an ever-increasing number of resources to deliver results for growing problem sizes in a reasonable amount of time. In the past 20 years, while the largest projects were able to afford expensive supercomputers, the smaller ones were forced to opt for cheaper resources such as commodity clusters or, more challenging to build, computational Grids. To program such large-scale distributed heterogeneous infrastructures, scientific workflows emerged as an attractive paradigm by allowing the programmers to focus on the composition of existing legacy code fragments to create larger and more powerful applications. Therefore, numerous efforts have been spent on researching and developing integrated programming and computing environments [1] to support the workflow lifecycle and meet scientists' needs.

Nowadays, *Cloud computing* proposes an alternative such that resources are no longer owned by the application scientists, but leased from large specialized data centers on-demand and in a cost-effective fashion according to temporal needs. This separation frees research institutions from the permanent costs of over-provisioning, operation, maintenance, and depreciation of resources. Existing workflow systems cannot

Cloud Services, Networking, and Management, First Edition.
Edited by Nelson L. S. da Fonseca and Raouf Boutaba.

senselessly take advantage of this new infrastructure without appropriate middleware support that often requires nontrivial extensions to the scheduling, enactment, resource management, and other runtime execution services. At the same time, existing Cloud providers such as Amazon recognized the importance of workflows to science and engineering and started to provide highly tuned solutions integrated into their native platforms [2] such as the *Amazon Simple Workflow (SWF)* service. Other platforms like OpenStack or CloudStack do not offer advanced services for workflows and require the used of external workflow engines for such executions. However, existing workflow systems [1] cannot immediately take advantage of this advanced support because of different, incompatible languages, interfaces and communication protocols. Another downside of SWF is that it requires applications to be written in Java and to implement specific interfaces, which is problematic for scientific workflows based on the composition of legacy code fragments. Using SWF requires scientists to learn a new development and execution platform in addition to the one they already regularly use but is very simple to get used to compared to most other scientific workflow environments.

To address this heterogeneity in workflow systems and underlying computing infrastructures, the SHIWA European project (http://www.shiwa-workflow.eu/) researched and developed the *Interoperable Workflow Intermediate Representation (IWIR)* [3] that enables fine-grained interoperability between workflow systems via transparent translation of workflows applications programmed in different languages. IWIR is a generic and system-neutral workflow representation able to sufficiently describe the large majority of existing workflow constructs. The common representation reduces the complexity of porting n workflow systems on m computing platforms from $O(m \cdot m)$ to $O(n + m)$. Additionally, it enables the integration of new workflow systems and new computing platforms with constant $O(1)$ complexity by implementing IWIR importers/exporters. This ensures not only interoperability across workflow systems but also enables workflows to be executed on new external foreign (or nonnative) computing infrastructures. IWIR provides additional tools and libraries to ease the development of language translators, and is currently supported by five major workflow systems: ASKALON (AGWL language) [4], Moteur (GWENDIA language) [5], WS-PGRADE (gUSE language) [6], Pegasus [7], and Triana (DAX representation) [8] (see Fig. 13.1).

In this chapter, we take advantage of IWIR and present a scalable software engineering solution that provides existing scientific workflows access to the Amazon

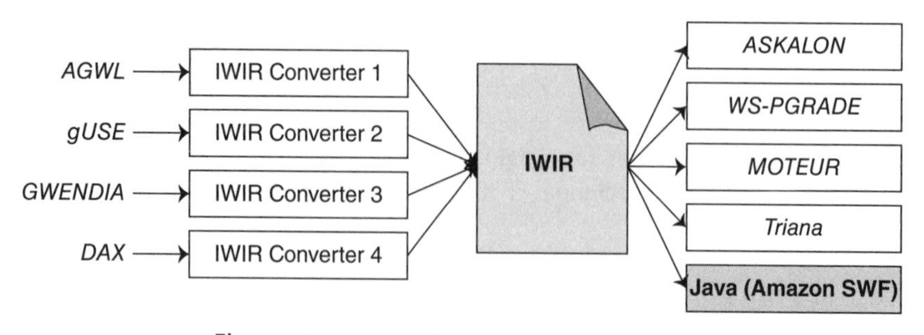

Figure 13.1. SHIWA fine-grained interoperability.

Elastic Compute Cloud (EC2) infrastructure. By designing and implementing one single IWIR-to-SWF converter, we automatically allow all IWIR-compliant workflow systems to benefit from the SWF features and to access the EC2 infrastructure with native performance. We present a method for automatically converting a scientific workflow specified in IWIR into Amazon SWF, and a supporting architecture for reusing and executing existing legacy code on EC2. We illustrate the integration and the advantages of our architecture with the help of a real-world scientific workflow originally programmed in the ASKALON integrated development and computing environment.

The chapter is organized as follows. We discuss related work in Section 13.3. Section 13.4 introduces the IWIR workflow model, followed by an introduction to Amazon SWF in Section 13.5. Section 13.6 introduces our pilot workflow application used for validation. Section 13.7 describes the conversion process of an IWIR workflow into an Amazon SWF workflow. Section 13.8 presents experimental results from porting our pilot application to SWF. Section 14.7 concludes the chapter.

13.2 BACKGROUND INFORMATION

Nowadays, scientific computing is an important part of research in most academic disciplines. Problems are getting more and more complex and finding solutions requires an ever-increasing amount of computation. Simulations for weather, earthquake, nuclear research, and material science are just a few examples of areas where there one can never have enough computation capacity available to make a realistic simulation without lots of model simplifications. Therefore, computer scientists are trying to find solutions to make scientific computing faster, easier, and more reliable on the available set of resources. Most applications developed by non-computer scientists are often hart to scale onto clusters or super computers and need support from the computer science community to scale them to nowadays clusters.

By introduction of cloud computing, a new resource type was added as possible platform to execute scientific applications. As this new technology is slowly adapted by computer scientists, it can be assumed that other fields of research that are relying on parallel computing for solving their problems have a even higher learning curve to include this new technologies into their everyday tool set. This gap between new technologies and need for computing needs to be closed by tools developed by computer scientists to allow easier adaptation of clouds for scientific computing.

13.3 RELATED WORK

Since the advent of Cloud computing, the scientific community showed interest in bringing scientific workflows on this new infrastructure. This trend increased with the availability of commercial Clouds featuring nearly the same performance as traditional Grid parallel computers [9]. There exist two major approaches in this community effort: pure Cloud and hybrid combining Grid and Cloud infrastructure.

FutureGrid [10] provides a Cloud test-bed that allows scientists explore the features of Cloud computing and experiment without charging real costs, as commercial providers do. [11] shows a proof-of-concept astrophysics workflow called Montage using the Pegasus Grid workflow system adapted for Clouds. The work in Ref. [12] shows a meteorological workflow executed in combined Grid and Cloud infrastructures using the ASKALON environment. In Ref. [13] the Pegasus Workflow Management System is used to execute a astrophysics workflow across multiple clouds and show how challanging this task still is. A hybrid approach for extending clusters with additional Cloud resources during peak usage for better throughput, transparent to the end-users is presented in Ref. [14]. [15] presents a similar approach using the Torque job manager. The work in Ref. [16] presents a workflow engine purposely developed for Clouds and extended Cloud federations. The Megha workflow system [17] provides a portal for submitting workflows to combined Grid and Cloud resources.

A drawback of all these efforts is that they provide custom non-interoperable solutions that isolate scientists on specific workflow system and Cloud infrastructures. In this chapter, we show how the IWIR-based approach opens the Amazon EC2 infrastructure and its SWF workflow system to the scientific community through one single IWIR-to-SWF translator. The idea of a single intermediate language has been explored in other domains, for example, by the UNiversal Computer Oriented Language (UNCOL) [18] proposed in 1958 by Conway as a solution for making compiler development economically viable. Following the UNCOL idea, the Architecture Neutral Distribution Format (ANDF) is a technology defined by the Open Software Foundation allowing common "shrink wrapped" binary programs be distributed for use on Unix systems running on different hardware platforms. Unfortunately, ANDF was never widely adopted either. IWIR is the first effort to investigate this idea on scientific workflows in distributed Grid and Cloud computing infrastructures.

13.4 IWIR WORKFLOW MODEL

In IWIR, a *workflow application* is represented by a composite activity $A = (I, O, G)$ consisting of n input ports $I = \bigcup_{i=1}^{n} \{I_i\}$, m output ports $O = \bigcup_{i=1}^{m} \{O_i\}$, and a directed acyclic graph (DAG) $G = (A, D)$, consisting of k activities $A = \bigcup_{i=1}^{k} \{A_i\}$, interconnected through data flow dependencies:

$D = \{(A_i, A_j, (O_{im}, I_{jn})) \mid (A_i, A_j) \in A \times A \wedge (O_{im}, I_{jn}) \in O_i \times I_j\}$, where (O_{im}, I_{jn}) represents a data transfer from the output port O_{im} of activity A_i to the input port I_{jn} of activity A_j. A data flow dependency between two activities implies a control flow precedence too. A pure control flow dependency between A_i and A_j has $D_{ij} = (A_i, A_j, \emptyset)$. We use $pred\,(A_i) = \{A_k | \,(A_k, A_i, (O_{km}, i_{in})) \in D \vee (A_k, A_i, \emptyset) \in D\}$ to denote the set of *predecessors* of activity A_i (i.e. activities to be completed before starting A_i). Figure 13.2 shows an detailed example of such a DAG and its components.

Compared to business workflows, the main difference to the scientific workflows we are focusing on is the high computational requirments of the activities and not buissnes processes in general.

There are two categories of activities in IWIR: atomic and composite. An *atomic activity*, represented by $A = (I, O, \emptyset)$, is characterized by an *activity type*, uniquely defined by a *name* and a *signature*. For example, activity names are `PrepareLM`,

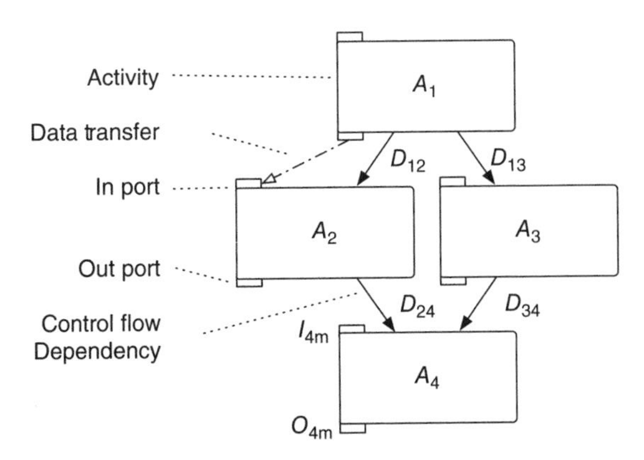

Figure 13.2. Example of a DAG with four activities.

`LinearModel`, `PostProcessSingle` and `PostprocessFinal` for our pilot workflow introduced in Figure 13.5 and Section 13.6. The signatures of `LinearModel` and `PostProcessSingle` are shown in lines 8 and 19 of Listing 13.1. A *composite activity*, represented by $A = (I, O, G)$, where $G \neq \emptyset$, can be of four kinds: conditional (`if`), sequential loop (`while`, `for`, `forEach`), parallel loop (`parallelFor`, `parallelForEach`), and sub-workflow (or DAG, added for modularity reasons).

13.5 AMAZON SWF BACKGROUND

Amazon SWF provides a high-level method for implementing workflow applications and for coordinating their synchronous and asynchronous task executions on multiple systems, which can be cloud-based, on-premises, or both. The architecture of Amazon SWF is displayed in Figure 13.3. SWF implements a work-stealing approach consisting of three parts: decider, task queues, and activity workers.

The *decider* implements the logic of the workflow. Unlike schedulers in scientific workflow systems, the decider only decides which activity to execute next based on the history of already executed tasks, and not where to execute it. However, one still has limited control by using several task queues where the decider puts the activities to be executed next.

The *task queues* hosted by Amazon are identified by their name and can be accessed via an HTTP API. There are two types of tasks. First, *decision tasks* are generated by Amazon SWF and executed by the decider every time a state change exists (e.g., start of workflow instance or activity task termination). The result of a decision task is usually a set of activity tasks that can be executed next. Second, *activity tasks* are executed by the activity workers and represent the individual pieces of work which comprise the workflow.

The *activity workers*, as shown in Figure 13.4, execute the individual workflow activities. The decider and the activity workers actively listen to one or more task queues and, when a task is received, execute the corresponding code and report back the execution status to Amazon SWF. All input values of an activity are contained in the task

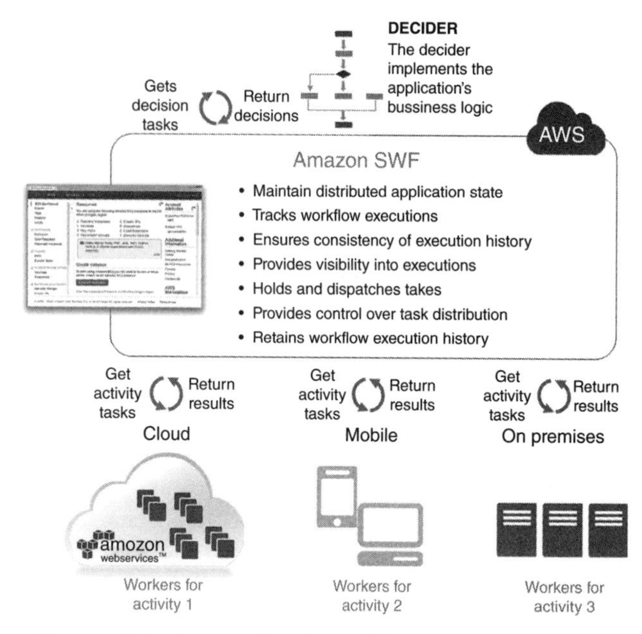

Figure 13.3. Amazon Simple Workflow Service architecture.

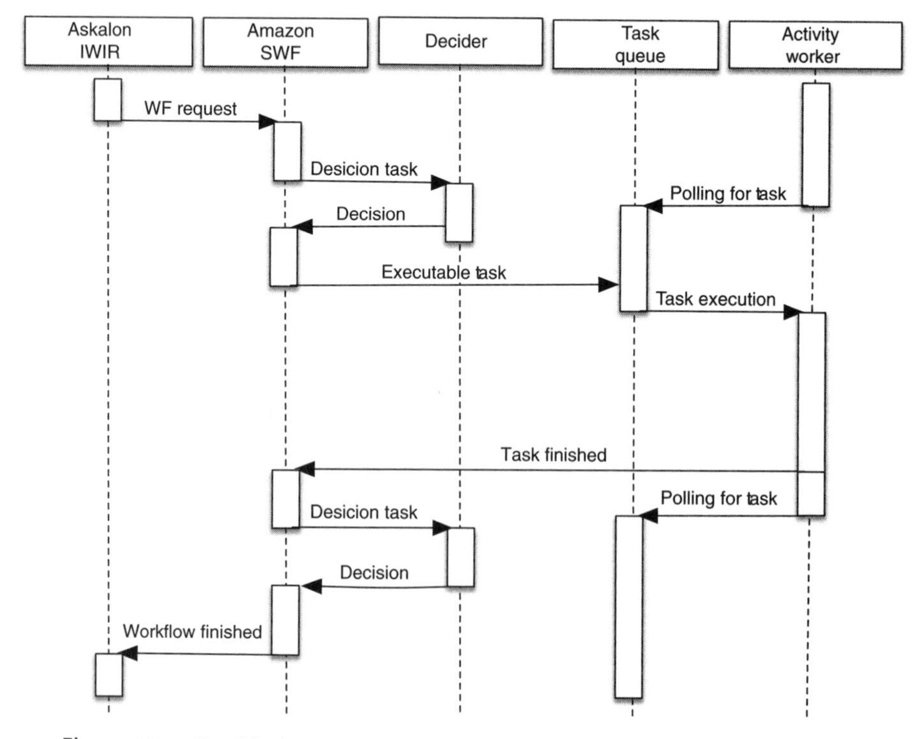

Figure 13.4. Simplified sequence diagram of execution of a workflow with one task.

request received from the task queue. Unlike traditional workflow systems, Amazon SWF provides no means to transfer files or prepare the execution environment. If an activity requires some input or produces some output files, it has to transfer them by itself. For Amazon SWF, workflow activities are simply remote asynchronous procedure calls.

Developing a workflow with Amazon SWF requires the following steps: (1) develop a decider implementing the logical workflow coordination, (2) develop activity workers implementing the individual activities, (3) register the workflow at Amazon SWF, (4) start the decider and activity workers and let them listen to the SWF endpoint, and (5) start the workflow.

AWS Flow Framework allows the development of Amazon SWF workflows via the AWS SDK for Java by specifying its coordination steps as a sequential Java program, where the workflow activities are represented as function calls. Functions representing activities and functions used to handle or manipulate data produced or consumed by activities need to have the `@Asynchronous` annotation (called in the following *asynchronous functions*), and their input arguments and return values need to be of type `Promise`. A `Promise` object acts as a handle to the actual data that will be available as soon as the corresponding asynchronous function has been executed. When used as input argument, a `Promise` object can also be used to represent data dependencies between several asynchronous functions. An asynchronous function, having a `Promise` object produced by another asynchronous function as input, will only be executed when the actual data referenced by the `Promise` object is available.

Amazon SWF executes a workflow application through repeated invocations of the decider program, which is executed every time a state change occurs (signaled via a *decider task*), and a history of all decider executions is recorded. To intercept all calls to asynchronous functions in the decider program, AWS uses AspectJ, a Java implementation of aspect-oriented programming [19]. An asynchronous function is instantiated only once during the entire workflow execution, and its return value is saved into the execution history. In every subsequent decider execution, the same function is not re-executed, but its result extracted from the execution history and returned as `Promise` object. An asynchronous function that has not been executed yet is put into a queue, and a `Promise` object with no actual data is returned. This data will be instantiated as soon as the corresponding asynchronous function has produced it. Before the decider finishes its execution, it examines all asynchronous functions in the queue, executes those whose dependencies are satisfied, and records their results in the execution history. The workflow execution finishes if there are no more nonexecuted asynchronous functions.

13.6 RAINCLOUD WORKFLOW

We introduce in this section the RainCloud workflow used in this chapter for illustrating and validating our approach. Raincloud is a meteorological workflow for investigating and simulating precipitations in mountainous regions using a simple numerical linear model of orographic precipitations [20]. The workflow has been developed in the ASKALON environment by the Institute of Meteorology of the University of Innsbruck to analyze certain meteorological phenomena by extending the linear model theory. The

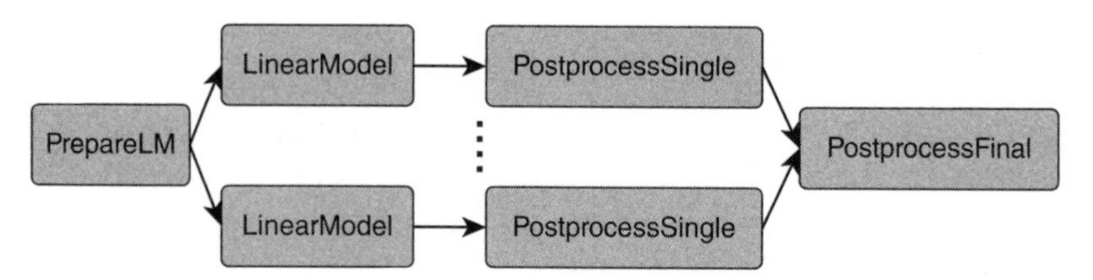

Figure 13.5. Simplified view of the RainCloud workflow.

workflow is also is used by the Tyrolean avalanche service (Tiroler Lawinenwarndienst) for their daily issued avalanche bulletin. We choose this applications as it is used on a daily baisis by scientists.

Figure 13.5 shows a simplified architecture of the RainCloud workflow. The first activity `PrepareLM` prepares and partitions the data for the linear model. Each partition is then processed in a parallel loop iteration by a pipeline of two activities: `LinearModel` and `PostprocessSingle`, the last one being optional. The number of parallel loop iterations can be configured by setting the appropriate input parameter. The last activity collects the output data and produces the final result. Listing 13.1 shows the specification of the `parallelForEach` loop in IWIR. Inside this loop, we first have the atomic activity `linearModel` (line 8), followed by an `if`-construct (line 14) containing the atomic activity `postProcessSingle` (line 19).

Listing 13.1 RainCloud's `parallelForEach` loop in IWIR

```
1  ...
2  <parallelForEach name="ParallelForEach_1">
3   <inputPorts><inputPort name="isPPS" type="boolean"/>
4    <loopElements><
5      loopElement name="PLMTars" type="collection/file"/>
6    </loopElements></inputPorts>
7   <body>
8    <task name="linearModel" tasktype="linearModel">
9     <inputPorts><inputPort name="PLMTar" type="file"/>
10    </inputPorts>
11    <outputPorts><outputPort name="LMTar" type="file"/>
12    <outputPort name="outfile" type="file"/></outputPorts>
13   </task>
14   <if name="DecisionNode_1">
15    <inputPorts><inputPort name="LMTar" type="file"/>
16    <inputPort name="isPPS" type="boolean"/></inputPorts>
17    <condition>isPPS = true</condition>
18    <then>
19     <task name="postProcessSingle" tasktype="postProcessSingle">
20      <inputPorts><inputPort name="LMTar" type="file"/>
21      </inputPorts>
22      <outputPorts><outputPort name="PPSTar" type="file"/>
23      </outputPorts>
24     </task>
25    </then>
26    <outputPorts><outputPort name="PPSlistTars" type="file"/></outputPorts>
27    <links>
28     <link from="DecisionNode_1/LMTar" to="postProcessSingle/LMTar"/>
29     ...
30    </links>
31   </if>
32  </body>
33  <outputPorts>
```

```
34      <outputPort name="PPSlistTars" type="collection/file"/>
35      <outputPort name="outfiles" type="collection/file"/>
36    </outputPorts>
37    <links>
38      <link from="ParallelForEach_1/PLMTars" to="linearModel/PLMTar"/>
39      ...
40    </links>
41  </parallelForEach>
42  ...
```

13.7 IWIR-TO-SWF CONVERSION

Figure 13.6 shows the architecture of our IWIR-to-SWF conversion solution, consisting of four parts: the decider, Amazon SWF, the legacy code execution service on each worker node, and the file storage. With Amazon SWF, the decider and workflow activities are individual Java programs, purposely designed for Amazon SWF. The goal of this chapter is to present a method for translating scientific workflows from the interoperable IWIR representation to Amazon SWF with as little effort for the programmer as possible. While the abstract workflow coordination can be automatically translated into an SWF decider Java program, there is no practical way to automatically convert the legacy code implementing the concrete workflow activities into an SWF-compatible Java program. To still achieve this goal with minimal programmer involvement, we implemented an execution service that interfaces with Amazon SWF and acts as a Java wrapper for existing legacy code.

The only requirement imposed by Amazon SWF on the worker nodes is an outgoing HTTP connection to Amazon SWF. This makes Amazon SWF easy to set up with no need of firewall reconfiguration. Technically, direct file transfers between worker nodes are possible, but this requires a corresponding service running on the worker nodes and the firewall to be reconfigured accordingly. As we did not want to loose the advantage of an easy setup, we decided to use an intermediate file storage for the file transfers, so that there is no need for incoming connections on the worker nodes. Currently, we support

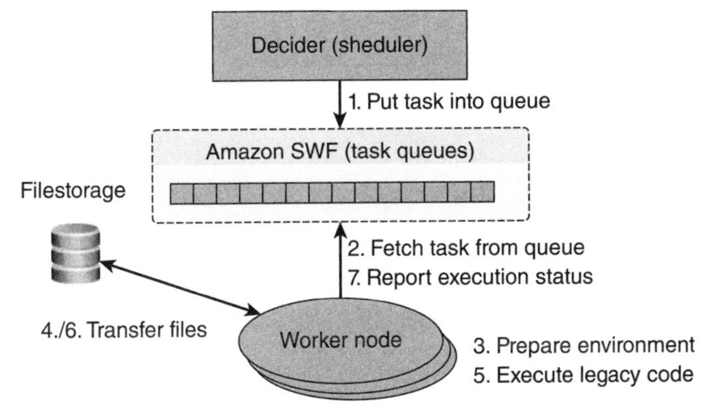

Figure 13.6. Architecture of a generated Amazon SWF workflow.

only Amazon S3 as an intermediate file storage, but other file storage technologies can be easily added as extensions.

13.7.1 Decider Generation

As presented in Section 13.4, an IWIR workflow is constructed from a top-level DAG activity that explicitly describes the data flow between its activities. Control flow constructs such as loops and conditionals are represented by composite activities. To convert an IWIR workflow into an Amazon SWF decider, we have to transform the data flow-driven IWIR DAGs and the semantics of the composite activity types into a control flow-driven Java program. Moreover, we also have to take care that the concepts of the AWS Flow Framework, namely asynchronous functions and semantics of the returned `Promise` objects, are correctly applied.

The basic principle of the conversion is that every atomic or composite activity is represented by its own *activity function* in the Java program. Listing 13.2 shows the generation process of the decider. The first step is the generation of a function representing the start of the workflow (line 3). The signature of this function represents the input and output ports of the workflow (line 10). In the function body, the top-level activity function is called with the appropriate input arguments (line 11). Afterwards, the results of the top-level activity function are presented to the user in an appropriate way (line 13). Every activity encountered during the conversion process with no activity function created yet is put into a queue (e.g., in line 12). After the workflow entry function has been generated, the algorithm iterates through this queue (line 4) and generates an activity function for each queue element (line 6).

Listing 13.2 SWF decider generation algorithm

Input: Scientific workflow: $A = (I, O, G)$
Output: SWF decider (Java program)
1: **function** GenDecider($A = (I, O, G)$)
2: $\quad$ $Queue \leftarrow \emptyset$
3: $\quad$ GenWfStart(A, $Queue$)
4: $\quad$ **while** Queue $\neq \emptyset$ **do**
5: $\quad\quad$ $A \leftarrow$ Pop($Queue$)
6: $\quad\quad$ GenActivityFunction(A, $Queue$)
7: $\quad$ **end while**
8: **end function**
9: **function** GenWfStart(A, $Queue$)
Input: $A = (I, O, G)$
10: $\quad$ GenWfStartProlog(I, O)
11: $\quad$ GenActivityFuntionCall(A, $Queue$)
12: $\quad$ Put($Queue$, A)
13: $\quad$ GenWfStartEpilog(O)
14: **end function**

13.7.2 Activity Function Generation

Listing 13.3 shows the generation of an activity function representing a workflow activity. The function signature of an activity function corresponds to the input and output ports, while the function body implements its semantic behavior, including any associated

DAG. For atomic activities, we only have to generate the function signature with the `Activity` annotation and an empty function body (lines 9–10). The AWS Flow Framework will then automatically generate function stubs, which allow us to communicate with the SWF task queue. For composite activities, we need to additionally generate, besides the function signature, a function body implementing the activity behavior (lines 4–6). In the following text, we describe in detail how the activity functions are generated. To facilitate understanding, we divided the code generation of the composite activity function bodies in three logical sections: (1) activity semantics, (2) DAG control flow, and (3) DAG data flow. However, these steps are not distinct, but interleaved with each other (e.g., the function call in line 5 generates not only the control flow but also the data flow).

Listing 13.3 Activity function generation algorithm

```
Input: Workflow activity: A = (I, O, G)
Output: Activity function (in Java)
 1: function GENACTIVITYFUNCTION(A, Queue)
Input: A = (I, O, G)
 2:     if G ≠ ∅ then                                  ▷ Composite activity
 3:         N ← GETACTIVITYNAME(A)
 4:         GENFUNCTIONPROLOG(N, A)
 5:         GENDAGCONTROLFLOW(G, Queue)
 6:         GENFUNCTIONEPILOG(A)
 7:     else                                           ▷ Atomic activity
 8:         N ← GETACTIVITYTYPENAME(N, I, O)
 9:         GENFUNCTIONSIGNATURE(N, I, O)
10:         GENEMPTYFUNCTIONBODY()
11:     end if
12: end function
```

13.7.2.1 *Function Signature.* The first step in generating an activity function is the function signature. The arguments of the activity function represent the input ports and the return value the output ports of the associated workflow activity. However, this representation has some inadequateness. In a workflow representation, the input and output ports of an activity are usually identified by their names, and the number of output ports is not limited. By contrast, the arguments of a Java function are identified by their order and the return argument is restricted to one. Moreover, returning values by call by reference does not work in an SWF program because the activity functions are executed asynchronously from the rest of the program (see Section 13.5). In practice, the first inadequateness can be neglected when generating the decider by consistently maintaining the same parameter order. However, this may pose a problem for the legacy wrapper service, as changes in the order of the input arguments cannot be automatically distributed to this service. To address this problem, we implemented a wrapper class for the input arguments of atomic activity functions with a field for the name of the input port and a field for the actual value. The legacy wrapper service can then assign the input values to the correct input port by looking at the name field. To address the second inadequateness, we implemented another wrapper class that stores several output values into an array that is returned by the activity functions.

For example, Listing 13.4 shows a function signature representing the atomic activity `linearModel` of RainCloud. Because the activity has more than one output port, the corresponding function returns an object of type `PortWrapperArray` encapsulating the output values. All input values have the type `PortWrapper` because the function represents an atomic activity and, therefore, needs to interface with the legacy wrapper service. The AWS Flow framework automatically generates a stub function for interfacing with the task queues declared as `asynchronous` and returning a `Promise` object.

Listing 13.4 Function signature of the atomic `linearModel` activity

```
1 @Activity(name="RainCloudActivities.linearModel")
2 public PortWrapperArray linearModel ( PortWrapper PLMTar )
```

Listing 13.5 contains another example of a function signature representing the composite activity `ParallelForEach_1`.

Listing 13.5 Function signature of the `ParallelForEach_1` activity

```
1 @Asynchronous
2 private Promise parallelForEach_1 ( Promise<Boolean> isPPS, Promise<String[]>
      PLMTars );
```

13.7.2.2 Activity Semantics. The next step is the generation of code that implements the semantics of the three types of composite activities: (1) container, (2) conditional, and (3) loop. *Container activities* only contain other activities without additional semantics. *Conditional activities* consist of an `if-else` construct and separate activity function control flows for the two branches. The conditional expression may contain input port values that can be easily referenced by specifying the appropriate function argument. *Loop activities* are the hardest to implement because of the several IWIR loop flavors: `while`, `for`, `forEach`, `parallelFor`, and `parallelForEach`. We exploited the asynchronous function invocation feature of SWF to implement parallel loops as simple sequential loops in the decider program. Because activity functions are executed asynchronously, the decider does not wait for an activity function to finish before starting the next loop iteration. To force sequential execution of activity functions inside a nonparallel loop, we have to introduce artificial dependencies between activity functions called in different iterations using `Promise`-objects.

Listing 13.6 shows an example of a function body representing the composite activity `ParallelForEach_1` of RainCloud. The number of loop iterations is first calculated in line 3. Lines 5–10 represent the actual `for` loop, whereas lines 12–13 deal with the construction of the return value.

Listing 13.6 `ParallelForEach_1` activity semantics

```
1 private Promise parallelForEach_1 (...) {
2   // Get number of elements.
3   int maxIter = PLMTars.get().length;
4   // Iterate over the given array.
5   for (int i = 0; i < maxIter; i++)  {
6     // Get current element
7     Promise<String> p = Promise.asPromise(PLMTars.get()[i];
```

```
8       // Activity function control flow goes here
9       . . .
10    }
11    // Build return value.
12    Promise[] retval = new Promise[2];
13    return Promise.asPromise( retval );
14 }
```

13.7.2.3 DAG Control Flow. The workflow activities of a given DAG are sorted according to their topological order that preserves the original data flow. In the topological order, a workflow activity can only be executed after all its predecessors have been completed and produced the required input data. As a workflow may consist of several DAGs, we calculate the topological order for each DAG independently.

For example, RainCloud's `ParallelForEach_1` loop calls the activity `linear-Model` whose results are fed into the activity `PostProcessSingle`, depending on the value of the input parameter `isPPS`. Listing 13.7 shows the equivalent Java activity with calls to the contained activity functions in lines 6 and 8. The `if` statement, which determines whether `PostProcessSingle` should be executed, is represented by the `decisionNode_1` function in line 8, with the missing parameter added in the data flow step (Listing 13.8, line 14).

Listing 13.7 Control flow inside `ParallelForEach_1` activity

```
1  private Promise parallelForEach_1 (...) {
2    int maxIter = PLMTars.get().length;
3    for (int i = 0; i < maxIter; i++) {
4      Promise<String> currEl = Promise.asPromise(PLMTars.get()[i];
5      // Call to atomic activity "LinearModel"
6      activityClient.linearModel(currEl);
7      // Call to composite if-activity
8      decisionNode_1(..., isPPS);
9    }
10   Promise[] retval = new Promise[2];
11   return Promise.asPromise( retval );
12 }
```

13.7.2.4 Data Flow. The last step in generating the body of an activity function is to introduce variables that model the data flow between the enclosed activity functions. To ease the variable handling, we use the *single static assignment* technique employed in compiler construction, which requires every variable be written once and not reused afterwards. Every value returned by an activity function is assigned to its own unique variable and passed as input to each activity function with a connected input port. The main idea is that the implementation of an activity function does not need to know how the preceding activity functions produced and stored their output values. This is also reflected in the activity function signatures (see Section 13.7.2.1) which only consists of the input arguments from the original workflow specification. Activity functions returning more than one value return a wrapper object (see Section 13.7.2). The individual values contained in this wrapper object need to be extracted before they are fed to a subsequent activity function. Unfortunately, `Promise` objects can only be accessed inside asynchronous functions, otherwise an exception will be thrown. To address this problem, we implemented several asynchronous helper functions for data manipulation and conversion.

Listing 13.8 presents the data flow of the activity function representing the composite activity `ParallelForEach_1`. First, an array for holding the results of each loop iteration is created for each activity in lines 4 and 6. The activities' output ports are directly connected to a corresponding output port of the surrounding composite activity. Then, the return value of each activity function is stored in its own variable in lines 10 and 14. Since the activity `linearModel` returns a wrapper object, we have to convert it (line 12) before using the actual return values (lines 14 and 16). At the end of each loop iteration, the values produced in the iteration are stored into the corresponding variables (lines 16 and 18). At the end of the function body, we construct the return object and convert the variables into a more suitable form (lines 22 and 24).

Listing 13.8 Data flow within the `ParallelForEach_1` composite activity

```
1  private Promise parallelForEach_1 (...) {
2    int maxIter = PLMTars.get().length;
3    // Holds output values of linearModel activity
4    Promise[] out1 = new Promise[maxIter];
5    // Holds output values of decisionNode_1 activity
6    Promise[] out2 = new Promise[maxIter];
7    for (int i = 0; i < maxIter; i++) {
8      Promise<String> p = Promise.asPromise(PLMTars.get()[i];
9      // Save linearModel return value of in lmo1
10     Promise<PortWrapperArray> lmo1 = activityClient.linearModel(p);
11     // Convert lmo1 in a format for further processing
12     Promise[] lmo2 = Utils.convertPWA2Pa(lmo1, 2);
13     // Input first value stored in lmo2; save return value into dno1
14     Promise dno1 = decisionNode_1(lmo2[0], isPPS);
15     // Store linearModel return value in a collection
16     out1[i] = lmo2[1];
17     // Store if return value in a collection
18     out2[i] = dno1;
19   }
20   Promise[] retval = new Promise[2];
21   // Convert collection to a suitable return format
22   retval[0] = Utils.convertAoP(out1);
23   // Convert collection to a suitable return format
24   retval[1] = Utils.convertAoP(out2);
25   // Return the output values
26   return Promise.asPromise(retval);
27 }
```

13.8 EXPERIMENTS

The goal of our experiments is to compare the performance of the RainCloud workflow in three configurations: automatically generated SWF workflow (using the technique described in Section 13.7), manually optimized SWF workflow, and original ASKALON version executed using the ASKALON middleware. To be able to interface with the EC2 infrastructure, we pragmatically extended the ASKALON middleware services such as security with Amazon credentials, information service with virtual machine image manipulation, and enactment engine with SSH-based job submission [21].

13.8.1 Setup

We run the experiments on 16 Amazon instances of type `m1.medium`. For the SWF workflow, we used S3 as intermediate file storage. We executed the SWF decider and the ASKALON scheduler on a dedicated host with an Intel i7-2600K quad-core processor running at 3.4 GHz and 8 GB of memory, outside of Amazon EC2. For ASKALON we used a just-in-time scheduler that maps the next ready activities on the machines delivering the earliest completion time, because it mostly resembles the SWF operation. This simple approach does not benefit from several optimizations normally used in workflow executions but the goal of this analysis was not to compare the features of the ASKALON worflow system with Amazon SWF. We executed the RainCloud workflow in two scenarios: *noncongested* with 16 parallel loop iterations and two problem sizes (small and large) and *congested* with 64 parallel loop iterations and a small problem size. The small problem size corresponds to a 18×18 simulation grid and the large one to a 36×36 grid. For each scenario and workflow version, we calculated the two metrics: *average total execution* time and *cumulative execution* time of all workflow activities plus the scheduling time. To get an understanding on the amount of overhead present in a workflow execution, we further split its cumulative execution time into *processing* time (performing actual computation), *scheduling* time, *waiting* time (in an engine internal queue) due to insufficient free resources, *queuing* time due to middleware and external load latencies, and *file transfer* time.

13.8.2 Results

Figure 13.7 shows the total execution times for the three workflow versions with 16 parallel iterations and small and large problem sizes in the non-congested scenario. The manually written SWF workflow is only marginally faster than the automatically generated version. We expected this result because the two versions only differ in the implementation of the decider, whose overhead is negligible compared to the total workflow execution time. Surprisingly, the ASKALON version suffers from significantly higher execution times due to the much higher overhead for transferring files between the worker nodes, as shown in Figure 13.8a. We found out that this overhead is caused by the Java CoG Kit [22] employed by ASKALON as a black-box library for interfacing

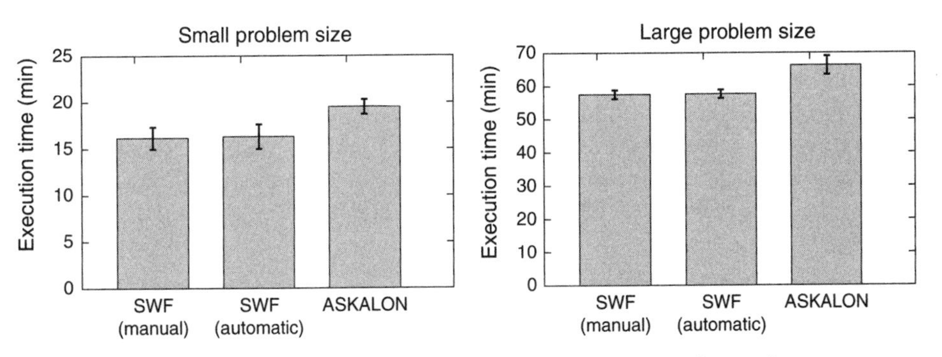

Figure 13.7. RainCloud execution time in noncongested scenario.

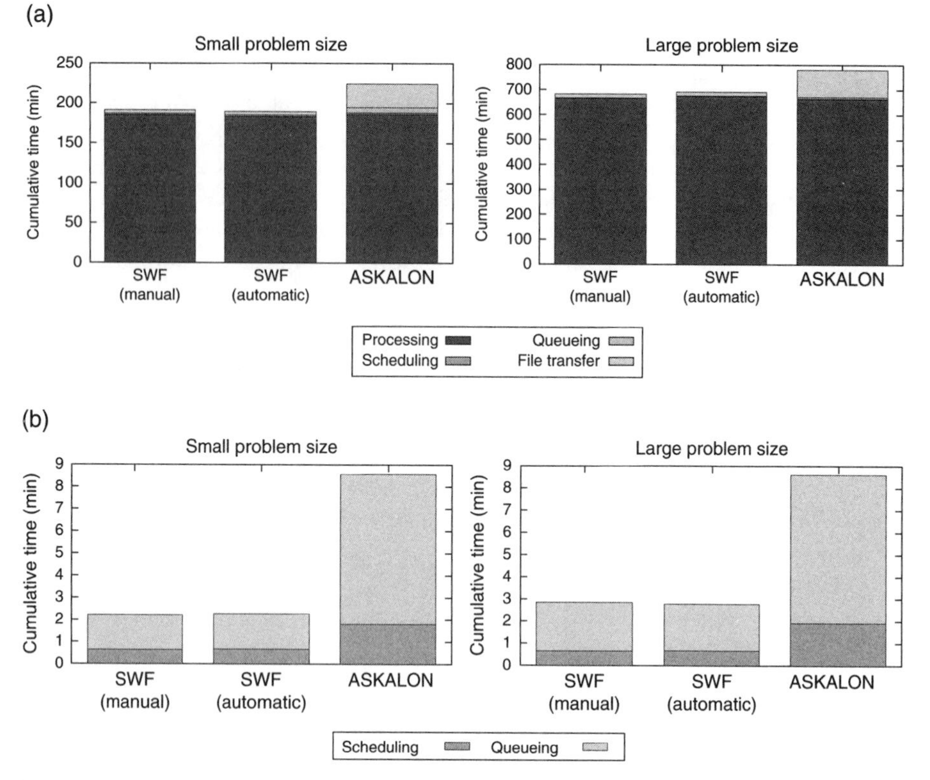

Figure 13.8. RainCloud cumulative times in noncongested scenario. (a) Cumulative times. (b) Cumulative overheads except file transfer.

with Grids (through Globus plugin) and Clouds (through SSH plugin), which uses an ASKALON middleware machine outside Amazon EC2 as an intermediary for transferring files between two remote machines. In the following text, we disregard the file transfer times to make the experiments more comparable.

The other reasons for ASKALON's performance losses are the scheduling and queuing overheads shown in Figure 13.8b. The scheduling overhead in ASKALON is approximately three times higher than in SWF because it is tuned for highly heterogeneous and distributed Grid infrastructures, as opposed to Clouds that tend to be more homogeneous and located within one data center. Because of this, the ASKALON Grid scheduler needs to interact with a resource manager for discovering the available shared resources which is not a requirement in static Clouds owned by a single organization. Moreover, the ASKALON scheduler also needs to evaluate the external load generated by scientists sharing a specific Grid resource, not required for dedicated Cloud resources. Finally, the ASKALON scheduler is also responsible for preparing the remote execution environments (and directories) through multiple SSH connections, not required for SWF that delegates the setup of the environment to the locally running legacy wrapper service. A more generic execution approach has more features resulting in overheads than a specialized platform dependent one.

Also, the workflow activities wait three times longer in the queue of ASKALON compared with SWF. The average overhead per workflow activity without file transfer is approximately 4–5 s for SWF and around 15 s for ASKALON, which is comparable for scientific workflows with long running activities. The queuing time is larger for ASKALON than for SWF because of the higher middleware stack required by ASKALON for supporting a broader range of heterogeneous infrastructures (i.e., clusters, Grids, and Clouds), as opposed to SWF tuned for running in the native EC2 infrastructure only. In addition, ASKALON actively pushes workflow activities to be executed onto the worker nodes which introduces higher overhead than the pull approach used by Amazon SWF where the worker nodes actively fetch tasks from a task queue.

Figure 13.9 shows the total execution times in the congested scenario. The SWF version performs slightly better for 64 parallel loop iterations than for 16; however, this improvement due to load imbalance on the 16 iteration parallel loop and coarse-grain activity sizes is still within the standard deviation. Using 64 iterations produces a finer grained parallelization and smaller activity sizes that enable a better schedule with smaller load imbalance overhead. The ASKALON version with 64 parallel iterations performs worse than with 16, but this is again within the standard deviation.

Figure 13.10 shows the cumulative execution time for 64 parallel loop iterations. As expected, the waiting times in the internal engine queue are extremely high because there are four times as many workflow activities ready to execute than worker nodes. Again, the queuing time of ASKALON is larger than of SWF because of the higher middleware stack and the batch-mode access to resources. The average overhead per workflow activity without the file transfer and waiting overheads is 17 s for SWF and 40 s for ASKALON, which is an increase by a factor of 3.4 for SWF and 2.7 for ASKALON compared to the previous scenario. The slight increase in execution time of the ASKALON version in the congested scenario is mainly caused by file transfer overheads.

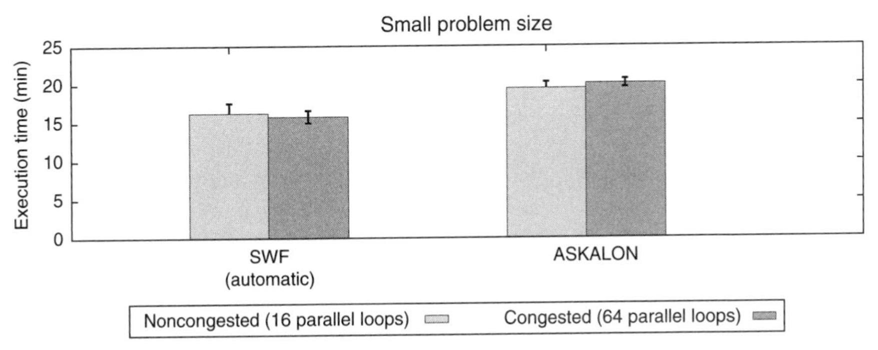

Figure 13.9. RainCloud execution time with 16 and 64 parallel loops.

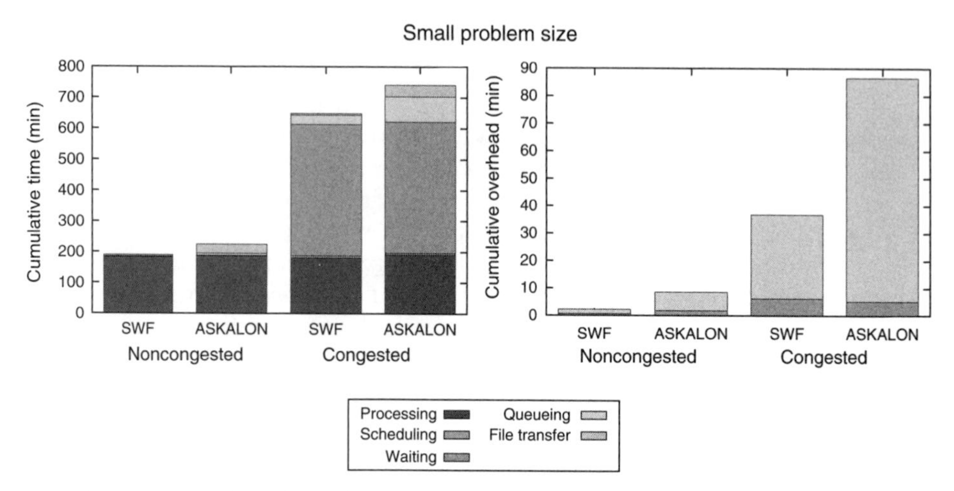

Figure 13.10. Cumulative RainCloud execution times with 16 and 64 parallel loops.

13.8.3 Discussion

To conclude, ASKALON has been designed to support a variety of heterogeneous and distributed computing environments, including Globus, gLite, EC2, and GroudSim-based [23]. This heterogeneity in the supported infrastructures is achieved through a modular architecture consisting of several layers and comprising complex services such as enactment engine, scheduling, file transfers, and resource management. Although we paid high attention at tuning the ASKALON overheads when building the Cloud plugins, we exhibit a performance drop due to the higher middleware stack compared to SWF, tuned for working with the local, simpler, and more homogeneous EC2 infrastructure. For this reason, SWF features a much simpler architecture where the decider only decides which workflow activities can be executed next and not on which resources. The tasks of preparing the execution environment and transferring local files from S3 need to be manually implemented by the programmer incurring a lower execution overhead, at the cost of a higher programming effort. Workflows consisting of numerous relatively short activities will mostly suffer from the larger ASKALON middleware overheads.

13.9 OPEN CHALLENGES

The approach shown in this chapter represents one possible solution for using cloud resources for scientific computing utilizing the workflow paradigm. Not all applications can utilize this approach to efficiently use the available resource pool. There are still open challenges for different types of scientific applications that are not covered with the shown solution:

Big Data: Some scientific domains rely on enormous amounts of data to be processed. The challenge for this applications is to transfer this data into the cloud for processing as transferring the processing power from the cloud to the data is technically not possible [24, 25]. For this class of application, a faster Internet connection will be needed; but when looking at physics experiments (like ATLAS from CERN), this fastest possible transfer speed might still be to low.

Security: Most scientific applications are self-written, open source or free to use. When utilizing commercial software there might be problems with licensing them to leased cloud hardware. Additionally the input data and results might also be restricted in their distribution (i.e., medical studies). To allow such applications to utilize cloud resources without violating copyrights or laws is still an open challenge [26, 27].

Super computing: Some applications are build for massive parallel architectures commonly only available in supercomputers. Cloud providers might have shown in demo applications that it is possible to build a setup that is fast enough to reach the TOP500 [28] speed wise but that is far away from a regular use case any scientist can deploy on the cloud everyday. For those applications the only solution is still having access to a supercomputer.

13.10 CONCLUSION

In this chapter, we proposed a method for automatic porting of scientific workflows to Amazon SWF, able to exploit the native performance of the EC2 infrastructure. The solution is based on the SHIWA fine-grained interoperability technology for translating workflows written across different languages and workflow systems through the common IWIR representation. This scalable software engineering solution enables five major workflow systems currently supporting the IWIR representation access the EC2 infrastructure through the SWF service: ASKALON, MOTEUR, Pegasis, Triana, and WS-PGRADE.

We presented in this chapter the difficulties we encountered in translating an data flow-oriented ASLALON workflow into a control flow-oriented SWF decider program. The method is based on an algorithm that automatically generates the SWF decider Java program and the underlying activity functions in four phases: function signature, activity semantics, DAG control flow, and data flow generation.

We presented experimental results for porting an original real-world ASKALON workflow to the EC2 infrastructure in two configurations: conversion to a Java SWF decider or execution through the ASKALON middleware connected to EC2 via an SSH plugin. The results demonstrate that the SHIWA fine-grained interoperability solution that translates an ASKALON workflow into an SWF version through the common IWIR representation is a promising alternative for porting workflows to a new infrastructure and able to exploit its native performance. Amazon SWF represents an attractive environment for running traditional workflow applications, especially those consisting of numerous relatively short activities affected by large middleware overheads when executed in traditional ways. This is demonstrated by the performance of the automatically generated SWF workflow, which is similar to the manually optimized version. By contrast, porting existing Grid workflow middleware environments such as ASKALON to the Cloud, although effective, have performance drawbacks compared to the translated SWF version. The reasons of performance losses lie in the high middleware stack required for supporting a wider range of distributed and heterogeneous cluster, Grid, and Cloud computing infrastructures and a more generic scheduling and execution approach.

A downside of SWF is its proprietary implementation hosted by a commercial vendor who charges costs and may abandon this service anytime if it is lacking success. Another difference to clusters and Grids is the pull-based assignment of tasks to an unknown number of activity workers that requires different scheduling methods.

ACKNOWLEDGMENTS

This work has been performed in the projects TRP 237-N23 funded by the Austrian Science Fund (FWF) and eRamp, co-funded by the grant 843738 from the Austrian Research Promotion Agency (FFG) and the ENIAC Joint Undertaking.

REFERENCES

1. I. J. Taylor, E. Deelman, D. Gannon, and M. Shields, Eds., *Workflows for e-Science. Scientific Workflows for Grids*. Springer, London, U.K., 2007.
2. J. Varia and S. Mathew, "Overview of amazon web services," 2012. Available: http://d36cz9bnwru1tt.cloudfront.net/AWS_Overview.pdf. Accessed on December 9, 2014.
3. K. Plankensteiner, J. Montagnat, and R. Prodan, "IWIR: A language enabling portability across grid workflow systems," in *Proceedings of the 6th Workshop on Workflows in Support of Large-Scale Science*, Seattle, WA. ACM, New York, 2011, pp. 97–106.
4. T. Fahringer, R. Prodan, R. Duan, J. Hofer, F. Nadeem, F. Nerieri, S. Podlipnig, J. Qin, M. Siddiqui, H. Truong *et al.*, "Askalon: A development and grid computing environment for scientific workflows," *Workflows for e-Science*, Springer, Berlin, 2007, pp. 450–471.
5. T. Glatard, J. Montagnat, D. Lingrand, and X. Pennec, "Flexible and efficient workflow deployment of data-intensive applications on grids with moteur," *International Journal of High Performance Computing Applications*, vol. 22, no. 3, pp. 347–360, 2008.
6. P. Kacsuk, "P-grade portal family for grid infrastructures," *Concurrency and Computation: Practice and Experience*, vol. 23, no. 3, pp. 235–245, 2011.
7. E. Deelman, G. Singh, M. Su, J. Blythe, Y. Gil, C. Kesselman, G. Mehta, K. Vahi, G. Berriman, J. Good *et al.*, "Pegasus: A framework for mapping complex scientific workflows onto distributed systems," *Scientific Programming*, vol. 13, no. 3, pp. 219–237, 2005.
8. I. Taylor, M. Shields, I. Wang, and O. Rana, "Triana applications within grid computing and peer to peer environments," *Journal of Grid Computing*, vol. 1, no. 2, pp. 199–217, 2003.
9. G. Juve, E. Deelman, K. Vahi, G. Mehta, B. Berriman, B. Berman, and P. Maechling, "Scientific workflow applications on amazon ec2," in *2009 5th IEEE International Conference on E-Science Workshops*. Oxford, U.K. IEEE, New York, 2009, pp. 59–66.

10. P. Riteau, M. Tsugawa, A. Matsunaga, J. Fortes, T. Freeman, D. LaBissoniere, and K. Keahey, "Sky computing on futuregrid and grid'5000," in *5th Annual TeraGrid Conference: Poster Session*, vol. 68, Pittsburgh, PA, 2010, p. 119.
11. C. Hoffa, G. Mehta, T. Freeman, E. Deelman, K. Keahey, B. Berriman, and J. Good, "On the use of cloud computing for scientific workflows," in *IEEE Fourth International Conference on eScience, 2008 (eScience'08).*, Indianapolis, IN. IEEE, New York, 2008, pp. 640–645.
12. G. Morar, F. Schueller, S. Ostermann, R. Prodan, and G. Mayr, "Meteorological simulations in the cloud with the ASKALON environment," in *Euro-Par 2012: Parallel Processing Workshops*. Springer, Berlin, Germany, 2013, pp. 68–78.
13. J.-S. Vöckler, G. Juve, E. Deelman, M. Rynge, and B. Berriman, "Experiences using cloud computing for a scientific workflow application," in *Proceedings of the 2nd International Workshop on Scientific Cloud Computing ser. ScienceCloud ('11, Boulder, CO)*. ACM, New York, 2011, pp. 15–24. Available: http://doi.acm.org/10.1145/1996109.1996114
14. M. De Assunção, A. Di Costanzo, and R. Buyya, "Evaluating the cost-benefit of using cloud computing to extend the capacity of clusters," in *Proceedings of the 18th ACM International Symposium on High Performance Distributed Computing*, Munich, Germany. ACM, New York, 2009, pp. 141–150.
15. P. Marshall, K. Keahey, and T. Freeman, "Elastic site: Using clouds to elastically extend site resources," in *Proceedings of the 2010 10th IEEE/ACM International Conference on Cluster, Cloud and Grid Computing*, Melbourne, Australia. IEEE Computer Society, Washington, DC, 2010, pp. 43–52.
16. D. Franz, J. Tao, H. Marten, and A. Streit, "A workflow engine for computing clouds," in *CLOUD COMPUTING 2011, the Second International Conference on Cloud Computing, GRIDs, and Virtualization*, Rome, Italy, 2011, pp. 1–6.
17. S. Pandey, D. Karunamoorthy, K. Gupta, and R. Buyya, "Megha workflow management system for application workflows," *IEEE Science & Engineering Graduate Research Expo*, Melbourne, Australia, 2009.
18. M. E. Conway, "Proposal for an uncol," *Communications of the ACM*, vol. 1, no. 10, pp. 5–8, Oct. 1958. [Online]. Available: http://dl.acm.org/citation.cfm?id=368928. Accessed on December 9, 2014.
19. G. Kiczales, J. Lamping, A. Mendhekar, C. Maeda, C. Lopes, J.-M. Loingtier, and J. Irwin, "Aspect-oriented programming," *ECOOP'97—Object-Oriented Programming*, Jyuäskylä, Finland, pp. 220–242, 1997.
20. I. Barstad and F. Schueller, "An extension of Smith's linear theory of orographic precipitation: Introduction of vertical layers," *Journal of the Atmospheric Sciences*, vol. 68, no. 11, pp. 2695–2709, 2011.
21. S. Ostermann, R. Prodan, and T. Fahringer, "Extending grids with cloud resource management for scientific computing," in *2009 10th IEEE/ACM International Conference on Grid Computing* Banff, Alberto, Canada. IEEE, New York, 2009, pp. 42–49.
22. G. von Laszewski, I. Foster, J. Gawor, and P. Lane, "A Java commodity Grid kit," *Concurrency and Computation: Practice and Experience*, vol. 13, no. 89, pp. 643–662, 2001.
23. S. Ostermann, K. Plankensteiner, and R. Prodan, "Using a new event-based simulation framework for investigating resource provisioning in Clouds," *Scientific Programming*, vol. 19, no. 2, pp. 161–178, 2011.

24. D. Agrawal, S. Das, and A. El Abbadi, "Big data and cloud computing: Current state and future opportunities," in *Proceedings of the 14th International Conference on Extending Database Technology*, Uppsala, Sweden. ACM, New York, 2011, pp. 530–533.
25. S. Chaudhuri, "What next?: A half-dozen data management research goals for big data and the cloud," in *Proceedings of the 31st Symposium on Principles of Database Systems*, Scottsdale, AZ. ACM, New York, 2012, pp. 1–4.
26. B. R. Kandukuri, V. R. Paturi, and A. Rakshit, "Cloud security issues," in *IEEE International Conference on Services Computing, 2009. (SCC'09)* Bangalore, India. IEEE, New York, 2009, pp. 517–520.
27. D.-G. Feng, M. Zhang, Y. Zhang, and Z. Xu, "Study on cloud computing security," *Journal of Software*, vol. 22, no. 1, pp. 71–83, 2011.
28. "Top500 homepage," Website, accessed on 06/2014, http://www.top500.org/system/10661.

14

INTERACTIVE MULTIMEDIA APPLICATIONS ON CLOUDS

Karine Pires and Simon Gwendal

Telecom Bretagne, Institut Mines-Telecom, Paris, France

14.1 INTRODUCTION

In less than 10 years, services over Internet have switched from static Web pages to interactive multimedia applications. To deal with the demand for more interactivity and more multimedia content, service providers have been forced to upgrade their infrastructure to offer their service. In the meantime, the benefits of virtualizing the infrastructure and of delegating the delivery to external companies have prevailed over the traditional architecture where the service provider owns a set of servers and delivers the content by itself. What is now referred to as the cloud is the combination of multiple actors tied by commercial agreements and orchestrated by the service provider.

In this chapter, we will focus on the service providers that offer massive, interactive, multimedia services. These service providers face challenging scalability and response time issues. We will describe the solutions that have been recently developed to address these issues. In particular, we will pay a close attention to three representative services:

1. *Cloud Gaming*: On-demand gaming, also known as cloud gaming, is a new video gaming application/platform. Instead of requiring end users to have sufficiently

Cloud Services, Networking, and Management, First Edition.
Edited by Nelson L. S. da Fonseca and Raouf Boutaba.

powerful computers to play games, on-demand gaming performs the intensive game computation, including the game graphics generation, remotely with the resulting output streamed as a video back to the end users. For game developers, shifting from traditional gaming platforms to cloud gaming means a better control on the delivery and easier software development and upgrade. On their side, end users benefit from platform independence, for example, to play computationally intensive games on portable devices that do not have the required hardware, and also to be able to play at any time on any available device including smart TVs and smartphones.

2. *Massive User-Generated Content (UGC) Live Streaming*: Anybody can become a TV provider. This promise, which has been floating in the air for almost 10 years, has led to a considerable effort from the research community [1]. The popularity of UGC live streaming aggregators has however not grown as fast as some expected. Yet, the past couple of years have seen a surge of interest for such services, pushed by new usages, including crowdsourced journalism [2] and e-sport [3]. The major actors of the area have been forced to take some measures to cope with traffic explosion. Typically, Twitch.tv, gaming branch of justin.tv, announced a significant increase in the delay,[1] while the new live service from YouTube was only offered to a subset of users.[2]
3. *Time-shifting On-Demand TV*: In time-shifted TV, a program broadcasted from a given time t is made available at any time from t to $t + \delta$ where δ can be potentially infinite. The popularity of TV services based on time-shifted streaming has dramatically risen [4]: *nPVR* (a personal video recorder located in the network), *catch-up TV* (the broadcaster records a channel for a shifting number of days, and proposes the content on demand), *TV surfing* (using pause, forward or rewind commands), and *start-over* (the ability to jump to the beginning of a live TV program). Today, to enjoy catch-up TV requires a digital video recorder (DVR) connected to the Internet. However, TV broadcasters need to protect advertisement revenue, whereas a DVR viewer can decide to fast forward through commercials. By controlling the TV stream, not only the broadcasters may guarantee that commercials are played, but they can also adapt them to the actual time at which the viewer watches the program. This calls for a cloud-based time-shifted TV service.

These three services make use of the same basic infrastructure components to offer their services in the best conditions. We will first describe in a generic way these delivery components. Then, we will study each of these services in an iterative manner with the ambition to reveal some of the unique characteristics of these services and the solutions that have been deployed so far.

[1] Twitch: The Official Blog http://is.gd/PdqlZI

[2] YouTube Live Introduction http://is.gd/Aw0yAx

14.2 DELIVERY MODELS FOR INTERACTIVE MULTIMEDIA SERVICES

14.2.1 Background

The interactive *response time* is defined as the elapsed time between when an action of the user is captured by the system and when the result of this trigger can be perceived by the user. For example in cloud gaming, which is one of the most demanding services on this aspect, the work in Ref. [5] demonstrates that a latency around 100 milliseconds is highly recommended for dynamic, action games, while response time of 150 milliseconds is required for slower paced games.

The overall interactive response time T of an application includes several types of delays, are defined as follows:

$$T = t_{\text{client}} + \overbrace{t_{\text{access}} + t_{\text{isp}} + t_{\text{transit}} + t_{\text{provider}}}^{t_{\text{network}}} + t_{\text{server}}$$

14.2.1.1 Hardware Latency. We define t_{client} as the *playout delay*, which is the time spent by the client to (1) send action information (e.g., in cloud gaming, initiating character movement in a game) and (2) receive and play the video. Only the client's hardware is responsible for t_{client}, but the software that runs at the client side is commonly provided by the service provider.

Additionally, we define t_{server} as the *processing delay*, which refers to the time spent by the server to process the incoming information from the client, to generate the corresponding video information, and to transmit the information back to the client. The service provider is mainly responsible for the processing delay.

Both playout and processing delays can be reduced with hardware changes and software development by the service provider.

14.2.1.2 Network Latency. The remaining contribution of total latency comes from the network. We further divide the network latency into four components: t_{access}, t_{isp}, t_{transit}, and t_{provider}.

First, t_{access}, is the data transmission time between the client's device and the first Internet-connected router. Three quarters of end users who are equipped with a DSL connection experience a t_{access} greater than 10 milliseconds when the network is idle [6], and the average access delay exceeds 40 milliseconds on a loaded link [7]. The behaviour of different network access technologies can greatly vary, as the latency of the access network can differ by a factor of 3 between different Internet Service Providers (ISPs) [7]. Additionally, the home network configuration and the number of concurrent active computers per network access can double access link latency [8]. Finally, when the network connection is through cellular networks, some other parameters can affect the delay, including the technologies at the base station and the underlying network protocol (the "generation" of the network).

The second component of network delay is t_{isp}, which corresponds to the transmission time between the access router and the peering point connecting the ISP network to the next hop transit network. During this phase, data travel exclusively within the ISP network. Although ISP networks are generally fast and reliable, major ISPs have reported congestion due to the traffic generated by new multimedia services [9].

The third component is t_{transit}, which is defined as the delay from the first peering point to the front-end server of the service provider. The ISP and provider are responsible for t_{transit}; however, the networks along the path are often owned by third-party network providers. Nonetheless, the ISP and the cloud provider is responsible for good network connectivity for their clients.

The fourth component, t_{provider}, is defined as the transmission delay between the front-end server of the service provider and the hosting server for the client. The provider is responsible for t_{provider}. This delay is however rarely significant. Network latencies between two servers in modern datacenters are typically below 1 milliseconds [10].

14.2.2 Introducing Delivery Models

Service provides have several options to build their "delivery cloud." Here is a short introduction to them.

14.2.2.1 Data Center. The most common way to deliver content is to use a datacenter (DC), which is basically a large set of servers [11]. The DC can be either owned or rented by the service provider. In the former case, the infrastructure is almost exclusively paid at the construction, however it has some fixed capacity limitations. In the latter case, the infrastructure can scale up and down on demand but the service provider has to deal with another actor (DC provider). Although DCs are attractive, easy-to-manage infrastructures, they do not enable low response time for a large population of users because they are located in one location (or few locations if the service provider deals with several DC). That is, the aforementioned network latency is too high for a vast fraction of the population because t_{isp} and $t_{transit}$ are large. Moreover, the monetary cost to transfer data is higher because the traffic should cross several networks until the content eventually reaches the users.

14.2.2.2 Peer to Peer. The challenge of delivering multimedia content on a large scale is essentially a problem related to the reservation of physical resources. To address this problem, the scientific community has advocated for years for a peer-to-peer (P2P)-based infrastructure, where users themselves contribute to the delivery by forwarding the content they received. A lot of algorithms have been designed to improve the delivery performances [1]. However, various constraints have limited the deployment of P2P systems for commercial purpose. First, firewalls and network address translator (NAT) still prevent many direct connection between users [12]. Second, P2P require users to install a program on their computers. Such a "technical," security-sensitive requirement can prevent users from using the service. Moreover, despite some new browser-based technologies (e.g., WebRTC), a P2P software depends on the configuration of the computer of end users, which is a cause of many development difficulties. Third, the service provider has a low control on the Quality of experience (QoE) of users since it does not directly control the performances. Last, the complexity of P2P system can increase the delay. Many initiatives have aimed at ensuring that peers connect preferentially with the other peers that are located in the same network [13], which make $t_{transit}$ null. However, a peer usually get data from multiple other peers. Even if the direct connection between two peers is short, aggregating data multiple from multiple peers requires synchronisation and buffering, which cause extra delay.

14.2.2.3 Content Delivery Network. In the recent years, content delivery networks (CDNs) have emerged as the privileged way for large-scale content delivery. CDN comprises three types of communication devices: a relatively small number of *sources*, which directly receive the content from the service producer, a medium-sized network of *reflectors*, and a large number of *edge servers*, which are deployed directly in the access networks, close to the users. The proximity between the end-users and the edge-servers makes that network latency is small.

For a decade, the CDN providers have met the demand of two families of players in the value chain of content delivery: service providers (because large-scale Internet services have to be distributed for redundancy, scalability, and low-latency reasons) and network operators (because minimizing inter domain traffic while still fulfilling their own users' requests is a business objective). CDNs have thus emerged as a new category of market players with a dual-sided business. They provide caching capacities "as a service" to network operators, and they provide a distributed hosting capacity to service providers. The CDN providers provide both scalability and flexibility, they deal with distribution complexities, and they manage multiple operator referencing—all of these services at a unique selling point.

The Works in Refs. such as [14, 15] confirm that edge-servers are not only used for serving static content. As studied in Ref. [1], current CDN infrastructure has the ability to serve millions of end users and is well-positioned to deliver game content and software [16]. However, CDN edge servers are generally built from commodity hardware that have relative weak computational capabilities and often lack GPUs.

14.2.3 Composing Hybrid Delivery Models

At the time this chapter was written, there was no clear consensus about the best solutions to deploy. Typically for video streaming, we observe that the main actors made different choices. To name a few:

- Google uses multiple DCs distributed over the globe to delivery their services, including YouTube [17].
- NetFlix uses a composition of multiple CDNs on its delivery chain [18].
- A composition of P2P assisted by CDN to improve viewers QoE was deployed on LiveSky [19].
- Justin.tv, one of the biggest live streaming service, uses private DC assisted by CDN [20].

A recent trend is to build *hybrid* delivery models that compose several of the aforementioned models. We depict in Figure 14.1 and list hereafter some frequent compositions, each one with its own pros and cons.

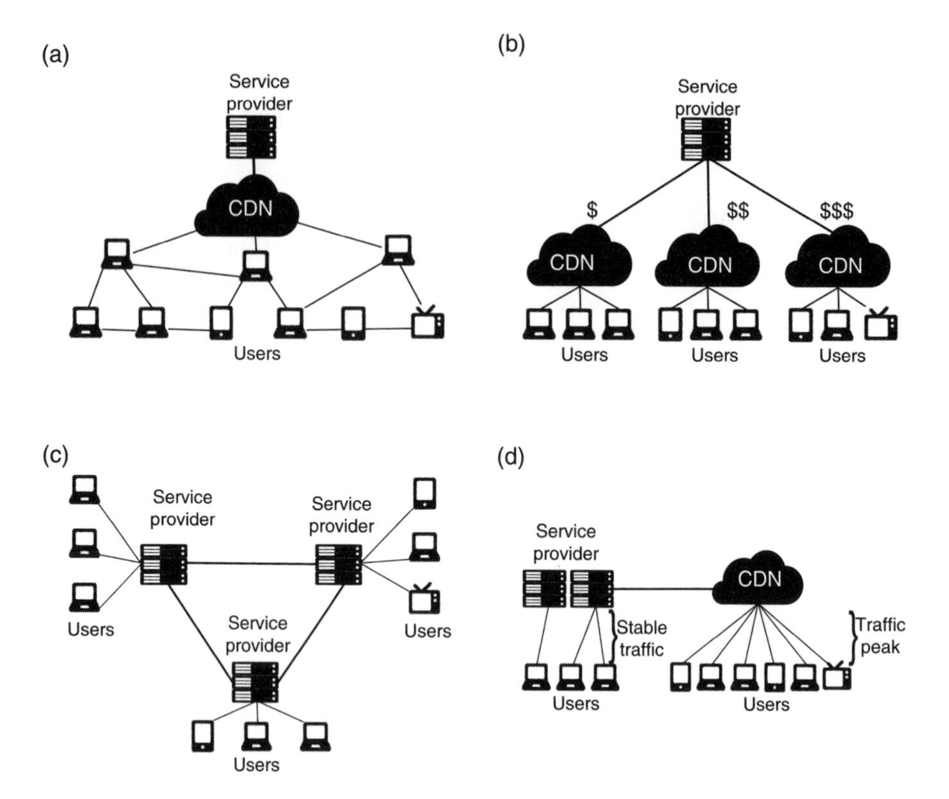

Figure 14.1. Hybrid delivery models compositions. (a) CDN-P2P, (b) Multi-CDN, (c) Multi-DC, and (d) DC-CDN.

14.2.3.1 CDN P2P. Such a composition is managed by either the service provider, as shown in Ref. [21] or the CDN provider, as shown in Ref. [22]. The CDN offers some guarantee on the QoE by offering a minimum amount of resources and by reducing the first response time. The CDN also allows users behind NAT to be properly served. On its side, the P2P system assists the CDN in case of traffic peak. The more users to be served by the system, the more resources in the system. The potential problem with such composition is that service providers want to control all parts of the delivery chain. Indeed, most profits come from a clear understanding of the demand from end users and a capacity to adapt the delivered content to every user (e.g., embedded advertisement). Another potential problem comes from the lack of guarantee of QoS. Finally CDN-P2P compositions suffer from the same drawback as P2P one, including the requirement of installing a software on users' computer.

14.2.3.2 Multi-CDN. The service provider is commonly the main manager of this composition. A typical example of such a composition has been thoroughly studied in Ref. [18]. The main idea is that the service provider relies on several CDNs to deliver the content. For each user, the service provider decides the CDN in charge of serving this user. The advantage in this composition is the possibility to achieve the best QoE for the viewers with the lowest cost from the multiple prices applied by each CDN. Another advantage is that the delivery is more robust since a downtime from one CDN provider can be mitigated by using another CDN. However, this delivery is only based on third-party actors, which means that even the consolidated background traffic is dealt in a pay-as-you-go way. Therefore, the overall price can be high.

14.2.3.3 Multi-DCs. Service providers that want to provide features beyond the basic delivery of the same content are interested in hosting the service in their own servers in a DC. However, response time requirements force service providers to deploy multiple DCs in order to serve the whole population with low response time [23]. In that case, it becomes crucial to manage the traffic such that the load is well balanced among the different DCs [24] and to manage the sharing of content over the multiple DCs [25]. The advantage is a lower cost rate per GBps than multi-CDN, the total control of the delivery chain, and a relatively low response time since every end-users should have a DC nearby (so $t_{transit}$ is reduced). The cons include substantial high cost for the initial deployment for multiple DCs.

14.2.3.4 DC CDN. In order to mitigate the disadvantages of the aforementioned models, it is frequent that video service providers deploy hybrid DC-CDN compositions [20]. DC-CDN hybrid composition is expected to combine the main advantages of both delivery solutions at a minimum cost. The high prices paid for CDN are minimized by using the CDN resources only when DC is out of capacity, normally at traffic peaks. The DC is dimensioned so that the consolidated background traffic (or valleys of usage) is dealt by the DC. In the cloud computing context, such composition is often called hybrid cloud where conventional DCs and cloud solutions are deployed together to aim the same combined advantage [26]. Various studies have indicated that it is not trivial to outsource tasks from the internal DCs to the external delivery infrastructure [27], typically due to security [28], QoS [29], and economic [27] reasons.

14.3 CLOUD GAMING

As said in Section 14.1, cloud gaming is a new paradigm that has the potential to change the video game industry. Attractive for both end-users and developers, cloud gaming faces two main technical challenges: latency and the need for servers with expensive, specialized hardware that cannot simultaneously serve multiple gaming sessions. By offloading computation to a remote host, cloud gaming suffers from

- encoding latency, that is, the time to compress the video output
- network latency, which is the delay in sending the user input and video output back and forth between the end user and the cloud.

Past studies [30–32] have found that players begin to notice a delay of 100 milliseconds [5]. Although the video encoding latency will likely fall with faster encoders, at least 20 milliseconds of this latency should be attributed to playout and processing delay [33]. It means that 80 milliseconds is the threshold above which network latency begins to appreciably affect user experience, among which a significant portion of network latency is unavoidable as it is bounded by the speed of light in fiber. Because of this strict latency requirement, servers are restricted to serving end users that are located in the same vicinity. This explains the inaptitude of a DC-only solution for cloud gaming. Even multi-DC hybrid solutions are inefficient for cloud gaming when the number of DC is too small. To validate this statement, we perform a large-scale measurement study consisting of latency measurements from PlanetLab and Amazon EC2 to more than 2,500 end users. These results are originally presented in Ref. [34].

In the following, we study the effectiveness of various infrastructures to offer on-demand gaming services. We focus on the network latency since the other latencies, especially the generation of game videos, have been studied in previous work [30, 35]. We evaluate in particular a multi-DC solution and a hybrid CDN-DC solution, which has been originally proposed in Ref. [36].

14.3.1 Measurement Settings

To determine the ability of today's cloud to provide the cloud gaming service, we conduct two measurement experiments to evaluate the performance and latency of cloud gaming services on existing cloud infrastructures in the United States. First, we perform a measurement campaign on the Amazon EC2 infrastructure during May 2012. Although EC2 is one of today's largest commercial clouds, our measurements show that it has some performance limitations. Second, we use PlanetLab [37] nodes to serve as additional DCs in order to estimate the behavior of a larger, more geographically diverse cloud infrastructure.

In our model, a DC (either Amazon EC2 or PlanetLab) is able to host all games and to serve all end-users that are within its latency range as it has a significant amount of storage and computational resources. This model is based on public information available regarding the peak number of concurrent end users using on-demand gaming today (less than 1800 [38]) and the size of modern cloud DCs (hundreds of thousands of servers [39]).

As emphasized in previous network measurement papers [6, 7], it is challenging to determine a representative population of real clients in large-scale measurement experiments. For our measurements, we use a set of 2,504 IP addresses, which were collected from 12 different BitTorrent[3] swarms. These BitTorrent clients were participating in popular movie downloads. Although 2,504 IP addresses represent a fraction of the total population in the United States, these IP addresses likely represent home users who are using their machines for entertainment purposes. Therefore, we believe that these users are a reasonable cross-section of those who use their computers for entertainment purposes, which includes gaming. We refer to this selected users as the *population*.

We choose BitTorrent as the platform for our measurement experiments since, in our beliefs, it provides a realistic representation of end users and their geographic

[3]http://www.bittorrent.com/

distribution. We use the *GeoIP* service to restrict our clients to the United States, which is the focus of this measurement study. Moreover, it allows us to determine the approximate geographical locations of our end users, which are used as a parameter for many of our measurement experiments. After determining the clients, we use TCP measurement probe messages to determine latency between servers and clients. Note that we measure the round-trip time from the initial TCP handshake, which is more reliable than a traditional ICM *ping* message and less sensitive to network conditions.

14.3.2 Measurement of a State-of-the-Art Multi-DC Infrastructure

The Amazon EC2 cloud offers three DC in the United States to its customers. We obtain a virtual machine instance in each of the three DCs. Every 30 min, over a single day, we measure the latency between each DC to all of the 2,504 clients. We use the median value from ten measurements to represent the latency between an end-host to a PlanetLab node or EC2. Figure 14.2 depicts the ratio of covered end users that have at least one network connection to one of the three DCs for a given latency target. Two observations can be made from the graph shown in Figure 14.2:

- *More than one-quarter of the population cannot play games from an EC2-powered cloud gaming platform.* The thin, vertical gray line in Figure 14.2 represents the 80 milliseconds threshold network latency yielding a 70% coverage.
- *Almost 10% of the potential clients are essentially unreachable.* In our study, unreachable clients are clients that have a network latency over 160 milliseconds, which renders them incapable of using an on-demand gaming service. Although we filter out the IP addresses that experienced highly variable latency results, we still observe that a significant proportion of the clients have a network latency

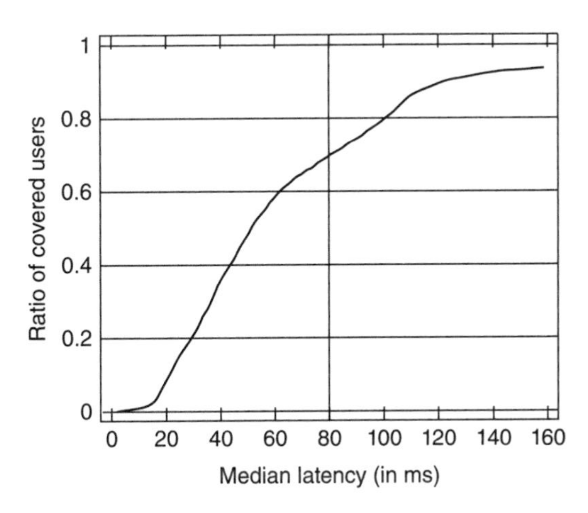

Figure 14.2. Population covered by EC2 cloud infrastructure as a function of the median latency.

over 160 milliseconds. This result confirms the measurements made by previous work, which identified that home gateways can introduce a significant delay on data transmission [7].

14.3.2.1 Effects of a Larger Cloud Infrastructure. An alternative to deploying a small number of large DCs is to instead use a large number of smaller DCs. The main providers have claimed to possess up to a dozen DCs within the United States [40, 41] in order to improve their population coverage. A large DC is generally more cost-efficient than a small DC; therefore, cloud providers should carefully determine if it is economically beneficial to build a new DC. In the following, we investigate the gain in population coverage when new DCs are added into the existing EC2 infrastructure.

We create a simulator that uses our collected BitTorrent latencies in order to determine how many users are able to meet the latency requirement for gaming. We use 44 geographically diverse PlanetLab [37] nodes in the United States as possible locations for installing DCs. We consider a cloud provider that can choose from the 44 locations to deploy a k-DC cloud infrastructure. We determine latencies between clients and PlanetLab nodes using the result of our measurement campaign. Afterwards, we determine the end user coverage when using PlanetLab nodes as additional DCs.

We design two strategies for deciding the location of DCs:

- *Latency-based strategy*: the cloud provider wants to build a dedicated cloud infrastructure for interactive multimedia services. The network latency is the *only* driving criteria for the choice of the DC locations. For a given number k, the cloud provider places k DCs such that the number of covered end users is maximal.
- *Region-based strategy*: the cloud provider tries to distribute DCs over an area. We divide the United States into four regions as set forth by the US Census Bureau: Northeast, Midwest, South, and West. Every DC is associated with its region. In every region, the cloud provider chooses *random* DC locations. For a given total number of DCs k, either $\lfloor \frac{k}{4} \rfloor$ or $\lceil \frac{k}{4} \rceil$ DCs are randomly picked in every region.

For cloud providers, the main concern is to determine the minimum number of DCs required to cover a significant portion of the target population. Figure 14.3 depicts the ratio of covered users as a function of the response time target for two targets network latencies: 80 and 40 milliseconds. The former 80 milliseconds network latency target come from previous works [5, 31, 42], which indicate that 100 milliseconds is the latency threshold that is required for realism and acceptable gameplay for action games. Because at least 20 milliseconds can be attributed to playout and processing delay [33], network latency can account for up to 80 milliseconds of the total latency. We select 40 milliseconds as a stricter requirement for games that require a significant amount of processing or multiplayer coordination.

We observe that a large number of DCs are required if one wants to cover a significant proportion of the population. Typically, a cloud provider, which gives priority to latency, reaches a coverage ratio of 0.85 with 10 DCs for a target latency of 80 milliseconds. Using the region-based strategy requires nine DCs to reach a 0.8 ratio. In all cases, a 0.9 coverage ratio with a 80 milliseconds response time is not achievable without a significant increase in the number of DCs (around 20 DCs). For more

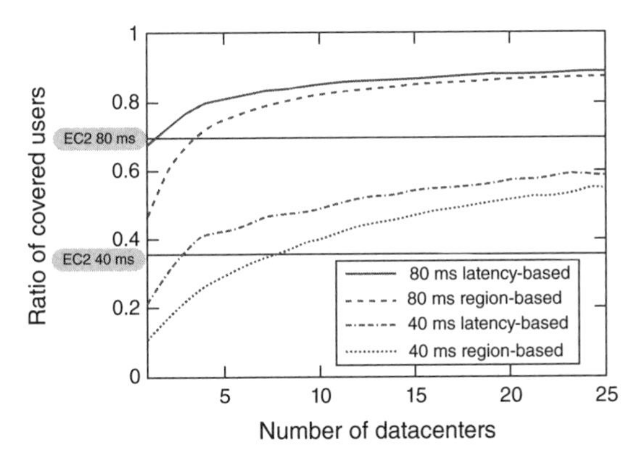

Figure 14.3. Coverage vs. the number of deployed DCs.

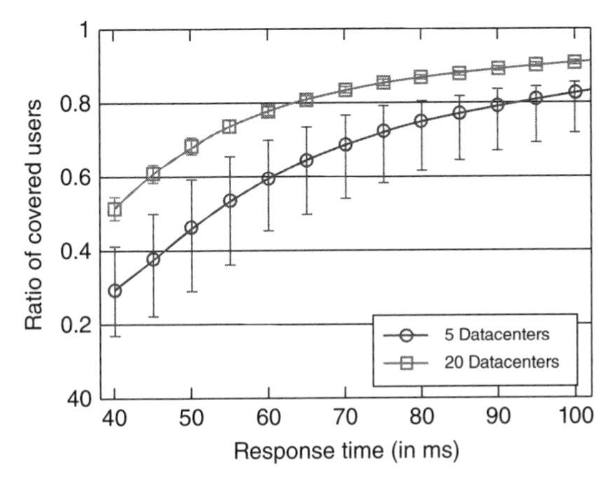

Figure 14.4. User coverage for a region-based DC location strategy (average with min and max from every possible set of locations).

demanding games that have a lower latency requirement (e.g., 40 milliseconds), we find that cloud provides exceedingly low coverage. Even if 20 DCs are deployed, less than half of the population would have a response time of 40 milliseconds. Overall, the gains in coverage are not significant with regard to the extra cost due to the increase in the number of DCs.

We then focus on the performance of two typical cloud infrastructures: a 5- and 20-DC infrastructure. We assume a region-based location strategy since it is a realistic trade-off between cost and performance. We present the ratio of covered populations for both infrastructures in Figure 14.4.

We observe that there can be significant performance gaps between a 5 and 20-DC deployment. Moreover, five DCs do not guarantee reliably good performance, despite the expectation that a region-based location strategy provides good coverage. Typically, a well-chosen 5-DC deployment can achieve 80% coverage for 80 milliseconds. However, a poorly chosen 5-DC deployment can result in a disastrous 0.6 coverage ratio. By contrast, a 20-DC deployment exhibits insignificant variances in the coverage ratio.

14.3.3 Hybrid DC-CDN Infrastructure

Since the multi-DC hybrid solution has some significant shortcomings. We explore in the following the potential of a hybrid DC-CDN infrastructure to meet the latency requirements of on-demand gaming end users. More details can be found in [36].

14.3.3.1 Experimental Settings. Out of the 2,504 IP addresses collected in our measurement study, unless otherwise specified, we select 1,500 of these IP addresses to serve as on-demand gaming end users. Of the remaining IP addresses, we select 300 of them to represent edge servers.

Client-to-client latency is determined as follows. Our simulator requires a latency matrix between all of our collected BitTorrent clients. A BitTorrent client may be used to represent either an edge server or an end-user. Since we do not have control of our collected BitTorrent clients, we estimate client-to-client latency by mapping a Client C_1 to its closest PlanetLab node, P. Suppose we wish to determine the latency between Client C_1 and C_2. This latency is the sum of P's latency to C_2 and a fuzzing factor that is between 0 and 15 milliseconds. We assume that client C_1 is located relatively near its closest PlanetLab node P; thus, the additional 0–15 milliseconds accounts for the latency between C_1 and P. Furthermore, an edge server may only serve an end user if it hosts the end user's demanded game.

We evaluate the effectiveness of a deployment or configuration by the number of end users that it is able to serve. In all of our experiments, we only model active users, and they are statically matched to either a datacenter or an available edge-server that has the requested game and meets the user's latency requirement. For our experiments, an edge server can only serve one end user at a time. An end user is served (or satisfied) if one of the following conditions are true:

- Its latency to a DC is less than its required latency.
- It is matched to an edge server that is within its latency requirement and hosts its requested game.

An end user may be unmatched if a DC cannot meet its latency requirement and all suitable edge servers are matched to other end users.

14.3.3.2 Determining the Size of the Augmented Infrastructure. We now focus on the 80 milliseconds target response time (for reasons that are described in Section 14.3.2.1), and we consider the factors that affect the performance when additional servers are added to the existing cloud infrastructure. Upon closer inspection, we are able determine whether clients are covered/served by EC2 or not. The *EC2-uncovered clients* can then also be differentiated between those who may be covered by a edge-server and those who are unreachable for a given response time.

In this experiment, each edge server hosts one game, and there is only one game in the system. Furthermore, we restrict edge-servers to serve a single user as opposed to many users. Figure 14.5 shows that approximately 10% of end users are unable to meet the 80 milliseconds latency target using EC2 or be served by edge servers. These

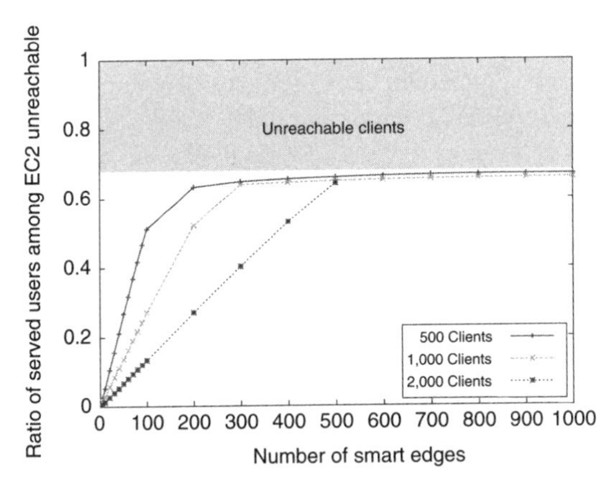

Figure 14.5. Ratio of served end users among the EC2-uncovered end users. Each edge-server can host one game, and there is only one game in the system. One edge-server can serve up to one on-demand gaming end user. The gray area indicates the percentage of end users that cannot be served by both edge-servers and EC2 datacenters.

end users exhibit excessive delay to all edge servers and datacenters, which is likely due to nonnetwork delays that are outside of our system's control. Therefore, the system's performance with respect to the ratio of covered end users is limited by this ceiling.

The results of our measurement study point to a hybrid DC-CDN infrastructure that combines existing DCs with CDN servers. Because CDN servers are in closer proximity to end users, they are able to provide lower latency for end users than distantly located cloud DCs. In addition, a hybrid DC-CDN infrastructure is more attractive than a multi-CDN because DC, which are less costly than CDN, can serve a significant fraction of users. Therefore, DC-CDN is attractive for such demanding interactive, multimedia service.

Yet, there are still many challenges that need to be addressed. One challenge is to determine the selection of edge servers that maximizes user coverage. Unfortunately, this is an instance of the facility location problem which is NP-hard. Furthermore, since edge servers cannot host an infinite number of games, due to physical limitations and cost considerations, another challenge is to strategically place games on edge servers in order to achieve a maximal matching between end users and edge servers. Solutions to these challenges are especially required in case of a growth of the number of concurrent gamers.

14.4 UGC LIVE STREAMING

Over-the-top (OTT) TV channels mimic regular TV channels, but instead of using traditional mass communication medium (e.g., broadcast, satellite and cable) they use the Internet to deliver live video stream to their audience. A key consequence of the development of OTT TV services is that anybody can be a TV provider. Crowdsourced news channel [2] and e-sport channels [3] are examples of the emerging usages that are enabled by TV delivery over an open medium. The user became an important source of content providing to the services continuously massive amount of information. The vast majority of works related to the delivery of live streams, both with P2P (see Ref. [1] for a survey) and CDN (see Refs. [43–45] for recent works).

The behaviors of contributors to video sharing platforms like YouTube has been extensively studied since [46]. For example, a study presented in Ref. [47] estimates that in May 2011 there was a total of roughly 500 millions YouTube videos, a minimum total storage needed for these videos was around 5 petabytes (PBs) and the network capacity to run YouTube ranged from 17 to 46 PBs/day. To the best of our knowledge, there is no similar measurements for UGC *live* streaming. A characterization of professional players broadcasting in twitch.tv (a branch of justin.tv exclusively for *gamecasting*) is presented in Ref. [3]. Another work [48] focusing on gamecasting community is about XFire, a social network for gamers featuring live video sharing. Live video sharing is also explored in Ref. [49], where authors analyzed 28 days of data from two channels associated with a popular Brazilian TV program aired in 2002. A study over a free-to-use P2P live streaming system, namely Zattoo, with provider side traces pointed out that it served over 3 million registered users across eight European countries with peaks of $60,000$ simultaneously users on a single channel [50]. During China 2008 Olympic Games, data were collected from the largest Chinese CDN [51], showing that the live nature of such events results in differences on access patterns compared to video on demand (VoD) and other UGC systems. However, none of these works has analyzed the behavior of contributors nor estimated the size of the delivery networks.

To understand the behavior of UGC live video-streaming services, we performed an extensive study over real traces of a major live streaming service, namely justin.tv.

14.4.1 Analysis of justin.tv UGC Live Streaming System

Justin.tv offers a free platform for publishing user-generated live video content. In the following, we distinguish *uploaders* and *viewers*. The uploaders are registered users that have been captured broadcasting one live video at least once during the months of our study. An uploader is the generator of only one given *channel* (a live video stream), so we will interchangeably use the terms channel and uploader hereafter. A channel can be either *online* at a given time, which means that it can be viewed by viewers, or *offline* when the user in charge is not uploading video on this channel. A channel can alternatively switch from offline to online and vice versa during our analysis. Viewers can *subscribe* to a channel so that they are notified every time the channel switches on.

We use justin.tv REST API with a set of synchronized computers to collect a global view of the justin.tv system every 5 min. We fetch information about the global popularity (total number of viewers in the system), total number of streams, channel's popularity (number of viewers by channel), and channel's metadata every five minutes. From the collected data, we target the months of August and November 2012. Supported by the results of our measurement campaign, we give two messages: *A large and international population* and *Uploaders guarantee a 24/7 TV-like service*.

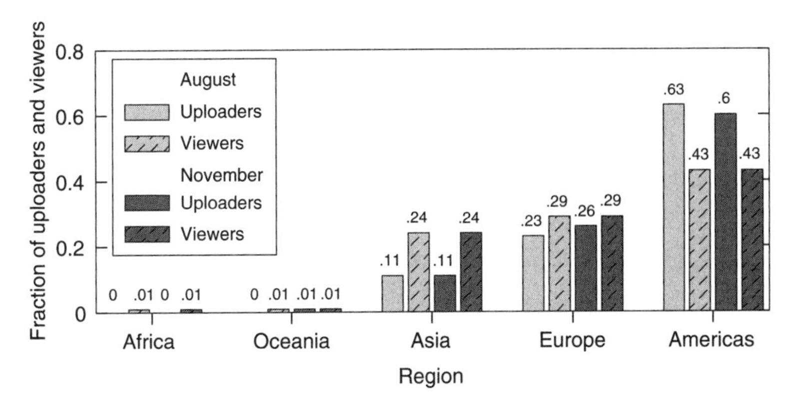

Figure 14.6. Each of five regions fraction of viewers and uploaders.

14.4.1.1 *A Large and International Population.* First we emphasize that justin.tv is (1) an international service, and (2) a service that is fueled by a large population of uploaders. We first use our data traces to get the origin of uploaders, which we associate to five regions (Africa, Americas, Oceania, Europe, and Asia). The origin of the viewers is not provided by justin.tv API, so we collected the estimated viewers information from Google Ad Planner service, which includes geolocalization data. Likewise, the viewers were grouped into five regions.

Our main observation in Figure 14.6 is that the viewers distribution conforms to the distribution of Internet users [52–54]. It is important to note that previous work related to P2P UGC live video systems (e.g., Ref. [55]) does not highlight such well-balanced distribution of viewers. We can also notice that in both months there is an over-representation of uploaders located in Americas. We suspect that uploaders do not pay full attention to their profile settings, the default country being America.

We then want to show how vast is the population of uploaders. We analyze both entire periods to measure the number of *distinct* channels that had been online. In average, there are around 2,000 simultaneous online channels. In August, we find that around 200,000 *distinct* uploaders have started channels during this one month period and almost 240,000 for the same period analyzed in November. This number demonstrates the massiveness of UGC live streaming system in comparison with traditional IPTV systems.

14.4.1.2 *Uploaders Guarantee a 24/7 TV-like Service.* Our second message is that justin.tv is an always-on service, thanks to its contributors. We have to recall that justin.tv differs from other UGC services like VoD in the sense that the service depends on the activity of uploaders *at every time*. There is a critical need of online channels. Fortunately, justin.tv has loyal uploaders, who manage to be more consistently active (here online) than on other typical UGC platforms. It thus guarantees service continuity.

We measure the number of online channels over the whole month, and then we compute the average numbers per hour of a day (respectively per day of a week), thus we

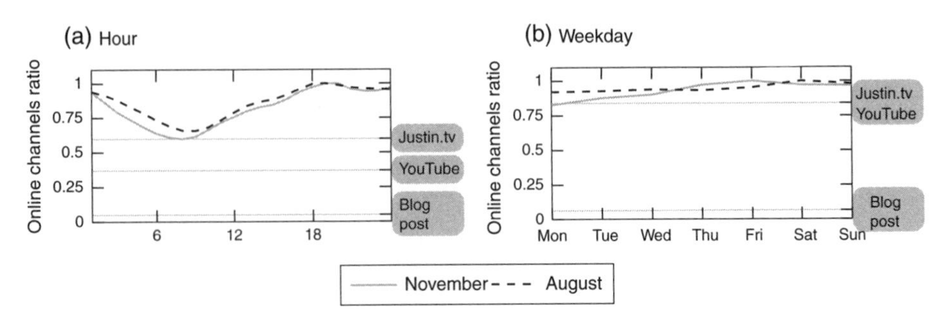

Figure 14.7. Normalized average of diurnal (a) and weekly (b) peak ratio of simultaneous online channels.

measure diurnal (respectively weekly) patterns. We normalize the results so that the peak of the number of online channels is equal to 1. We show our results in Figure 14.7.

To demonstrate the continuity of the live service over any time of the day we explored the diurnal pattern of concurrent online channels. The diurnal pattern has a traditional shape (daylight), but the main important point to note is that this pattern is low in comparison with other platforms. We draw with thin lines the same lowest popularity in a scale to 1 for two other UGC platforms: YouTube (discussed in Ref. [56] and [57]) and blogposts ([58] described it in 2009). It is noteworthy that justin.tv lowest global popularity in a day is more than 0.65 of its peak (noted 0.65:1), which means that there are many online channels all along the day. On YouTube, the number of uploaded videos is significantly less important at some day time than other (nearly 0.37:1). If justin.tv followed the same pattern as YouTube, there would be some day time without enough channels to guarantee a large enough choice of channels. Finally, blogposts have a gigantic diurnal pattern according to Ref. [58] (around 0.05:1).

The same observation holds for weekdays. The difference between its lowest and peak global popularity is not significant on justin.tv for the month of August (0.92:1) and rather interesting for the month of November (0.83:1). In other words, there are online channels all along the week. These results are comparable with YouTube (0.84:1) and outperform blogposts (0.06:1).

14.4.2 Motivations for a Hybrid DC-CDN Delivery

We now discuss a selection of insights to justify the usage of a hybrid DC-CDN delivery model in the case of UGC live streaming systems. First, it is well known that a small number of contributors of UGC systems represents the vast majority of the global popularity of these platforms. Such distribution simplifies the management of CDN infrastructures. The provider is also interested in delegating to the CDN the channels with the highest resolution, which can throttle the limited bandwidth capacity of DCs. Finally, channels that are stable over time are easier to manage in CDN, with less configuration of edge servers. We are interested in measuring such facts for justin.tv.

14.4.2.1 Most of the Traffic Comes from a Tiny Proportion of Uploaders. This is our main observation, and it is important to understand that these

TABLE 14.1. Number of channels for top categories

	Top				
	10	20	30	40	50
Aug. # channels	559	1086	1499	1830	2166
Nov. # channels	458	922	1342	1670	1985

special uploaders are not online simultaneously. They have alternatively been online and offline; but at every time, the subset of online channels out of this tiny subset of uploaders represents most of the traffic.

Every 5 minutes, we collect the k most popular channels. To simplify, we focus here on values of k in $\{10, 20, 50\}$. Please recall that there are around 2,000 simultaneous online channels, so these top channels represent a small fraction of all uploaders. Overall, for each month, we gathered more than 8,500 different lists (one new list every five minutes) of top-k channels.

We show that a small number of distinct channels occurs in these top-channel lists over the whole months. In Table 14.1, we give the number of different channels (# channels) having at least one occurrence in these lists over the whole months. Only 559 uploaders (0.3% of monthly total) have occurred in the top-10 channels in August. It means that 559 uploaders have occupied the over 85,000 "spots" that were available in the month. This result is even stronger in November for which the number of distinct uploaders is only 458 (0.2% of monthly total) although the overall number of distinct channels is larger than in August, as discussed in Section 14.4.1.1.

We then measure the popularity of these channels and calculate the footprint of top channels on the overall traffic of justin.tv by collecting the bitrates given in the API. We can thus extrapolate the total bandwidth in justin.tv system. This information is depicted in Figure 14.8. First, as can be expected, the popularity of top channels decreases fast. The gap between top-10 and top-20 channels is around 10% of the overall traffic, and also small between top-20 and top-50 channels. Second, the peak of global popularity of justin.tv can be exclusively credited to the top-10 channels. We see a direct correlation between peak of overall popularity and peak in top-10 channels. A third remarkable observation is that November peak accounted for 1 Tbps of uplinked data. Such enormous bandwidth makes the case for interfacing justin.tv with CDN. A fourth remark is regarding the usage of a hybrid DC-CDN model on this scenario. For example, with a DC provisioned with 100 Gbps of bandwidth capacity, all the peak traffic would have been sent to CDN and the DC capacity would have being almost fully used almost every other time of the month.

14.4.2.2 The Most Popular Channels are in the Highest Resolutions. Another noteworthy observation is that the ratio of traffic generated by the aforementioned top channels is bigger than for the ratio of viewers. During peaks, almost 98% of traffic comes from the very small subset of top channels that are online, while viewers account at most 70%. The reason for such difference between the ratio of viewers and

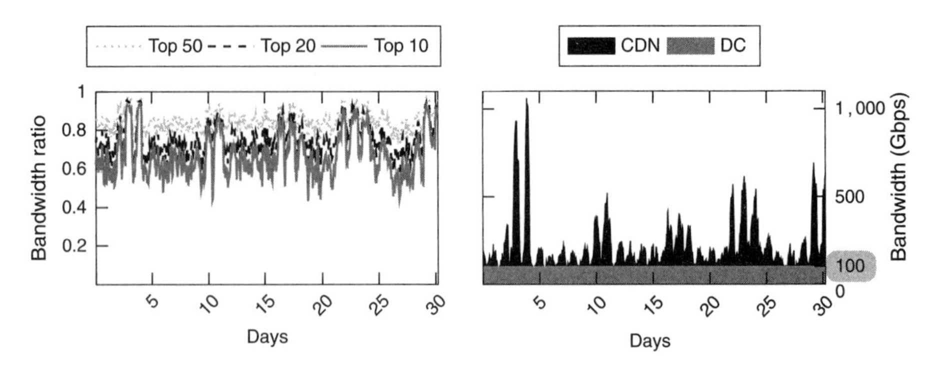

Figure 14.8. Maximum of bandwidth usage ratio by each hour per top 10, 20, and 50 channels, and total bandwidth of November.

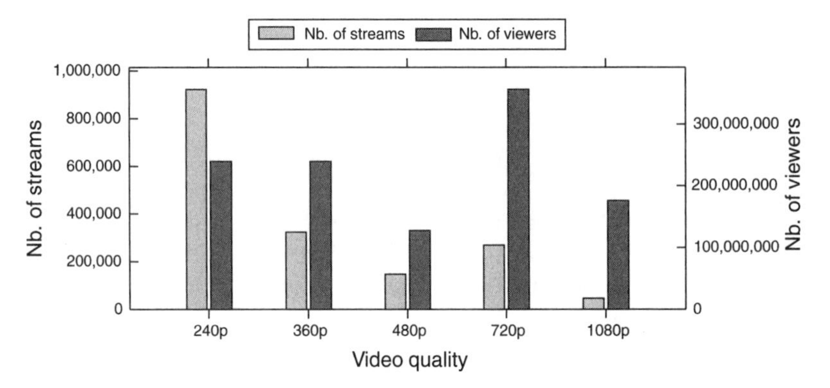

Figure 14.9. Total number of streams and viewers for each video quality in November.

the ratio of traffic for top channels is revealed in Figure 14.9. We associate each range of bitrates with a determined video quality, based on the values of YouTube LiveStream Guide and what we get from the API. As can be noted, videos with better quality are more popular (720p being the resolution for which videos are the most popular) although these resolutions represent a small portion of total streams.

Overall, these observations are significant in the perspective of integrating justin.tv into a hybrid CDN-DC architecture. CDN are efficient to handle a small number of very popular content. Based on our findings, we claim that it is easy to integrate justin.tv into a CDN. Since a relatively small number of uploaders (around one thousand) can (at least) halve the burden on DCs, justin.tv platform should focus on these uploaders and ensure that they get handled by the CDN as soon as they switch on their channels.

14.4.2.3 The Number of Simultaneously Online Popular Channels is Stable. In the scenario where a CDN manages top-k channels, a question is the number of uploaders that are *simultaneously* online at a given time. As previously said, CDN knows how to manage a small number of channels. We measure the number of online channels out of the overall population of top-k channels, every 5 minutes. We present in

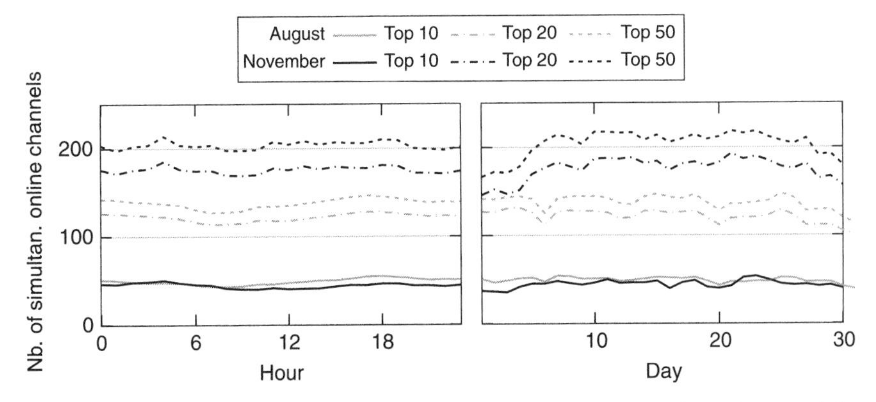

Figure 14.10. Average number of simultaneously online channels per hour and day.

Figure 14.10 the results, where the average number every hour over all days in the month is given in the first graphic of Figure 14.10, while the evolution during the month is given in the second one.

The number of simultaneously online uploaders out of the population of CDN-friendly uploaders is both stable and small. Typically for the set of thousand uploaders that occurred at least once in the top-20 channels, the number of online channels is between 100 and 130 for the month of August, which is a range that a CDN can handle without problem. To conclude, we claim that *justin.tv platform can easily interface with a hybrid DC-CDN model because a small and stable population of uploaders is responsible of the traffic peaks.*

14.5 TIME-SHIFTING VIDEO STREAMING

As stated in Section 14.1, time-shifted TV is a core element of a number of potential *killer apps* of the connected TV. We emphasize below some of the most critical differences between VoD services and time-shifted TV services:

- Time-shifted services allow end users to time-shift a program that is still on air (typically *via* the popular pausing feature of Personal DVR). Studies have shown that most time-shifted requests are for the ongoing TV program [59], so delivery models that do not consider *simultaneous* ingestion and delivery of content do not meet the demand from the end users. Typically, catch-up TV services, where every program is proposed separately after it has been fully broadcast and recorded, do not provide the interactivity expected by most users.
- The length of a TV stream is several orders of magnitude longer than a typical movie in VoD. While a movie can be considered as one unique object, the stream of a time-shifted video is a series of *portions*, which are not uniformly popular. The popularity of video portions in a time-shifted streaming system is complex because it depends on multiple parameters including the popularity of the TV program

associated with a given portion and also the time at which the portion has been broadcasted. Moreover, the popularity of a given portion varies with time ; it usually tends to decrease with time but sometimes events that were unnoticed at the broadcasting time can become popular later due to, for example, social networks.

- The volatility of viewers is more important than in VoD. In Ref. [60], a peak has been identified at the beginning of each program, where many clients start streaming the content, while the spikes of departure occur at the end of the program. More than half of the population quits during the first 10 min of a program in average, and goes to another position in the history [61]. In a same session, a user of time-shifted TV systems (hereafter called *shifter*) is interested in several distinct portions, which can be far from each other in the stream history.

The characteristics of time-shifted streaming services make the delivery especially challenging. In particular, DC-based solutions have some serious weaknesses because current servers do not meet all the requirements. First, conventional disk-based VoD servers cannot massively ingest content, and keep pace with the changing viewing habits of subscribers, because they have not been designed for concurrent read and write operations. Second, client-server delivery systems are not cost-efficient in the case of applications where clients require distinct portions of a stream. They can indeed not use group communication techniques such as multicast protocols. As a matter of facts, current time-shifted services managed by TV broadcasters are restricted to a time delay ranging from 1 to 3 hours, despite only 40% of shifters watch their program less than three hours after the live program [59].

Due to lack of space, we will not enter here into the details of the different proposals for delivering time-shifted streams. The most complete overview of the literature is in [62]. Previous work has highlighted the problems met by time-shifted systems based on a DC infrastructure [63–65]. New server implementations are described in [65]. Cache replication and placement schemes are extensively studied by the authors of [63]. Such a solution corresponds to a hybrid CDN-DC infrastructure. A different option is to opt for DC-based solutions such as Ref. [64]. When several clients share the same optical Internet access, a patching technique is used to handle several concurrent requests so that the server requirement is reduced.

The delivery model that appears to be the most attractive is a hybrid P2P-DC solution. In such architecture, the main motivation is that the most popular video portions at a given time are usually the video portions that just aired a few minutes ago. The idea is thus to cache these fresh downloaded portions in the viewers' computer or home gateway—this is the P2P part—while the older and less popular video portions are stored in the DC using cost-effective storage systems—this is the DC part. Hybrid P2P-DC solutions have been the topic of several papers [66–69].

In Table 14.2, we summarized the results of simulations that we conducted on a set of synthetic traces from [68]. These results indicate the percentage of video portions that are served by either the P2P delivery models or the DC. As can be seen, results can differ a lot among the presented solutions. The DC is used when either the portion is not available in the P2P system (typically because it has not been stored in the user computer) or the P2P system does not have enough capacity to serve the users.

TABLE 14.2. Ratio of video portions from P2P vs. DC

	from P2P	*from DC*	
		missing portion in P2P	*not enough capacity in P2P*
PACUS [68]	75.2%	0.1%	24.7%
Turntable [67]	78.5%	0%	21.5%
P2TSS-Rand [66]	11.2%	23%	65.8%
P2TSS-Live [66]	2.8%	22.8%	74.4%

14.6 OPEN CHALLENGES

The management of interactive multimedia service is still considered a challenging task. The solutions that have been described throughout this chapter fix some of the most prevailing challenges and allow today's services to be used all over the world. But there are still some open challenges, which will require a significant effort from the scientific community in the next years. We would like to highlight three topics, which, in our opinion, will matter in the near future.

- *Economics of networks*: Behind the services that everybody enjoys everyday, there is a complex value chain where multiple actors interact to provide components of the service (to name some of the most important actors, the CDN provider, the ISP, the transit network operator and the content producer). For a given service, each actor should be profitable (its revenues should exceed the cost of the infrastructure it provides for the said service) and aims to maximize their profits. In the case of multimedia services, it is frequent that a decision taken by one actor has implications on the context of another actor. Such an interplay between actors makes the design of services even harder. To study the behavior of rationale actors and the consequences of their actions on the global system, scientists use theoretical models combining game theory and discrete optimization [70]. The recent disputes between major Internet actors (for example Netflix and Comcast[4]) have highlighted the complexity of wide-scale multimedia services and the stress these services impose on the infrastructure. Despite the relative youth of network economics as a scientific domain, we believe that future works related to CDN and content provider should take into account the economical drivers for these actors.
- *Virtualization for Intensive Multimedia Tasks*: Multimedia services have a high demand for specialized resources, for example, Graphics Processing Unit (GPU). The migration from private DCs (with dedicated hardware) to the cloud (with virtual machines, shared resources, and standard hardware) is a long, still ongoing journey. The elasticity of DCs has the potential to convince service providers to migrate their most intensive tasks but some of these tasks are difficult to migrate

[4]http://blog.streamingmedia.com/2014/02/heres-comcast-netflix-deal-structured-numbers.html

because the commoditized hardware cannot accomodate the requirements of specialized software (e.g. a game engine requires GPU), and because these software have been designed to maximize the utilization of hardware although DC management requires smart resource sharing. Among the advanced solutions, the development of virtual DC is expected to offer well-configured computing infrastructure in shared data-center [71]. Virtual DCs are however still in their infancy. More generally, although some vendors claim that some tasks can now be run in the cloud (see the Amazon Elastic Transcoding offer), we believe that scientists dealing with system and network management will find in the next years a lot of open problems related to the hosting of multimedia software services in shared hardware resources.

- *Improvement of Adaptive Streaming*. Dynamic adaptive streaming technologies have been recently adopted by a majority of streaming vendors and service providers. The standardization efforts at MPEG has allowed various key advancements in the technologies, but they also reveal the multiple open problems that still need proper solutions. We emphasize three topics. First the work related to Server and network-assisted DASH Operations (SAND) at MPEG is key for those who call for a better integration of network operators in the adaptive process. Today's solutions are only based on the client side. As it has been shown in various papers (e.g., Ref. [72]), a client-only adaptive systems has serious weaknesses. A better collaboration between every actor in the chain would be beneficial, while the technology must keep its current simplicity, which is part of the reasons for its widespread adoption. The second important topic is the implementation of low-latency live streaming. Some papers (e.g. Ref. [73]) have started studying live adaptive streaming more carefully with the goal of offering the same level of adaptivity as for regular stored video although the stream is generated on the fly. Finally, a third topic which requires extra-attention is the pre-delivery phase in multimedia services. The decision of how to encode the stream that has to be delivered (the number of representations, the bit-rates, the resolutions) is typically critical because the whole delivery infrastructure has to address the consequences of these decisions. Preliminary works have studied some of the problems in a formal way [74], but much more has to be done.

14.7 CONCLUSION

Interactive multimedia services have become a key component of the Internet. In this chapter, we highlighted three of them: cloud gaming, UGC live streaming and time-shifted TV. These services are however extremely challenging to implement, deploy and manage. The delivery infrastructure, which is referred to as the cloud, is far more complex than for typical static websites.

One of the main messages we conveyed in this chapter is that hybrid delivery architecture feature attractive characteristics to address the challenges of interactive multimedia services. Their management is however difficult. Moreover, there is no "one-fits-all" solution. We exhibited in this chapter that, for each service, a different hybrid architecture is the most appropriate.

The management of hybrid architectures is a tremendously promising research area. In particular, recent works have shown that cost savings by an order of magnitude can be achieved by the implementation of smart hybrid architecture instead of a more conventional DC-only or CDN-only infrastructure. A lot of opportunities exist typically in exploring content delivery with optimization approaches, in applying data analysis techniques to large-scale services, and in leveraging new multimedia technologies to improve the QoE of mobile users.

REFERENCES

1. Andrea Passarella. A survey on content-centric technologies for the current internet: Cdn and p2p solutions. *Computer Communications*, 35(1):1–32, 2012.
2. Usama Mir, Houssein Wehbe, Loutfi Nuaymi, Aurelie Moriceau, and Bruno Stevant. The zewall project: Real-time delivering of events via portable devices. In *Proceedings of the 77th IEEE vehicular Technology Conference, VTC Spoing 2013*, June 2–5, Dresden, Germany. IEEE, 2013.
3. Mehdi Kaytoue, Arlei Silva, Loïc Cerf, Wagner Meira Jr., and Chedy Raïssi. Watch me playing, i am a professional: A first study on video game live streaming. In *Proceedings of the 21st world wide web Conference, www 2012*, April 16–20, Lyon, France. ACM, 2012.
4. Nielsen Company. *Three Screen Report Q1*, June 2010.http://www.nielsen.com lusien linsignts/reports/2010/three-screen-report-91-2010. html
5. Michael Jarschel, Daniel Schlosser, Sven Scheuring, and Tobias Hoßfeld. Gaming in the clouds: Qoe and the users' perspective. *Mathematical and Computer Modelling*, 57: 2883–2894, 2013.
6. Marcel Dischinger, Andreas Haeberlen, P. Krishna Gummadi, and Stefan Saroiu. Characterizing residential broadband networks. In *Proceedings of the 7th ACM SIGCOMM Conference on Intenent Measuremer, 2007*, October 22–26, San Diego, CA 2007.
7. Srikanth Sundaresan, Walter de Donato, Nick Feamster, Renata Teixeira, Sam Crawford, and Antonio Pescapè. Broadband internet performance: A view from the gateway. In *Proceedings of the ACM SIGCOMM 2011 Conference on Applications, Technologies, Architectures, and Protocals for Computer Communications*, August 15–19, Toronto, ON, 2011.
8. Lucas DiCioccio, Renata Teixeira, and Catherine Rosenberg. Impact of home networks on end-to-end performance: Controlled experiments. In *Sigcomm Workshop on Home Networks*, New Delhi 2010.
9. Stacey Higginbotham. Smart TVs cause a net neutrality debate in S. Korea. Giga OM, Febuary 2012. http:// gigaom. com/2012/02110/smart-trs-cause-a-net-neturality-debate-in-s-korea/
10. Stephen M. Rumble, Diego Ongaro, Ryan Stutsman, Mendel Rosenblum, and John K. Ousterhout. It's time for low latency. In 13th Workshop opn Hot Topics in Operating Systems, HotOS.XIII May 9–11, Napa, CA, 2011.
11. Luiz André Barroso and Urs Hölzle. The datacenter as a computer: An introduction to the design of warehouse-scale machines. *Synthesis Lectures on Computer Architecture*, 4(1): 1–108, 2009.

12. Adele Lu Jia, Lucia D'Acunto, Michel Meulpolder, and Johan A. Pouwelse. Modeling and analysis of sharing ratio enforcement in private bittorrent communities. In *Proceedings of the IEEE International Conference on Communication, ICC 2011*, June 5–9, Kyoto, Japan, 2011.
13. Jan Seedorf, Sebastian Kiesel, and Martin Stiemerling. Traffic localization for p2p-applications: The alto approach. In *Proceedings P2P 2009, Nineth International Conference on Peer-to-Peer Computing*, September 9–11, Seattle, WA, 2009.
14. Avraham Leff and James T. Rayfield. Alternative edge-server architectures for enterprise javabeans applications. In *Proceedings of the 5th ACM/IFIP/USENIX International Conference on Middleware*, Middleware '04, pages 195–211, New York, 2004. Springer-Verlag, New York.
15. Mikael Desertot, Clement Escoffier, and Didier Donsez. Towards an autonomic approach for edge computing: Research articles. *Concurry and Computation: Practice Experceince*, 19(14):1901–1916, September 2007.
16. Kris Alexander. Fat client game streaming or cloud gaming. Akamai Blog, Aug. 2012. https://blogs.akamai.com/2012/08/part-2-fat-client-game-streaming-or-cloud-gaming.html.
17. Vijay Kumar Adhikari, Sourabh Jain, Yingying Chen, and Zhi-Li Zhang. Vivisecting youtube: An active measurement study. In *INFOCOM, Orlando, FL*. IEEE, New York, 2012.
18. Vijay Kumar Adhikari, Yang Guo, Fang Hao, Matteo Varvello, Volker Hilt, Moritz Steiner, and Zhi-Li Zhang. Unreeling netflix: Understanding and improving multi-cdn movie delivery. In *INFOCOM, Orlando, FL*. IEEE, New York, 2012.
19. Hao Yin, Xuening Liu, Tongyu Zhan, Vyas Sekar, Feng Qiu, Chuang Lin, Hui Zhang, and Bo Li. Design and deployment of a hybrid cdn-p2p system for live video streaming: Experiences with livesky. In *Proceedings of the 17th International Confernece on Multimedia 2005*, October 19–24, Vancouver, BC. ACM, 2009.
20. Todd Hoff. Gone fishin': Justin.tv's live video broadcasting architecture. High Scalability blog, November 2012. http://is.gd/5ocNz2.
21. Pietro Michiardi, Damiano Carra, Francesco Albanese, and Azer Bestavros. Peer-assisted content distribution on a budget. *Computer Networks*, 56(7):2038–2048, 2012.
22. Paarijaat Aditya, Mingchen Zhao, Yin Lin, Andreas Haeberlen, Peter Druschel, Bruce Maggs, and Bill Wishon. Reliable client accounting for p2p-infrastructure hybrids. In *NSDI, San Jose, CA*. USENIX, Berkeley, CA, 2012.
23. Ricardo A. Baeza-Yates, Aristides Gionis, Flavio Junqueira, Vassilis Plachouras, and Luca Telloli. On the feasibility of multi-site web search engines. In *Proceedings of the ACM CIKM*, Hong Kong, Chince 2009.
24. Jimmy Leblet, Zhe Li, Gwendal Simon, and Di Yuan. Optimal network locality in distributed virtualized data-centers. *Computer Communications*, 34(16):1968–1979, 2011.
25. Nikolaos Laoutaris, Michael Sirivianos, Xiaoyuan Yang, and Pablo Rodriguez. Inter-datacenter bulk transfers with netstitcher. In *Proceedings of the ACM SIGCOMM 2011 Conference on Application, Technologies, and Protocols for Computer Communications*, August 15–19, Toronto, ON. ACM, 2011.
26. Michael Armbrust, Armando Fox, Rean Griffith, Anthony D. Joseph, Randy H. Katz, Andy Konwinski, Gunho Lee, David A. Patterson, Ariel Rabkin, Ion Stoica, and Matei Zaharia. A view of cloud computing. *Communications of the ACM*, 53(4):50–58, 2010.
27. Ruben Van den Bossche, Kurt Vanmechelen, and Jan Broeckhove. Cost-optimal scheduling in hybrid iaas clouds for deadline constrained workloads. In *IEEE, International Conference on Cloud Comuting, CLOUD 2010*, July 5–10, Miami, FL, 2010.

28. Michael Smit, Mark Shtern, Bradley Simmons, and Marin Litoiu. Partitioning applications for hybrid and federated clouds. In *Center for Advanced Studies on Collaborative Research CASCON'12*, November 5–7, Toronto, ON. IBM / ACM, 2012.
29. Hui Zhang, Guofei Jiang, Kenji Yoshihira, Haifeng Chen, and Akhilesh Saxena. Intelligent workload factoring for a hybrid cloud computing model. In *2009 IEEE Congress on Services, Part I, SERVICESI2009*, July 6–10, Los Anseles, CA. IEEE, 2009.
30. Michael Jarschel, Daniel Schlosser, Sven Scheuring, and Tobias HoÃŸfeld. An evaluation of qoe in cloud gaming based on subjective tests. In *Proceedings of the Fifth International Conference on Innovative Mobile and Internet Services in Ubiquitous Computing, 1MIS 2011*, June 30–July 2, Seoul, Korea, 2011.
31. Mark Claypool and Kajal T. Claypool. Latency and player actions in online games. *Communications of the ACM*, 49:40–45, 2006.
32. Mark Claypool and Kajal Claypool. Latency can kill: Precision and deadline in online games. In *MMSys*, Phoenix, AZ 2010.
33. Sean K. Barker and Prashant Shenoy. Empirical Evaluation of Latency-sensitive Application Performance in the Cloud. In *MMSys*, Phoenix, AZ, 2010.
34. Sharon Choy, Bernard Wong, Gwendal Simon, and Catherine Rosenberg. The brewing storm in cloud gaming: A measurement study on cloud to end-user latency. In *Proceedings of ACM NetGames*, Venice, Italy 2012.
35. Kuan-Ta Chen, Yu-Chun Chang, Po-Han Tseng, Chun-Ying Huang, and Chin-Laung Lei. Measuring the latency of cloud gaming systems. In *ACM Multimedia*, Scottsdale, AZ, 2011.
36. Sharon Choy, Bernard Wong, Gwendal Simon, and Catherine Rosenberg. A hybrid edge-cloud architecture for reducing on-demand gaming latency. *Multimedia Systems Journal*, 20(2), 503–519 March 2014.
37. Andy Bavier, Mio Bowman, Brent Chun, Daird Culler, Scott Karlin, Steve Muir, harry. Peterson, Jimothy. Roscoe, Jommo. Spalink, and Mike Wawrzoniak. Operating system support for planetary-scale network services. In *1st Symposium on Networks System Design and Implementation (NSDI 2004)*, March 29–31, 2004, San Francisico, CA, 2004.
38. Xav de Matos. Source: Onlive averaged 1800 concurrent users, ceo promised to protect patents against gaikai. http://www.joystiq.com/2012/08/17/source-onlive-ceo-showed-no-remorse-when-announcing-layoffs/.
39. Albert Greenberg, James Hamilton, David A. Maltz, and Parveen Patel. The cost of a cloud: Research problems in data center networks. *SIGCOMM Comput. Commun. Rev.*, 39(1):68–73, December 2008.
40. Gaikai will be fee-free, utilize 300 data centers in the us. http://www.joystiq.com/2010/03/11/gaikai-will-be-fee-free-utilize-300-data-centers-in-the-us/.
41. Gdc09 interview: Onlive founder steve perlman wants you to be skeptical. http://www.joystiq.com/2009/04/01/gdc09-interview-onlive-founder-steve-perlman-wants-you-to-be-sk.
42. Lothar Pantel and Lars C Wolf. On the impact of delay on real-time multiplayer games. In *Proceedings of the 12th International Workshop on Network and Operating Systems Support for Digital Audio and Video*, pages 23–29. May 12–14, Miani Beach, FL ACM, 2002.
43. Micah Adler, Ramesh K. Sitaraman, and Harish Venkataramani. Algorithms for optimizing the bandwidth cost of content delivery. *Computer Networks*, 55(18):4007–4020, 2011.

44. Jiayi Liu, Gwendal Simon, Catherine Rosenberg, and Géraldine Texier. Optimal delivery of rate-adaptive streams in underprovisioned networks. *IEEE Journal on Selected Areas in Communications*, 32: 706–713, 2014.

45. Jiayi Liu and Gwendal Simon. Fast near-optimal algorithm for delivering multiple live video streams in cdn. In *22nd International Conference on Computer Communication and Networks, ICCCN 2013*, July 30–August 2, Nassau, Bahamas. 2013.

46. Meeyoung Cha, Haewoon Kwak, Pablo Rodriguez, Yong-Yeol Ahn, and Sue B. Moon. I tube, you tube, everybody tubes: Analyzing the world's largest user generated content video system. In. ACM, 2007.

47. Jia Zhou, Yanhua Li, Vijay Kumar Adhikari, and Zhi-Li Zhang. Counting youtube videos via random prefix sampling. In *Proceedings of the 11th ACM SIGCOMM Conference on Internet Measurement, IMC'11*, November 2, Berlin Gemany. ACM, 2011.

48. Siqi Shen and Alexandru Iosup. XFire online meta-gaming network: Observation and high-level analysis. In *The 4th International Workshop on Massively Multiuser Virtual Environments at IEEE International Symposium on Audio-Visual Environments and Games (HAVE 2011)* October 15, Hebel, China, 2011.

49. Eveline Veloso, Virgílio A. F. Almeida, Wagner Meira Jr., Azer Bestavros, and Shudong Jin. A hierarchical characterization of a live streaming media workload. *IEEE/ACM Transactions on Networking*, 14(1):133–146, 2006.

50. Hyunseok Chang, Sugih Jamin, and Wenjie Wang. Live streaming performance of the zattoo network. In *Proceedings of the 9th ACM SIGCOMM Conference on Intennet Measuments* November 4–6, 2009, Chicago, IL. ACM, 2009.

51. Hao Yin, Xuening Liu, Feng Qiu, Ning Xia, Chuang Lin, Hui Zhang, Vyas Sekar, and Geyong Min. Inside the bird's nest: Measurements of large-scale live vod from the 2008 olympics. In *Proceedings of the 9th ACM SIGCOMM Conference on Internet Measurement 2009*, November 4–10, Chicago, IL. ACM, 2009.

52. Yuan Ding, Yuan Du, Yingkai Hu, Zhengye Liu, Luqin Wang, Keith W. Ross, and Anindya Ghose. Broadcast yourself: understanding youtube uploaders. In *Proceedings of the 11th ACM SIGCOMM Conference on Intenent Measurement, 1MG'11*, November 2, Berlin, Germmay. ACM, 2011.

53. Sunghwan Ihm and Vivek S. Pai. Towards understanding modern web traffic. In *Proceedings of the 11th ACM SIGCOMM Conference on Internet Measurement, 1MC'11*, November 2, Berlin, Germany, 2011.

54. Zi Hu, John Heidemann, and Yuri Pradkin. Towards geolocation of millions of ip addresses. In *Proceedings of the 12th ACM SIGCOMM Conference on Internet Measurement, IMC'12* November 14–16, Boston, MA, 2012.

55. Xiaojun Hei, Chao Liang, Jian Liang, Yong Liu, and Keith W. Ross. A measurement study of a large-scale p2p iptv system. *IEEE Transactions on Multimedia*, 9(8):1672–1687, 2007.

56. Meeyoung Cha, Haewoon Kwak, Pablo Rodriguez, Yong-Yeol Ahn, and Sue B. Moon. Analyzing the video popularity characteristics of large-scale user generated content systems. *IEEE/ACM Transactions on Networking*, 17(5):1357–1370, 2009.

57. Gloria Chatzopoulou, Cheng Sheng, and Michalis Faloutsos. A first step towards understanding popularity in youtube. In *INFOCOM Workshops, San Diego, CA*. IEEE, 2010.

58. Lei Guo, Enhua Tan, Songqing Chen, Xiaodong Zhang, and Yihong Eric Zhao. Analyzing patterns of user content generation in online social networks. In *Proceedings of the 15th ACM*

SIGKDD International Conference on Knowledge Discovery and Data Mining, June 28–July 1, Paris, France. 2009.

59. Nielsen Company. *How DVRs Are Changing the Television Landscape*, April 2009. http://www.nielsen.comlus/en/insigntslnews/2009/how-drrs-are-changing-the-television-lanscape.html
60. Tim Wauters, Wim Van de Meerssche, Filip De Turck, Bart Dhoedt, Piet Demeester, Tom Van Caenegem, and E. Six. Management of time-shifted IPTV services through transparent proxy deployment. In *Proceedings of the Global Telecommunications Conference, GLOBE Com'ob*, November 7–December 1, Franscisco, CA, pages 1–5, 2006.
61. Xiaojun Hei, Chao Liang, Jian Liang, Yong Liu, and Keith W. Ross. A measurement study of a large-scale P2P IPTV system. *IEEE Transactions on Multimedia*, 9(8):1672–1687, December 2007.
62. Niels Bouten, Steven Latré, Wim Van de Meerssche, Bart De Vleeschauwer, Koen De Schepper, Werner Van Leekwijck, and Filip De Turck. A multicast-enabled delivery framework for qoe assurance of over-the-top services in multimedia access networks. *Journal of Network and Systems Management*, 21: 1–30, 2013.
63. Juchao Zhuo, Jun Li, Gang Wu, and Su Xu. Efficient cache placement scheme for clustered time-shifted TV servers. *IEEE Transactions on Consumer Electronics*, 54(4):1947–1955, November 2008.
64. Wei Xiang, Gang Wu, Qing Ling, and Lei Wang. Piecewise patching for time-shifted TV over HFC networks. *IEEE Transactions on Consumer Electronics*, 53(3):891–897, August 2007.
65. Cheng Huang, Chenjie Zhu, Yi Li, and Dejian Ye. Dedicated disk I/O strategies for IPTV live streaming servers supporting timeshift functions. In *Seventh International Conference on Computer and Infomation Technology (C11 2007)*, October 16–19, 2007, University of Aizu, Fukusnima, Japan, 2007.
66. Sachin Deshpande and Jeonghun Noh. P2tss: Time-shifted and live streaming of video in peer-to-peer systems. In *IEEE International Conference on Multimedia and Expo*, Hannover, Germany, June 2008.
67. Yaning Liu and Gwendal Simon. Distributed delivery system for time-shifted streaming systems. In *Proceedings of the 35th Annual IEEE Confernence on Local Computer Networks, LCN 2010*, October 10–14, Denver, Co, 2010.
68. Yaning Liu and Gwendal Simon. Peer-assisted time-shifted streaming systems: Design and promises. In *Proceedings of IEEE International Conference on Communication, ICC 2011*, June 5–9, Kyoto, Japan, 2011.
69. Fabio Victora Hecht, Thomas Bocek, Richard G Clegg, Raul Landa, David Hausheer, and Burkhard Stiller. Liveshift: Mesh-pull live and time-shifted p2p video streaming. In *Proceeding of IEEE LCN*, Bonn, Germany 2011.
70. Patrick Maillé and Bruno Tuffin. *Telecommunication Network Economics: From Theory to Applications*. Cambridge University Press, Cambridge 2014.
71. Mohamed Faten Zhani, Qi Zhang, Gwendal Simon, and Raouf Boutaba. VDC Planner: Dynamic migration-aware virtual data center embedding for clouds. In *2013 IFIP/IEEE International Symposium on Integrated Network Management (IM 2013)*, May 27–31, Ghent, Belgium, 2013.
72. Rémi Houdaille and Stéphane Gouache. Shaping http adaptive streams for a better user experience. In *Proceedings of the Third ACM SIGMM Conference on Multimedia System, MMSYS 2012*, Febuary 22–24, Chapel Hill, NC, 2012.

73. Cyril Concolato, Nassima Bouzakaria and Jlan. Le Feuvre. Overhead and performance of low latency live streaming using mpeg-dash. In *The 5th International Conference on Information, Intelligence, Systems and Applications, 11SA 2014*, July 7–9, Chania, Crete, 2014.
74. Laura Toni, Ramon Aparicio Pardo, Gwendal Simon, Pascal Frossard, and Alberto Blanc. Optimal set of video representations in adaptive streaming. In *Multimedia Systems Conference 2014, MMSYS'14*, March 19–21, Singapore, 2014.

15

BIG DATA ON CLOUDS (BDOC)

Joseph Betser and Myron Hecht

The Aerospace Corporation, El Segundo, CA, USA

15.1 INTRODUCTION

Big data is the term for a collection of data sets so large and complex that it becomes difficult to process using on-hand database management tools or traditional data processing applications. The challenges include capture, curation, storage, search, sharing, transfer, analysis, and visualization [1]. This chapter focuses on big data on clouds (BDOC). In fact, an excellent overview of the state-of-the-art and research challenges for the management of cloud computing enterprises is presented in Ref. [2]. Indeed, the main thesis of that paper is that heterogeneity and scale are the driving forces of many of the research challenges for the management of cloud computing systems. BDOC further exacerbate both the scale and heterogeneity of the resulting enterprises. It is the thesis of this chapter that hybrid management, involving disciplined and innovative site reliability Engineering (SRE), is the enabling operations paradigm by which to successfully tackle these growing, emerging challenges. By hybrid management we mean a combination of an increasing level of automated, autonomic management, and fully engaged dynamic human SRE organizations. The SREs provide operations oversight, as well as develop increased automation and insight, in order to afford yet greater scale, heterogeneity, overall enterprise capability, and business performance.

Cloud Services, Networking, and Management, First Edition.
Edited by Nelson L. S. da Fonseca and Raouf Boutaba.

The business appetite for big data continues to grow as cloud computing continues to emerge as the dynamic vessel by which to supply the ever-growing demand for ubiquitous online and mobile services. Social networks, technical computing, ever-growing global communities, and heterogeneous enterprises are the key drivers for ever-growing cloud computing systems. The successful management of these challenging global networks and computing resources is important to successful business performance and high quality-of-service delivery across the globe. This chapter articulates some of the success enablers for deploying BDOC, in the context of some historical perspectives and emerging global services. We consider cloud and mobile applications, complex heterogeneous enterprises, and discuss big data availability for several commercial providers. In addition, we offer some legal insights for successful deployment of BDOC. In particular, we highlight the emergence of emerging hybrid BDOC management roles, the development and operations (DevOps), and SRE. Last, we highlight science, technology, engineering, and mathematics (STEM) talent cultivation and engagement, as an enabler to technical succession and future success for global enterprises of BDOC.

15.2 HISTORICAL PERSPECTIVE AND STATE OF THE ART

This section covers the historical perspective and then discusses some existing solutions to some technical challenges presented by BDOC.

Cloud computing evolved over time, as connectivity of computer networks steadily increased in the advent of the Internet. The Internet itself started as an Advanced Research Projects Agency (ARPA, currently known as DARPA) research project which sought to connect computer systems, and thus achieve greater availability in 1969 [3] at University of California, Los Angeles (UCLA). The greatest invention that propelled the Internet from a research, e-mail, and ftp infrastructure to an everyday utility was the Mosaic [4] Web browser, which was created at the University of Illinois in 1993. This enabled a plethora of new applications based on the higher connectivity and easier access via the web browser. In fact, Zhang et al. [2] argue very well that most of the technologies that enable cloud computing are not new. It is the heterogeneity and scale of today's growing enterprises that call for innovative research for successful management of BDOC.

15.2.1 From Application Service Provider to Cloud Computing

One of the initial services that emerged was the use of the Application Service Provider (ASP) business model. This model used the Web browser as the primary user interface, and the service provider would run the application on a server. The initial level of sophistication of these client-server architectures was low to moderate, and some of these applications are reviewed next.

15.2.1.1 E-mail, Search, and E-Commerce. E-mail is a service that existed in research environments from the 1970s, but did not become a common household technology until the 1990s. As simple as e-mail appears today, it took considerable efforts of the Internet Engineering Task Force [5], in order to achieve the necessary communication protocol standardization that enabled various platforms and operating systems to be able to seamlessly interoperate using (Simple Mail Transfer Protocol (SMTP). The Internet governance model of *rough consensus and running code* [6] speaks volumes in terms of achieving interoperability over the Internet, as well as over BDOC. Things have to work and work properly for the end user, or the user will go elsewhere by a click of the mouse.

Search: Once Internet browsers became available, one of the earlier services offered was Internet search. Some of the early companies (AltaVista, etc.) are no longer in business in this very competitive space. It is now dominated by Google, Bing, and Yahoo, which together own over 90% of the search market [7]. Battelle gives an excellent review of the evolution of the search market. Updated information can be found at Battelle's media blog [8].

Once search engines gained considerable capability, e-commerce was born. With the instant ability to identify merchandise items of interest with a click of a mouse, brick and mortar stores became obsolete, and an increasing volume of business moved to the cyber domain. In December 2013, both UPS and FedEx were overwhelmed with the volume of packages being shipped, and experienced significant delays during holiday deliveries!

Overall, these emerging trends, boosted by affordable increasing computational power, as well as by network bandwidth availability, brought about the concept of "The world is flat" [9].

15.2.1.2 Grid Computing, and Open Grid, Global Grid. Grid computing addresses loosely coupled computers, typically owned by different research organizations, which collaborate on various computing tasks. The management of such grids is looser, and grid computing middleware provides the interface for these tasks. Most of the computing performed by such grids is scientific and technical computing.

The Open/Global Grid Forum [10] is the global forum that provides for the international collaboration among the researchers and scientists, in order to provide the interoperable middleware that enables grid computing.

15.2.1.3 Openstack. Openstack [11] is the virtual organization that promotes the open standardization of cloud technologies. Since BDOC requires all components to interoperate efficiently for high-performance computing, it is critically important to maintain open interfaces, so that technologies developed by global collaborators could be well integrated and interoperate smoothly.

15.2.1.4 Apple iCloud, Yahoo, Google, Amazon, and DropBox. These commercial cloud service providers (CSPs) provide commercial cloud services to the global consumer community. This is a fiercely competitor marketplace, and there are other entrants into this space with perhaps less name recognition, but novel capabilities and unique price performance. It is anticipated that considerable consolidation will continue to take place going forward. It is important to note that the cloud providers who also offer content and other associated services are in a stronger position than pure storage providers and/or computing providers. For example, Apple provides access to i-tunes and many apps, and Google provides Gmail, Maps, Docs, News, and dozens of other popular apps.

15.2.2 State of the Art and Available Technical Solutions

This subsection presents some recent technical capabilities that are available to the BDOC enterprise management community. These contributions are described herewith, and some of them are referenced throughout the chapter. It should be noted that one of the challenges of BDOC is the heterogeneous nature of the hardware, software, user demands, and geographically distributed nature of both the cloud components and user community. In fact, a very good overview of cloud computing is presented in Ref. [2]. Some important areas identified for promising research include: Automated service provisioning, virtual machine (VM) migration, server consolidation, energy management, traffic management and analysis, software frameworks, storage technologies and data management, and novel cloud architectures. The discussion is based on the classical cloud architecture of physical layer, infrastructure-as-a-service (IaaS) layer, platform-as-a-service (PaaS) layer, and finally the software-as-a-service (SaaS) layer that runs the APP that the end user interacts with.

In this chapter, we chose to focus on service availability, data security, business considerations, SRE, and STEM talent considerations. For completeness, we mention here some recent research.

15.2.2.1 Performance Enhancement—Rhea: Automatic Filtering for Unstructured Cloud Storage. Microsoft Research developed performance enhancements in order to expedite filtering of BDOC storage [12]. This technique helps co-locate data and processing whenever possible, thus enhancing performance. They have shown that this technology can expedite searches and reduce cost by 2x–13x.

15.2.2.2 Dynamic Service Placement in Geographically Distributed Clouds. This chapter provides performance enhancement by developing dynamic algorithms using game theory and control techniques in order to enhance performance [13]. They clearly demonstrate that such global optimizations work far better than local optimizations of the subsystems.

15.2.2.3 MemC3: Compact and Concurrent MemCache with Dumber Caching and Smarter Hashing. This chapter authored by Carnegie Mellon and Intel develops caching and Cuckoo hashing schemes that enhance performance for read-mostly workloads [14]. This is an example of a specific strategy to enhance performance under a specific load pattern.

Additional references that are specific to the areas discussed in the forthcoming sections are embedded within those sections.

15.3 CLOUDS—SUPPLY AND DEMAND OF BIG DATA

The explosive growth in the prevalence of CSPs is driven by the plethora of online services, as well as by the growing communities that consume these services. This section reviews some of these trend-setting phenomena.

15.3.1 Social Networks

Social networks started their explosive commercial growth in the early 2000s with companies such as Facebook, Twitter, Google+, Tmblr, and others. These social networks grew in break neck speed, and Facebook in 2014 is offering service to over a billion users worldwide. This kind of scale and heterogeneity are unprecedented, and in fact connect some 18% of the world population on many types of servers and edge devices. Other social networks are growing rapidly, and the infrastructure needed to support them is indeed BDOC based.

15.3.2 Communities

Online communities exist in many areas. Some of the communities are social network based, others are based on professional activities and interests, and still others are based on hobbies, travel, and so on. Online communication is continually taking over the papyrus based communication since 1993. The most up-to-date professional publications are online publications. The same is true for many other types of information and expertise for many communities of interest.

15.3.3 New Business Models

The online communities that continue to expand present a business audience to many innovative companies. This new audience is used for advertisement and marketing campaigns. The business models for these e-business campaigns is quite novel, and is rapidly taking market share from traditional advertisement media such as TV, radio, and newspapers. The TV advertising market is $70B versus $50B of the online advertising Market [15]. Hence, we are quickly approaching the tipping point where online advertisement will take over TV advertisement. This trend is similar to other e-commerce trends, in that the internet business engagement is quickly overtaking the brick-and-mortar traditional commerce. This will be discussed in further detail in the next section.

Additional details can also be found in *Architecting the Enterprise via Big Data Analytics* [16].

15.4 EMERGING BUSINESS APPLICATIONS

Cloud computing and the Internet have completely revolutionized the business world. The instant access to people, information, computing, and network resources indeed make our world "flat." BDOC computing is the enabling resource that makes all this possible. This section will examine a number of dimensions of these emerging business models of BDOC and the successful management of these global resources.

15.4.1 Growing Global Enterprises

When one examines the global resources of some of the CSPs, it becomes clear that these enterprises exhibit an unprecedented scale, size, heterogeneity, and scope of operations.

Google has cloud data center facilities in places such as Finland, Oklahoma, Oregon, and South Carolina, to name only a few of the cloud hosting locations. Since these are energy consuming behemoth, it makes sense to place them near energy sources, such as hydro-electric and geo-thermal locations. On the other hand, most of the talent managing this vast cloud is located where talent is concentrated, that is, near universities and major metropolitan centers, where Google has technical engineering offices. This will be discussed further within the "Site Reliability Engineers" section.

15.4.2 Technical Computing

Technical computing is not as big as business computing, but did spearhead the development of the BDOC technologies that now enable vast business enterprises. Technical computing is mostly focused on scientific and technical tasks of high computational complexity. Examples include high energy physics, oil field exploration and simulation, aerodynamic simulation, jet engine simulation, traffic simulation for transportation system, discrete event simulation for networks and communication switches, electromagnetic field simulation, finite element structures analyses, finite difference fluid flow simulations, and so on. Overall, even though the sophistication of these technical disciplines is high, the scope of these activities is relatively small. They are done mostly by specific organizations using super computers or clusters, and the level of cloud computing utilization is not high. To their credit, many of these technical and research activities are crucial for the invention and development of novel technologies, including BDOC computing. Since the scope of business activities on BDOCS computing is considerably larger, we focus our discussion on business applications.

15.4.3 Online Advertising

As indicated earlier, online advertising is the foundational business model and driving force of many of the innovative companies that experience very high growth. Companies such as Google, Facebook, Twitter, Tumbler, and others generate most of their revenue stream from targeted advertisement, which in turn generates sales for the advertisers and customers of the BDOC companies. The loss of any production application for any of these companies result in immediate and substantial revenue loss, as well as service interruption for the customers and users, which reduces satisfaction, and in extreme cases can cause users and customers to shift their interest and investment of both time and money to competitive online services.

15.4.3.1 E-Commerce. Some companies are dedicated to online commerce or e-commerce. Examples that come to mind are Amazon, e-Bay, Google, and others. In addition to these companies, which are exclusively online companies, many brick and mortar companies establish successful online presence. Such companies include Walmart, Fry's, Best Buy, and others. Related infrastructure companies include shipping companies like FedEx and UPS, and other companies involved in supply chain management. The better the integration of the supply chain management companies and the e-commerce companies, the better the service, delivery speed, and ultimate customer experience. Some fulfillment centers of the e-commerce companies are collocated with shipping hubs of the shipping companies in order to expedite service. In addition the BDOC enterprise systems of these collaborators enjoy considerable interoperability, such that customers can track packages from the e-commerce companies, and the companies are better able to predict in at the time of sale the delivery times of their shippers.

15.4.4 Mobile Services

Mobile is big and growing bigger fast. In addition to the convenience of online access availability 24 hours a day, 7 day a week (24/7), mobile access can readily provide geo-location of the mobile device, hence the location of the end user. This information is extremely useful to the BDOC providers, as they can fine tune advertisement placement for optimal sales and service capture. This dynamic capability makes the BDOC ASP more nimble and at the same time more complex to design, develop, and operate. On the positive side, the SRE team supporting this BDOC ASP, can use mobile devices to support the smooth operation of the applications (APPs).

15.4.5 Site Reliability Engineers

SREs are the professionals who work together with the development team in making sure that the BDOC app is up and running at all times. It is a novel BDOC APP management approach to have such strong collaboration among the development teams and the SRE teams. Google is one of the pioneers to take this new approach [17], but other application service providers soon followed. This is driven by the notion that the huge scale of BDOC ASP requires very high reliability.

It is critically important to develop software that will automate as much as possible of the site reliability engineering capability. In a sense, the successful SRE strives to "automate the human out of manual tasks." Achieving this enables the humans to focus on the creative aspects of system reliability engineering, by building intelligent tools that will fix the system automatically as much as possible, or issue an alert/trouble ticket to a human SRE. In those exceptional cases, the SRE works the issues. If necessary, the SRE engages the development teams. In all cases, a post mortem is written, in order to trace all outages to root causes. Ultimately, it is the role of the SRE and the developers to fix the root cause.

It is important to notice that the SRE operates at a high abstraction and semantic level of the actual app or service. Unlike the traditional network operations center (NOC) that deals with links, packets, and nodes, the SRE is focused at the level of the BDOC APP. Attention might need to address lower level layers such as IaaS, PaaS, and SaaS, but the end user or the customers do not care about anything else, as long as their transparent service or APPs are up. Hence, that is the focus of the SRE team. The SRE teams specialize in specific APPs, and continually strive to improve their reliability and quality of service. As the scale and heterogeneity of the BDOC enterprise grow, more and more of the enterprise management services are automated. This allows the enterprise to continually grow is scope, heterogeneity, and capability. It is the role of the SRE team and the DevOps team and the development team to work together on improving overall performance for the APPs that they are responsible for. Overall, the concept of "Site" for the SRE does not mean a physical location or any of the layers above it. It means the actually web enabled service that supports the APP, whether mobile or wired. It is all about providing quality service to the customer, and a growing revenue stream to the service provider.

With that in mind, the next section will examine BDOC ASP service availability, and the following section will examine legal aspects associated with BDOC ASP apps and services. We will then return to the role of the SRE, and offer strategies to grow the SRE team availability and capabilities.

15.5 CLOUD AND SERVICE AVAILABILITY

Public cloud big data platform offerings [18] provide compelling pricing, outsourcing of support resources, and no capital budgeting. Thus, they are likely to be the dominant platform for big data computing. In this construct, the availability of a big data cloud resident application is dependent on (i) uptime under normal circumstances, which we call operational availability, and (ii) disaster tolerance. The following subsections discuss each of these topics.

15.5.1 Operational Availability

The operational availability of the big data cloud application is dependent on

a. the likelihood that computing resources will be available upon demand,
b. the availability of communication networks for the transfer of data and results to and from the cloud application,
c. the probability of the platform and infrastructure resources of the CSP being operational throughout the data analysis operation,
d. The probability of successful operation of the big data application itself.

Operational availability requires that all of these conditions be met. This can be represented mathematically as follows:

$$A_{\mathrm{op}} = A_a A_b A_c A_d \tag{15.1}$$

where A_{op} is the operational availability and the terms on the right-hand side correspond to the four points listed earlier. For the purposes of quantitative prediction, availabilities A_{a}, A_{b}, and A_{c} are determined by the service-level agreements (SLAs) of the platform and Internet service providers. However, a method for computing A_{a} based on stochastic Petri nets (SPNs) was documented by Khazaei et al. [19]; a model for computing A_{c} was described by Longo et al. [20] Both models were developed from the perspective of the service provider rather than the data owner.

One availability issue of big data implementations on cloud computing arises from network bottlenecks for which several solutions such as Camdoops [20] and FlowComb [21] have been proposed. A second can arise from high workloads imposed by a large number of users for which one reported effective solution is a large-scale implementations of memcached at Facebook [22]. Another arises from the actual size of the store data which, by virtue of its volume, adds to the likelihood of failure. An example of an approach to address this problem is Scalus for HBase [23]. Other failure causes in cloud computing platforms are no different than for other computing systems: hardware failures, software programming errors, data errors, network errors, system power failures, application protocol errors, procedural errors, and redundancy management. Architectures and system management practices for maintaining uptimes are well known and documented elsewhere [24].

Performing large-scale computation is difficult. To work with this volume of data requires distributing parts of the problem to multiple machines to handle in parallel using approaches such as MapReduce. A MapReduce program consists of four functions: map, reduce, combiner, and partition. The input data are split into chunks and, assuming with approximately single chunks stored per server. Usually, a chunk is no larger than 64 Mbytes is used to increase parallelism and improve performance if tasks need be rerun [25]. As the number of number of machines used in cooperation with one another increases, the probability of failures rises. Big data platforms will handle such failures in various mechanisms. The following description for Hadoop is illustrative [1].

The failure detection and recovery scheme of Hadoop is based on three entities: tasks, the tasktracker (which monitors tasks), and the jobtracker (which monitors jobs). When the jobtracker is notified of a task attempt that has failed (by the tasktracker's heartbeat call or a runtime exception), it will reschedule execution of the task. The jobtracker will try to avoid rescheduling the task on a tasktracker where it has previously failed. Furthermore, if a task fails four times (or more), it will not be retried further. This value is configurable: the maximum number of attempts to run a task is controlled by the mapred.map.max.attempts property for map tasks and mapred.reduce.max.attempts for reduce tasks. By default, if any task fails four times (or whatever the maximum number of attempts is configured to), the whole job fails.

Child tasks failures are detected either through the absence of heartbeats or runtime exceptions. If a child task throws a runtime exception (either due to user code in the map or a reduce task exception), the child JVM reports the error back to its parent tasktracker, before it exits. The error ultimately makes it into the user logs. The tasktracker marks the task attempt as *failed*, freeing up a slot to run another task. In Streaming tasks, if the Streaming process exits with a nonzero exit code, it is marked as failed. Another failure mode is the sudden exit of the child JVM—perhaps there is a JVM bug that causes the JVM to exit for a particular set of circumstances exposed by the MapReduce user code. In this case, the tasktracker notices that the process has exited and marks the attempt as failed.

Hanging tasks failures are detected by means of a failure of a progress update for a while and proceed to mark the task as failed. If a tasktracker has not received updates after an expiration period, the child JVM process is killed. The timeout period is normally 10 minutes and can be configured on a per-job basis (or a cluster basis) by setting the mapred.task.timeout property to a value in milliseconds. Setting the timeout to a value of zero disables the timeout. This measure should be avoided because the hanging slot will not be freed; and over time, there may be cluster slowdown as a result.

The maximum percentage of tasks that are allowed to fail without triggering job failure can be set for the job. Map tasks and reduce tasks are controlled independently, using the mapred.max.map.failures.percent and mapred.max.reduce.failures.percent properties.

A task attempt may also be killed, because it is a speculative duplicate or because the tasktracker it was running on failed, and the jobtracker marked all the task attempts running on it as killed. Killed task attempts do not count against the number of attempts to run the task (as set by mapred.map.max.attempts and mapred.reduce.max.attempts), since it wasn't the tasks fault that an attempt was killed. Users may also kill jobs or fail task attempts using the Web UI or the command line.

If a tasktracker fails by crashing, or running very slowly, it will stop sending heartbeats to the jobtracker (or send them very infrequently). The jobtracker detect a tasktracker failure through a timeout (default is 10 minutes, configured via the mapred.tasktracker.expiry.interval property, in milliseconds) and remove it from its pool of tasktrackers to schedule tasks on. The jobtracker arranges for map tasks that were run and completed successfully on that tasktracker to be rerun if they belong to incomplete jobs, since their intermediate output residing on the failed tasktracker's local file system may not be accessible to the reduce task. Any tasks in progress are also rescheduled. A tasktracker can also be blacklisted by the jobtracker, even if the tasktracker has not failed. A tasktracker is blacklisted if the number of tasks that have failed on it is significantly higher than the average task failure rate on the cluster. Blacklisted tasktrackers can be restarted to remove them from the jobtrackers blacklist. Failure of the jobtracker is the most serious failure mode. Currently, Hadoop has no mechanism for dealing with failure of the jobtracker—it is a single point of failure—so in this case the job fails. However, this failure mode has a low chance of occurring, since the chance of a particular machine failing is low.

15.5.2 Disaster Tolerance

Disaster tolerance, also referred to as business continuity, addresses measures to resume operations after damage from "force majeure" events such as fire, flood, atmospheric electrical discharge, solar induced geomagnetic storm, wind, earthquake, tsunami, explosion, nuclear accident, volcanic activity, biological hazard, civil unrest, mudslide, and tectonic activity. These events are generally out of scope of the availability considerations described earlier—replication or restart will not be effective if the physical building housing the cloud data center is flooded.

A necessary condition for disaster tolerance is a partial or complete replication of the data and system resources in an alternate geographical location that is sufficiently distant that it is unlikely to be affected by the event which damaged or destroyed the primary location. Disaster tolerance requires planning. Such planning includes procedures for establishing organizational contacts for the purposes of decision-making, activating the remote site (if it is not already activated), transferring the most recent data (if possible and if not already at the remote site), and changing IP addresses at the appropriate routers.

The value of business continuity depends on the impact of the loss of the analysis function to the enterprise. For example, disruption of continuously running big data operations which are business critical (e.g., fraud detection, information system log monitoring, click stream monitoring, or weather prediction) can have a high organizational impact and justify the expenditure of considerable expenditures.

The key performance metrics in business continuity are the time needed to recover and resume and the amount of tolerable degradation in service once operations are resumed at the alternate location—the higher the value of either metric, the greater the cost. The business case for business continuity measures is that the cost of these measures (both initial nonrecurring and recurring) are less than the expected value of the damage or impact of the loss of the operations. This condition can be summarized by the following equation:

$$\sum_j [C_j(t_r) + \text{NPV}[R_j(t_r)]] \leq \sum_i p_i \text{NPV}[D_i(t_r)] \quad (15.2)$$

where C_j is the capital (nonrecurring) cost of the jth continuity measure, t_r is the resumption time associated with that capability, NPV is the net present value function; R_j is the recurring cost of the jth continuity measure over the time period under consideration, p_i is the probability of the ith disaster of business continuity loss event (cumulative for the entire time period under consideration), and D_i is economic value of the damage or impact to the enterprise associated with the ith disaster of business continuity loss event—which is a function of t_r, the resumption time

The assumptions of this equation are as follows:

1. Transition from the primary to alternate data center occurs with 100% success.
2. Downtime associated with this transition has insignificant cost (or in the alternative, that it occurs in 0 time).

A more complete model that relaxes these assumptions has been created [26].

The left-hand side of this equation (recurring and nonrecurring costs of business continuity measures) includes not only the costs of the primary resources necessary for resumption but also dependencies such as processes, applications, business partners, and third party service providers. In many cases, the left-hand side might simply be the cost of creating additional replicas in other geographically diverse data centers and establishing a periodic data update procedure.

The right-hand side (expected value of the loss) includes both the probability of the event and the economic impact of the loss. The probability of disruptive events is dependent on the geographic location of the CSP and the physical measures it undertakes to protect the facility. These probabilities can be reduced by avoiding locations subject to high probability environmental risks, implementing strong security measures, and housing centers in structures most able to withstand flood, earthquake, winds, and other environmental forces. The economic value of the impacts resulting from planned or unplanned disruptions depend on the duration of the disruption and may also vary over time (e.g., credit card fraud detection during peak shopping seasons). If the alternative site offers less than a full-service capability, the value of the loss of functionality in degraded must needs to also be considered.

Next we discuss a number of security issues that affect BDOC.

15.6 BDOC SECURITY ISSUES

The importance of data security is related to the consequences of loss of integrity, availability, or confidentiality of the data. For example, Hadoop provides no security model, nor safeguards against maliciously inserted data; it cannot detect a man-in-the-middle attack between nodes [27]. If the data used in or result from the big data application are sensitive, cloud computing should be approached carefully with due consideration to that sensitivity. The cloud used in the big data deployment might be entirely under the control of the customer, utilize the platform of the cloud provider, or use a cloud implementation of a service entirely. Thus, any of the three NIST models of cloud services: IaaS, PaaS, or SaaS) might apply. The implementation might reside on an organizations own cloud (internal cloud), a cloud provided by a third-party provider (external cloud), or a combined cloud (hybrid cloud).

If the cloud is entirely under organizational control (a private cloud), security concerns, assurance processes, and practices are defined by general IT security guidelines and standards [28, 29] as well as domain specific standards [30–32]. However, public cloud big data platform offerings [18] provide compelling pricing, outsourcing of support resources, and no capital budgeting. Thus, they are likely to be the dominant platform for big data computing. Thus, we will assume for the remainder of this section that there are two separate organizational entities: the customer, also known as the tenant, that owns the big data and associated analytical applications and VM templates, and the CSP, which provides the hardware and software platform upon which the customer's applications run. A public Internet network cloud is used to move data to the CSP and return results to the big data owner.

The "outsourcing" of the computing platform to a multitenant cloud provider from the resource owned and controlled by a data owner represents a significant paradigm shift from the conventional norms of an organizational data center to an infrastructure without an organizationally controlled security perimeter thereby more open to exploitation by potential adversaries.

15.6.1 Threats and Vulnerabilities

The security challenges of big data on a multitenant CSP are formidable. General classes of vulnerabilities in the cloud computing platforms used in big data processing include the following [33, 34].

- *Session riding and hijacking*: Web application technologies must overcome the problem that, by design, the HTTP protocol is a stateless protocol, whereas Web applications require some notion of session state. Many techniques implement session handling and are vulnerable to session riding and session hijacking [29].

- *Erosion of encryption algorithms*: Encryption is currently relied upon as the primary defense against data breaches [35]. However, technical advances as well as faster processors are rendering an increasing number of cryptographic mechanisms less secure as novel methods of breaking them are discovered. In addition, flaws exist in cryptographic algorithm implementations, which can turn strong encryption into weak encryption (or sometimes no encryption at all). For example, cryptographic vulnerabilities might exist if the abstraction layer between the hardware and OS kernel has flawed mechanisms for tapping that entropy source for random number generation, or having several VM environment son the same host might exhaust the available entropy, leading to weak random number generation [29].
- *Limited system monitoring*: CSPs offer limited system monitoring for the purposes of performance and availability monitoring, but do not provide the complete traffic n network monitoring, logging, and intrusion detection that is often used in internal information system installations [27].
- *Configuration management and control*: One of the most common vulnerabilities in IT systems is incomplete change and configuration management and resultant outdated and incomplete system documentation and organizational policies for configuration management and control (including system documentation) may be in place for the internal IT system, but they may be difficult or impossible to enforce on the external CSP.
- *Inability to sanitize storage media*: Policies for disk reformatting, degaussing, or even destruction that might be in place to prevent malware propagation or to mitigate data spillage cannot be readily applied or enforced in public clouds, where the hardware is under the control of the service provider [27, 29].

Cloud-based big data installations have all the weaknesses of standard IT installations. In addition, there are unique weaknesses including the following:

- *Breaching tenant boundaries*: an attacker might successfully escape from the boundaries on which public clouds rely to separate tenants including VMs [36], big data services, data base management systems, and communication infrastructures [37]. In 1 year, VMware had released 14 security advisories [38]. IBM found that that more than 50% of the 80 VM vulnerabilities it found in 2010 could compromise the administrative VM or lead to hypervisor escapes [39].
- *Vulnerability to disaffected insiders*: Disaffected insiders are not unique to cloud computing; they are a threat to any organization. However, the damage they can cause is can do is—even if they are unable to defeat the account control and access privileges of the infrastructure itself. For example, In February 2011, a terminated IT administrator at a pharmaceutical company used a service account to create an unauthorized installation of VMware vSphere to delete 88 virtual servers [40]. While this incident involved a private cloud, it could also affect a public cloud

and could result in the loss of massive amounts of data or the deletion of machine images that contain specific configuration, encryption key, or other information that could be difficult to replace.

- *User authentication defects*: Many widely used authentication mechanisms are weak. For example, usernames and passwords can be compromised by insecure user behavior (weak passwords, reused passwords) or the inherent limitations of one-factor authentication—even encryption is used for the remote login process. Use of multifactor authentication and role-based access by means of LDAP or Active Directory can be complicated by the need to maintain multiple account files in different servers because the organizational private directory servers cannot be integrated into the public cloud infrastructure.
- *Configuration stability of VM environments*: Cloud elasticity and metering as well as live migration help organizations harness the power of virtualization and make the processing environment extremely dynamic.
- *Propagation of flawed or vulnerable VM templates*: Vulnerable VM template images cause OS or application vulnerabilities to spread over many systems. An attacker might be able to analyze configuration, patch level, and code in detail using administrative rights by renting a virtual server as a service customer, and thereby gaining knowledge helpful in attacking other customers images. Other attacks can use side channels from a co-resident VM [18, 41]. The use of others specialized big data implementations because of purported superior properties might be taken from an untrustworthy source and have been manipulated so as to provide back-door access for an attacker. Data leakage by VM replication is a vulnerability that's also rooted in the use of cloning for providing on-demand service. Cloning leads to data leakage problems regarding machine secrets: certain elements of an OS—such as host keys and cryptographic salt values—are meant to be private to a single host. Cloning can violate this privacy assumption [29].
- *Presence of large amounts of unencrypted data*: Big data that needs to be decrypted for processing by a framework such as Hadoop is exposed as the analysis is being done, and then the results are transferred back in to some traditional data warehouse or business intelligence framework. A capable attacker would probably not move or destroy thousands of terabytes of data to avoid detection. However, such data would be attractive to an attacker might be looking for patterns that match a credit card number or a Social Security number regardless of size [42].

15.6.2 Mitigation Approaches

Mitigating the risks and addressing the threats defined above involves various approaches and tasks including the following:

- security planning
- ensuring network security and interoperability

- addressing the unique security strengths and weaknesses of VM infrastructures in cloud computing
- ensuring tenant separation
- identity and access management

These measures are discussed in the following subsections.

15.6.2.1 Security Planning and Risk Assessment for Big Data Processing on Cloud Computing Platforms. Planning is necessary to address these and other threats and to maximize the security of the computing environment. Risk assessment is necessary to weigh the cost of implementing this security versus the sensitivity of the data. Factors influencing planning include organizational policies (including commitments to conformance to specific security standards), contractual commitments on confidentiality and non-disclosure, and legal requirements with respect to privacy and security. These should be documented in an Information Security Management Plan (or an equivalent document). Among the aspects of the plan that affect use of big data platforms on clouds are the following [43]:

- Risk management
- Security policy
- Organization of information security and incident response
- Information asset management
- Communications and operations management
- Access control
- Information systems acquisition, development, and maintenance.

The products of such planning would include specific requirements and conformance criteria, contractual requirements on the service providers, design and configuration measures to be undertaken by the cloud users, and processes and procedures.

15.6.2.2 Ensuring Networking Communications Security and Interoperability. The CSP and the data owner share responsibility for security of big data while it is being transferred from the data owner to the cloud. The CSP also has additional responsibilities to monitor and safeguard its network perimeter and to prevent the introduction of rogue devices into its facility. The following are relevant practices and procedures:

1. Standardized network protocols: The CSP should provide secure (e.g., nonclear text and authenticated) standardized network protocols for the import and export of data and to manage the service. Documentation for the data owners describing the protocols should be sufficient to enable the data owner organization to be able to create and configure network links of adequate circuitry.

2. Network and system monitoring: The CSP network environments and virtual instances should be designed and configured to control (and restrict if necessary) network traffic, reviewed at planned intervals, supported by documented business justification for use of all services, protocols, and ports allowed, including rationale or compensating controls implemented for those protocols considered to be insecure.
3. Response to attacks: The cloud service provide should have the capability to detect attacks (e.g., deep packet analysis, anomalous ingress or egress traffic patterns) and defend the perimeter (e.g., traffic throttling, and packet black-holing) for detection and timely response to network-based attacks MAC (e.g., spoofing and ARP poisoning, and distributed denial-of-service or DDoS attacks)
4. Documentation: Network architecture diagrams must clearly identify high-risk environments and data flows that may have legal, statutory, and regulatory compliance impacts.
5. Configuration of boundary devices: Policies and procedures shall be established, and supporting business processes and technical measures implemented, to protect environments, including the following:

   - Perimeter firewalls implemented and configured to restrict unauthorized traffic,
   - Security settings enabled with strong encryption for authentication and transmission, replacing vendor default settings (e.g., encryption keys, passwords, and SNMP community strings),
   - User access to network devices restricted to authorized personnel,
   - The capability to detect the presence of unauthorized (rogue) network devices for a timely disconnect from the network.

15.6.2.3 Addressing VM Security. The VM infrastructure within a cloud installation consists of are executable software and as such, provide an additional attack surface. There are additional infrastructure and management layers to protect as well as the hypervisor itself. Virtual systems aren't unique and are just as vulnerable as any other system running code. If it runs code, someone can compromise it. The NIST's Guide to Security for Full Virtualization Technologies provides vendor-agnostic guidance on securing virtual environments [44]. The CSP is responsible for configuration, monitoring, and control of the virtual infrastructure. However, depending on the sensitivity of the data, the responsibility for ensuring that the safeguards are appropriate to the risk is the responsibility of the data owner. The following are specific requirements generated by the Cloud Service Alliance [34].

1. The CSP shall inform data owner (tenant) of policies, procedures, supporting business processes and technical measures implemented, for timely detection of vulnerabilities within organizationally-owned or managed (physical

and virtual) applications and infrastructure network and system components, applying a risk-based model for prioritizing remediation through change-controlled, vender-supplied patches, configuration changes, or secure software development for the organization's own software.
2. The provider should use well-known virtualization platforms and standard virtualization formats (e.g., OVF) to help ensure interoperability. Customized changes made to any hypervisor should be available for data owner (tenant) review.
3. The provider should ensure the integrity of all virtual machine images. Any changes made to virtual machine images must be logged and an alert rose regardless of their running state (e.g., dormant, off, or running). The results of a change or move of an image and the subsequent validation of the image's integrity should be reported immediately to data owners.
4. Each operating system should be hardened to provide only necessary ports, protocols, and services to meet business needs.
5. Virtual machines should include antivirus, file integrity monitoring, and logging as part of their baseline operating build standard or template.

15.6.2.4 Ensuring Tenant Separation. Segregation of tenants in data centers owned or managed by CSP is a concern affecting (physical and virtual) applications, and infrastructure system and network components. These should be designed, developed deployed and configured such that provider and data owner (tenant) user access is appropriately segmented from other tenant users, based on the following considerations:

- Established policies and procedures
- Isolation of business critical assets and/or sensitive user data and sessions that mandate stronger internal controls and high levels of assurance
- Compliance with legal, statutory and regulatory compliance obligations

15.6.2.5 Identity and Access Management. Identity and access management is a joint concern between the CSP and the big data user. The data owner is responsible for the identity and access management of its cloud users (or attackers who have misappropriated its credentials) whereas the CSP is responsible for regulating access of its own staff as well as other tenants to system infrastructure, monitoring, and configuration resources as well as to log data that could be misused by an attacker.

Policies, procedures, supporting business processes, and technical measures must be established and documented for ensuring appropriate identity, entitlement, and access management for (i) data owner (tenant) users to their data and applications and (ii) service provider staff to its owned or managed (physical and virtual) application interfaces and infrastructure network and systems components. These policies, procedures, processes, and measures should address the following [34]:

- Roles and responsibilities for provisioning and de-provisioning user account entitlements following the rule of least privilege based on job function;

- Criteria for higher levels of assurance and multifactor authentication secrets (e.g., management interfaces, key generation, remote access, segregation of duties, emergency access, large-scale provisioning or geographically distributed deployments, and personnel redundancy for critical systems);
- Access segmentation to sessions and data in multitenant architectures by any third party (e.g., provider and/or other customer (tenant));
- Identity trust verification and service-to-service application (API) and information processing interoperability (e.g., single sign on (SSO) and federation);
- Account credential lifecycle management from instantiation through revocation;
- Account credential and/or identity store minimization or re-use when feasible;
- Authentication, authorization, and accounting (AAA) rules for access to data and sessions (e.g., encryption and strong/multifactor, time-limited, nonshared authentication secrets);
- Permissions and supporting capabilities for customer (tenant) controls over AAA rules for access to data and sessions;
- Adherence to applicable legal, statutory, or regulatory compliance requirements;
- Access to, and use of, audit tools that interact with the organization's information systems shall be appropriately segmented and restricted to prevent compromise and misuse of log data;
- User access to diagnostic and configuration ports shall be restricted to authorized individuals and applications;
- Management of identity information about every person who accesses IT infrastructure and to determine their level of access;
- Control access to network resources based on user identity;
- Control of user access based on defined segregation of duties to address business risks associated with a user-role conflict of interest.

15.6.2.6 Data Security. Defenses such as data encryption and access control are essential because (i) systems that collect sensitive data such as consumer information are attractive targets; (ii) they may be required contractual terms and laws; and (iii) release of such data may significantly damage the organization. Policies and procedures business processes and technical measures should be defined and implemented for the following:

- Use of encryption protocols for protection of sensitive data in storage and in transmission. These measures have to balance the need for security against the overhead of decryption and re-encryption as the data are processed in frameworks such as Hadoop.
- Key management and usage. Keys should not be stored at the CSP, but maintained by the cloud consumer or trusted key management provider. Key management and key usage should be separated, but the extent of this separation requires the balance between throughput and security.

- Access control, input, and output integrity routines (i.e., reconciliation and edit checks) to detect manual or systematic processing errors, corruption of data, or misuse.
- Labeling, handling, and the security of data and objects which contain data. Mechanisms for label inheritance shall be implemented for objects that act as aggregate containers for data.
- Data exchanged between one or more system interfaces—particularly when affected by legal, statutory and regulatory compliance obligations

15.6.3 Incident Response

Incident response in the event of a breach of a big data application and store on a remote cloud provider's platform is more complicated than a breach of a single organization because of organizational boundaries (and resultant non-aligned interests, contractual issues, and segregated system management structures. These must be planned for in advance and documented in an incident response plan. Such plans vary by industry and by circumstance. A general example is available from the American Bar Association [45].

Issues that must be addressed are as follows:

- Points of contact in both the CSP and the data owner for applicable regulation authorities, national and local law enforcement, and other legal jurisdictional authorities for compliance issues and to be prepared for a forensic investigation requiring rapid engagement with law enforcement.
- Policies, procedures, business processes and technical measures to triage security-related events and ensure timely and thorough incident management. Coordination between the data owner and the CSP to agree on the priority and severity of breaches is necessary—particularly if the breach affects multiple tenants at the CSP site.
- Coordination on follow-up actions and investigations after an information security incident requires legal action, including preservation of evidence, documenting chains of custody, and other actions needed to support potential legal action subject to the relevant jurisdiction

The next section will cover legal aspects of BDOC.

15.7 BDOC LEGAL ISSUES

The information in this section is for informational purposes only and not for the purpose of providing legal advice. You should contact your attorney to obtain advice with respect to any particular issue of problem.

The two central legal issues are (1) management of the risks and liabilities of the data owner for the data it relegates to the CSP which relate to privacy and governance and (2) the contractual terms and provisions (including the SLA) between the service provider and the data owner. The following subsections discuss these issues.

15.7.1 Privacy and Governance

The privacy of data is a large concern and one that increases in the context of big data. Perhaps the most obvious example is location data [46], but others include health, browsing, or purchasing activity [47]. Big data privacy and governance concerns fall primarily on the data owner but affect the relationship with the CSP. As such, there are both internal organizational issues and supplier contractual issues that need to be considered. This section focuses on the legal responsibilities and potential liabilities of the data owner.

The data owner should evaluate its data and assess its risk of massive amounts of data in order to determine its approach to moving data onto cloud platforms. The general approach to this task is to

- Enumerate its legal, statutory, and regulatory compliance obligations associated with (and mapped to) sites where its data are stored. Examples of such data include consumer data that would be affected by state privacy breach laws, health data that would be affected by Federal HIPAA privacy regulations, or credit card data whose storage and security requirements are affected by the Payment Card Industry Data Security Specification (PCI DSS). Some of these obligations will affect whether data organization can allow its data to be processed externally, others may affect its contractual requirements with the CSP.
- Classify its data and objects containing data shall be based on data type, jurisdiction of origin, jurisdiction domiciled, context, legal constraints, contractual constraints, value, sensitivity, criticality to the organization, third-party obligation for retention, and prevention of unauthorized disclosure or misuse.

The data owner establish policies and procedures together with records of performance and compliance to assert a defense against claims of negligence and non-conformance by governmental agencies, contracting parties, or individuals affected by data breaches.

With this insight, the data owner can establish information security requirements for the prospective service providers. Compliance with security baseline requirements must be reassessed at least annually unless an alternate frequency has been established and established and authorized based on business need higher levels of assurance are required for protection, retention, and lifecycle management of audit logs and other evidence of statutory or regulatory compliance obligations.

15.7.2 Considerations for Contracting with a CSP

The most common form of a contracting with a CSP is an agreement with a standard set of terms and conditions that is nonnegotiable. These terms and conditions in general promise a best effort level of service and security but offer no guarantees (other than a de minimus refund or discount) and disclaim responsibility from legal consequences and damages from both outages and data breaches. The benefit of these "take it or leave it" arrangements is low cost of use; and if the legal risks and liabilities of outages or data breaches are low, the terms are quite appropriate. However, for applications where these risks are significant, a negotiated contract with terms and conditions tailored to the needs of the data owner are appropriate. This section discusses the consideration that goes into a discussion of what the considerations are from the perspective of the data owner.

15.7.2.1 General Contractual Terms. Contractual agreements between providers and customers (tenants) should specify at least the following mutually agreed-upon provisions and terms:

- Parties: contract must specify the following:
  - Service providers (there may be more than one, depending on if the analytical service provider is distinct from the infrastructure platform service provider),
  - The service consumer—in this case, the data owner,
  - Third parties—the task of these third parties may vary from measuring service parameters to taking actions on violations as delegated by either the service provider or service consumer.
- Scope of business relationship and services offered (e.g., customer (tenant) data acquisition, exchange and usage, feature sets and functionality, personnel and infrastructure network and systems components for service delivery and support, roles and responsibilities of provider and customer (tenant) and any subcontracted or outsourced business relationships, physical geographical location of hosted services, and any known regulatory compliance considerations).
- Expiration of the business relationship and disposition of the data owner (tenant) data at the end of the agreement.
- Customer (tenant) service-to-service application (API) and data interoperability and portability requirements for application development and information exchange, usage, and integrity persistence.
- Policies, procedures, and terms for service-to-service application (API) and information processing interoperability, and portability for application development and information exchange, usage, and integrity persistence.
- Configuration management and control of IT infrastructure network and systems components including change management policies and procedures prior to deployment, provisioning, or use and authorization prior to relocation or transfer of hardware, software, or data.
- File formats for structured and unstructured data between the data owner and the cloud service provider.

15.7.2.2 Service-Level Agreements. SLAs nearly always specify levels of availability (with availability frequently being defined in terms of response times); in some cases they may specify security services, and liabilities for security breaches. These issues are discussed in Section 15.7.2.4.

SLAs define technical and legal responsibilities of the data owner and the service provider. These responsibilities depend on the cloud service model (IaaS, PaaS, or SaaS) and the cloud deployment model (private, hybrid, or public). If the big data platform is a private cloud, then SLA enforcement is an internal organizational matter. If the deployment model is hybrid or public, then the SLA assumes legal significance and is subject to the interpretation and enforcement of the judiciary. The less responsibility and control the data owner has, the more it relies on the provider and the more critical the terms of the SLA between those two parties becomes. The greater the complexity of the interaction between systems owned by the third party provider and those owned by the data owner, the greater the need to explicitly assign responsibilities and specify liabilities for breaches.

In the case of SaaS, the SLA must address service levels, security, governance, compliance, and liability, expectations of the service and provider are contractually stipulated; managed to; and enforced by the SLA. In the case of PaaS or IaaS, it is the responsibility of the consumer's system administrators to effectively manage the same, with some offset expected by the provider for securing the underlying platform and infrastructure components to ensure basic service availability and security. It should be clear in either case that one can assign/transfer responsibility by the SLA. To a limited extent, accountability, can also be transferred to the service provider (e.g., by means of indemnification clauses)—if the service provider is willing to accept. However, governmental laws particularly with respect to breaches of privacy may make it impossible for the data owner to completely transfer liability to the service provider.

At a minimum, the following items, taken Web SLA (WSLA) framework [48], should be defined in an SLA:

1. SLA parameters: SLA parameters metrics define how service parameters can be measured. These include the following:

    - Resource metrics are retrieved directly from the provider resources. In the case of the big data cloud, these could include, data bandwidth, transaction count, uptime, RAM, infrastructure resources (processing capacity, platform middleware, storage capacity) resources, middleware resources (including Lucene, Solr, Hadoop, and HBase), and other required resources.
    - Composite metrics represents a combination of several resource metrics, calculated according to a specific algorithm. For example, transactions per hour combines the raw resource metrics of transaction count and uptime. Composite metrics may be necessary to characterize higher level big data metrics such as velocity, volume, and variety.
    - Business metrics that relate SLA parameters to financial terms specific to a service customer. These include the cost of the services.

2. Service-level objectives (SLOs): WSLA, these are a set of formal expressions in the form of an if-then structure. The antecedent (if) contains conditions and the nine consequent (then) contains actions. An action represents what a party has agreed to perform when the conditions are met. In the context of a plain language

legal document, these conditions would include the cost of service (when metrics are met) and the consequences of not meeting those metrics.

3. Data recording and analysis requirements to establish conformance with availability and performance parameters
4. Terms specifying the penalties or monetary damages for outages. In standard SLAs from service providers such as Amazon, the maximum liability for outages is refunds or additional time at low cost, and liability for damages due to the consequences of security breaches and cyber attacks are specifically excluded.

15.7.2.3 Disaster Tolerance and Business Continuity. General considerations disaster tolerance were identified in Section 15.5.2 including the basis for deciding on the locations and extent of the standby resources. An SLA for the standby resources should be established. In addition, the requirements for standby datacenter security, readiness, and monitoring should be established. The requirements placed on the CSP should be consistent with the business continuity/disaster tolerance plan. The terms should cover utilities services and environmental conditions (e.g., water, power, temperature and humidity controls, telecommunications, and internet connectivity). Terms should be described the extent and how they are secured, monitored, maintained, and tested (e.g., inspection intervals, and power backups at the remote site, and failover testing.)

15.7.2.4 Information Security Requirements. The level of security and privacy controls and supplying the evidence of their implementation and effectiveness is usually established by the terms and conditions of the contract or SLA with the CSP [49]. The terms and condition must be consistent with the information security management plan (see Section 15.6.2.1) and require coordination between both the contractor negotiators and the data owner's information security experts.

The legal consequences of data breaches resulting in the release of data are the same as for conventionally structured and stored data. Liability, notification requirements, and penalties for releases of data items that constitute Personally Identifiable Information (PII) and health related information are often governed by national, provincial or state statutes [50, 51]. Additional consequences may result from industry agreements such as the Payment Card Industry Data Security Standard (PCI-DSS) [52] as well as specific non-disclosure agreements between the data set owner and third parties

The specific information security measures should be consistent with the risk of loss of integrity, confidentiality, or availability of the data at the remote site. Too few provisions would shift the liability to the data owner from the service provider; too much might make the cloud implementation cost prohibitive.

The following terms and provisions from the Cloud Service Alliance should be considered for the inclusion in the information security clauses of negotiated contracts between service providers and data owners:

- Provider and data owner (tenant) primary points of contact for the duration of the business relationship;

- References to detailed supporting and relevant business processes and technical measures implemented to enable effectively governance, risk management, assurance and legal, statutory and regulatory compliance obligations by all impacted business relationships;
- Responsibility for disposition of the data upon termination of the business relationship and treatment of customer (tenant) data impacted;
- Notification and/or pre-authorization of any system configuration or procedural changes in CSP resources;
- Timely notification of a security incident (or confirmed breach) to all customers (tenants) and other business relationships impacted (i.e., up- and down-stream impacted supply chain);
- Assessment and independent verification of compliance with agreement provisions and/or terms (e.g., industry-acceptable certification, attestation audit report, or equivalent forms of assurance), without posing an unacceptable business risk of exposure to the organization being assessed;
- Review of the risk management and governance processes of their partners to ensure that practices are consistent and aligned to account for risks inherited from other members of that partner's cloud supply chain;
- Oversight of third-party service provider information security programs, service definitions, and delivery-level agreements included in third-party contracts.
- Efforts in support of follow-up actions concerning a person or organization after an information security incident. These may include forensic procedures for gathering evidence suitable for admission to legal proceedings, including chain of custody and preservation. Upon notification, customers (tenants) and/or other external business relationships impacted by a security breach shall be given the opportunity to participate as is legally permissible in the forensic investigation.

15.8 ENABLING FUTURE SUCCESS—STEM CULTIVATION AND OUTREACH

This section presents the STEM management strategy that enables the proper execution of the BDOC attributes discussed in this chapter. Combining business needs, availability and reliability, heterogeneous issues, legal issues, security issues, and scaling issues. It is the SRE organization, which is charged with the management of vast and complex BDOC enterprises.

Clearly, the successful execution of BDOC ASP depends on very capable and robust technical professionals for both the development, as well as for the ongoing SRE function, and for the technical understanding of the legal aspects associated with SLAs and all other aspects of BDOC enterprise.

This section addresses the supply of talent in the areas of STEM. These skills are absolutely critical to the successful execution of BDOC management activities. This applies to both the development as well as to the operations of such BDOC ASP enterprises.

15.8.1 Criticality of Creating and Growing a STEM Pipeline and Engagement Ideas

There has been a continuing shortage in the supply of STEM talent in the United States over recent years. In 2013 the computer science (CS) field had some 50,000 undergraduate degrees awarded annually, against the growing demand of over 100,000 CS jobs in the United States alone. While this is wonderful for those graduates who obtain CS degrees, it exacerbates the challenges that the BDOC ASP industry is facing. In 2014 LinkedIn showed hundreds of job openings in the SRE category alone. This includes openings at Facebook, VMWare, Google, SalesForce, A9, Microsoft, Tmblr, Akamai, BestBuy and many others.

15.8.2 Computer Science, Networking, and Computer Engineering

SREs or DevOps are the fastest growing job type at this time [53]. This is astounding, given that this kind of job title did not even exist in 2009. Hence, the pace of change in the landscape teaches us that the most important part of STEM education, is to teach our students to become life-long learners, so that they can continually build on the technical foundations that we educate them.

Even though the United States has been battling declining in STEM interest at the K-12 levels, it appears that the job prospects for positions such as SRE and DevOps are pushing an increase in both enrollment and graduation rates in STEM and in particular in CS. Some of this work is done by industry sponsored capstone projects as has been very successful at Harvey Mudd College since the 1960s for Engineering, and from the 1990s in Computer Science [54]. Similar capstone programs are being created by accreditation requirements promulgated by the Accreditation Board for Engineering and Technology (ABET) 2020 study [55] and its member Computer Science Accreditation Board (CSAB), founded in 1985.

The following section will address some of the open-research areas. A major challenge entails the management of SRE talent growth, and the associated STEM education, which drives the SRE talent supply.

15.9 OPEN CHALLENGES AND FUTURE DIRECTIONS

This section discusses some of the open challenges and new directions of BDOC which include workforce issues, privacy and security, resource utilization, and integration with the Internet of things.

15.9.1 SRE and DevOps Professions

Going forward, we face some exciting challenges, which will afford the community an opportunity to acquire a deeper technical understanding in a number of areas. Furthermore, the community stands to benefit from developing new approaches and enabling new technologies and solutions in order to address these emerging challenges. One growth area for which there has been only limited work to date, is the emerging profession of DevOps and SRE professionals, in the context of the entire BDOC enterprise that is being managed.

This topic is not always viewed as a "hard" technical topic that technologists study, but rather a "soft" personnel topic. However, organizational management experts correctly point out that "the soft stuff is the hard stuff". Solving these challenges, within the context of the United States, will not be easy. In fact, it is a daunting challenge, in that the societal views of technologists is not high, and "geeks" are not considered "cool" by their peers during the most critical and formidable years of K-12 education. The late astronaut Sally Ride served on a research team that determined that "We lose the boys away from STEM in 8th grade, and we lose the girls in 5th grade" [56]. This hostile environment to technology during the early educational experience of the K-12 of children educated in the United States creates a cascading effect that results in an insufficient supply of STEM-capable students at the college level. Even if college students realize that becoming an engineer or computer scientist affords great job opportunities, they simply lack the STEM background in order to compete within the college level STEM curriculum.

While this is the most critical challenge that is the most crucial to address, there are additional challenges is building the SRE and DevOps workforce, that can be addressed, provided that the STEM talent supply issue is improved. Running a strong SRE organization requires considerable management talent, as well as a strong mentoring and training culture. This culture is a strong team culture, in which seasoned SREs teach incoming SREs the tradecraft. These positions include both operational responsibilities, as well as development activities. The development activities look to address the scale of BDOC enterprises. The only way to do that effectively is to develop the methodologies and automation technologies that will "automate the human manager out of the manual operational activities." Since the SRE position is rather new, it has not been studied adequately by the research community. This delicate balance needs to be studied further. Trade-offs between the benefits of automation and the cost of developing automated alert and troubleshooting systems must be better understood. The organizational challenges of creating and leading strong SRE organizations must be quantified and further studied. The focus of SRE teams on specific applications represents a semantically higher level of responsibility, in contrast with traditional management of networks and other lower stack layer semimanual operations.

The complexity associated with the technical challenges associated with availability, reliability, legal issues, and the emerging business models demands nothing less than a very adaptable SRE workforce, which is schooled in these complex areas. The operational responsibility of the SRE team encompasses multiple emerging disciplines and applications. It is anticipated that these challenges will continue to present multiple research challenges, which will be mutually beneficial to all stakeholders.

While the highest leverage would be achieved by major contributions to the STEM supply, and by achieving SRE organizational enhancements, the following subsections and the "Conclusions" section offer more traditional research areas that would contribute to BDOC management solutions going forward.

15.9.2 Information Assurance (Confidentiality, Integrity, and Availability)

This is a fertile area for future research. When many users are involved, the consequences of data breaches (i.e., the 2013 Target Data Breach) are substantial. Disaster response and transaction verification offer considerable research opportunities. For intelligent highway outages, the loss of integrity of the data from the sensors feeding the cloud-resident applications and from the cloud resident applications to the users can be particularly serious. In many cases, a portion of the data path from the sensors to the cloud will include wireless links. New methods for rapid authentication for a large number of users (possibly hundreds of thousands simultaneously), ensuring data integrity (including overcoming "man-in-the-middle" attacks) over wireless communications for high volume traffic are also necessary.

15.9.3 More Efficient Use of Cloud Resources

While not all cloud–resident databases may reach the level of exabytes or petabytes in the immediate future, the trend is certainly in that direction. Research is necessary in order to more efficiently utilize storage and develop algorithms to reduce the costs of data collection, integration and transformation, data analyses (searches and queries), storage, and disposal. Specific design issues include the following:

- Tailoring DBMSs for cloud computing including the tradeoffs between atomicity, consistency, isolation, and durability (ACID) in traditional SQL databases and less rigorous basically available soft-state and eventually consistent (BASE) models.
- *Data access*: whether to have data repositories located closer the user than the provider.
- *Consistency of replicated data*: If data are moved closer to the users, then replication will be necessary. Along with the replication comes issues of currency, consistency, and completeness.
- Deployment of BDOC installations including the use of staging for data collection processing.

15.9.4 Big Data and the Internet of Things

Information gathering, processing, and computing of massive amounts of data generated from and delivered to highly distributed devices (e.g., sensors and actuators) create new challenges, especially for interoperability of services and data. These requirements will impact the underlying cloud infrastructure requiring efficient management of very large sets of globally distributed nonstructured or semistructured data that could be produced at very high rates (i.e., big data). A multicloud service platform supported by broadband networks needs to handle all these challenges and appear to the application environment as one uniform platform.

15.10 CONCLUSIONS

This chapter reviewed BDOC from a number of perspectives. A historical perspective was provided as to the evolution of computing Web services, and how the current prevailing information architecture is the result of a steady process of increase in connectivity, mobility, and adoption of applications such as those offered by Google, Facebook, and Amazon. We discussed a number of BDOC enterprise management challenges such as SRE/DevOp training, availability and reliability, legal and security aspects, and the STEM educational challenges that must be addressed. Further work is recommended in each of these areas, in order to accommodate further growth and enable future capabilities, especially in the mobile space.

It is anticipated that other promising research areas for BDOC would continue to include [2] automated service provisioning, VM migration, server consolidation, energy management, traffic management and analysis, software frameworks, storage technologies and data management, and novel cloud architectures.

It is clear that BDOC is the wave of the future, and that these applications will continue to further benefit the user communities and many other stakeholders.

REFERENCES

1. Tom White, *Hadoop: The Definitive Guide*, 2nd Edition. Sebastopol, CA: O'Reilly Media, 2012.
2. Qi Zhang, Lu Cheng, and Raouf Boutaba, "Cloud computing: state-of-the-art and research challenges." *Journal of Internet Services and Applications* 1 (2010): 7–18
3. Leonard Kleinrock, "Models for computer networks." Proceedings of the International Conference on Communications, 1969, http://www.lk.cs.ucla.edu/index.html. Accessed November 19, 2014.
4. Marc Andreseen, Mosaic—the first global web browser, NCSA Technical Report. 1992, http://www.livinginternet.com/w/wi_mosaic.htm and http://en.wikipedia.org/wiki/Mosaic_(web_browser). Accessed November 19, 2014.
5. Internet Engineering Task Force. This is where the Request for Comments (RFCs) are maintained, www.ietf.org. Accessed November 19, 2014.
6. David Clark "We reject: kings, presidents, and voting. We believe in: rough consensus and running code." 24th IETF Meeting, July 1992, Cambridge, MA, 1992.
7. John Battelle, *The Search: How Google and Its Rivals Rewrote the Rules of Business and Transformed Our Culture*. New York: Portfolio, 2005.
8. John Battelle, Battelle's media blog, www.battellemedia.com. Accessed November 19, 2014.
9. Thomas L. Friedman, *The World is Flat [Updated and Expanded]: A Brief History of the Twenty-First Century*. New York: Farrar, Straus and Giroux, 2006.
10. Open Grid Forum (OGF), www.ogf.org. Accessed November 19, 2014.
11. OpenStack, the virtual organization that promotes the open standardization of cloud technologies, http://www.openstack.org/. Accessed November 19, 2014.
12. Christos Gkantsidis, Dimitrios Vytiniotis, Orion Hodson, Dushyanth Narayanan, Florin Dinu, and Antony Rowstron, "Rhea: automatic filtering for unstructured cloud storage." Presented

as part of the 10th USENIX Symposium on Networked Systems Design and Implementation. USENIX, 10th USENIX Symposium on Networked Systems Design and Implementation, April 2–5, 2013, Lombard, IL, 2013.

13. Qi Zhang, Quanyan Zhu, Mahamed Faten Zhani, Raouf Boutaba, and Joseph L. Hellerstein, "Dynamic service placement in geographically distributed clouds," *IEEE Journal on Selected Areas in Communications* 31 (2013): 762–772.
14. Bin Fan, David G. Andersen, and Michael Kaminsky, "MemC3: compact and concurrent memcache with dumber caching and smarter hashing," Proceedings of the 10th USENIX NSDI, 10th USENIX Symposium on Networked Systems Design and Implementation, April 2–5, Lombard, IL, 2013.
15. Adage.com, "70 Billion TV to Digital Ads", 2013. http://adage.com/article/media/70-billion-tv-ad-market-eases-digital-direction/244699/. Accessed November 19, 2014.
16. Joseph Betser and David Belanger, "Architecting the enterprise via big data analytics," in *Big Data and Business Analytics*, Jay Liebowitz, ed. Boca Raton, FL: CRC Press, 2013.
17. Underwood, Todd, Google, Usenix Panel, 2013, https://www.usenix.org/sites/default/files/conference/protected-files/underwood.pdf. Accessed November 19, 2014.
18. Amazon Elastic Map Reduce Web page, 2014. http://aws.amazon.com/elasticmapreduce/. Accessed November 19, 2014.
19. Hamzeh Khazaei, Jelena Mišic, Vojislav B. Mišic, and Nasim Beigi Mohammadi, "Availability analysis of cloud computing centers," 2012 IEEE Global Communications Conference (GLOBECOM), Anaheim, CA, pp. 1957–1962, http://ieeexplore.ieee.org/xpl/login.jsp?tp=&arnumber=6503402&url=http%3A%2F%2Fieeexplore.ieee.org%2Fxpls%2Fabs_all.jsp%3Farnumber%3D6503402. Accessed November 19, 2014.
20. Francesco Longo, Rahul Ghosh, Vijay K. Naik, and Kishor S. Trivedi, "A scalable availability model for infrastructure-as-a-service cloud," 2011 IEEE/IFIPS Conference on Dependable Systems and Networks, International, Hong Kong, pp. 335–346, http://ieeexplore.ieee.org/xpl/articleDetails.jsp?arnumber=5958247&navigation=1. Accessed November 19, 2014.
21. Anupam Das, Cristian Lumezanu, Yueping Zhang, Vishal Singh, Guofei Jiang, Curtis Yu, "Transparent and flexible network management for big data processing in the cloud," Proceedings of the 5th USENIX Worskhop on Hot Topics in Cloud Computing (Hotcloud '13), 2013, http://www.nec-labs.com/ lume/files/flowcomb-hotcloud13.pdf. Accessed November 19, 2014.
22. Rajesh Nishtala, Hans Fugal, and Steven Grimm, "Scaling Memcache at Facebook," Proceedings of the 10th USENIX Symposium on Networked Systems Design and Implementation, April 2013, pp. 356–370, https://www.usenix.org/conference/nsdi13/technical-sessions/presentation/nishtala. Accessed November 19, 2014.
23. Yang Wang, Manos Kapritsos, Zuocheng Ren, Prince Mahajan, Jeevitha Kirubanandam, Lorenzo Alvisi, and Mike Dahlin, "Robustness in the Salus scalable block store," Proceedings of the 10th USENIX Symposium on Networked Systems Design and Implementation, April 2013, pp. 356–370, https://www.usenix.org/conference/nsdi13/technical-sessions/presentation/wang_yang. Accessed November 19, 2014.
24. Eric Bauer and Randee Adams, "Analyzing cloud reliability and availability," in *Reliability and Availability of Cloud Computing*. New York: Wiley-IEEE, 2012, p. 84.
25. Paolo Costa, Austin Donnelly, Antony Rowstron, and Greg O'Shea, " Camdoop: exploiting in-network aggregation for big data applications," in 9th USENIX Symposium on Networked

Systems Design and Implementation (NSDI'12), 2012, http://research.microsoft.com/apps/pubs/default.aspx?id=163081. Accessed November 19, 2014.

26. Bruno Silva, Paulo Maciel, Eduardo Tavares, and Armin Zimmermann, "Dependability models for designing disaster tolerant cloud computing systems," 2013 43rd Annual IEEE/IFIP International Conference on Dependable Systems and Networks (DSN), Budapest, Hungary, http://www.computer.org/csdl/proceedings/dsn/2013/6471/00/06575323-abs.html. Accessed November 19, 2014.
27. Apache Hadoop tutorial, http://developer.yahoo.com/hadoop/tutorial/module1.html. Accessed November 19, 2014.
28. ISO 27001, "Information security management system," International Standards Organization, October 2005, http://www.27000.org/iso-27001.htm. Accessed November 19, 2014.
29. Joint Task Force Transformation Initiative, NIST SP 800-53 Rev. 4., "Security and privacy controls for federal information systems and organizations", http://nvlpubs.nist.gov/nistpubs/SpecialPublications/NIST.SP.800-53r4.pdf. Accessed November 19, 2014.
30. HIPAA/HITECH Act, 45 CFR 164, February 2013, http://www.hipaasurvivalguide.com/hipaa-regulations/hipaa-regulations.php. Accessed November 19, 2014.
31. North American Electric Reliability Council (NERC) Critical Infrastructure Protection Committee, "Security Guidelines." http://www.nerc.com/comm/CIPC/Pages/Security-Guidelines.aspx. Accessed November 19, 2014.
32. Thomas Ristenpart, Eran Tromer, Hovav Shacham, and Stefan Savage, "Hey, you, get off of my cloud: exploring information leakage in third-party compute clouds," CCS '09: Proceedings of the 16th ACM Conference on Computer and Communications Security, November 9–13, 2009, Chicago IL, 2009.
33. Hassan Takabi, James B.D. Joshi, and Gail-Joon Ahn, "Security and privacy challenges in cloud computing environments," *IEEE Security & Privacy* 8 (2010): 24–31.
34. Kui Ren, Cong Wang, and Qian Wang, "Security challenges for the public cloud," *IEEE Internet Computing* 16 (2012): 69–73.
35. Rick Holland, Stephanie Balaouras, John Kindervag, and Kelley Mak, *The CISO's Guide To Virtualization Security*. Forrester Research Inc., 2012, http://www.forrester.com/The+CISOs+Guide+To+Virtualization+Security/fulltext/-/E-RES61230. Accessed November 19, 2014.
36. Ellen Messmer, "VMware strives to expand security partner ecosystem," Network World, August 31, 2011, https://www.networkworld.com/news/2011/083111-vmware-security-partners-250321.html. Accessed November 19, 2014.
37. Bernd Grobauer, Tobias Walloschek, and Elmar Stöcker, "Understanding cloud computing vulnerabilities," *IEEE Security & Privacy* 9 (2011): 50–57.
38. Advisories & Certifications, VMware, 2014. http://www.vmware.com/security/advisories. Accessed November 19, 2014.
39. IBM X-Force, "IBM X-Force 2010 Trend and Risk Report," IBM, 2010. https://www.ibm.com/services/forms/signup.do?source = swg-spsm-tiv-sec-wp&S_PKG = IBM-X-Force-2010-Trend-Risk-Report. Accessed November 19, 2014.
40. "Former Shionogi employee sentenced to Federal Prison for hack attack on company computer servers," The United States Attorney's Office, District of New Jersey press release, December 9, 2011, http://www.justice.gov/usao/nj/Press/files/Cornish,%20Jason%20Sentencing%20News%20Release.html. Accessed November 19, 2014.
41. Yanpei Chen, Vern Paxson, and Randy H. Katz, "What's new about cloud computing security?" University of California at Berkeley Technical Report UCB/EECS-2010-5, http://www.eecs.berkeley.edu/Pubs/TechRpts/2010/EECS-2010-5.html. Accessed November 19, 2014.

42. Linda L. Briggs, Q&A: New Approaches for Tackling Big Data Security Issues, http://tdwi.org/articles/2011/09/20/big-data-security-2.aspx. Accessed November 19, 2014.
43. Cloud Security Alliance, "Cloud Controls Matrix (CCM) Version 3.0.1, July 10, 2014," https://cloudsecurityalliance.org/research/ccm/#_new. Accessed November 19, 2014.
44. Karen Scarfone, Murugiah Souppaya, and Paul Hoffman, *Guide to Security for Full Virtualization Technologies*. National Institute of Standards and Technology (NIST), January 2011, http://csrc.nist.gov/publications/nistpubs/800-125/SP800-125-final.pdf. Accessed November 19, 2014.
45. Christopher Wolf, "Requirements and security assessment procedures," 2012, http://www.americanbar.org/content/dam/aba/administrative/litigation/materials/sac_2012/22-15_intro_to_data_security_breach_preparedness.authcheckdam.pdf. Accessed November 19, 2014.
46. Marta C. González, César A. Hidalgo, and Albert-László Barabási, "Understanding individual human mobility patterns". *Nature* 453, 779–782 (2008).
47. Alexandros Labrinidis and H. Jagadish, "Challenges and opportunities with big data," Purdue University Cyber Technical Reports, Purdue University Cyber Center Technical Reports, 2012, http://docs.lib.purdue.edu/cctech/1/. Accessed November 19, 2014.
48. Alexander Keller and Heiko Ludwig, "The WSLA framework: Specifying and monitoring service level agreements for web services." *Journal of Network and Systems Management* 11 (2003), 57–81.
49. National Institute of Standards and Technology Public Cloud Computing Security Working Group, "Challenging security requirements for US Government cloud computing adoption," 2012, http://collaborate.nist.gov/twiki-cloud-computing/pub/CloudComputing/CloudSecurity/Challenging_Security_Requirements_for_US_Government_Cloud_Computing_Adoption_v6-WERB-Approved-Novt2012.pdf. Accessed November 19, 2014.
50. Miriam Russom, Robert Sloan, and Richard Warner, "Legal concepts meet technology: a 50-state survey of privacy laws", Proceedings of the 2011 Workshop on Governance of Technology, Information, and Policies, ACM, December 6, 2011, Orlando, FL, 2011.
51. Electronic Frontiers Foundation, "Privacy: statutory provisions," 2011, https://ilt.eff.org/index.php/Privacy:_Statutory_Protections. Accessed November 19, 2014.
52. Payment Card Industry, Data Security Standard 3.0 Requirements and Security Assessment Procedures, November 2013, https://www.pcisecuritystandards.org/documents/PCI_DSS_v3.pdf. Accessed November 19, 2014.
53. Mashable.com, "Fastest growing jobs," 2013, http://mashable.com/2013/11/13/fastest-growing-jobs/. Accessed November 19, 2014.
54. Joseph Betser, "Knowledge management (KM) and e-learning (EL) growth for industry and university outreach activities via capstone projects: case studies and future trends," Jay Liebowitz and Michael Frank, eds. *Knowledge Management and E-Learning*. London: Taylor and Francis Press, pp. 305–321, 2010.
55. ABET 2020: Face the Future, "Proceedings of the 2003 ABET Annual Meeting," Minneapolis, October 30–31, 2003, www.abet.org. Accessed November 19, 2014.
56. Carol B. Muller, Sally M. Ride, Janie Fouke, Telle Whitney, Denice D. Denton, Nancy Cantor, Donna J. Nelson, et al. "Gender differences and performance in science," *Science* 307 (2005): 1043.

INDEX

Note: Page numbers in *italics* refer to Figures; those in **bold** to Tables

activity function generation, workflow activity
- activity semantics, 322, **322–3**
- algorithm, **321**, 321–4
- DAG control flow, 323, **323**
- data flow, 323–4, **324**
- function signature, 321–2, **322**

Actual MAC (AMAC), 92
Alfredo, 159, 161, 162
Amazon Dynamo, 12
Amazon EC2, 11, 19, 25, 99, 140, 224, 232, 296
Amazon Simple Workflow (SWF) service
- activity workers, 315–17, *316*
- architecture, 315, *316*
- asynchronous functions, 317
- decider, 315, 317
- task queues, 315

Amazon Virtual Private Cloud (VPC), 11, 93–4
Amazon Web Services (AWS), 35, 36, 39, 317, 320
Apache Hadoop, 158
Apache Software Foundation project, 38–40
Apple iCloud, BDOC, 363
application platforms, 17, 129, 130, 132, 150, 333
application programming interface (API), 7, 31, 138, 244
architectures
- cloud computing, *7*
  - application layer, 8
  - hardware layer, 7
  - infrastructure layer, 8
  - platform layer, 8
- FlowIPS security design, *279*
- Interoperable Workflow Intermediate Representation (IWIR), 319, *319*
- mobile cloud computing (MCC), 158–161
- OpenDaylight, 145, *146*
- OpenStack, 139, *139*
- RainCloud workflow, 318, *318*
- scheduling, 199, *199*
- security, 271
- software-defined networking (SDN), 132–3, *133*
- virtual machine (VM) migration, *50*, 50–51

3-ary CamCube topology, 80, *81*
automated migration, 67, 134
autonomic computing, 220–221
autonomic management, 41, 219, 361
autonomic manager, 220, 233, 240

bag-of-tasks (BoTs), 249, 255
BCMS *see* Bounded Congestion Multicast Scheduling (BCMS)
BCube, hybrid switch/server topologies, 79–80, *80*
big data on clouds (BDOC), 19–20
- application service provider to cloud computing, 362–3

Cloud Services, Networking, and Management, First Edition.
Edited by Nelson L. S. da Fonseca and Raouf Boutaba.

big data on clouds (BDOC) (*cont'd*)
business appetite, 362
business applications
global enterprises, 365–6
mobile services, 367
online advertising, 366–7
site reliability engineers (SREs), 367–8
technical computing, 366
cloud and service availability
disaster tolerance, 370–372
operational availability, 368–70
cloud computing, 19–20, 362
cloud resources, 387
clouds—supply and demand, 364–5
contracting with cloud service providers (CSPs)
contractual agreements, 381
disaster tolerance and business continuity, 383
information security requirements, 383–4
service-level agreements (SLAs), 381–3
DevOps professions, 385–6
execution *see* science, technology, engineering and mathematics (STEM)
incident response, 379
information assurance, 387
and Internet, 387
legal issues, 380–384
privacy and governance, 380
search, 362–3
security issues *see* security
site reliability engineering (SRE), 361, 385, 386
technical capabilities
dynamic service placement, 364
MemC3, 364
performance enhancement, 364
bisection bandwidth constrains, 77
blocking switch port, 285
Bluetooth, 161, 162, 171, 176
Bounded Congestion Multicast Scheduling (BCMS), 91
broker, performance management and monitoring, 222
browsers, weaknesses of, 166
buzzword, 4

canonical three-tiered tree-like topology, 77, *77*
capacity constraints, 112
CDMI *see* Cloud Data Management Interface (CDMI)
CDNs *see* content delivery networks (CDNs)
Ceilometer project in OpenStack, 150
centralized monitoring, cloud data centers, 41
Chroma, 159, 161
CIMI *see* Cloud Infrastructure Management Interface (CIMI)
client server technique, MCC
Alfredo, 159
Apache Hadoop, 158
Chroma, 159
CMCVR, 160
Cuckoo, 160
Hyrax, 158–9
MWSMF, 159
Spectra, 159
VMC, 160
CloneClouds, 161, 162, 177, 181
cloud computing *see also* resource management
Amazon Web Services (AWS), 243
architecture, *7*, 7–8
big data on clouds (BDOC), 19–20
characteristics, 4–5
cloud networking, 11–12
cloud services, 8–10
computing models and, 6–7
data center networks and relevant standards, 15–16
data center virtualization, 11
data storage and management, 12
definition, 4–5, 243
energy management, 13–14, 18–19
Google Application Engine (GAE), 243–4
interactive multimedia applications, 20–21
interdata center networks, 16
key driving forces, 5–6
MapReduce programming model, 12–13

mobile cloud computing, 17
multi-clouds, autonomic performance management, 18
National Institute of Standards and Technology (NIST), 4–5
OpenFlow and, 16–17
and OpenStack *see* OpenStack
performance management and monitoring, 220
privacy, 14
resource management, 13, 17–18
scheduling, 17–18
scientific applications *see* scientific applications on clouds
security, 14, 19
software-defined networking (SDN) technology, 12, 16–17
survivability and fault tolerance, 19
see also survivability in cloud
virtualization, 14–15
VM migration, 11, 15
Cloud Data Management Interface (CDMI), 33–4
cloud gaming
Amazon EC2 infrastructure, 340–341
BitTorrent, 340–341
challenges, 339
definition, 333–4
hybrid DC-CDN infrastructure, 344–5, *345*
measurement settings, 340–341
multi-DC infrastructure
cloud providers, 342–3
EC2 cloud infrastructure, 341, *341*
latency-based strategy, 342
region-based DC location strategy, 343, *343*
region-based strategy, 342
offloading computation, 339
Cloud Infrastructure Management Interface (CIMI), 33
Cloudlets, 160–161, 179, 181
Cloud-Mobile Convergence for Virtual Reality (CMCVR), 158, 160
CloudMonkey, Python tool, 38
cloud networking, 11–12, 27–8, *28*
cloud platform, 25, 26
cloud providers, 8–9, 11
cloud gaming, 342–3
scientific applications on clouds, 312
service providers, 3
virtual machines (VMs), 25
cloud resource management, tools and systems
CloudStack, 38–9
Deltacloud, 39–40
Eucalyptus, 35–6
Libcloud, 39
Libvirt, 40
OpenNebula, 36–7
OpenStack, 37–8
cloud service providers (CSPs)
big data on clouds (BDOC), contract with
contractual agreements, 381
disaster tolerance and business continuity, 383
information security requirements, 383–4
service-level agreements (SLAs), 381–3
to cloud computing, 362–3
cloud slice, 25
CloudStack, 38–9, 147, 280, 312
CMCVR *see* Cloud-Mobile Convergence for Virtual Reality (CMCVR)
Code-Oriented eXplicit multicast (COXcast), 91
communication networks, 164, 166–7, 197–8, 368
communication protocols
Bluetooth, 162
3G/4G, 162
Wi-Fi, 161–2
communications infrastructure, 209–10
communication-to-computation ratio (CCR), 259
community clouds, 10, 157
computing server power consumption, 195–6, *196*
configuration management and control, 373, 381
Content Addressable Network (CAN), 80
content delivery networks (CDNs), 337
see also hybrid delivery models
cooperative monitoring, cloud data centers, 41

COXcast *see* Code-Oriented eXplicit multicast (COXcast)
Crossroads, datacenters, 94–5
Cuckoo, 160, 162, 364
Cuckoo, client-server based framework, 160, 162, 364

datacenter (DC), 11, 336 *see also* hybrid delivery models
data center infrastructure efficiency (DCIE), 194
datacenter networks (DCNs)
 3-ary CamCube topology, 80, *81*
 bisection bandwidth constrains, 77
 canonical three-tiered tree-like topology, 77, *77*
 commodity off-the-shelf (COTS) switches, 76
 Content Addressable Network (CAN), 80
 emerging technologies, 93–4
 equal-cost multipath (ECMP), 76
 expansion, efficient and incremental, 96–7
 full address space virtualization, 95
 hybrid switch/server topologies, 79–80, *80*
 incremental expansion, 77
 load balancing across multiple paths, 98
 multi-rooted tree, 76
 multitiered topology, 77
 name and locator separation, 94–5
 network expansion, 83–5
 requirements, 96
 routing, 89–93
 server only topology, 80–82
 sharing and performance guarantees, 97
 switch-oriented topologies, *78*, 78–9, *79*
 tenants, address flexibility, 97–8
 topologies, 76–82
 traffic *see* traffic, datacenter networks (DCNs)
 valiant load balancing (VLB), 76
data integrity, 166, 174
data replication, cloud computing
 applications, 205
 cost-based, 206
 databases, 206–7
 downlink bandwidth requirements, 207, *207*
 energy and residual bandwidth, 207–8, *208*
 optimal location, 205–6
 policy maker, 205
data security, 166, 181, 184, 372, 378
DC *see* datacenter (DC)
DCell, hybrid switch/server topologies, 79–80, *80*
Deadline-Markov decision process (MDP) algorithm, 260–261
delivery cloud, interactive multimedia applications
 content delivery networks (CDNs), 337
 datacenter (DC), 336
 peer-topeer (P2P)-based infrastructure, 336–7
Deltacloud, 39–40
deterministic packet marking (DPM), 276
development and operations (DevOps) professions, 20, 385–6
disaster tolerance
 business continuity, 371
 definition, 370
 disruptive events probability, 371–2
 planning, 370–371
 primary resources, 371
Distributed Management Task Force (DMTF), 32
Distributed Overlay Virtual Ethernet (DOVE), 65, 66
DOVE *see* Distributed Overlay Virtual Ethernet (DOVE)
DPM *see* deterministic packet marking (DPM); dynamic Power Management (DPM)
DropBox, BDOC, 363
Dryad computational model, 12–13
DVFS *see* dynamic voltage and frequency scaling (DVFS)
dynamic Power Management (DPM), 194, 195, 198, 209
dynamic voltage and frequency scaling (DVFS), 194, 195, 209
dynamic voltage scaling (DVS) links, 197

ECG data analysis software, 171, *171*, 176
ECMP *see* equal-cost multipath (ECMP)

e-commerce, 3, 23, 243, 362–3, 366–7
edge-core clouds (scenario), 225
electric bills, energy-efficient design
  data center subset selection, 116
  demand profile, 118, *119*
  demand response (DR) component, 115
  dynamic pricing, 115
  IDC demands, 115
  IDC workload sharing, 116–18, **118**
  Opex savings, *119*, 119–20
  physical link cost assignment, 116
  smart grid, 115
  virtual link cost assignment, 115–16
e-mail, 23, 362–3
embedding, network virtualization, 24, *301*, 301–2, **306**
EMC^2, storage provider enterprise, 67
encryption algorithms erosion, 373
energy consumption in data centers
  challenges, 210–211
  communication networks, 197–8
  computing servers and switches, 195–6
  energy-efficiency
    communications infrastructure, 209–10
    computing servers, 197
    data replication, 205–8
    load balancing, 200–205
    networking switches, 196, 197
    scheduling, 198–201
    virtual machines placement, 208–9
  Interactive Data Corporation (IDC), 194
  next-generation user devices, 193
equal-cost multipath (ECMP), 76, 85, 89, 98
e-STAB scheduling policy
  queue-size, 203–4, *204*
  racks and modules, 203, *203*
  server selection, 204, *204*
  steps, 202
Eucalyptus, 35–6, 244, 245
European Network and Information Security Agency (ENISA), 10
Execution Engine, 227
extended semishadow images (ESSI), 161

fabric manager, 92
failure tolerance in cloud
  hard disk failures, 299
  load balancers (LBs), 299
  Microsoft data centers, 299
  NetPilot, 298–9
  raid controller failures, 299
faulty backup mechanisms, 166
FireCol, flooding DDoS attacks detection solution, 275
flow conservation constraints, 112
FlowIPS security design
  architecture, *279*
  cloud cluster, 280
  controller, 281
  Open vSwitch (OVS), 280–281
  processing flow, 281–2, *282*
  Snort, 281
  Snort/Iptables IPS *vs.*, 282–4, *283*
  system components, 280–281
Forwarding Rules manager (FRM), 145

generic routing encapsulation (GRE), 25, 141
global enterprises, BDOC, 365–6
Google, 3, 5, 8, 185, 363, 366, 367
Google Glass, 185, 193
grid computing, 6–7, 363

Hadoop distributed file system (HDFS), 158, 159
hardware latency, 335
Heat project in OpenStack, 150
Hedera, 89–90
HEFT *see* Heterogeneous earliest finish time (HEFT)
heterogeneous clouds
  heterogeneous monitoring systems, 239
  performance management and monitoring, 238–9
  public cloud environment, 239
  rapid reaction, 239
  traditional monitoring techniques inaccuracy, 239
Heterogeneous earliest finish time (HEFT), 259
heuristic solution, IDC network virtualization, 113–15, *114*
high-availability virtual infrastructure management framework (Hi-VI), 302–3

Hi-VI *see* high-availability virtual infrastructure management framework (Hi-VI)
Host-based intrusion detection system, 275
hybrid cloud, 9–10, 217, 224–5, 247
hybrid cloud optimization cost (HCOC) algorithm, 259
hybrid delivery models
  CDN-P2P compositions, 338
  compositions, 337, *338*
  DC CDN, 339
  multi-CDN, 338–9
  multi-DCs, 339
hybrid switch/server topologies, 79–80, *80*
hypervisor, 25, 50–51
Hyrax, client-server approach
  computational resources, 158–9, 177
  fault tolerance mechanism, 162
  VMC, 160

IDS/IPS *see* intrusion detection and prevention systems (IDPS)
infrastructure as a service (IaaS), 8, *9*
  performance management and monitoring, 217
  provider, 8
input/output (I/O) virtualization
  drawback, 57
  hardware-assisted network, 55, *56*
  modes, 55, *55*
  Network Plug-In Architecture (NPA/NPIA), 56
  Peripheral Component Interconnect Express (PCIe), 55
  single root I/O virtualization (SR-IOV), 55
  techniques, 57, **58**
  VM Device Queues (VMDq), 55
  VM live migration, 56, 57
insecure/incomplete data deletion, 166
Interactive Data Corporation (IDC), 194
interactive multimedia applications on clouds
  adaptive streaming, 354
  delivery cloud, 336–7
  Graphics Processing Unit (GPU), 353–4
  hardware latency, 335
  hybrid delivery models, 337–9, *338*
  Massive User-Generated Content (UGC) Live Streaming, 334
  multimedia tasks virtualization, 353–4
  network latency, 335–6
  networks economics, 353
  on-demand gaming *see* cloud gaming
  response time, 335
  time-shifting on-demand TV, 334
  time-shifting video streaming, 351–3, **353**
  user-generated content (UGC) live streaming, 345–351
inter-data-center (IDC) networks
  electric bills, energy-efficient design, 115–20
  energy efficiency impacts, 107
  energy *vs.* resilience trade-off, 123–4
  heterogeneous, 105, *106*
  optical, 106–7
  penalties, design, 120–123
  public telecom network, 105–6
  schemes, 125, **126**
  virtualization *see* virtualization
internal clouds, 9, 372
Internet Group Management Protocol (IGMP), 93–4
Interoperable Workflow Intermediate Representation (IWIR)
  atomic activity, 314–15
  composite activity, 315
  directed acyclic graph (DAG), 314–15, *315*
  IWIR-to-SWF conversion
    activity function generation algorithm, **321**, 321–4
    architecture, 319, *319*
    control flow constructs, 320
    HTTP connection, 319–20
    principle, 320
    SWF decider generation algorithm, **320**
  workflow application, 314
intrusion detection and prevention systems (IDPS), 19, 269, 274
  in cloud virtual networking environments, 279
  design challenges, 272

latency, 272
Network Intrusion Detection System (NIDS) mode, 274
network reconfigurations, 272
resource consumption, 272
SDN-based cloud security solutions, 277–8
Snort, 272, 274
solutions, limitation of, 278–9
Suricata, 272
IPS *see* intrusion detection and prevention systems (IDPS)
IWIR *see* Interoperable Workflow Intermediate Representation (IWIR)

Java EE web application, 232–3, 235, 240
Jellyfish, *84*, 84–5
justin.tv analysis *see* user-generated content (UGC) live streaming

Knowledge Store component, 226, 227

label switched paths (LSPs), 25
legacy IP networks, 105
legislative/organizational risks, MCC, 168–9, 176
Legup, 83, *83*
Libcloud, 39
Libvirt, 40
limited system monitoring, 373
Link Aggregation Control Protocol (LACP), 91, 92
link protection and restoration, 304
Linux bridging, 142
live migration
big data on clouds (BDOC), 374
definition, 50
Internet Protocol (IP) address, 52
memory migration, 52–3
offline and, 53, **54**
page fetching, 53
post-copy, 53
pre-copy, 53
retransmission of dirty pages, 52
strategies, 53
transmission control protocol (TCP), 52
virtual CPU (vCPU), 52
virtual machines, 66, 145, 225
load balancing
electricity cost, 204–5
e-STAB scheduler, 202–4, *203*, *204*
Ethernet standards, 202
idle servers, 200–201
stochastic power reduction scheme (SAVE), 205
virtual machines, 204
Locator/Identifier Separation Protocol (LISP), 66–7

malicious insider, 166, 270, 279
Management Logic implementation, *226*, 226–7
manycast constraints, 112–13
MAPE-k loop, XCAMP components
Execution Engine, 227
information aggregation service, 225–6
Knowledge Store component, 227
Management Logic implementation, *226*, 226–7
notification engine, 225–6
Plugin Engine, 226
MapReduce
cloud computing and, 6
programming model, 12–13
resource management, 13
MAUI framework, 158, 159, 161, 162
MCC *see* mobile cloud computing (MCC)
Microsoft Hyper-V, 93–4
MobiCloud, 161, 162
mobile ad hoc networking (MANET), 161
mobile cloud computing (MCC)
abstraction and virtualization, 155
application partitioning/client server, 158–60
benefits, 156
characteristics, 155
client server, 157
complexity and dynamism, 154
definition, 155, 156
description, 153
frameworks and application models
architectures, 158–161
communication protocols, 161–2
risk management strategies, 162
hybrid approach, 157
peer to peer approach, 157
popularity, 156

mobile cloud computing (MCC) (*cont'd*)
risk management *see* risks in mobile cloud computing (MCC)
services, 155–6
VM technology, 160–161
mobile services, 17, 177, 362, 367
Mobile Web Services Mediation Framework (MWSMF), 158, 159
MPLS *see* multiprotocol label switching (MPLS) networks
multiclouds, 18, 217–19, 222, 225, 387
Multiple Spanning Tree Protocol (MSTP), 91
multiprotocol label switching (MPLS) networks, 25, 90, 105
multitiered topology, 77
MWSMF *see* Mobile Web Services Mediation Framework (MWSMF)

natural disasters, risks, 166
NETCONF, management protocols, 28, 44
NetLord, full address space virtualization, 95, *95*
NetPilot, failure characterization, 298–9
network latency, 335–6, 342
Network Plug-In Architecture (NPA/NPIA), 56
network reconfiguration (NR), 19
actions, **284**
blocking switch port, 285
filtering, 285
mechanism, 286
QoS adjustment (QA), 284, 286–7
quarantine, 285
selection policy, **287**, 287–8
traffic isolation (TI), 284–5
traffic redirection (TR), 284–6
network security, 166, 271, 277
network virtualization, 4, 11, 24, 27, 42, 59, 108–115
Net-Zero Energy Data Center, 14
NR *see* network reconfiguration (NR)

offline migration, 52–3, **54** *see also* live migration
OFRHM *see* OpenFlow Random Host Mutation (OFRHM)
on-demand gaming *see* cloud gaming
online advertising, BDOC, 366–7
Open Cloud Computing Interface (OCCI), 31–2, 36, 37
Open Cloud Networking Interface (OCNI), 32
OpenDaylight
architecture, 145, *146*
cloud computing and SDN interaction, *139, 146*, 147
description, 145
Forwarding Rules manager (FRM), 145
Open-Vswitch database (OVSDB), 148
service provider, 147
virtual infrastructures management, architecture for, 148, *148*
virtualization edition, 147, *147*
Virtual Tenant Network (VTN) service, 147
VN embedding problem, 148
OpenFlow
components, 135, **135**
controller, 90, 95
elements, 134
Ethernet destination and source addresses, 135–6
flow entries removal from entries, 136
identifier, 137
matching of packets, 136, *136*
message types, 137
Open Networking Foundation, 134–5
ports, 136
QoS implementation, 137
switch and controller, 134, *135*
technology, 16–17
transport layer security (TLS), 137
OpenFlow Management Infrastructure (OMNI), 65–6
OpenFlow Random Host Mutation (OFRHM), 278, 279
open/global grid forum, 363
OpenNebula, 36–7, 39
open source cloud management platforms
CloudStack, 38–9
Eucalyptus, 35–6
OpenNebula, 36–7
OpenStack, 37–8
OpenStack, 37–8
application programming interface (API), 138

architecture, 139, *139*
BDOC, 363
data center deployment, 140–141, *141*
Identity service (Keystone), 140
interrelated services, 138–9
Neutron networking service, 140
nova compute service, 140
tenant and provider networks, 141–2, *142*
tenant resource, 140
virtual machines (VM), 138, *138*
Open Virtualization Format (OVF), 32–3, 377
Open Virtual Switch (OVS) cloud environment, 279
Open vSwitch (OVS), 19, 142–4, *143*
Open-Vswitch database (OVSDB), 148
operational availability, big data on clouds (BDOC)
Camdoops, 369
child tasks failures, 369
FlowComb, 369
hanging tasks failures, 369–70
Internet service providers, 368
MapReduce program, 369
map tasks and reduce tasks, 370
Scalus for HBase, 369
service-level agreements (SLAs), 368
tasktracker failure, 370
opportunistic redundancy pooling (ORP), *300*, 300–301
optical interconnection networks, 209–10
optical switching architecture (OSA), 78–9, *79*
ORP *see* opportunistic redundancy pooling (ORP)
OVF Packages, 32–3

packet marking technique, IP traceback, 275–6
pattern-based deployment service (PDS), 229–231, 237
peer-to-peer (P2P)-based infrastructure, 336–7
Pegasus Workflow Management System, 314
penalties, inter-data-center (IDC) networks
elastic optical network (EON) backbone, 120
ij, virtual link, 120
minimum outage probability in cloud, 121
resource saving minimum outage probability in cloud (RS-MOPIC), 121–3, *122*
upstream data center demand, 121
virtual path (VP), outage probability, 120
performance management and monitoring
autonomic computing, 220–221
autonomic manager, 220
broker, 222
cloud computing, 220
heterogeneous clouds, 238–9
hybrid clouds, 217
implementation
Amazon EC2, 233
analysis and planning, 229–30
execution, 229
Java EE web application, 232–3, 235
Misure, 230
monitoring components, 229
pattern-based deployment service (PDS), 230–231
scaling experiment, measurements from, 233, *234*
scaling test, measurements, 234, *235*
infrastructural flexibility, 220
infrastructure-as-a-service (IaaS)-style, 217
Management Logic, design and implementation, 222, 233, 235–7
multiclouds, 217, 222
ownership and access, 219–20
pay-as-you-go model, 220
private clouds, 221
X-Cloud Application Management Platform (XCAMP), 222–9
performance risks
complexity, 168
data availability, 167, 175
data location, 167
data segregation/isolation, 167, 175
MCC technology and services, 167
network constraints, 168
portability, 167
quality-of service-(QoS) parameter, 174–5
reliability, 168, 175

performance risks (*cont'd*)
resource exhaustion, 168
service availability, 167, 175
Peripheral Component Interconnect Express (PCIe), 55
platform as a service (PaaS), 8, *9*, 25, 155, 244, 245, 270, 364
Plugin Engine, 226
Portland, data center network architectures, 12, 92, 99
port-switching-based source routing (PSSR), 90, *90*, 98, 99
power usage effectiveness (PUE), 13–14, 194
private cloud, 9, 10, 25, 221, 247
probabilistic packet marking (PPM) method, 276
ProtoGENI (federation), 43
Pseudo IP (PIP), 95
Pseudo MAC (PMAC), 92
PSSR *see* Port-switching-based source routing (PSSR)
public cloud, 9, 10, 25, 246, 368
PUE *see* power usage effectiveness (PUE)
pyOCNI, OCNI reference implementation, 32

QoS adjustment (QA), 284, 286–7
Quantum project, OpenStack, 37
quarantine, network reconfiguration (NR), 285

RabbitMQ, 140
raid controller failures, 299
RainCloud workflow, 324–8
architecture, 318, *318*
ASKALON
black-box library, 325–6
environment, 317–18
heterogeneous and distributed computing environments, 328
scheduler, 326
execution time with 16 and 64 parallel loops, 327, *327–8*
file transfer time, 325
noncongested scenario, 325, *325–6*
parallelForEach loop in IWIR, 318, **318–19**
processing time, 325
queuing time, 325
scheduling time, 325
SWF version, 327
waiting time, 325
random graph-based topologies, 85
RBridges (routing bridges), 92
remote connectivity, 157
RESERVOIR (European Union's Seventh Framework Programme (FP7) project), 36
resilient provisioning with minimum power consumption in cloud (RPMPC), 123–4, *125*
resource management
allocation, 17–18, 248
applications
bag-of-tasks (BoTs), 249
characteristics, 248
data transfers, 249
service-oriented computing, 248–9
variations, 249
workflow task, 249, *250*
Big Data, 262–3
charging models and service-level agreements (SLAs), 247
cloud computing definition, 244
cloud service models, 244–6, *245*
cloud types, 246–7
greeness, 263
hybrid clouds and uncertainty, 263–4
infrastructure, 251
multiple workflows scheduling, 263
optimization techniques, 253
scheduler and VM allocation cooperation, 262
scheduling in clouds *see* scheduling
service level agreements (SLAs), 246, *246*, 251, *252*
user tasks submission, 249, *251*
VM allocation, 252–3
resource saving minimum outage probability in cloud (RS-MOPIC), 121–3, *122*
Rewire, 84
risks in mobile cloud computing (MCC)
application mobility, 170
CloneCloud and Cloudlets, 179, 181
context-awareness, 170

cost and resource savings, 154
definition, 163
device mobility, 170
ECG data analysis software, 171, *171*
effectiveness, 181
factors, 172, **172–4**
framework analysis, 181, **182–3**
hierarchy, 179–181, *180*
Hyrax and MobiCloud, 162, 177, 179
identification, analysis and treatment, 163
legislative/organizational, 168–9, 176
"metering of services", 170
mobile-specific, 176
monitoring and control, 163
performance, 167–8, 174–5
physical risks, 170
resource limitation, mobile devices, 170
scenarios, 170, 176–7, **178**
security and privacy, 164–7
Spectra and Alfredo, 159
Routelet, 66
routing
bridges, 91–2
layer 2
Actual MAC (AMAC), 92
fabric manager, 92
Link Aggregation Control Protocol (LACP), 91, 92
Multiple Spanning Tree Protocol (MSTP), 91
Portland, 92
Pseudo MAC (PMAC), 92
RBridges, 92
routing bridges, 91–2
Smart Path Assignment in Networks (SPAIN), 92
Spanning Tree Protocol (STP), 91
Transparent Interconnect of Lots of Links (TRILL), 91
Virtual LANs (VLANs), 91
layer 3
Bounded Congestion Multicast Scheduling (BCMS), 91
Code-Oriented eXplicit multicast (COXcast), 91
equal-cost multipath (ECMP), 89
Hedera, 89–90
multiprotocol label switching (MPLS), 90
OpenFlow controller, 90
Port-switching-based source routing (PSSR), 90, *90*
valiant load balancing (VLB), 89
RPMPC *see* resilient provisioning with minimum power consumption in cloud (RPMPC)
RS-MOPIC *see* resource saving minimum outage probability in cloud (RS-MOPIC)

SAVI two-tier cloud testbed, 237, *238*
scheduling
in clouds *see also* resource management
allocation and, 248
definition, 252, *252*
dependent tasks, 257–261
elasticity management entity, *256*, 256–7, *257*
independent tasks, 254–6, *255*
output, 249–50, *251*
submission, 250–251
VM allocation, 261–2
data centers
computing servers, 198
congestion/hotspots, 199
DENS methodology, 199–200
DPM-like power management, 198
Gigabit Ethernet (GE) interfaces, 198
server load and communication potential, 200, *201*
three-tier data center architecture, 199, *199*
science, technology, engineering and mathematics (STEM)
criticality, 385
DevOps professions, 385–6
engagement ideas, 385
site reliability engineering (SRE), 385, 386
scientific applications on clouds
Amazon, cloud providers, 312
Amazon Simple Workflow (SWF) service, 312, 315–17
Architecture Neutral Distribution Format (ANDF), 314
big data, 328–9

scientific applications on clouds (*cont'd*)
cloud computing, 313
Elastic Compute Cloud (EC2) infrastructure, 313
FutureGrid, 313–14
Interoperable Workflow Intermediate Representation (IWIR), 312, 314–15
IWIR-to-SWF conversion
activity function generation, 320–324
decider generation, 320
Megha workflow system, 314
Pegasus Grid workflow system, 314
RainCloud workflow, 317–19, 324–8
security, 329
SHIWA European project, 312, *312*
super computing, 329
UNiversal Computer Oriented Language (UNCOL), 314
SDN *see* software-defined networking (SDN)
secure virtual environment
access control, 62
authentication and authorization, 62
availability and isolation, 62
confidentiality, 62
integrity, 62
nonrepudiation, 62
replay resistance, 63
vulnerabilities
access control policy, 64
covert channel, 64
intruders, 64
loopholes in migration module, 65
side channel attack, 63–4
system administrator, 63
unprotected channel transmission, 64–5
security
algorithms, 290
in big data *see* big data on clouds (BDOC)
big data on clouds (BDOC)
data encryption and access control, 378–9
identity and access management, 377–8
networking communications, 375–6
planning and risk assessment, 375
tenant separation, 377
threats and vulnerabilities, 372–4
VM security, 376–7
cloud computing, 270
data integrate, 270
DDoS attack, 290
IDS/IPS cloud security solutions *see* intrusion detection and prevention systems (IDPS)
malicious insider, 270
network characteristics *see* network reconfiguration (NR)
OpenFlow-based IDPS solution *see* FlowIPS security design
performance comparison, 288–90
robust network architecture design, 271
SDN-based cloud security solutions *see* software-defined networking (SDN)
signature-and anomaly-based detection, 290
software-defined networking (SDN), 273–4
synchronization, 290
traditional non-SDN solutions, 275–6
virtualization hijacking, 270–271
server only topology
3-ary CamCube topology, 80, *81*
benefits and limitations, 80, **81**
CamCube, 80, 82
Content Addressable Network (CAN), 80
DCell and BCube, 82
Fat-Tree, 81
Forwarding Information Base (FIB), 80
optical switching architecture (OSA), 82
VL2 scales, 81
service-oriented computing and virtualization, 129
service providers, 25 *see also* cloud service providers (CSPs)
cloud providers, 3
Internet, 368
lease resources, 3
OpenDaylight, 147
virtualization management in cloud, 25
session riding and hijacking, 372–3
Simple Mail Transfer Protocol (SMTP), 363
Simple Network Management Protocol (SNMP), 28–9

single root I/O virtualization (SR-IOV), 55
site reliability engineers (SREs), 20, 367–8
small to medium enterprises (SMEs) survey on cloud computing, 10
Smart Path Assignment in Networks (SPAIN), 92
SMTP *see* Simple Mail Transfer Protocol (SMTP)
SNMP *see* Simple Network Management Protocol (SNMP)
Snort
  FlowIPS security design, 281–4, *283*
  intrusion detection and prevention systems (IDPS), 272, 274
software as a service (SaaS), 8, *9*, 18, 25, 155, 244, 364, 372
software-defined infrastructures (SDI)
  cloud and network controllers, 149
  programmable hardware resources, 149
  resource management system (RMS), 149, *149*
  SAVI project, *149*, 149–50
software-defined networking (SDN) *see also* OpenStack
  automated scaling, 150
  Ceilometer project in OpenStack, 150
  for cloud computing
    inter-data center networking, 145
    Linux bridging, 142
    networking requirements, 144
    Open vSwitch (OVS), 142–4, *143*
    physical interfaces (PIFs), 142
    virtual interfaces (VIFs), 142
  cloud security solutions, 269
    actuating trigger module, 277
    ALARMS, 276
    AvantGuard, 277
    based security solutions, 276–7
    CONA, content-oriented networking architecture, 277
    CPRecovery component, 277
    FortNox, Security Enforcement Kernel, 276–7
    intrusion detection and prevention system (IDS/IPS), 277–8
    OpenFlow-enabled solutions, 273, *273*, 276–7
    OpenFlow switch (OFS)'s flow tables, 273–4
    OpenSafe, 276
    QuagFlow, 276
    replication mechanism, 277
  definition, 132
  flexible and customizable networking, 133
  Heat project in OpenStack, 150
  layered architecture, 132–3, *133*
  management systems, scalability, 151
  paradigm, 95
  technology, 12, 16–17
Spanning Tree Protocol (STP), 91
Spectra framework, 159, 161
standardized generic interfaces, 31
stochastic power reduction scheme (SAVE), 205
storage media, inability to sanitize, 373
Storage Networking Industry Association (SNIA), 34
storage virtualization, 26–7
survivability in cloud *see also* failure tolerance in cloud
  availability, 297–8, **298**
  cloud computing fundamentals, 296–7
  embedding schemes
    comparison, **306**
    high-availability virtual infrastructure management framework (Hi-VI), 302–3
    opportunistic redundancy pooling (ORP), *300*, 300–301
    survivable mapping, 299–300
    survivable virtual network embedding (SVNE), 304–5, *305*
    Vailability-aware EmbeddiNg framework In Cloud Environments (VENICE), *301*, 301–2
    worst-case survival (WCS), *303*, 303–4
  fault domain and tolerance, 297, 298
  reliability, 297
survivable virtual network embedding (SVNE), 304–5, *305*
switch-oriented topologies
  Clos-based topologies, 78, *78*
  optical switching architecture (OSA), 78–9, *79*
System Resource, CIMI, 33

Technical Working Group (TWG), 34
time-shifting video streaming
  characteristics, 352
  hybrid P2P-DC solution, 352
  video portions ratio, 352, **353**
  VoD services and, 351–2
TPM *see* trusted platform module (TPM)
traffic, datacenter networks (DCNs)
  asymmetry, 86
  bandwidth guarantees, 88
  categories, 85
  flow arrival patterns, 87
  flow size, duration and number, 87
  hot spots, 87
  intra-and inter-application communication, 86
  link utilization, 87
  location and exchange, 86
  management, 87–9
  nature, 86
  packet losses, 87
  properties, 86–7
  proportional sharing, 88
traffic isolation (TI), 284–5
traffic redirection (TR), 284–6
Transparent Interconnect of Lots of Links (TRILL), 91
transport layer security (TLS), 40, 137
"TrustCloud", security and privacy risks, 184
trusted computing base (TCB), 50–51
trusted platform module (TPM), 14

unauthorized access, risk, 165, 174
user-generated content (UGC) live streaming
  hybrid DC-CDN delivery
    CDN infrastructures management, 348
    channels top categories, 349, **349**
    online popular channels stability, 350–351, *351*
    popular channels, 349–50, *350*
  justin.tv analysis
    international service, 347
    24/7 TV-like service, 347–8
    uploaders, 346–7, *347*
    viewers, 346–7, *347*
  over-the-top (OTT) TV channels, 345–6
  video sharing platforms, 346
utility-based services, 155

Vailability-aware EmbeddiNg framework In Cloud Environments (VENICE), *301*, 301–2
valiant load balancing (VLB), 89
virtual data centers (VDCs), 11 *see also* survivability in cloud
Virtual eXtensible Local Area Network (VXLAN), 93–4, 141
Virtual Grid Application Development Software Project (VGrADS), 35
virtualization
  cloud computing, 14–15
  inter-data-center (IDC) networks
    backbone, 108, *109*
    cloud customers, 107
    constraints, 111–12
    erbium-doped fiber amplifiers (EDFAs), 110
    heuristic solution, 113–15
    infrastructure as a service (IaaS), 108
    infrastructure design, 109–10
    mathematical formulation, 110–113
    mixed integer linear programming (MILP), 108–9
    notation, 110, **111**
    objective, 108–9
    physical infrastructure, 108
    power consumptions, 110
virtualization management in cloud
  cloud services, 25
  cloud tenants, 41–2
  embedding, 24
  energy efficiency, 42
  fault management, 42
  federation, 43–4
  infrastructures, 23
  interfaces for, 30–34
  monitoring, 41
  multiple customers, 23–4
  operations, 29–30
  private clouds, 25
  resource configurations, 24
  scalability, 40–41
  security, 43
  service providers, 25

standard management protocols and information models, 44
tools and libraries *see* cloud resource management, tools and systems
virtualized elements
computing, 26
management, 28–9
networking, 27–8
storage, 26–7
VMmonitor, hypervisor, 25
Virtual LANs (VLANs), 91–3, 141
Virtually Clustered Open Router (VICTOR), 66
virtual machines (VMs)
in cloud computing, 130
cloud providers, 25
computing resources, 26
Deltacloud, 40
Machine Resources, 33
migration, 11, 15
automated migration, 67
confidentiality, 67–8
hypervisor, 50–51
input/output (I/O) virtualization, *55*, 55–7, *56*, **58**
isolation, access control, and availability, 65–66
live migration, 50, 52–3, **54**
network connections, 66–7
offline migration, 52–3, **54**
process migration, 50
relocation, 49
security *see* secure virtual environment
storage, 67
trusted computing base (TCB), 50–51
virtual network migration without packet loss, 59–61
VM downtime, 66
Xen-based virtualization architecture, *50*, 50–51
XenFlow virtual topology migration, 51–3, *52*
operations, 29–30
OVF Package, 32–3
SDN *see* software-defined networking (SDN)
security issues, 43
server virtualization technologies, 26
virtual mobile computing (VMC), 160, 161
virtual network migration without packet loss
control plane directives, 59
local area network (LAN), 59
network topology, 59
OpenFlow switched networks, 59
XenFlow, 59
virtual networks (VNs), 27–8, *28*, 33
Virtual Private Cloud (VPC), 11, 93–4, 99
Virtual Tenant Network (VTN) service, 147
VLB *see* valiant load balancing (VLB)
VL2, datacenters, *94*, 94–5
VMC *see* virtual mobile computing (VMC)
VM Device Queues (VMDq), 55
VMM-MIB module, management protocols, 44
VR-MIB module, management protocols, 44

WCS *see* worst-case survival (WCS)
Web browsing use case, 131, *131*
Web services, security defects, 165
wireless backhaul networks, 105
wireless sensor networks (WSNs), 105
wireline local area networks (LANs), 105
worst-case survival (WCS), *303*, 303–4

X-Cloud Application Management Platform (XCAMP)
components, 222, *223*
deployment stage examination, 237
deployment view, 227–8
edge-core clouds (scenario 2), 225
hybrid clouds (scenario 1), 224–5
hypothesis evaluation, 237–8
information abstraction, 228
initial exploratory runs, 237
Java EE application, 240
Management Logic, 228–9
MAPE-k loop, 222, 225–7
SAVI two-tier cloud testbed, 237, *238*
Xen-based cloud environment, 279
XenFlow and VM migration techniques, 15

Yahoo, 363

IEEE Press Series on

Networks and Services Management

The goal of this series is to publish high quality technical reference books and textbooks on network and services management for communications and information technology professional societies, private sector and government organizations as well as research centers and universities around the world. This Series focuses on Fault, Configuration, Accounting, Performance, and Security (FCAPS) management in areas including, but not limited to, telecommunications network and services, technologies and implementations, IP networks and services, and wireless networks and services.

---

1. *Telecommunications Network Management into the 21st Century*
   Edited by Thomas Plevyak and Salah Aidarous
2. *Telecommunications Network Management: Technologies and Implementations: Techniques, Standards, Technologies, and Applications*
   Edited by Salah Aidarous and Thomas Plevyak
3. *Fundamentals of Telecommunications Network Management*
   Lakshmi G. Raman
4. *Security for Telecommunications Network Management*
   Moshe Rozenblit
5. *Integrated Telecommunications Management Solutions*
   Graham Chen and Qinzheng Kong
6. *Managing IP Networks: Challenges and Opportunities*
   Thomas Plevyak and Salah Aidarous
7. *Next-Generation Telecommunications Networks, Services, and Management*
   Edited by Thomas Plevyak and Veli Sahin
8. *Introduction to IT Address Management*
   Timothy Rooney
9. *IP Address Management: Principles and Practices*
   Timothy Rooney
10. *Telecommunications System Reliability Engineering, Theory, and Practice*
    Mark L. Ayers
11. *IPv6 Deployment and Management*
    Michael Dooley and Timothy Rooney
12. *Security Management of Next Generation Telecommunications Networks and Services*
    Stuart Jacobs
13. *Cable Networks, Services, and Management*
    Mehmet Toy
14. *Cloud Services, Networking, and Management*
    Edited by Nelson L. S. da Fonseca and Raouf Boutaba